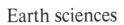

Review of research on modern problems in geochemistry

International Association for Geochemistry
and Cosmochemistry

Edited by F. R. Siegel

unesco

Published in 1979 by the United Nations Educational,
Scientific and Cultural Organization,
7 Place de Fontenoy, 75700 Paris
Printed by Ceuterick, Leuven

ISBN 92-3-101577-X

Preface

The last three decades have seen important changes of emphasis in the Earth sciences. During this relatively short span, the new concepts of global tectonic evolution and a much broader understanding of the early phases of crustal evolution during Pre-Cambrian times have gained universal acceptance. At the same time, new methods developed for field and laboratory work, and modern data processing have greatly enhanced the volume and precision of available information. Geochemistry has been no exception to this rapid evolution and, in the process, it has emerged as a key discipline, progressively increasing its relevance to many other branches of science. Thus, for theoretical reasons alone, a modern review of geochemistry would be justified.

There are, however, still other reasons for this endeavour. In view of the rapidly rising demand for agricultural products, minerals and energy to satisfy a fast-growing world population which is tending to become an essentially industrialized, raw-material-consuming society, it is appropriate to ascertain the natural resources of the earth which must provide the basic materials for such a development. To answer this challenging question, we need to know the particular conditions which were responsible for the formation of our natural resources—both those already exploited and those not yet discovered—within the context of the general evolution of the Earth. Moreover, geochemistry will be needed in the struggle to ensure that our air, soil and water, i.e. man's natural environment, are maintained in an ecological balance.

Thanks to the assistance of an able scientific organization, the ambitious task of addressing the specialist in geochemistry, the interested scientist in neighbouring disciplines, and the public at large, has been successfully undertaken. Unesco is indebted to the members of the International Association for Geochemistry and Cosmochemistry who volunteered to write the contributions presented in this review. They were written before 1975 and the Organization wishes to thank the authors for their efforts, and also for their patience during the various stages of this publication prior to its appearance. A special acknowledgement is due to the editor, whose enthusiasm made it possible to bring together the team of authors who finally produced the present version of this text. While the idea of addressing as many readers as possible with comprehensive reviews of important scientific disciplines is fully endorsed by Unesco, it should, of course, be understood that the responsibility for the scientific content of this volume rests entirely with the respective authors.

Contents

Introduction

Frederic R. Siegel

Washington, D.C., United States of America

This *Review of Research on Modern Problems in Geochemistry* and reviews of other related disciplines have two facets: one theoretical and one applied. In our contemporary society of developed nations, nations in stages of development, and undeveloped nations, stress on the environment is increasing and it is recognized that man and his living environment are rapidly approaching a critical mass with respect to survival potential on our planet. Research on recent past and ancient past events and environments permits a hindsight evaluation of causes and 'cures' (of results of such events) which when coupled with studies on contemporary problems should allow for futuristic modelling and result in foresight in planning.

Geochemists have accurately and precisely determined the abundances of the elements (and their isotopes) in the accessible phases of the Earth, in extraterrestrial matter fallen on the Earth, and in lunar samples brought to the Earth. Extrapolation of geochemical data to the past and from the past is essential for establishing baseline data and the natural geochemical concentration variance in the component phases of environments studied. The accuracy and precision of geochemical analyses have improved markedly during the past quarter century especially in so far as minor and trace element analyses are concerned. This is the result of the recognition of the importance of representativeness and care of sampling, the proper handling and storage of samples during and after collection, and the technological advances in instrumentation which have pushed the threshold of detection to remarkably low limits. With a sophisticated data base thus developed and inter-element or element-environmental factor relations made on physico-chemical principles and biological concepts, the geoscientist can confidently (but not overly so) project which are the factors that influence and/or control the fractionation, migration, deposition and distribution of the elements in the rock classes of the lithosphere, and extend this knowledge and interpretative capability to include the hydrosphere, atmosphere and biosphere.

That limits to the 'carrying capacity' of our ecosystem do indeed exist—and could suddenly be exceeded with disastrous results—can best be illustrated by the example of a guppy tank capable of supporting a population of 1,000 (Ehrlich *et al.*, 1973). If the initial population of 2 guppies has a doubling time of 1 month, 8 doubling times (8 months) will pass before about half the carrying capacity of the tank is reached. The population seems secure and there is no stress on the living environment. However, danger signs appear when the population is well beyond half of the tank's capacity, i.e. within the last month and when the last 100 guppies are added in less than 5 days. Thus, there have been 265 days of environmental health and security until about 90 per cent of the tank's capacity is achieved; then disaster strikes in less than a week. How many more doubling times can humanity pass through before pressures on the system (water and food, for example) exceed the Earth's ability (technology notwithstanding) to oppose them—1, 2, 3, 4—8 billion[1] population, 16 billion, 32 billion, 64 billion—A.D. 2010, 2045, 2080, 2115?

We study the past and present environments and when concepts derived from this work are considered, it is obvious that the past and present are the keys to projecting for the future. The geochemistry discipline now includes subfields of hydrogeochemistry, biogeochemistry, pedogeochemistry, lithogeochemistry, etc.; atmogeochemistry, for example, is in the forefront of interdisciplinary studies that will generate the data and concepts which, together with those from other major disciplines and subdisciplines, will allow us to evaluate where we have been, where we are and where we must go to preserve the quality of existence as we know it—and to develop this quality for peoples who strive' to achieve it—this, in a rational, well-planned manner.

Geochemical knowledge and understanding of individual and interdependent Earth systems are essential factors to be assessed in planning for 'man's' continued

---

1. One billion equals 1,000 million.

'good' existence on our planet. Survival within limits of the ecosystem requires complex evaluations and decisions on water, food, energy and mineral resources, in terms of quality of life. To what limits can the water–food–energy triplet be extended in light of industrialization, pollution and rapid population increase in a fragile ecosystem such as we inhabit? If disaster strikes an environment, it can recover, but at what cost in terms of life and suffering. Why plan on disaster? It is far better to know and defend against the cause than to wait for or effect a cure.

The *Review of Research on Modern Problems in Geochemistry* is, among other things, directed towards the role of geoscience in assessing the state and health of the environment. It builds upon a fundamental knowledge of the composition of the lithosphere and associated ecospheres (hydrosphere, biosphere and atmosphere). With compositional norms established, it is possible to follow processes of weathering and the development of the soil cover essential for food growth to man and the other fauna and flora of our living system. Man and his 'associates', inhabitants of the solid, liquid and gaseous phases of the environment, require chemical elements in their food supplies for proper nutrition. Recognition of these chemical life requirements and the limits of 'allowable ingested concentration' (or maximum permissible concentration) is basic to the planning for and maintenance of our food supply chain. How excesses and deficiencies in rocks and, hence, in soils derived from them, have affected and could affect plant and animal nutrition must be more precisely known so that danger signals can be recognized early.

The makings of foodstuffs for plants and animals is based on energy (ultimately from the sun) and water. Both the quality and quantity of water, which is a recyclable commodity, are of concern to those projecting earth requirements for the present and future. Although only about 18 per cent of the world's cultivated lands are now irrigated, they supply from 40 to 50 per cent of the world's food; only a fraction of 1 per cent of the planet's water is available for irrigation and hence its quality must be preserved. About 25,000 people die daily (or over 9 million yearly) from the ingestion of contaminated water. Where the potential for polluting the fresh-water supplies exists, it must be stopped; and where pollution has taken hold of an environment, it must be halted, reversed and eliminated. Geochemical measurements define the quality or pollution levels of waters, highlight potential for pollution in an environment and can specify pollution sources. Geochemical concepts can be used with those of other disciplines to resolve and, ultimately, remove pollution problems. What is true in studies and evaluations of non-marine systems is true of marine systems, especially those in close proximity to the land environment: estuaries and marine coastal zones. The coastal marine environments (here including estuaries) are frequently areas close to major population concentrations and provide a major

food source for these and inland populations. About 50 per cent of the ocean's productivity is from three areas of nutrient upwelling (the west coast of Africa, the coast of California and the coast of Peru) which are thought to represent slightly more than about 0.1 per cent of the world oceans—the remaining 50 per cent is from other coastal zones. If these zones are increasingly subjected to any one of a great number of pollutants at present being injected into them, a major food source for the Earth's population will undoubtedly be adversely affected.

As essential as food and water are to life, air is temporally more important. A human can live without food for about 40 days, without water for about 4 days, but without oxygen in the atmosphere for only a few minutes. The atmosphere has naturally occurring pollutants, but human activity has discharged many unnatural and unacceptable contaminants into the atmosphere. These ultimately find their way into other parts of the ecosystem. As with other media, the geochemist works to establish background levels for atmospheric gaseous and particulate components, finds where pollutants exist, attempts to pin-point the origins of these pollutants and works on theoretical and applied methods for lessening the input of the sources for the life sinks.

The preservation of our earth environment and key to the resolution of many of our existing and projected problems lies in the proper evaluation, development and use of mineral resources; because of national aspirations or driven by multinational corporation profit motives, such propriety is often not observed. Any evaluation of or exploration for mineral resources must be based on a fundamental understanding of how these minerals deposits originated. Once this is defined and specific rock-type/ore-type relations indicated, geoscientists can use regional and local geological data bases and the concepts of element mobility as a function of geological environment parameters to begin geochemical prospecting. This prospecting can guide explorationists to broad areas of mineralization potential and target propitious zones within the regional areas.

In a final assessment, after food, water and shelter requirements have been met by our planetary life population, progress or maintenance of the quality of life is dependent upon the existence of metalliferous and non-metalliferous resources, and energy sources for their successful exploitation. This reduces ultimately to (a) the economics of energy costs for commodity extraction versus (b) a market's ability to pay for the final product.

Finally, such problems concerning many disciplines (scientific, economic, social and political) are complex and intertwined, and though for clarity not always so treated in this volume must always be recognized as such. For example, in a very real and practical sense, the nations of the world must come to specific agreements on legal problems involving the marine environment. Food, mineral and energy resources are present outside existing, artificially-designated 'limits of national sovereignty' now commonly taken as about 200 miles from land. These resources belong to all nations but there exists the question as to whether their development for the benefit of mankind be delayed or denied due to international, legal (questionable), non-agreements—or whether they should be developed as soon as technologically and economically feasible (with agreements established in the light of development patterns). And while considering the contamination of the environment through ocean 'dumping' beyond the 'national sovereignty' limits, what rights of mankind are not being considered that should be considered? Or we could consider the morality of injecting radioactive matter into the atmosphere with the full knowledge that there will be radioactive cloud movement across international boundaries with the fall-out of the cloud burden, poisoning all living things not for a moment but for the many half-lives of decay necessary for radioactivity to achieve 'maximum permissible concentrations'—an onerous term since we really do not know what maximum permissible concentrations are—such acts are immoral and must be stopped; not limited, but stopped, now and for ever.

We inhabit a fragile system and in fragility there is resiliency but only if the limits of resiliency are not breached.

'Serás lo que debas ser, o sino no serás nada.'—General José de San Martín

Reference

EHRLICH, P. R.; EHRLICH, A. H.; HOLDREN, J. P. 1973. *Human Ecology*. San Francisco, Calif., Freeman. 304 p.

Composition of the Earth's crust and distribution of the elements

James E. Mielke

Marine and Earth Sciences,
Library of Congress,
Washington, D.C., United States of America

Types of rocks and the problems of averaging rock analyses to establish a frame of reference

A discussion of the composition of the Earth's crust and distribution of elements in the rock types can only be based on the best estimates and calculations that can be made with limited evidence. The present chapter is a general review of some of these estimates and observations regarding the distribution of the elements.

It has been suggested that we probably know less about the composition of the crust of the Earth than about the composition of chondritic meteorites (Turekian, 1968). With respect to the capability for representative sampling, this is undoubtedly true. Only a fraction of the Earth's crust is exposed or accessible, either by drilling or in deep mines. Estimates of the rock types at depth can be made from geophysical evidence and extrapolations of deep structural features. The most easily investigated areas are the near-surface portions of the continents. However, the continents cover less than a third of the Earth's surface. Seismic evidence indicates that the average thickness of the continental crust is 35 km and that of the oceanic crust 5–6 km thick. Consequently, in terms of the total volume of Earth's crust, there is direct evidence of the composition of less than 30 per cent.

The greatest area of uncertainty regarding the chemical composition of the Earth's crust is the composition of the deep rocks of the crust and the 'basaltic shell' of the continental crust. Tatel and Tuve (1955) describe two types of seismic behaviour in the lower crust: a gradual increase in velocity with depth, with a marked increase at the Mohorovicic discontinuity of the upper mantle; and fairly constant velocity to 20 km, followed by a transition zone through which the velocity increases to that of the upper mantle. The latter especially suggests there may be an intermediate layer between the crust and mantle in some areas.

The average composition of the Earth's crust is essentially the same as the average composition of igneous rocks. This is because the total amount of sedimentary and metamorphic rocks in the crust is relatively insignificant compared with the mass of igneous rocks. Furthermore, the average composition of all sedimentary and metamorphic rocks is basically similar to the average igneous rock. To follow this approach would leave only the problem of calculating some average igneous rock. Several methods have been proposed in the past. In addition, methods based on other approaches have also been published.

The first attempt to calculate the chemical composition of the Earth's crust was made by Clarke (1889) based on the composition of 880 igneous rocks from several countries. This tabulation was later revised (Clarke and Washington, 1924) to include 5,159 'superior' analyses grouped geographically (Table 1). These analyses represent the continental crust since no sampling of the ocean floor was included.

The first attempt to include a quantitative distribution of rocks in a calculation of the average composition of the Earth's crust was made by Knopf (1916). He weighted the average compositions of Daly (1910) for Cordillerian and Appalachian rocks by the total area occupied by each type. This weighting was then used to calculate the composition of the average igneous rock (Table 1).

Another approach was the method used by Goldschmidt (1933). He based his average on 78 glacial and post-glacial clays from Norway, reasoning that the composition of the clays presented a representative sampling of the rock types in the region (Table 1). However, this average may be low in calcium and sodium and high in water due to leaching and hydration of the clays.

Although these averages, produced by three different methods, are in fairly close agreement, each has been the subject of some criticism. Among the most commonly mentioned problems are: (a) the over-weighting

13

TABLE 1. Average chemical composition of the Earth's crust (in percentage)

Oxide	1	2	3	4	5	6	7
SiO_2	59.14	61.64	59.19	59.4	55.2	61,9	59.3
Al_2O_3	15.34	15.71	15.82	15.5	15.3	15.6	15.8
Fe_2O_3	3.08	2.91	6.99	2.3	2.8	2.6	2.6
FeO	3.80	3.25		5.0	5.8	3.9	4.4
MgO	3.49	2.97	3.30	4.2	5.2	3.1	4.0
CaO	5.08	5.06	3.07	6.7	8.8	5.7	7.2
Na_2O	3.84	3.40	2.05	3.1	2.9	3.1	3.0
K_2O	3.13	2.65	3.93	2.3	1.9	2.9	2.4
H_2O	1.15	1.26	3.02				
TiO_2	1.05	0.73	0.79	1.2	1.6	0.8	0.9
P_2O_5	0.30	0.26	0.22	0.2	0.3	0.3	0.2
MnO	0.12	0.16		0.1	0.2	0.1	0.2

Column 1, Clarke and Washington, 1924.
Column 2, Knopf 1916.
Column 3, Goldschmidt, 1933.
Column 4, Poldervaart, 1955 continental crust.
Column 5, Poldervaart, 1955 entire lithosphere.
Column 6, Ronov and Yaroshevsky, 1969 continental crust.
Column 7, Ronov and Yaroshevsky, 1969 entire lithosphere.

by Clarke and Washington of rare and unusual rock types which, as curiosities, tended to be analysed to greater extents; (b) the over-weighting of certain regions that tended to be better mapped and analysed than other regions; (c) the lack of weighting or difficulty in weighting rock types according to volume; and (d) the difficulty in extrapolating analyses from only one region to the average igneous rock, or certainly the entire crust.

Many of these objections, as Mason (1958) and others have pointed out, may not be as serious as previously thought. The individual averages of each region tend to agree so that better analysed regions can be averaged with sparser data from other areas. Unusual rock types tend to cover the complete suite of rock compositions so that they too average out. Weighting of rock types according to volume is not possible without complete data at depth so that attempts at this weighting based on surface exposures are not likely to be more accurate. Furthermore, these averages tend to come out much alike as is shown in Table 1.

Another approach is that of Poldervaart (1955) based on averaging data from each of four major blocks or regions of the crust: continental shields; young folded belts; continental platforms and slopes; and deep oceanic regions. He then weighted the average from each region by the relative mass of each to arrive at average compositions for the continental crust and the entire lithosphere (Table 1). Since this average for the entire lithosphere included the oceanic regions, it is more representative of the entire crust of the Earth than some of the earlier estimates. It indicates the more basaltic and less alkaline nature of the oceanic crust.

A recent calculation of the average chemical composition of the Earth's crust was given by Ronov and Yaroshevsky (1969, 1972) following the basic method of Poldervaart but using different data and calculation methods. They consider three types of crust: continental, including platforms and geosynclines; subcontinental, encompassing the continental shelf and slope; and oceanic, consisting of the sediments of seismic layers I and II. Within each of these types of crust they averaged the compositions of each of the commonly occurring types of rocks (Tables 2 and 3). For a comparison with other estimates, their averages for the crust as a whole and the continental crust are also included in Table 1. Their averages are more granitic than Poldervaart's, indicating a granite to basalt ratio of roughly 2:3. Wedepohl (1969) estimated that greater than or equal to 50 per cent of the Earth's crust—including the lower continental crust and oceanic crust—consists of gabbro or basaltic rocks. The upper portion of the continents are generally considered to be granitic. This is the portion of the crust that has been sampled in greatest detail and many rock types have been identified. Compositions of some of the more common igneous rocks have been calculated by Nockolds (1954) and are presented in Tables 4 to 8. These compositions have served as the basis of several estimates of the composition of the Earth's crust.

More recently, Turekian and Wedepohl (1961) assembled a table of the elemental composition of some major units of the Earth's crust. They chose the best data available for each of ten groups of rocks. These data are presented in Table 9.

Many of the early averages of rock types included estimates of the abundance of trace elements. However, owing to considerable improvements in analytical methods, much of the early data on trace elements are now obsolete. One compilation of rock analyses—based on the best methods at the time—was made by Rankama and Sahama (1950). Their tabulation of the abundance of igneous rocks is presented in Table 10. Another more recent estimate of trace element abundances in the continental crust and the Earth's crust as a whole was presented by Ronov and Yaroshevsky (1972), based on data of Taylor (1964) and Vinogradov (1962). These data compared with the average composition of chondrites are presented in Table 11.

Taylor (1964) tried a different approach to estimate the composition of the Earth's crust by using the rare-earth compositions of shales to derive a ratio of granite and basalt in the crust. More recent data indicate that there is little difference in the distribution patterns of rate-earths in granites and continental basalts, as compared with shales (Haskin et al., 1966).

Computer processing of geochemical data is becoming more widespread. Programmes have been devised to show correlations or trends and to calculate geochemical material balances and element abundances using data from many sources (Horn and Adams, 1966; Miesch,

TABLE 2. Distribution of volumes and masses of magmatic and metamorphic rocks of continental, subcontinental, and oceanic types of the crust and chemical composition of rocks

Types of crust	Shells	Volume (10^6 km³)	Mass (10^{24} g)	Types of rocks	Abundance, percentage of total volume of shell	Mass of rock types (10^{24} g)	SiO_2	TiO_2	Al_2O_3	Fe_2O_3	FeO	MnO	MgO	CaO	Na_2O	K_2O	P_2O_5	$Corg$	CO_2	S	Cl	H_2O^+
Continental and subcontinental	Sedimentary	710	1.81	Traps and plateau-basalts of platforms	1.1	0.02	49.22	1.53	15.74	3.33	8.02	0.18	6.71	10.00	2.51	0.73	0.18	0.01	—	0.03	0.005	1.81
				Basalts of geosynclines	9.2	0.20	49.04	1.36	15.69	5.38	6.37	0.31	6.17	8.94	3.11	1.52	0.45	0.01	—	0.03	0.005	1.62
				Andesites of geosynclines	7.4	0.16	59.57	0.77	17.30	3.33	3.13	0.18	2.75	5.79	3.58	2.04	0.26	0.01	—	0.03	0.005	1.26
				Rhyolites of geosynclines	1.8	0.03	72.74	0.33	13.47	1.45	0.88	0.08	0.38	1.20	3.38	4.45	0.08	0.03	—	0.04	0.02	1.47
	'Granitic'	3,590	9.81	Granites	18.1[1]	1.75	72.33	0.31	14.00	0.95	1.50	0.05	0.53	1.36	3.14	5.07	0.15	0.03	—	0.04	0.02	0.61
				Granodiorites, diorites	19.9[1]	1.93	65.66	0.55	15.72	1.54	2.63	0.10	1.73	3.71	4.10	3.05	0.23	0.03	—	0.04	0.02	0.98
				Syenites, nepheline syenites	0.3[1]	0.03	57.73	0.66	18.70	2.51	2.02	0.15	1.77	3.80	4.85	6.97	0.20	0.01	—	—	—	0.61
				Gabbro	3.7[1]	0.38	48.73	1.17	16.80	2.93	7.09	0.21	7.38	11.13	2.35	0.70	0.24	0.01	—	0.03	0.005	1.24
				Dunites, periodites	0.1[1]	0.01	43.54	0.81	3.99	2.51	9.84	0.21	34.02	3.46	0.56	0.25	0.05	—	—	—	—	0.76
				Gneisses	37.6[2]	3.66	69.6	0.5	14.7	1.7	2.2	0.1	1.3	2.1	3.2	3.7	0.2	—	—	—	—	0.7
				Crystalline schists	9.0[2]	0.88	62.0	0.9	18.5	2.7	4.8	0.1	2.8	1.5	1.9	3.8	0.2	—	—	—	—	0.8
				Marbles	1.5[3]	0.15	12.38	—	1.18	0.55	—	—	17.89	26.74	0.03	0.05	0.11	—	40.71	—	—	0.36
				Amphibolites	9.8[2]	1.02	49.6	1.6	15.5	3.5	7.7	0.2	6.9	9.4	2.9	1.1	0.3	0.03	—	0.04	0.02	1.3
				Average composition of magmatic rocks of 'granitic' shell	42.1	4.10	66.91	0.50	15.07	1.41	2.54	0.09	1.78	3.34	3.52	3.72	0.19	0.03	—	0.04	0.02	0.84
				Average composition of metamorphic rocks of 'granitic' shell	57.9	5.71	63.72	0.74	15.12	2.13	3.48	0.11	2.91	3.88	2.87	3.21	0.22	0.01	1.05	—	—	0.56
	'Basaltic'	3,740	10.85	Acid magmatic and metamorphic ortho- and para-rocks	50.0	5.24																
				Basic magmatic and metamorphic ortho- and para-rocks	50.0	5.61																
Oceanic	Basalts plus volcanic rocks of seismic layer II	1,785	5.18	Oceanic tholeiitic basalts	99.0	5.13	49.74	1.36	16.57	2.33	6.95	0.17	7.54	11.50	2.79	0.19	0.13	—	—	—	—	0.73
				Alkaline differentiates of oceanic basalts	1.0	0.05	47.83	2.84	16.39	4.13	7.50	0.16	5.22	9.07	3.71	1.87	0.48	—	—	—	—	0.80

Content of components (in percentage by weight)

1. The estimate of relative abundance of magmatic rocks is based on the results of measurement on the maps of Ukrainian and Baltic shields and crystalline basement of Russian platform (acid, 38.3 per cent; base, 3.8 per cent), as well as the data obtained by Fleischer and Chao (1960) (Daly's figures) on the relations of granites, granodiorites, and syenites in the acid group of rocks and basic and ultrabasic rocks in the basic group.
2. The percentage of marbles was reduced to one-third the amount of carbonate rocks in geosynclines, because of their less widespread occurrence in Early Archean rocks of the Pre-Cambrian.
3. A quantitative estimate of the abundances of different types of metamorphic rocks is obtained on the basis of the concept of the geosyncline origin of metamorphic 'para-' rock series. The second assumption is that the metasediments and 'ortho-' rocks have nearly the same volumetric abundances.
Source: Ronov and Yaroshevsky, 1972.

TABLE 3. Volume, mass, and average chemical composition of sedimentary and volcanic rocks of continents and sediments of oceans

Types of crust	Large structural units of crust and layers	Volume, 10^6 km³	Average thickness[1] (km)	Mass (10^{24} g)	Types of rocks and sediments	Abundance percentage of volume on continents and percentage of area in oceans	SiO₂	TiO₂	Al₂O₃	Fe₂O₃	FeO	MnO	MgO	CaO	Na₂O	K₂O	PO₅	Corg	CO₂	SO₃	Cl[2]	H₂O⁺
Continental	Platforms	135	1.8 (1.2)	0.35	Sands	23.5	75.75	0.49	6.90	2.55	1.31	0.06	1.43	3.40	0.58	1.83	0.16	0.30	2.75	0.29	0.09	2.11
					Clays	49.3	55.09	0.86	16.30	4.17	1.87	0.05	2.46	4.75	0.75	3.01	0.11	0.99	3.92	0.43	0.12	5.16
					Carbonates	21.0	9.80	0.18	2.54	0.85	0.54	0.06	6.83	38.93	0.24	0.76	0.07	0.33	35.05	2.36	0.08	1.40
					Evaporites: sulfates, 50 per cent salts, 50 per cent	2.3	0.30	—	0.07	0.04	—	—	0.31	18.07	24.92	0.31	—	—	0.66	24.95	28.74	8.11
					Volcanics	3.9	49.22	1.53	15.74	3.33	8.02	0.18	6.11	10.00	2.51	0.73	0.18		0.01	0.03	0.005	1.81
					Average composition of sedimentary series of platforms	100.0	49.12	0.64	10.86	2.97	1.65	0.06	3.36	12.17	1.23	2.11	0.11	0.63	10.00	1.34	0.76	3.26
	Geosynclines	365	10.0	0.94	Sands	18.7	62.93	0.52	12.12	2.30	3.20	0.11	2.28	5.69	1.92	1.69	0.11	0.25	4.22	0.12	0.03	2.47
					Clays and shales	39.4	55.76	0.71	17.56	3.61	3.35	0.08	2.52	4.08	1.27	2.76	0.15	0.78	2.80	0.11	0.03	4.37
					Carbonates	16.3	13.30	0.14	2.70	0.43	0.94	0.09	2.92	42.40	0.56	0.49	0.10	0.35	34.44	0.03	0.02	1.05
					Evaporites: sulfates, 50 per cent salts, 50 per cent	0.3	0.30	—	0.07	0.04	—	—	0.31	18.07	24.92	0.31	—	—	0.66	24.95	28.74	8.11
					Volcanics	25.3	55.62	1.00	16.12	4.17	4.52	0.24	4.22	6.91	3.33	2.02	0.34		0.01	0.03	0.007	1.46
					Average composition of sedimentary series of geosynclines	100.0	50.00	0.65	13.69	2.98	3.21	0.13	2.97	11.49	1.86	2.00	0.18	0.42	7.51	0.14	0.03	2.75
	Total	500	4.2 (3.4)	1.29			49.76	0.65	12.93	2.98	2.79	0.11	3.05	11.67	1.69	2.03	0.16	0.48	8.18	0.46	0.21	2.89
Subcontinental	Shelf and continental slope	210	3.2	0.52		*Composition considered to be similar to that of sedimentary rocks of continents*																
Oceanic	Sediments of seismic layer I	120 (70)[3]	0.4	0.19	Terrigenous	7.3	52.92	0.81	15.73	5.04	1.92	0.33	3.82	4.68	1.59	3.09	0.08	0.30	3.35	—	—	6.34
					Calcareous	41.5	18.84	0.29	6.09	2.55	0.46	0.19	1.98	35.61	0.70	1.03	0.11	0.28	28.90	—	—	2.97
					Siliceous	17.0	62.80	0.66	13.21	4.92	1.15	0.14	3.02	1.47	1.52	2.70	0.16	0.26	1.08	—	—	6.91
					Red deep-sea clays	31.2	54.51	0.80	16.40	6.93	0.84	1.05	3.41	2.02	1.27	2.78	0.21	0.22	2.45	—	—	7.11
					Volcanogenic sediments	3.0	44.30	2.48	12.41	5.83	6.40	0.41	8.85	9.79	1.88	0.78	0.20	0.32	3.36	—	—	2.99
					Average composition of pelagic sediments	100.0	40.73	0.61	11.41	4.60	0.97	0.47	2.94	16.29	1.13	2.01	0.15	0.26	13.27	—	—	5.17
	Sediments of seismic layer II	265	0.9	0.66		*Composition considered to be similar to that of sediments of seismic layer I*																
	Sediments of oceanic crust	385 (335)[3]	1.3	0.85		*Composition considered to be similar to that of sediments of seismic layer I*																
	Total for sedimentary shell including volcanics	1095 (1045)[3]	2.2 (2.0)[4]	2.65			46.63	0.64	12.44	3.50	2.21	0.22	3.01	13.15	1.51	2.02	0.16	0.41	9.81	0.46	0.21	3.62
							1.240	0.017	0.331	0.093	0.059	0.006	0.080	0.350	0.040	0.054	0.004	0.011	0.261	0.012	0.006	0.096
	Total for sedimentary shell excluding volcanics	900[3]	2.25				46.62	0.62	10.84	3.24	2.00	0.18	2.71	13.62	1.28	2.09	0.13	0.49	11.60	0.53	0.27	3.78
							1.049	0.014	0.244	0.073	0.045	0.004	0.061	0.306	0.029	0.047	0.003	0.011	0.261	0.012	0.006	0.085

1. Thicknesses including areas of shields are shown in parentheses.
2. The sum of analyses exceeds 100 per cent by O=Cl₂.
3. Recalculated for consolidated sediment with a density of 2.5.
4. In calculating average values, the content of SO₃ and Cl in marine sediments was taken to be the same as in continental ones.
Source: Ronov and Yaroshevsky, 1972.

TABLE 4. Averages of granite rocks (in percentage)

Element	Alkali granites (48)[1]	Alkali rhyolites (21)	Granites (72)	Rhyolites (22)	Quartz monzonites (121)	Quartz laites (58)	Grano-diorites (137)	Rhyodacites (115)	Quartz diorites (58)	Dacites (50)
SiO_2	73.86	74.57	72.08	73.66	69.15	70.15	66.88	66.27	66.15	63.58
TiO_2	0.20	0.17	0.37	0.22	0.56	0.42	0.57	0.66	0.62	0.64
Al_2O_3	13.75	12.58	13.86	13.45	14.63	14.41	15.66	15.39	15.56	16.67
Fe_2O_3	0.78	1.30	0.86	1.25	1.22	1.68	1.33	2.14	1.36	2.24
FeO	1.13	1.02	1.67	0.75	2.27	1.55	2.59	2.23	3.42	3.00
MnO	0.05	0.05	0.06	0.03	0.06	0.06	0.07	0.07	0.08	0.11
MgO	0.26	0.11	0.52	0.32	0.99	0.63	1.57	1.57	1.94	2.12
CaO	0.72	0.61	1.33	1.13	2.45	2.15	3.56	3.68	4.65	5.53
Na_2O	3.51[2]	4.13	3.08	2.99	3.35	3.65	3.84	4.13	3.90	3.98
K_2O	5.13[2]	4.73	5.46	5.35	4.58	4.50	3.07	3.01	1.42	1.40
H_2O^+	0.47	0.66	0.53	0.78	0.54	0.68	0.65	0.68	0.69	0.56
P_2O_5	0.14	0.07	0.18	0.07	0.20	0.12	0.21	0.17	0.21	0.17
qz[3]	32.2	31.1	29.2	33.2	24.8	26.1	21.9	20.8	24.1	19.6
or	30.0	27.8	32.2	31.7	27.2	26.7	18.3	17.8	8.3	8.3
ab	29.3	35.1	26.2	25.1	28.3	30.9	32.5	35.1	33.0	34.1
an	2.8	2.0	5.6	5.0	11.1	9.5	16.4	14.5	20.8	23.3
c	1.4	—	0.8	0.9	—	—	—	—	—	—
$CaSIO_3$	—	0.1	—	—	—	0.2	—	1.3	0.3	1.3
$MgSIO_3$	0.6	0.3	1.3	0.8	2.5	1.6	3.9	3.9	4.9	5.3
$FeSIO_3$	1.1	0.6	1.7	—	2.2	0.8	2.9	1.3	4.1	2.8
ac	—	—	—	—	—	—	—	—	—	—
mt	1.2	1.9	1.4	1.9	1.9	2.5	1.9	3.0	2.1	3.3
il	0.5	0.3	0.8	0.5	1.1	0.8	1.1	1.4	1.2	1.2
ap	0.3	0.2	0.4	0.2	0.5	0.3	0.5	0.3	0.5	0.3

1. Number of analyses used for finding averages (in parentheses).
2. Pegmatites mainly differ in averages by their higher potassium and slightly lower sodium contents (6.3 per cent K_2O), etc.
3. The following abbreviations are used for normative minerals: qz = quartz; or = K-feldspar; ab = albite; an = anorthite; c = corundum; ac = acmite; mt = magnetite; il = ilmenite; ap = apatite.
Source: Nockolds, 1954.

1969). These may be of significant value if consideration is given to the limitations of their construction. As Turekian (1968) pointed out, the method of programming a computer to choose between the maximum and minimum values of element analyses to find the best solution to balance equations, may not produce the representatively best value for a given rock type. Turekian further suggested that tables that were numerical constructs of the total composition of the Earth's crust or any of its members were less useful than actual rock and sediment data. Green (1972) discussed other problems and values of geochemical data processing by machine.

Wedepohl (1971) pointed out the value of the average composition of the crust as a frame of reference for the study of the distribution of elements among different rocks. Average compositions, including trace element data, can be used as an indication of secondary rock-forming processes. In the same manner, crustal abundances of minerals can be useful. These can be calculated from element analyses or abundances of rock types. The ten most abundant minerals are listed in Table 12.

Distribution of the elements

In discussing the chemical distribution of the elements, some ordering or tabulation of similar properties is convenient. One of the most fundamental and useful arrangements is the periodic diagram (Fig. 1) first proposed by Mendeleev in 1869. This system is primarily based on valence, which is a reflection of electronic structures of the atoms. Many geochemical parameters can be explained on the basis of this tabulation. These include lattice types and packing densities, heats of formation, thermodynamic properties, compressibility, density, heat of vaporization, ionization potential, atomic volumes, ionic radii, abundance of isotopes, melting point, ore-element assemblages, and liquid-state structure. In addition, there are other factors which are inter-

TABLE 5. Averages of intermediate rocks (in percentage)

Element	Alkali syenites (25)[1]	Alkali trachytes (15)	Syenites (18)	Trachytes (24)	Monzonites (46)	Latites (42)	Monzo-diorites (56)	Latite andesites (38)	Diorites (50)	Andesites (49)
SiO_2	61.86	61.95	59.41	58.31	55.36	54.02	54.66	56.00	51.86	54.20
TiO_2	0.58	0.73	0.83	0.66	1.12	1.18	1.09	1.29	1.50	1.31
Al_2O_3	16.91	18.03	17.12	18.05	16.58	17.22	16.98	16.81	16.40	17.17
Fe_2O_3	2.32	2.33	2.19	2.54	2.57	3.83	3.26	3.74	2.73	3.48
FeO	2.63	1.51	2.83	2.02	4.58	3.98	5.38	4.36	6.97	5.49
MnO	0.11	0.13	0.08	0.14	0.13	0.12	0.14	0.13 ·	0.18	0.15
MgO	0.96	0.63	2.02	2.07	3.67	3.87	3.95	3.39	6.12	4.36
CaO	2.54	1.89	4.06	4.25	6.76	6.76	6.99	6.87	8.40	7.92
Na_2O	5.46	6.55	3.92	3.85	3.51	3.32	3.76	3.56	3.36	3.67
K_2O	5.91	5.53	6.53	7.38	4.68	4.43	2.76	2.60	1.33	1.11
H_2O^+	0.53	0.54	0.63	0.53	0.60	0.78	0.60	0.92	0.80	0.86
P_2O_5	0.19	0.18	0.38	0.20	0.44	0.49	0.43	0.33	0.35	0.28
qz^2	1.7	—	2.0	—	—	0.5	2.0	7.2	0.3	5.7
or	35.0	32.8	38.4	43.9	27.8	26.1	16.7	15.6	7.8	6.7
ab	46.1	54.0	33.0	28.8	29.3	27.8	31.9	29.9	28.3	30.9
an	4.2	3.3	10.0	9.7	15.8	19.2	21.1	22.2	25.8	27.2
ne	—	0.6	—	2.0	—	—	—	—	—	—
$CaSiO_3$	3.0	2.1	3.0	4.2	6.3	4.5	4.5	4.1	5.6	4.2
$MgSoO_3$	2.4	1.6	5.0	3.2	8.0	9.7	9.9	8.5	15.3	10.9
$FeSiO_3$	2.1	—	2.1	0.5	4.1	2.4	5.4	3.0	8.5	5.3
Mg_2SiO_4	—	—	—	1.4	0.8	—	—	—	—	—
Fe_2SiO_4	—	—	—	0.2	0.4	—	—	—	—	—
mt	3.3	3.3	3.3	3.7	3.7	5.6	4.9	5.3	3.9	5.1
il	1.2	1.4	1.5	1.2	2.1	2.3	2.1	2.4	2.9	2.4
ap	0.5	0.4	1.0	0.5	1.0	1.2	1.0	0.8	0.8	0.7

1. Number of analyses used for finding average (in parentheses).
2. qz = quartz; or = K-feldspar; ab = albite; an = anorthite; ne = nepheline; mt = magnetite; il = ilmenite; ap = apatite.
Source: Nockolds, 1954.

related and may have a major influence. For example, in the isomorphic substitution in a mineral of one element for another of similar properties, the ease of substitution depends not only on the relative thermodynamic stabilities of the resulting compound and the exactness of 'fit' in the crystal site but also on the relative abundance of both elements.

Only eight elements account for the composition of nearly 99 per cent of the Earth's crust. Therefore, the occurrence in minerals and distribution of most elements is controlled by crystal chemistry and their geochemical properties. While not denying the general usefulness of the periodic grouping, Krauskopf (1967) cautions that it cannot alone explain the distribution of the elements.

Another significant classification for the distribution of the elements was proposed by Goldschmidt in 1923. His classification was based in part on observations of: (a) the distribution of metal, silicate, and sulphide phases in smelting; (b) the composition of meteorites; and (c) the composition and geologic occurrence of rocks and ores. He divided the elements into four groups: siderophile, chalcophile, lithophile and at-

mophile. Siderophile elements are those related to iron and are concentrated in the Earth's core. Chalcophile elements associate with sulphur and tend to form sulphide ore deposits. Lithophile elements tend to form silicates, and atmophile elements are primarily gases. A tabulation of Goldschmidt's classification is presented in Table 13.

In addition to the tendencies of the elements to be distributed in igneous rocks as formulated by Goldschmidt, certain rules or generalizations can be made concerning the behaviour and substitution of one element for another. These can be stated as follows:
Ions whose radii are similar may co-substitute if their charges do not differ by more than one unit,
Differences in ionic radii greater than about 15 per cent generally prevent significant substitution,
In competing for a site in a lattice structure, stronger bonds are formed by the smaller ion or the ion with the higher charge,
Differences in the degree of covalency of the bonds formed by two elements may affect their ability to co-substitute even though they may be of similar size and charge.

TABLE 6. Averages of gabbroic-basaltic rocks (in percentage)

Element	Gabbros (160)[1]	Tholeiitic basalts (137)[1]	Alkali olivine basalts (96)[1]
SiO_2	48.36	50.83	45.78
TiO_2	1.32	2.03	2.63
Al_2O_3	16.81	14.07	14.64
Fe_2O_3	2.55	2.88	3.16
FeO	7.92	9.00	8.73
MnO	0.18	0.18	0.20
MgO	8.06	6.34	9.39
CaO	11.07	10.42	10.74
Na_2O	2.26	2.23	2.63
K_2O	0.56	0.82	0.95
H_2O^+	0.64	0.91	0.76
P_2O_5	0.24	0.23	0.39
qz^2	—	3.5	—
or	3.3	5.0	6.1
ab	18.9	18.9	18.3
an	34.2	25.9	24.7
ne	—	—	2.3
$CaSiO_3$	8.0	10.3	10.8
$MgSiO_3$	14.0	15.8	7.1
$FeSiO_3$	7.4	11.2	2.9
Mg_2SiO_4	4.3	—	11.5
Fe_2SiO_4	2.5	—	5.0
mt	3.7	4.2	4.6
il	2.4	3.8	5.0
ap	0.6	0.5	1.0

1. Number of analyses used for finding averages.
2. qz = quartz; or = K-feldspar; ab = albite; an = anorthite; ne = nepheline; mt = magnetite; il = ilmenite; ap = apatite.
Source: Nockolds, 1954.

TABLE 7. Averages of periodic and anorthositic rocks (in percentage)

Element	Periodite (23)[1]	Anorthosite (9)[1]
SiO_2	43.54	54.54
TiO_2	0.81	0.52
Al_2O_3	3.99	25.72
Fe_2O_3	2.51	0.83
FeO	9.84	1.46
MnO	0.21	0.02
MgO	34.02	0.83
CaO	3.46	9.62
Na_2O	0.56	4.66
K_2O	0.25	1.06
H_2O^+	0.76	0.63
P_2O_5	0.05	0.11
qz^2	—	1.4
or	1.7	6.7
ab	4.7	39.3
an	7.5	45.9
ne	—	—
$CaSiO_3$	3.9	0.3
$MgSiO_3$	14.8	2.1
$FeSiO_3$	2.6	1.2
Mg_2SiO_4	49.1	—
Fe_2SiO_4	9.6	—
mt	3.7	1.2
il	1.5	0.9
ap	0.1	0.3

1. Number of analyses used for finding averages.
2. qz = quartz; or = K-feldspar; ab = albite; an = anorthite; ne = nepheline; mt = mangetite; il = ilmenite; ap = apatite.
Source: Nockolds, 1954.

These general rules provide insight into the characteristic behaviour of many elements, e.g. why some trace elements which readily substitute for major elements are commonly dispersed in other minerals, or why some trace elements which are dissimilar to major elements tend only to form independent minerals.

Metamorphism frequently causes the redistribution of elements in minerals but may not alter the whole rock analysis. Some of the more mobile trace elements may be lost under higher grades of metamorphism or if the metamorphic process introduces leaching by solutions. If partial melting is reached, elements will crystallize in much the same patterns as in igneous rocks.

The formation of sedimentary rocks offers another opportunity for redistribution of elements and the grouping of new mineral assemblages. In sedimentary processes several factors influence the element distribution patterns observed. Such factors are chemical precipitation, solubility, resistance to weathering, specific gravity of resistates, biological concentration, reactivity with organic compounds, ionic substitution in clays or sedimentary minerals and absorption in clays and shales. In general, trace elements which are not concentrated in resistates (minerals resistant to weathering) are enriched in clays and shales relative to sandstones. Trace elements similar to calcium are enriched in carbonate rocks. Organic rich sediments are generally enriched in several trace elements.

Two general references present excellent reviews and bibliographies of the distribution of the elements. Discussions of the distributions of most of the elements can be found in the *Handbook of Geochemistry*, Volume II, Nos. 1, 2 and 3, 1969–73, edited by K. H. Wedepohl. A somewhat older text, Rankama and Sahama (1950), still represents a good summary of the manner of occurrence of the elements.

Figure 1 in Green's paper (1959) presents an excellent summary of element properties and element distribution and provides a ready and concise reference to important geochemical information.

HYDROGEN (H)

Hydrogen is covalently bound to oxygen (or sometimes

TABLE 8. Averages of alkalic rocks (in percentage)

Element	Nepheline syenites (80)[1]	Phonolites (47)	Essexites (15)	Nepheline tephrites (8)	Leucite tephrites (31)	Ijolites (11)	Olivine nephelinites (21)	Olivine leucitites (11)	Olivine melilitites (10)
SiO_2	55.38	56.90	46.88	44.82	47.05	42.58	40.29	43.64	37.08
TiO_2	0.66	0.59	2.81	2.65	1.54	1.41	2.90	2.54	3.31
Al_2O_3	21.30	20.17	17.07	15.42	16.05	18.46	11.32	10.82	8.08
Fe_2O_3	2.42	2.26	3.62	4.28	3.49	4.01	4.87	5.11	5.12
FeO	2.00	1.85	5.94	6.61	5.78	4.19	7.69	5.89	7.23
MnO	0.19	0.19	0.16	0.16	0.17	0.20	0.22	0.15	0.18
MgO	0.57	0.58	4.85	7.27	6.20	3.22	13.28	13.86	16.19
CaO	1.98	1.88	9.49	10.32	10.80	11.38	12.99	10.66	16.30
Na_2O	8.84	8.72	5.09	5.30	2.35	9.55	3.14	2.16	2.30
K_2O	5.34	5.42	2.64	1.26	5.38	2.55	1.44	4.09	1.36
H_2O^+	0.96	0.96	0.97	1.56	0.60	0.55	1.08	0.72	1.89
P_2O_5	0.19	0.17	0.48	0.35	0.59	1.52	0.78	0.63	0.96
CO_2	0.17	—	—	—	—	0.38	—	—	—
Cl	—	0.23	—	—	—	—	—	—	—
SO_3	—	0.13	—	—	—	—	—	—	—
or[2]	31.1	31.7	15.6	7.8	22.2	10.0	—		—
ab	32.0	36.2	14.7	12.6	—	—	—	6.9	—
an	2.8	1.7	16.1	14.5	17.5	—	12.8	6.1	7.5
lc	—	—	—	—	7.4	3.9	6.5	13.8	6.5
ne	23.3	18.7	15.3	17.3	10.8	43.7	14.2	9.9	10.5
Ca_2SiO_4	—	—	—	—	—	—	1.6	—	12.8
$CaSiO_3$	2.1	2.9	11.6	14.3	13.6	18.5	17.2	17.8	10.7
$MgSiO_3$	1.2	1.4	8.2	10.4	9.3	8.0	13.1	14.5	8.6
$FeSiO_3$	0.8	0.9	2.4	2.5	3.2	2.4	2.2	1.1	0.8
Mg_2SiO_4	0.1	—	2.8	5.5	4.3	—	14.1	14.1	22.3
Fe_2SiO_4	0.1	—	0.8	1.5	1.7	—	2.7	1.2	2.5
mt	3.5	3.3	5.3	6.3	5.1	5.8	7.2	7.4	7.4
il	1.4	1.2	5.3	5.0	2.9	2.7	5.5	4.9	6.2
ap	0.4	0.3	1.2	0.8	1.3	3.6	1.8	1.5	2.3
cc	0.4	—	—	—	—	0.9	—	—	—

1. Number of analyses used for finding averages (in parentheses).
2. or = K-feldspar; ab = albite; an = anorthite; lc = leucite; ne = nepheline; mt = magnetite; il = ilmenite; ap = apatite; cc = calcite.
Source: Nockolds, 1954.

nitrogen atoms) in essentially all mineral structures in which hydrogen is present. Other modes of occurrence of hydrogen are of little or no geochemical importance with the exception of petroleum or coal hydrocarbons. Water molecules, H_2O, or hydroxyl ions, OH^-, are the most common forms of occurrence of hydrogen in minerals. Hydrogen is also present in some micas, amphiboles, and clay minerals as hydronium ions, H_3O^+. Ammonium ions, NH_4^+, are present in some minerals.

LITHIUM (Li)

The major characteristic of the alkali metals, of which lithium is a member, is a single valence electron surrounding a stable noble gas configuration. This electron is readily removed to leave a stable, singly charged cation, Li^+. Most of the alkali metals have similar properties but a few differences are worth noting. Because of its small size, Li substitutes more readily with the alkaline earth metal, magnesium, than with other alkali metals. The Li/Mg ratio in minerals increases with increasing differentiation. The ratio increases through the series orthopyroxene, clinopyroxene, amphibole, mica, in which the Li concentration also generally increases. Lithium may become concentrated in pegmatites to the point where it forms independent minerals generally containing F, Cl, P, or Mn. Some of these minerals include spodumene, lepidolite, petalite, lithium tourmaline and amblygonite. The average Li concentration in the upper third of the continental crust is 20 p.p.m. Lithium concentrations in granites average about 30 p.p.m. and gabbros about 10 p.p.m. (Heier and Adams, 1964). Heier and Billings (1970) found that marine shales averaged 76 p.p.m. Li.

TABLE 9. Distribution of the elements in the Earth's crust (in parts per million)[1]

Atomic No. and Element		Ultrabasic	Basaltic rocks	Granitic rocks		Syenites	Shales	Sandstones	Carbonates	Carbonate	Clay
				High calcium	Low calcium						
1 Hydrogen	H	A	A	A	A	A	A	A	A	A	A
2 Helium	He	B	B	B	B	B	B	B	B	B	B
3 Lithium	Li	0.X	17	24	40	28	66	15	5	5	57
4 Beryllium	Be	0.X	1	2	3	1	3	0.X	0.X	0.X	2.6
5 Boron	B	3	5	9	10	9	100	35	20	55	230
6 Carbon	C	A	A	A	A	A	A	A	A	A	A
7 Nitrogen	N	6	20	20	20	30	A	A	A	A	A
8 Oxygen	O	A	A	A	A	A	A	A	A	A	A
9 Fluorine	F	100	400	520	850	1,200	740	270	330	540	1,300
10 Neon	Ne	B	B	B	B	B	B	B	B	B	B
11 Sodium	Na	4,200	18,000	28,400	25,800	40,400	9,600	3,300	400	20,000	40,000
12 Magnesium	Mg	204,000	46,000	9,400	1,600	5,800	15,000	7,000	47,000	4,000	21,000
13 Aluminium	Al	20,000	78,000	82,000	72,000	88,000	80,000	25,000	4,200	20,000	84,000
14 Silicon	Si	205,000	230,000	314,000	347,000	291,000	73,000	368,000	24,000	32,000	250,000
15 Phosphorus	P	220	1,100	920	600	800	700	170	400	350	1,500
16 Sulphur	S	300	300	300	300	300	2,400	240	1,200	1,300	1,300
17 Chlorine	Cl	85	60	130	200	520	180	10	150	21,000	21,000
18 Argon	Ar	B	B	B	B	B	B	B	B	B	B
19 Potassium	K	40	8,300	25,000	42,000	48,000	26,600	10,700	2,700	2,900	25,000
20 Calcium	Ca	25,000	76,000	25,300	5,100	18,000	22,100	39,100	302,300	312,400	29,000
21 Scandium	Sc	15	30	14	7	3	13	1	1	2	19
22 Titanium	Ti	300	13,800	3,400	1,200	3,500	4,600	1,500	400	770	4,600
23 Vanadium	V	40	250	88	44	30	130	20	20	20	120
24 Chromium	Cr	1,600	170	22	4.1	2	90	35	11	11	90
25 Manganese	Mn	1,620	1,500	540	390	850	850	X0	1,100	1,000	6,700
26 Iron	Fe	94,300	86,500	29,600	14,200	36,700	47,200	9,800	3,800	9,000	65,000
27 Cobalt	Co	150	48	7	1.0	1	19	0.3	0.1	7	74
28 Nickel	Ni	2,000	130	15	4.5	4	68	2	20	30	225
29 Copper	Cu	10	87	30	10	5	45	X	4	30	250
30 Zinc	Zn	50	105	60	39	130	95	16	20	35	165
31 Gallium	Ga	1.5	17	17	17	30	19	12	4	13	20
32 Germanium	Ge	1.5	1.3	1.3	1.3	1	1.6	0.8	0.2	0.2	2
33 Arsenic	As	1	2	1.9	1.5	1.4	13	1	1	1	13
34 Selenium	Se	0.05	0.05	0.05	0.05	0.05	0.6	0.05	0.08	0.17	0.17
35 Bromine	Br	1	3.6	4.5	1.3	2.7	4	1	6.2	70	70
36 Krypton	Kr	B	B	B	B	B	B	B	B	B	B
37 Rubidum	Rb	0.2	30	110	170	110	140	60	3	10	110
38 Strontium	Sr	1	465	440	100	200	300	20	610	2,000	180
39 Yttrium	Y	0.X	21	35	40	20	26	40	30	42	90
40 Zirconium	Zr	45	140	140	175	500	160	220	19	20	150
41 Niobium	Nb	16	19	20	21	35	11	0.0X	0.3	4.6	14
42 Molybdenum	Mo	0.3	1.5	1.0	1.3	0.6	2.6	0.2	0.4	3	27
43 Technetium	Tc	C	C	C	C	C	C	C	C	C	C
44 Ruthenium	Ru	D	D	D	D	D	D	D	D	D	D
45 Rhodium	Rh	D	D	D	D	D	D	D	D	D	D
46 Palladium	Pd	0.12	0.02	0.00X	0.00X	D	D	D	D	D	D
47 Silver	Ag	0.06	0.11	0.051	0.037	0.0X	0.07	0.0X	0.0X	0.0X	0.11
48 Cadmium	Cd	0.X	0.22	0.13	0.13	0.13	0.3	0.0X	0.035	0.0X	0.42
49 Indium	In	0.01	0.22	0.0X	0.26	0.0X	0.1	00X	0.0X	0.0X	0.08
50 Tin	Sn	0.5	1.5	1.5	3	X	6.0	0.X	0.X	0.X	1.5
51 Antimony	Sb	0.1	0.2	0.2	0.2	0.X	1.5	0.0X	0.2	0.15	1.0
52 Tellurium	Te	D	D	D	D	D	D	D	D	D	D
53 Iodine	I	0.5	0.5	0.5	0.5	0.5	2.2	1.7	1.2	0.05	0.05

James E. Mielke

TABLE 9—*contd*

Atomic No. and Element		'Igneous' rocks					Sedimentary rocks			Deep-sea sediments	
		Ultrabasic	Basaltic rocks	Granitic rocks		Syenites	Shales	Sandstones	Carbonates	Carbonate	Clay
				High calcium	Low calcium						
54 Xenon	Xe	B	B	B	B	B	B	B	B	B	B
55 Cesium	Cs	0.X	1.1	2	4	0.6	5	0.X	0.X	0.4	6
56 Barium	Ba	0.4	330	420	840	1,600	580	X0	10	190	2,300
57 Lanthanum	La	0.X	15	45	55	70	92	30	X	10	115
58 Cerium	Ce	0.X	48	81	92	161	59	92	11.5	35	345
59 Praseodymium	Pr	0.X	4.6	7.7	8.8	15	5.6	8.8	1.1	3.3	33
60 Neodymium	Nd	0.X	20	33	37	65	24	37	4.7	14	140
61 Promethium	Pm	C	C	C	C	C	C	C	C	C	C
62 Samarium	Sm	0.X	5.3	8.8	10	18	6.4	10	1.3	3.8	38
63 Europium	Eu	0.X	0.8	1.4	1.6	2.8	1.0	1.6	0.2	0.6	6
64 Gadolinium	Gd	0.X	5.3	8.8	10	18	6.4	10	1.3	3.8	38
65 Terbium	Tb	0.X	0.8	1.4	1.6	2.8	1.0	1.6	0.2	0.6	6
66 Dysprosium	Dy	0.X	3.8	6.3	7.2	13	4.6	7.2	0.9	2.7	27
67 Holmium	Ho	0.X	1.1	1.8	2.0	3.5	1.2	2.0	0.3	0.8	7.5
68 Erbium	Er	0.X	2.1	3.5	4.0	7.0	2.5	4.0	0.5	1.5	15
69 Thulium	Tm	0.X	0.2	0.3	0.3	0.6	0.2	0.3	0.04	0.1	1.2
70 Ytterbium	Yb	0.X	2.1	3.5	4.0	7.0	2.6	4.0	0.5	1.5	15
71 Lutetium	Lu	0.X	0.6	1.1	1.2	2.1	0.7	1.2	0.2	0.5	4.5
72 Hafnium	Hf	0.6	2.0	2.3	3.9	11	2.8	3.9	0.3	0.41	4.1
73 Tantalum	Ta	1.0	1.1	3.6	4.2	2.1	0.8	0.0X	0.0X	0.0X	0.X
74 Tungsten	W	0.77	0.7	1.3	2.2	1.3	1.8	1.6	0.6	0.X	X
75 Rhenium	Re	D	D	D	D	D	D	D	D	D	D
76 Osmium	Os	D	D	D	D	D	D	D	D	D	D
77 Iridium	Ir	D	D	D	D	D	D	D	D	D	D
78 Platinum	Pt	D	D	D	D	D	D	D	D	D	D
79 Gold	Au	0.006	0.004	0.004	0.004	0.00X	0.00X	0.00X	0.00X	0.00X	0.00X
80 Mercury	Hg	0.0X	0.09	0.08	0.08	0.0X	0.4	0.03	0.04	0.0X	0.X
81 Thallium	Tl	0.06	0.21	0.72	2.3	1.4	1.4	0.82	0.0X	0.16	0.8
82 Lead	Pb	1	6	15	19	12	20	7	9	9	80
83 Bismuth	Bi	D	0.007	D	0.01	D	D	D	D	D	D
84 Polonium	Po	E	E	E	E	E	E	E	E	E	E
85 Astatine	At	E	E	E	E	E	E	E	E	E	E
86 Radon	Rn	E	E	E	E	E	E	E	E	E	E
87 Francium	Fr	E	E	E	E	E	E	E	E	E	E
88 Radium	Ra	E	E	E	F.	E	E	E	L	E	E
89 Actinium	Ac	E	E	E	E	E	E	E	E	E	E
90 Thorium	Th	0.004	4	8.5	17	13	12	1.7	1.7	X	7
91 Protactinium	Pa	E	E	E	E	E	E	E	E	E	E
92 Uranium	U	0.001	1	3.0	3.0	3.0	3.7	0.45	2.2	0.X	1.3
93 Neptunium	Np	F	F	F	F	F	F	F	F	F	F
94 Plutonium	Pu	F	F	F	F	F	F	F	F	F	F

1. In some cases, only order of magnitude estimates could be made. These are indicated by the symbol X.

A : These elements are the basic constituents of the biosphere, hydrosphere and atmosphere. Oxygen is also the most important element of the lithosphere, whereas carbon is important in sedimentary rock.

B : The rare gases occur in the atmosphere in the following amounts (volume per cent: He, 0.00052; Ne, 0.0018; A, 0.93; Kr, 0.0001; Xe, 0.000008. He is produced by radioactive decay of U and Th but is also lost to outer space. A^{40} is produced by the radioactive potassium 40 and is the major isotope of argon in the atmosphere. The argon and helium contents of rocks will vary with their age owing to the effect of radioactive decay.

The estimated rare-gas contents of igneous rocks are (in cc per gm of rock): He, 6×10^{-5}; Ne, 7.7×10^{-8}; A, 2.2×10^{-5}; Kr, 4.2×10^{-9}; Xe, 3.4×10^{-10}.

C : These elements do not occur naturally in the Earth's crust.

D : The data for these elements are missing or unreliable.

E : All these elements are present as radioactive nuclides in the decay schemes of U and Th.

F : These elements occur naturally only as a consequence of neutron capture by uranium.

Source: Turekian and Wedepohl, 1961.

TABLE 10. Abundance of the elements in igneous rocks, in meteorites, and in the sun's atmosphere

Atomic No.	Element	Igneous rocks		Meteorites		Sun's atmosphere	
		g/ton	Atoms per 100 Si	g/ton	Atoms per 100 Si	100 mg·m^{-2}	Atoms per 100 Si
1	H	Present		Present		31,600	88,500
2	He	0.003	0.0000076	Present		4,000	2,800
3	Li	65	0.091	4	0.010	0.0006*	0.00024*
4	Be	6	0.0067	1	0.0020	0.0006	0.00019
5	B	3†	0.0028	1.5	0.0024	Probably present††	
6	C	320†	0.27	300	0.33	400	93.3
7	N	46.3	0.033	Present		1,300*	260*
8	O	466,000	296	323,000	347	15,900	2,782.5
9	F	600–900	0.32–0.48	28	0.021	Probably present	
10	Ne	0.00007	0.000000035	Present††		Absent	
11	Na	28,300	12.4	5,950	4.42	400	48.7
12	Mg	20,900	8.76	123,000	87.24	500	57.6
13	Al	81,3000	30.5	13,800	8.79	1,000	100
14	Si	277,200	100	163,000	100	1,000	100
15	P	1,180	0.38	1,050	0.58	0.3*	0.027*
16	S	520	0.16	21,200	11.4	16*	1.4*
17	Cl	314	0.09	1,000–1,500*	0.4–0.6*	Absent	
18	A	0.04	0.00001	Present††		Absent	
19	K	25,900	4.42	1,540	0.69	250*	17.9*
20	Ca	36,300	9.17	13,300	5.71	200	14.0
21	Sc	5	0.0011	4	0.0015	0.2	0.0124
22	Ti	4,400	0.92	1,320	0.47	8	0.47
23	V	150	0.030	39	0.013	5	0.27
24	Cr	200	0.039	3,430	1.13	25	1.35
25	Mn	1,000	0.18	2,080	0.66	40	2.05
26	Fe	50,000	9.13	288,000	89.1	1,000	50.13
27	Co	23	0.004	1,200	0.35	25	1.19
28	Ni	80	0.014	15,680	4.60	60	2.86
29	Cu	70	0.011	170	0.046	6	0.26
30	Zn	132	0.020	138	00.036	5	0.21
31	Ga	15	0.0022	4.2	0.00084	0.006*	0.00024*
32	Ge	7	0.00095	79	0.0188	0.08	0.0031
33	As	5	0.00067	Present		Absent	
34	Se	0.09	0.000012	7	0.0015	Absent	
35	Br	1.62	0.00020	20	0.0043	Absent	
36	Kr					Absent	
37	Rb	310	0.036	3.5	0.00068	0.004*	0.00013*
38	Sr	300	0.035	20	0.0040	0.16	0.005
39	Y	28.1	0.00307	4.72	0.000974	0.03	0.0009
40	Zr	220	0.026	73	0.0139	0.03	0.0009
41	Cb	24	0.0026	0.41	0.000076	0.001*	0.00003*
42	Mo	2.5–15	0.0003–0.0016	5.3	0.00095	0.0025	0.00007
43	Tc	Presence unsettled				Presence unsettled	
44	Ru	Present		2.23	0.00036	0.005	0.00014
45	Rh	0.001	0.00000010	0.80	0.00013	0.0003	0000008
46	Pd	0.010	0.0000009	1.54	0.00025	0.0013	0.00003
47	Ag	0.10	0.000009	2.0	0.00032	0.001	0.000026
48	Cd	0.15	0.000013	Present		0.016	0.0004
49	In	0.1	0.000007	0.15	0.000023	0.0001*	0.0000026*
50	Sn	40	0.00343	20	0.00291	Probably present	
51	Sb	1*	0.000083*	Present		0.0008*	0.000018*
52	Te	0.0018*		0.1*		Absent	
53	I	0.3	0.000024	1	0.000136	Absent	
54	Xe					Absent	
55	Cs	7	0.00053	0.08	0.000010	Absent	
56	Ba	250	0.018	6.9	0.00083	0.25	0.005
57	La	18.3	0.00128	1.58	0.000208	0.008	0.00017

James E. Mielke

TABLE 10—*contd*

Atomic No.	Element	Igneous rocks		Meteorites		Sun's atmosphere	
		g/ton	Atoms per 100 Si	g/ton	Atoms per 100 Si	100 mg·m^{-2}	Atoms per 100 Si
58	Ce	46.1	0.00321	1.77*	0.000232*	0.04	0.0008
59	Pr	5.53	0.000389	0.75	0.0000964	0.0006*	0.000012*
60	Nd	23.9	0.00162	2.59	0.000331	0.016	0.0003
61	Pm	Presence unsettled				Presence unsettled	
62	Sm	6.47	0.000419	0.95	0.000115	0.005	0.000093
63	Eu	1.06	0.000068	0.25	0.000028	0.004*	0.000074*
64	Gd	6.36	0.000394	1.42	0.000165	0.002*	0.000036*
65	Tb	0.91	0.000056	0.45	0.000052	Probably present	
66	Dy	4.47	0.000269	1.80	0.000203	0.006*	0.0001*
67	Ho	1.15	0.000068	0.51	0.000057	Presence unsettled	
68	Er	2.47	0.000144	1.48	0.000163	0.0002*	0.000003*
69	Tm	0.20	0.0000115	0.26	0.000029	0.0005*	0.000008*
70	Yb	2.66	0.000149	1.42	0.000150	0.0016*	0.000026*
71	Lu	0.75	0.000037	0.46	0000048	0.0016*	0.000026*
72	Hf	4.5	0.00030	1.6	0.00015	0.0004	0.000006
73	Ta	2.1	0.00012	0.30	0.000028	Probably present	
74	W	1.5–69	0.000082–0.0038	15	0.00145	0.0003	0.0000046
75	Re	0.001	0.000000054	0.0020	0.00000018	Absent	
76	Os	Present		1.92	0.000174	0.0006*	0.0000088*
77	Ir	0.001	0.00000005	0.65	0.000058	0.0001	0.0000014††
78	Pt	0.005	0.00000027	3.25	0.000287	0.007	0.00010
79	Au	0.005	0.00000026	0.65	0.000057	Probably present	
80	Hg	0.077–0.5	0.0000039–0.000025	Present		Absent	
81	Tl	0.3–3	0.000015–0.0015	Present		Absent	
82	Pb	16	0.00080	11	0.00091	0.003	0.00004
83	Bi	0.2	0.000009	Present		Absent	
84	Po	0.0000000003	$14 \cdot 10^{-15}$	0.00000000003	$2 \cdot 10^{-15}$	Absent	
85	At	Present				Presence unsettled	
86	Rn	Present		Present		Absent	
87	Fa	Present				Presence unsettled	
88	Ra	0.0000013	$58 \cdot 10^{-12}$	0.0000001	$76 \cdot 10^{-13}$	Absent	
89	Ac	0.0000000003	$13 \cdot 10^{-15}$	0.00000000002	$15 \cdot 10^{-16}$	Absent	
90	Th	11.5	0.00050	0.8	0.000059	Present	
91	Pa	0.0000008	$35 \cdot 10^{-12}$	0.00000006	$45 \cdot 10^{-13}$	Absent	
92	U	4	0.00016	0.36	0.000023	Presence unsettled	
93	Np	Probably present				Presence unsettled	
94	Pu	Present				Presence unsettled	
95	Am	Probably present				Presence unsettled	
96	Cm	Probably present				Presence unsettled	

* Uncertain value.
† Considerably higher content in sedimentary rocks.
†† More doubtful value.
Source: Rankama and G. Sahama, 1950.

SODIUM (Na)

Sodium is the most abundant alkali metal. In nature, sodium co-ordinates only with oxygen and halogen atoms. In rock-forming minerals most Na is located in feldspars but it is also important in amphiboles, pyroxenes, zeolites and micas. Sodium is a major element in most igneous rocks and is relatively mobile during metamorphic processes. An important solid solution series exists between Na and Ca in plagioclases, clinopyroxenes and amphiboles. Substitution between Na and K also occurs. In clastic sedimentary rocks most Na is located in clay minerals and detrital minerals, mainly feldspars. In carbonate rocks some additional Na may enter from sea-salt and skeletal material. Sodium is a major constituent of marine evaporite deposits in the form of halite, NaCl and related salts and of fresh-water evaporites such as sodium carbonates. The Na_2O concentrations in shales averages 0.8 per cent (Heier and Billings, 1970).

TABLE 11. Abundances of chemical elements in the Earth's crust and chondrites (in parts per million)

Element	Continental crust	Crust as a whole	Chondrites	Element	Continental crust	Crust as a whole	Chondrites
H	(1,530)[1]	(1,520)	—	Rh	—	—	—
Li	20	18	3	Pd	0.015	0.015	1
Be	3	2	3.6	Ag	0.075	0.08	0.1
B	10	9	2	Cd	015	0.16	0.1
C	200 (5,070)	180 (5,220)	400	In	0.25	0.24	0.01
N	19	19	1	Sn	2.3	2.1	1
O	461,000 (472,650)	456,000 (468,780)	350,000	Sb	0.2	0.2	0.1
F	585	544	28	Te	—	—	0.5
Na	23,550 (22,060)	22,700 (21,370)	7,000	J	0.45	0.46	0.04
Mg	23,300 (18,500)	27,640 (23,400)	140,000	Ca	3	2.6	0.1
Al	82,300 (80,210)	83,600 (80,950)	13,000	Ba	425	390	6
Si	281,500 (279,380)	273,000 (277,970)	180,000	La	39	34.6	0.3
P	1,050 (1,040)	1,120 (960)	500	Ce	66.5	66.4	0.5
S	350 (530)	340 (400)	20,000	Pr	9.2	9.1	0.1
Cl	145 (620)	126 (500)	70	Nd	41.5	39.6	0.8
K	20,850 (23,620)	18,400 (19,420)	850	Sm	7.05	7.02	0.2
Ca	41,500 (40,000)	46,600 (49,960)	14,000	Eu	2.0	2.14	0.08
Sc	22	25	6	Gd	6.2	6.14	0.4
Ti	5,650 (4,380)	6,320 (5,040)	500	Tb	1.2	1.18	0.05
V	120	136	70	Dy	—	—	0.35
Cr	102	122	2,500	Ho	1.3	1.26	0.07
Mn	950 (1,110)	1,060 (1,240)	2,000	Er	3.5	3.46	0.2
Fe	56,300 (46,810)	62,200 (50,880)	250,000	Tu	0.52	0.5	0.04
Co	25	29	800	Yb	3.2	3.1	0.2
Ni	84	99	13,500	Lu	—	—	0.03
Cu	60	68	100	Hf	3.0	2.8	0.5
Zn	70	76	50	Ta	2.0	1.7	0.02
Ga	19	19	3	W	1.25	1.2	0.15
Ge	1.5	1.5	10	Re	0.0007	0.0007	0.0008
As	1.8	1.8	0.3	Os	—	—	0.5
Se	0.05	0.05	10	Ir	—	—	0.5
Br	2.4	2.5	0.5	Pt	—	—	2
Rb	90	78	5	Au	0.004	0.004	0.17
Sr	370	384	10	Hg	0.085	0.086	3
Y	33	31	1	Tl	0.85	0.72	0.001
Zr	165	162	30	Pb	14	13	0.2
Nb	20	20	0.3	Bi	0.0085	0.0082	0.003
Mo	1.2	1.2	0.6	Th	9.6	8.1	0.04
Ru	—	—	—	U	2.7	2.3	0.015

1. The abundances of the main elements obtained by a quantitative method based on the volume measurements are shown in parentheses.

Source: Ronov and Yaroshevsky, 1972.

TABLE 12. Crustal abundance of minerals

Mineral	Volume percentage
Plagioclase	42
Potash feldspar	22
Quartz	18
Amphibole	5
Pyroxene	4
Biotite	4
Magnetite, ilmenite	2
Olivine	1.5
Apatite	0.5

Source: Wedephl, 1971.

TABLE 13. Goldschmidt's geochemical classification of the elements

Siderophile	Chalcophile	Lithophile	Atmophile
Fe Co Ni Ru	Cu Ag (Au)[1]	Li Na K Rb Cs Be	H N (C) (O)
Rh Pd Re Os	Zn Cd Hg Ga	Mg Ca Sr Ba B Al	(F) (Cl) (Br) (I)
Ir Pt Au Mo	In Tl (Ge) (Sn)	Sc Y R.E.[2] (C) Si Ti	Inert gases
Ge Sn C P	Pb As Sb Bi S	Zr Hf Th (P) V Nb	
(Pb) (As) (W)	Se Te (Fe)	Ta O Cr W U (Fe)	
	(Mo) (Re)	Mn F Cl Br I (H)	
		(Tl) (Ga) (Ge) (N)	

1. Parentheses round a symbol indicate that the element belongs primarily in another group, but has some characteristics that relate it to this group.
2. Rare-earth metals.

Group Period	I	II														III	IV	V	VI	VII	O
1	H 1																				He 2
2	Li 3	Be 4			Transition metals											B 5	C 6	N 7	O 8	F 9	Ne 10
3	Na 11	Mg 12	IIIA	IVA	VA	VIA	VIIA		VIII			IB	IIB			Al 13	Si 14	P 15	S 16	Cl 17	Ar 18
4	K 19	Ca 20	Sc 21	Ti 22	V 23	Cr 24	Mn 25	Fe 26	Co 27	Ni 28	Cu 29	Zn 30				Ga 31	Ge 32	As 33	Se 34	Br 35	Kr 36
5	Rb 37	Sr 38	Y 39	Zr 40	Nb 41	Mo 42	(Tc) 43	Ru 44	Rh 45	Pd 46	Ag 47	Cd 48				In 49	Sn 50	Sb 51	Te 52	I 53	Xe 54
6	Cs 55	Ba 56	* 57-71	Hf 72	Ta 73	W 74	Re 75	Os 76	Ir 77	Pt 78	Au 79	Hg 80				Tl 81	Pb 82	Bi 83	Po 84	(At) 85	Rn 86
7	(Fr) 87	Ra 88	** 89																		

Rare-earth metals	La 57	Ce 58	Pr 59	Nd 60	(Pm) 61	Sm 62	Eu 63	Gd 64	Tb 65	Dy 66	Ho 67	Er 68	Tm 69	Yb 70	Lu 71
Actinide metals	Ac 89	Th 90	Pa 91	U 92	(Np) 93	(Pu) 94	(Am) 95	(Cm) 96	(Bk) 97	(Cf) 98	(E) 99	(Fm) 100	(My) 101	(No) 102	(Lw) 103

FIG. 1. Periodic classification of the elements. (Elements whose symbols are enclosed in parentheses do not occur in nature, but have been prepared artificially by nuclear reactions.)

POTASSIUM (K)

The third member of the alkali metal family is potassium. Three isotopes occur naturally, ^{39}K, ^{40}K, and ^{41}K. Of these ^{40}K is radioactive and decays to ^{40}Ca and ^{40}Ar. This radioactive decay forms the basis for the K–Ar age-dating technique. Potassium occurs naturally only associated with oxygen and halogens. Because of their similar size, Rb easily substitutes for the more abundant K. The elements Pb and Ba also substitute for K in trace amounts. Although a great many minerals contain K in varying amounts, it is especially important in the magmatic formation of feldspars, micas, leucite and nepheline. Most of the potassium in the earth's crust is contained in alkali feldspar in which some Na and Ca is also associated. Potassium appears to be mobile under medium- to high-grade metamorphic conditions. Potassium in sedimentary rocks is incorporated into clay minerals, detrital feldspars, K-micas, and glauconite. The major K mineral in evaporite deposits is sylvite, KCl, although many others are found. Shales averaged 2.45 per cent K_2O by weight (Heier and Billings, 1970).

RUBIDIUM (Rb)

Rubidium does not form minerals of its own in nature but is dispersed in potassium minerals, mainly micas and feldspars. Rubidium occurs naturally in two iso- topes, ^{85}Rb and ^{87}Rb. The radioactive decay of ^{87}Rb and ^{87}Sr forms another important age-dating technique. Some association of Rb^+ and Tl^+ is found in muscovite and microcline; rubidium also replaces Cs to a limited extent. Ratios of K/Rb have been used to characterize rock types and have been a valuable means of studying the geochemistry of rubidium. In rock series progressing from mafic to felsic rocks, the Rb concentration increases twice as much as the K concentration. Granitic rocks average about 196 p.p.m. Rb (Hurley et al., 1962). In shales Rb is found to be concentrated relative to K. Rubidium concentrations in shales averaged 164 p.p.m. Rb (Heier and Billings, 1970).

CAESIUM (Cs)

Caesium is the heaviest member of the stable alkali metal family. Cs^+ is similar in size to Rb^+ and Tl^+; however, because of the greater abundance of K-minerals, Cs is usually found associated with K. Consequently, caesium is found in micas and feldspars. Ratios of K/Cs appear to decrease with increasing differentiation. Pegmatite minerals, especially muscovite and lepidolite, may contain high concentrations of Cs. The average Cs concentration in the continental crust is 3 p.p.m. (Heier and Billings, 1970). Granites averaged 3 to 6 p.p.m. Cs. Caesium is adsorbed on clay minerals and is associated with these in sedimentary rocks. Horstman (1957) found average Cs concentration in shales to be 5 p.p.m..

BERYLLIUM (Be)

The common oxidation state of beryllium in minerals is Be^{2+}. Geochemically, Be correlates with Li, B, and F. Despite the valence difference, Be^{2+} also chemically resembles Al^{3+} to some degree. Both Be and Al are strongly lithophilic. The only stable isotope of beryllium is ^9Be. Beryllium is found as a trace element in most rock-forming minerals. In addition, Be forms a number of minerals of its own, mostly silicates with some borates, carbonates, phosphates and oxides. The most important beryllium mineral is beryl, $Al_2Be_3Si_6O_{18}$. In basalts and gabbros Be contents generally range from 0.3 to 2.0 p.p.m. and average 0.6 p.p.m. In granitic rocks Be content varies from 1 to 30 p.p.m. The element is concentrated in pegmatites where most independent beryllium minerals are found. Whole rock concentrations in pegmatites can reach several hundred p.p.m. Metamorphic rocks are generally low in beryllium, averaging 3.5 p.p.m. (Hörmann, 1969). Higher concentrations of Be are found in clay-bearing sedimentary rocks (1 to 3 p.p.m.) than in sandstones. Hörmann (1969) estimated that bauxites contain 5 p.p.m. Be.

MAGNESIUM (Mg)

Three stable isotopes of magnesium occur in nature: ^{24}Mg, ^{25}Mg and ^{26}Mg, of which the lightest is the most abundant. Much of the geochemistry of magnesium is related to the great similarity of the Mg^{2+} and Fe^{2+} ions and their subsequent ability to substitute for each other in crystal lattices. During magmatic differentiation, MgO is concentrated in early precipitates, such as high-temperature olivines and pyroxenes, but may appear as amphiboles or micas at high oxygen and water fugacity. The major mineral occurrences of Mg in metamorphic rocks are biotite, hornblende and pyroxenes. Carbonates such as dolomite, Mg-calcite, and magnesite and associated silicates, talc and chondrodite, are also relatively common in marbles and calc-silicate skarns. Salt deposits contain small amounts of Mg in minerals such as carnallite, kainite and kieserite, but the major concentrations of magnesium in sediments are in carbonates such as dolomite and Mg-calcite and silicates such as chlorite and glauconite.

CALCIUM (Ca)

Calcium is the fifth most abundant element in the crust and forms many common rock-forming minerals. It is often substituted for by Na^+, Mg^{2+}, Mn^{2+} and Ce^{3+} or forms solid solutions with Na in plagioclase feldspars. Calcium carbonate and calcium sulphate are of significant commercial importance, Other calcium minerals include phosphates, borates and nearly every type of silicate. In magmatic differentiation, Ca tends to remain in the melt while Mg-rich olivenes form, but Ca enters into pyroxenes, amphiboles and calcic plagioclase as temperatures begin to fall. Calcium concentrations in igneous rocks average 36,300 p.p.m. (Rankama and Sahama, 1950). In sedimentary rocks calcium carbonate is the major constituent of limestones. Evaporite deposits also contain calcium minerals.

STRONTIUM (Sr)

In minerals, strontium always occurs in the +2 valency state. Strontium has four stable isotopes of which the isotope ^{87}Sr is formed by the decay of ^{87}Rb, an event that is useful in geochronology. Several strontium minerals are known. Strontium is one of the most abundant trace elements in the upper crust and follows Ca very closely. Strontium also substitutes for K^+ and to a lesser extent Ba^{2+} and Pb^{2+} in minerals formed in igneous rocks. Feldspars are the most common host for Sr followed by apatite, pyroxenes, amphiboles and micas. In igneous rocks, independent Sr minerals are rare and found mainly in pegmatites and hydrothermal deposits. The Sr concentrations in igneous rocks average about 150 p.p.m. (Rankama and Sahama, 1950). The most important Sr minerals are strontianite, $SrCO_3$, and celestite, $SrSO_4$, which are found in hydrothermal deposits and more commonly in sedimentary deposits. Most strontium minerals are formed as either primary or secondary minerals in weathering and depositional processes.

BARIUM (Ba)

Barium forms an ion of +2 oxidation state and has 7 stable isotopes. It frequently replaces Pb^{2+}, Sr^{2+} and occasionally K^+, Ca^{2+} and Ra. Barium forms a number of minerals including silicates, oxides, carbonates, a sulphate and a nitrate. In igneous rocks Ba usually enters into other silicates such as feldspars and micas. Apatite and calcite also may contain Ba. Gabbros and continental tholeiitic basalts average 246 p.p.m. Ba, alkali basalts 613 p.p.m. Ba, and oceanic tholeiitic basalts 14.5 p.p.m. Ba (Puchelt, 1972). High Ca granitic rocks are high in Ba, averaging 800 to 900 p.p.m. Puchelt (1972) found that granites averaged 732 p.p.m. Ba. The barium content of metamorphic rocks varies widely. The concentration of Ba in sediments is controlled by formation of $BaSO_4$ and absorption of Ba^{2+} by clays, hydroxides and organic matter. Consequently, shales commonly vary from 250 to 800 p.p.m. Ba, averaging around 600 p.p.m.; carbonates range from 1 to 10,000 p.p.m. Ba, averaging near 100 p.p.m.; and sandstones vary from 5 to 900 p.p.m., averaging a little over 300 p.p.m. Ba (Puchelt, 1967, 1972). Manganese nodules may contain up to 20,000 p.p.m. Ba.

BORON (B)

Boron has an oxidation state of +3 and is nearly always bound to oxygen to form anions such as BO_3^{3-} or BO_4^{5-}.

Hydrated borate minerals such as borax, $Na_2B_4O_7 \cdot 10\,H_2O$, and kernite, $Na_2B_4O_7 \cdot 4\,H_2O$, are the most common mode of B occurrence. These minerals are found in arid region salt deposits formed in closed basins. The average B concentration in the Earth's crust is less than 10 p.p.m. In igneous rocks boron is concentrated in late-stage magmatic differentiates. Some B is taken into biotites and amphiboles in earlier stages but most B appears in pegmatitic tourmalines. Fumaroles and volcanic emanations frequently contain boric acid or other boron compounds. Igneous rocks roughly average 3 p.p.m. B. In sedimentary rocks B is mainly incorporated into clay minerals or may be in retained porewater in marine sediments. Marine evaporites may also be high in B.

ALUMINIUM (Al)

Aluminium is found in the $+3$ oxidation state. Next to oxygen and silicon, aluminium is the third most abundant element in the Earth's crust. It normally occupies two types of structural positions: within the SiO_4 tetrahedra replacing Si; or outside the Si–O framework replacing Mg or Fe. Aluminium frequently forms solid solutions with many cations, notably Si^{4+}, Fe^{3+}, Cr^{3+}, Mg^{2+} and Sc^{3+}, and it is a major constituent in most rock-forming minerals. Aluminium tends to become concentrated in the first rocks to crystallize from a magma; hence, gabbros have somewhat higher Al content than granitic rocks. The feldspars contain the greatest amount of the aluminium in the Earth's crust. Of these, anorthite contains twice as much Al as albite and K-feldspars. Corundum, Al_2O_3, is relatively rare in igneous rocks. Most independent Al minerals are found in pegmatites and related rocks. In sedimentary rocks, the aluminium content is highest in shales and lowest in carbonates. This obviously reflects the mineral compositions where most Al is found in clay minerals and detrital silicates. For example, among the clay minerals, kaolinite contains nearly 21 per cent Al by weight, illite 15 per cent Al, and montmorillonite roughly 10 per cent Al.

SCANDIUM (Sc)

Scandium is dispersed primarily in ferromagnesium minerals of igneous rocks. Scandium behaves somewhat similarly to the lanthanides, although not so closely as the second 4 d transition element, yttrium. Commonly existing only in the Sc^{3+} ionic state, it generally occupies structural sites of Fe^{3+}, Al^{3+}, Cr^{3+} and Ti^{4+}. Scandium is found mainly in pyroxenes, amphiboles and biotite in igneous rocks or in zircon, titanium garnets and monazite in igneous and metamorphic rocks. Scandium is frequently found in pegmatites and hydrothermal deposits. Granites average in the order of 2–10 p.p.m. Sc (Norman and Haskin, 1968). In sedimentary rocks Sc is mainly adsorbed by clay minerals or hydrous Al or Fe oxides. Relatively high values have been found in some bauxites, phosphorites, laterites and placer sands.

YTTRIUM AND THE LANTHANIDES

Yttrium is included in a discussion of the lanthanide elements because of its similar geochemical behaviour. The III-B elements of the periodic table include lanthanium, cerium, praseodymium, neodymium, samarium, europium, gadolinium, terbium, dysprosium, holmium, erbium, thulium, ytterbium and lutetium. These elements are frequently referred to as the 'rare-earth' elements. However, in terms of actual crustal abundance they are more abundant than Co, Pb or Sn. More than 200 minerals contain trace amounts of the lanthanides; however, the highest concentrations occur in bastnaesite, monazite and cerite. Lanthanides are enriched in pegmatites up to thirty times their average concentrations in igneous rocks. Hermann (1970) calculated average concentrations in granitic rocks for Y and lanthanides combined of 290 p.p.m. which is three times their average concentration in basalts. The lanthanides primarily occur in the dark-coloured, calcium-bearing minerals such as pyrochlore, loparite and sphene. Little is known about the redistribution of these elements in metamorphic rocks. There appears to be little fractionation between lighter and heavier lanthanides during metamorphism. In sedimentary rocks some phosphates appear to concentrate lanthanide elements during diagenesis.

CARBON (C)

Carbon has two stable isotopes, ^{12}C and ^{13}C. In addition, radioactive ^{14}C is used in age-dating of organic material. The $^{12}C/^{13}C$ ratios are useful to identify formational processes of rocks and sediments. The most important carbon-containing minerals are the carbonates. Some bicarbonates, graphite and diamonds are also found. Organic carbon is important in coal, oil and gas in geologic deposits. Carbon concentrations in igneous rocks average between 100 and 200 p.p.m. The volatiles CO_2 and CO are both important constituents of magmatic gases and hydrothermal fluids. Carbonates are rare in pegmatites. Hoefs (1969) calculated mean values of 600 p.p.m. CO_2 as carbonate and 200 p.p.m. elementary C for granites. Carbon dioxide is released from carbonate minerals during metamorphism. Generally, higher-grade metamorphic rocks are lower in carbon content than lower-grade metamorphic rocks. Carbon is a major constituent of limestones and other carbonate rocks. Clastic sediments are generally enriched in organically derived carbon.

SILICON (Si)

Silicon is the second most common element in the earth's crust. Silicon–oxygen bonds form the common tetrahedral framework unit from which most silicate minerals are derived, either as isolated tetrahedra, corner-shared tetrahedra or, rarely, edge-shared fibrous SiO_2. Silicon also rarely co-ordinates octahedrally, as in stishovite and some high-pressure feldspars. Another mineral form of silicon is carborundum, SiC. Isolated tetrahedra form common minerals such as olivines and garnets. Double tetrahedra such as epidote and rare triple tetrahedra are also found. Another class of silicates is the single chains of which enstatite and augite are examples. Double chains form the framework of amphiboles such as hornblende and tremolite. Tetrahedral ring minerals include beryl and tourmaline. Tetrahedral sheet structures form the micas and clay minerals. Examples of tetrahedral networks are quartz and the feldspars. In sediments, the most abundant silicon minerals are quartz, feldspars, micas and clay minerals. Cherts and silica-cemented sandstones can reach nearly 100 per cent silica. In limestones silica, in the form of quartz, is usually present to some extent.

TITANIUM (Ti)

Titanium mainly occurs in minerals in the Ti^{4+} state. It is common as a minor constituent in many minerals and forms several minerals of its own. The most important titanium minerals are ilmenite, titanomagnetite, rutile and sphene. Among the elements titanium replaces in minerals are Al^{3+}, Fe^{3+}, Nb^{5+}, Ta^{5+} and Mn^{3+}. Ti^{4+} may also substitute for Si^{4+} in the absence of Al^{3+} and Fe^{3+}. Titanium follows iron in magmatic crystallization and is enriched in the early forming stages as ilmenite and titanomagnetite. Titanium is generally enriched most in alkalic rocks. Titanium remains in resistates during weathering and its concentration is highest in sedimentary rocks containing these. Bauxites and laterites may contain up to 4 per cent Ti.

ZIRCONIUM AND HAFNIUM (Zr, Hf)

Zirconium is only found as Zr^{4+} in minerals. It is lithophilic and resembles Si and Ti in many respects. Zirconium tends to form its own silicates such as zircon, $ZrSiO_4$, which accounts for most of the Zr found in igneous rocks. More complex zirconium silicates are also found, particularly in pegmatites. Some Zr enters iron-magnesium minerals in igneous rocks. In magmatic differentiation Zr differs from Ti in becoming concentrated in the later stage rocks such as granites. Zircon is very stable during chemical and physical weathering and consequently zirconium concentrations are highest in sediments containing resistates.

Hafnium behaves very similarly to zirconium and is strongly lithophilic. Most zircons contain an average of about 1 per cent HfO_2. Since hafnium forms no minerals of its own, its abundance is much less than that of Zr. Most Hf is found in Zr minerals and consequent-

ly Hf tends to follows Zr through the geochemical cycle.

THORIUM (Tn)

Thorium is only found in the +4 valency state. Geochemically it is very similar to uranium. Only radioactive isotopes of Th exist in nature although both ^{232}Th and ^{230}Th have very long half-lives. Thorium is useful in several age-dating methods involving U and Pb. In various respects Th chemically resembles U, Zr, Hf, Y and the lanthanides. Thorium minerals are extremely rare but Th commonly occurs as a trace substitute in other minerals including many rock-forming minerals. Much of the Th in the upper crust is incorporated into zircon structures. Thorium concentrations may average between 10 and 20 p.p.m. in granitic rocks and between 0.5 and 2 p.p.m. in basaltic rocks (Rogers and Adams, 1969). Thorium contents of metamorphic rocks are highly variable. Thorium is concentrated in shales, (10–13 p.p.m.) due to adsorption into clays; it is also high in bauxites and bentonites, averaging 49 p.p.m. and 24 p.p.m., respectively (Rogers and Adams, 1969), due to its retention in heavy resistate minerals. Its concentration in limestones is low.

NITROGEN (N)

The most common stable isotope of nitrogen in nature is ^{14}N, although ^{15}N is also found. The ratio of these two isotopes compared with that in the atmosphere has been measured in many rock systems and in coal, gas and petroleum deposits. These data are analysed for clues to the conditions or the origin of formation of these deposits. In mineral compounds of the earth's crust, nitrogen is mainly present as ammonium, NH_4^+. Ammonium can substitute for Na^+ and K^+ in feldspars and frequently enters layer-lattice silicates, particularly montmorillonite, by chemisorption or physical adsorption; ammonia commonly enters zeolites. Elemental nitrogen is frequently found in fluid or gas inclusions in minerals. Nitrogen is also trapped in ring structure silicates. Gases from volcanic vents generally contain nitrogen. Volcanic rocks, averaging 37 p.p.m. N (Wlotzka, 1972), frequently have higher N contents than other igneous rocks probably due to incomplete degassing of the magma during rapid cooling. Granites average 21 p.p.m. N (Wlotzka, 1972). Low N values (averaging 11 p.p.m.) are often found in gabbroic rocks. Nitrogen may be released from rock systems by outgassing during metamorphism. The main source of nitrogen in sedimentary rocks is from organic material. Nitrate salts can form in arid regions and cave deposits where leaching does not occur.

PHOSPHORUS (P)

The only phosphorus compounds known to occur in nature are phosphates. Because of rapid hydrolysis of

James E. Mielke

the P–O–P bond in water, all naturally occurring phosphates are monophosphates. These range from single, octahedral structures to octahedral frameworks. Phosphorus can replace Si in silicates, especially monosilicates, but most P in igneous rocks is found in independent phosphorus minerals such as apatite, $Ca_5(F, Cl, OH)(PO_4)_3$. Phosphorus tends to follow Ti in magmatic differentiation, not because of chemical similarity, but because each of their independent minerals, ilmenite, sphene and apatite, tend to separate out at the same time. In sedimentary rocks most P is found enriched in shales and clays two and a half times the average in igneous rocks. Much of the P in sediments is in resistate minerals, marine phosphate beds, and phosphate concretions containing many elements such as Zn, Cd, In, and Bi.

VANADIUM (V)

In magmas vanadium occurs in the +3 oxidation state although a wide range of valence is possible. Geochemically, it is similar in several ways to Fe^{3+} and is often a trace element in magnetite, pyroxene, amphibole and biotite. Vanadium also occurs in association with Ti, P and As. Vanadium is one of the more abundant trace elements and very rarely forms independent minerals in igneous rocks. It tends to become enriched in basic rocks along with P and Ti and reaches concentrations of 1,400 p.p.m. in apatite-bearing, titaniferous iron ores in Sweden (Landergren, 1948). A number of rare secondary vanadium minerals are found in sedimentary rocks. These minerals often contain Ca, Mn, Fe, Bi, U, Pb, Cu and Zn. Vanadium enters into clay minerals during weathering. Consequently, organic clays and shales are generally higher in V than other sedimentary rocks; laterites and bauxites are also frequently high in vanadium.

COLUMBIUM AND TANTALUM (Cb, Ta)

Columbium and tantalum are very similar geochemically. Columbium and tantalum substitute for Ti in independent titanium minerals and rock-forming minerals and for Zr in zircons. Several independent Cb and Ta minerals are found in pegmatites. Both Cb and Ta are lithophilic elements and are concentrated in late stages during magmatic differentiation. However, Cb is enriched to a greater extent in syenites and nepheline syenites, whereas Ta is more concentrated in granites. Both are concentrated to the greatest extent in pegmatites. The average abundance of Cb in igneous rocks is 24 p.p.m. and the average for Ta is 2.1 p.p.m. (Rankama, 1944). Minerals of Cb and Ta are generally fairly resistant to weathering and may form placer deposits. Some Cb and Ta is incorporated into clay, especially kaolinite. Bauxites are also frequently enriched in Cb

and Ta. Deep-sea manganese nodules may be enriched in Cb by several times its concentration in igneous rocks.

OXYGEN (O)

Oxygen is the most abundant element in the earth's crust. Of the stable isotopes of oxygen, ^{16}O is the most abundant with ^{18}O and ^{17}O of relatively minor importance. Oxygen isotope ratios, $^{18}O/^{16}O$, are of interest geochemically; their fractionation in nature represents a variety of thermodynamic conditions and their interpretation yields information on such problems as phase relations, source, etc. Oxygen contents of rocks and minerals are generally calculated from oxide analyses. For most rock-forming minerals and, consequently, for most igneous rocks the average oxygen content is 40–50 per cent by weight. Most metamorphic rocks fall in the same range. In sedimentary rocks the oxygen content is generally in the range of 40–53 per cent, pure quartzites being at the high end of the range. Oxygen is almost absent in coal.

CHROMIUM (Cr)

Two valency states of chromium are found in nature, Cr^{3+} and Cr^{6+}. Trivalent chromium replaces Al^{3+}, Mg^{2+}, Fe^{3+}, Fe^{2+} and Ti^{4+}. Hexavalent chromium is found in minerals containing chromate, CrO_4^{2-}, or dichromate, $Cr_2O_7^{2-}$. Chromium is strongly oxyphilic in the upper crust and is found only in oxide and silicate minerals. Independent Cr minerals are relatively rare; chromite, Cr_2O_3, is the most important chromium mineral. Most Cr in the upper crust is incorporated into other silicates, mainly augite, hornblende and magnesian olivines. Chromium follows Ti and P in separating into early crystallates during magmatic differentiation; at lower magma temperatures Fe^{3+} is replaced by Cr^{3+}. It will also replace Al^{3+} in minerals containing octahedral aluminium groups, whereas it will not replace Al or Si in tetrahedral co-ordination. Chromium concentrations in granitic rocks in southern Lapland average 2 to 6.8 p.p.m. (Sahama, 1945). Sedimentary rocks containing resistates and clays are highest in chromium. Bauxites may contain concentrations of 2,500 p.p.m. Cr.

TUNGSTEN (W)

Five stable isotopes of tungsten occur in nature. In igneous rocks tungsten is more concentrated in mica than other rock-forming minerals. Tungsten is also found in iron oxides and titanium minerals. The isotope W^{6+} can also substitute for Nb^{5+} and Ta^{5+} in niobium and tantalum minerals. Tungsten is lithophilic in character and is slightly more concentrated in granitic rocks (1.5 p.p.m. average) than in mafic rocks (0.5–1 p.p.m. average) (Krauskopf, 1970). Tungsten is similar to Mo in many respects. In hot spring deposits, manganese oxides may be greatly enriched in W. Essentially all independent tungsten minerals are oxides, the most common being scheelite and wolframite. Tungsten tends to favour the water-rich phases in magmatic differentiation. The element is less concentrated in carbonate rocks than clastic sediments and about equally concentrated in shales and sandstones where it averages 1–2 p.p.m. (Krauskopf, 1970).

MOLYBDENUM (Mo)

Molybdenum is oxyphilic and is concentrated in the last crystallates during magmatic differentiation. Granites may contain appreciable amounts of molybdenum in the form of molybdenite, MoS_2. Concentrations of Mo in silicic rocks may average 2.5 p.p.m. (Sandell and Goldich, 1943). Molybdenum is similar to W although it has a greater chalcophile tendency. Molybdenite and wulfenite, $Pb(MoO_4)$, are the most important Mo minerals. Several independent Mo minerals are found, many of which are formed by secondary processes. Molybdenum accumulates in sediments under reducing conditions.

URANIUM (U)

Uranium is strongly lithophilic and its geochemistry is closely related to that of thorium. All uranium isotopes are radioactive. The uranium decay series are important in several methods of age-dating. Uranium forms a variety of its own minerals mainly containing UO_2^{2+} ions; uraninite, UO_2–U_3O_8, is the most important uranium mineral. Uranium is also a common trace element in most rock-forming minerals and many accessory minerals. The average U concentration in the continental crust is in the range of 1–4 p.p.m.; concentrations are generally higher in intrusive than extrusive rocks. There is a tendency for uranium to become concentrated in the later members of a magmatic differentiation series. Uranium is probably quite mobile in metasomatic reactions but data are sparse. The average U content of shales is roughly the same as for the continental crust as a whole. Average uranium concentrations are high for bauxites—11 p.p.m. (Rogers and Adams, 1969)—presumably due to U retention in resistate minerals. The average U content in carbonate rocks is of the order of 2 p.p.m.

MANGANESE (Mn)

Manganese ions involved in mineral formation are Mn^{2+}, Mn^{3+} and Mn^{4+} of which the Mn^{2+} ion is most common in silicates. Other types of manganese minerals include simple and complex oxides, hydroxides, sulphides, tellurides and selenides. Manganese is a relatively abundant trace element and is strongly lithophilic geochemically. Manganese is related to iron in many respects. In addition to Fe^{2+}, Mn^{2+} can replace Mg^{2+},

Zn^{2+} and Ca^{2+}. Although a number of independent manganese minerals are found in igneous rocks they are relatively rare. Most manganese minerals are found in metamorphic or sedimentary rocks. Most Mn in igneous rocks is found in the structures of other rock-forming minerals. Manganese is concentrated in the late differentiates of magmatic crystallization such as pegmatites and hydrothermal deposits. Rankama and Sahama (1950) found an average of 965 p.p.m. Mn in granites. In sedimentary rocks, manganese is found associated with organic shales and clays formed in reducing environments. Manganese can be enriched in shells by replacing calcium. Deep-sea nodules having high Mn contents are also common.

RHENIUM (Re)

Rhenium has been found in all oxidation states from $+7$ to -1. Two isotopes occur naturally, ^{185}Re and ^{187}Re. The isotope ^{187}Re is radioactive and is used in Re–Os age-dating. Rhenium occurs only as a trace metal in other minerals. The chemistry of Re is similar to that of technetium, molybdenum and to a much lesser extent, manganese. The major host mineral of Re is molybdenite where concentrations as high as 1.88 per cent Re have been found (Magak'yan et al., 1963). The average abundance of Re in igneous rocks is approximately 0.5 p.p.b. (Morris and Short, 1969). Rhenium is concentrated in the early formed sulphide phase during magmatic differentiation, and may reach relatively high concentrations, 50–700 p.p.m., in some sedimentary rocks, particularly uranium–copper sandstones.

IRON (Fe)

Iron normally occurs in two oxidation states in nature, $+2$ and $+3$. Native iron is rare in the Earth's crust. The geochemistry of iron is frequently determined by the conditions under which it shifts from one valency state to the other or forms complexes with inorganic ions and organic chelating agents. Iron oxides, sulphides, carbonates and silicates all play important roles as ore minerals. In addition iron and magnesium together form a number of magmatic silicates such as olivines, pyroxenes, amphiboles and micas. Ferrous oxide forms solid solutions with MgO and Fe_2O_3 commonly substitutes for Al_2O_3 in silicate lattices or forms separate oxide minerals such as hematite and magnetite. Titanium is frequently present in iron oxides such as ilmenite. Iron compounds are deposited by hot springs and fumaroles. In addition to iron oxides and sulphides, sediments also may contain authigenic iron silicates, mainly glauconite. Sedimentary siderite, $(Fe, Mg, Mn)CO_3$, is generally associated with fresh-water swamp environments, high in organic material and iron and low in sulphate. Shales are higher in Fe than sandstones and both contain more iron than limestones. Pelagic clays are commonly enriched with iron.

COBALT AND NICKEL (Co, Ni)

Both cobalt and nickel are siderophilic elements although cobalt less so than nickel. In the upper lithosphere both elements have a chalcophilic and lithophilic tendency. However, cobalt tends to remain in the residual silicate phase during magmatic differentiation, whereas Ni associates with the sulphide phase. While small quantities of Co and Ni are found in common sulphide minerals such as pyrite, pentlandite and pyrrhotite, the bulk of Co and Ni in igneous rocks is bound into the structure of silicate minerals. Both Co and Ni are more concentrated in basic rocks than granitic rocks. In sedimentary rocks both cobalt and nickel are associated with relatively fine-grained organic-rich deposits. This could be linked to their affinity for sulphur, which leads to the formation of sulphides under reducing conditions.

PLATINUM METALS

The platinum metals include ruthenium, rhodium, palladium, osmium, iridium and platinum. Data on platinum metals are relatively sparse. Concentrations of the Pt metals are generally low—in the p.p.b. range—in most rocks and minerals. Platinum metals are associated with sulphides, selenides, tellurides, antimonides, arsenides and bismuth. In gabbroic Cu–Ni sulphide deposits, chalcopyrite, pyrrhotite and pentlandite are the most likely hosts for Pt metals. In the Bushveld Complex (South Africa) Pt metals are also found in pyrite and pyrrhotite. Platinum appears to be concentrated in chromite or chromite-bearing rocks and rare earth minerals. Platinum metals, particularly Pt and Pd, are evidently concentrated in residual fluids of sulphur-rich magmas. Crocket (1969) obtained weighted averages of 21 p.p.b. Pd and 30 p.p.b. Pt for basalts and gabbros. Few analyses of platinum metals in metamorphic and sedimentary rocks are available.

COPPER (Cu)

Copper is found in nature in its native, $+1$, and $+2$ valency states. Copper occurs as a native element or as an alloy with other metals, particularly As and Ag and in numerous simple and complex sulphides, in selenides and tellurides, and as oxides, hydroxides, and in oxygen-containing minerals. The majority of the copper in the Earth's crust is found in sulphur-containing minerals. In silicates, Cu may replace Fe^{2+} and Mg^{2+} to a limited extent. The average concentration of Cu in acidic igneous rocks is 16 p.p.m. and in basic igneous rocks, 150 p.p.m. (Sandell and Goldich, 1943). In sediments, copper is incorporated into clay minerals and may be precipitated as sulphides. Deep-sea manganese nodules are frequently enriched in copper.

SILVER (Ag)

Although native silver is found, the most common valency state of Ag is $+1$. Geochemically Ag resembles Cu and Au in many respects. Silver is a chalcophilic element, most commonly forming simple and complex sulphide minerals. In magmatic crystallization, silver is concentrated in the late differentiates. Silver may be especially enriched in minerals associated with hydrothermal tin deposits. Silver commonly occurs in galena —which may contain up to 2 per cent Ag (Rankama and Sahama, 1950). In sedimentary rocks, Ag concentrations are highest in clays, shales and organic sandstones.

GOLD (Au)

Gold is strongly siderophilic and consequently tends to occur in the metallic state or in metal alloys. Gold minerals usually are tellurides or mixtures of Au with Ag, Pd, Rh, Cu, Hg, Bi and Sb. Gold tends to be concentrated in late crystallates of a magmatic differentiation series and is found mostly in hydrothermal veins and pegmatites. In sedimentary rocks Au is concentrated in resistates and, due to its high specific gravity, tends to form rich placer deposits. Gold is also found in manganese nodules.

ZINC (Zn)

Zinc is found only in the $+2$ valency state. Five stable isotopes of Zn exist in nature. In addition to forming a great many minerals, zinc tends to replace Fe^{2+} and Mg^{2+} in rock-forming silicates and oxides, reaching concentrations of up to a few hundred p.p.m. Zinc concentrations in gabbros and basalts average 80–120 p.p.m., whereas granitic rocks average slightly lower at 30–70 p.p.m. In metamorphic rocks, the geochemistry of Zn is similar to igneous rocks via replacement of ferrous iron and magnesium in biotite, chlorite, amphiboles, staurolite, garnets and magnetite. Zinc is readily adsorbed by clay minerals and also accumulates in sediments containing organic material. Consequently, in black organic shales Zn concentrations are generally higher than in other shales which average around 100 p.p.m. (Wedephol, 1972). Greywackes also average around 100 p.p.m. from detrital Zn containing minerals, whereas quartz sandstones and limestones are lower. Zinc is also slightly concentrated in bauxite deposits, averaging approximately 150–300 p.p.m.

CADMIUM (Cd)

Cadmium exhibits only a $+2$ valency state in nature and has eight stable isotopes. Although Cd forms independent sulphides, carbonates and oxides, it most commonly substitutes for Zn^{2+} in zinc minerals and is nearly always associated with Zn ores. Although similar in size and charge, Cd apparently does not substitute for Ca to any great extent. This may be due to the relatively powerful covalency of the Cd–O bond. Cadmium is a strongly chalcophilic element and tends to become concentrated in hydrothermal rocks and minerals. Granites average 0.1–0.2 p.p.m. Cd (Wakita and Schmitt, 1970). In sedimentary rocks Cd is incorporated into bituminous shales; high concentrations are also found in manganese nodules.

MERCURY (Hg)

All seven stable isotopes of mercury are found in minerals. Mercury is found as a metal and in Hg^+ and Hg^{2+} ions. Metallic mercury is unusual by being in a liquid state in nature. The major mercury minerals are sulphides of which cinnabar, HgS, and metacinnabar, (HG, Zn, Fe)(S, Se), are the most important. Many minerals contain mercury, chiefly those containing Cu, Ag, Au, Zn, Cd, Bi, Pb, Ba and Sr all of which have similar radii. The average Hg content of igneous rocks in the upper crust is 20 p.p.b. (Ehmann and Lovering, 1967). Mineral mercury deposits are currently forming in some hot springs. Mercury is concentrated in marine sediments by adsorption on clay minerals. Shales averaged 400 p.p.b. Hg (Turekian and Wedepohl, 1961).

GALLIUM (Ga)

Two stable isotopes of gallium, ^{69}Ga and ^{71}Ga, occur in nature. Gallium tends to enter into feldspars, micas, amphiboles and magnetite. In rock-forming minerals Ga commonly tends to associate with Al^{3+} and, to a lesser extent, Fe^{3+} and Cr^{3+}. The two independent gallium minerals, a sulphide and a hydroxide, are rare. Alkaline pegmatite minerals are generally higher in gallium; however, the abundance of gallium throughout most igneous rocks is fairly uniform. Gallium has a chalcophillic tendency and is frequently found in sulphide minerals, mainly associated with Zn in sphalerite. Gallium concentrations in metamorphic rocks are generally similar to those in igneous rocks. In sedimentary rocks some investigators have found gallium generally lower in marine deposits than fresh-water deposits (Degens *et al.*, 1957; 1958). However, other factors than salinity apparently have an influence on gallium concentration in sediments and may reverse such a trend. Carbonate rocks are generally low in gallium, and ferro-manganese nodules relatively high.

INDIUM (In)

The more abundant of the two isotopes of indium is radioactive and has a long half-life. Indium is a strongly chalcophilic element and is rare in rock-forming minerals. It is usually found as a trace element in sulphide and tin minerals such as sphalerite, chalcopyrite and cassiterite. Indium frequently associates with Fe, Mn,

Sn, Zn, Cu and Pb. The few independent indium minerals are rare. Generally, In is concentrated in late stage magmatic rocks. Granitic and basaltic rocks average less than 100 p.p.b. indium (Linn and Schmitt, 1972). Indium averages 50–70 p.p.b. in clays and shales.

THALLIUM (Tl)

Thallium has two stable isotopes, ^{203}Tl and ^{205}Tl. The geochemistry of Tl is related to its close association with K and Rb. Thallium minerals are rare and generally associated with late stage hydrothermal deposits and are concentrated to an even greater extent than In in tin-bearing rocks. Thallium is present as a trace element primarily in biotites, potassium feldspar and amphiboles. Concentrations of Tl are generally higher in K feldspars and muscovites from pegmatites. Granitic rocks are enriched with Tl (0.6–3.5 p.p.m.) compared with mafic rocks (de Albuquerque and Shaw, 1972). Thallium is highly mobile in metasomatic processes. In sedimentary rocks Tl concentrations are highest in clastic materials, generally in the 1–3 p.p.m. range. The Tl^+ can be adsorbed into clay minerals in the same manner as Cs^+, Rb^+ and K^+. Sulphides in coal deposits are enriched in thallium. Ferromanganese nodules often concentrate Tl (up to 600 p.p.m.).

GERMANIUM (Ge)

Five stable isotopes of germanium occur in nature. Germanium substitutes for Si and Fe^{3+} in crystal lattices. The Ge concentration in most silicate rocks, igneous, metamorphic and sedimentary, is in the range 0.5–3 p.p.m. The Ge content of rock-forming minerals increases from basaltic to granitic rocks and is highest in pegmatite minerals, especially topaz and garnets. Germanium also tends to be more concentrated in granitic rocks of low calcium content. Granite pegmatites average 10 p.p.m. Ge (Hörmann, 1963). Germanium exhibits a chalcophilic tendency and associates with Cu, Pb and Zn sulphides in hydrothermal mineralizations. In sedimentary rocks Ge concentrations reflect the mineral composition. For example, clay minerals, especially kaolinite and illite, are high in Ge, and quartz and carbonate minerals low. In diagenetic processes Ge is chemically combined with or adsorbed to organic matter, especially in coal deposits.

TIN (Sn)

Tin is found in the native metallic state, Sn^{2+} and Sn^{4+} ions. There are relatively few minerals containing more than trace amounts of tin. Of these, cassiterite, SnO_2, is the most important. Of all the elements, tin has the largest number (10) of stable isotopes. Tin is generally found associated with oxygen in high temperature minerals but has an increased affinity for sulphur at lower temperatures. It tends to become concentrated in later stages of magmatic differentiation until cassiterite is formed. The Sn^{4+} may substitute for Fe^{3+}, Se^{3+} and Tl^{4+}, and Sn^{2+} may replace Ca^{2+} in silicate minerals. Tin may also substitute for columbium, tantalum and tungsten—leading to concentrations in wolframite of up to 1 per cent Sn. Tin is found in common accessory minerals such as apatite, ilmenite, rutile, sphene and magnetite. Relatively high Sn contents have been found in granitic micas. Hamaguchi et al. (1964) reported averages of 3.6 p.p.m. Sn in silicic rocks and 0.9 p.p.m. in mafic rocks. In general, clays do not show any appreciable enrichment of tin over concentrations in igneous rocks.

LEAD (Pb)

Metallic lead is rarely found in nature. The most common Pb mineral is galena, PbS. Three of the four stable Pb isotopes are formed radiogenically by decay of U or Th. This fact has given rise to the systems of age-dating involving U, Th and Pb. As a trace element in igneous rocks, Pb is found to the greatest extent in acidic differentiates where it averages 20 to 30 p.p.m. (Rankama and Sahama, 1950). The Pb^{2+} may replace K^+ in potassium feldspars and Ca^{2+} in minerals such as apatite, micricline, pyroxenes and aragonite. Lead is concentrated in resistates and clay minerals and enters sedimentary carbonate rocks, where it may occupy spaces of Ca^{2+} and Sr^{2+}

ARSENIC (As)

Only a single isotope of arsenic, ^{75}As, is found in nature. Two valency states, As^{3+} and As^{5+}, and the native element occur naturally. In ionic radius and charge, As is similar to and may replace Si^{4+}, Al^{3+}, Fe^{3+} and Ti^{4+} in trace amounts in rock-forming minerals. Arsenic is frequently present in magnetite and ilmenite. Arsenic minerals include simple and complex sulphides, arsenides, arsenates, arsenites, oxides and the native element. The average As concentration in igneous rocks was 1.5 p.p.m. (Onishi, 1969). The As is likely to migrate or be lost under metamorphic conditions. In sedimentary rocks As is present in clay minerals, iron sulphides and organic matter. Onishi (1969) reported that As concentrations in shales averaged 13 p.p.m.

ANTIMONY (Sb)

Two isotopes of antimony occur in nature, ^{121}Sb and ^{123}Sb. Antimony commonly occurs in the +3 and +5 oxidation states and is also found as the native metal. In rock-forming and accessory minerals, such as high magnesian olivines and ilmenite, Sb can substitute for Fe. In addition to the native metal, antimony forms simple and complex sulphides, antimonides and oxides, most of which are found in hydrothermal deposits. The most important antimony mineral is stibnite, Sb_2S_3.

Tetrahedrite, $Cu_{12}(Sb, As)_4Sb_{13}$, pyrargirite, Ag_3SbS_3, and stephanite, Ag_5SbS_4, are also important. Granitic rocks average approximately 0.2 p.p.m. Sb (Onishi, 1969) and may be slightly higher in Sb than basaltic rocks. The Sb in metamorphic rocks would be found primarily in sulphide minerals. Of the common sedimentary rocks, shales are highest in Sb, averaging 1–2 p.p.m., as are some bauxites and phosphorite rocks.

BISMUTH (Bi)

Only one isotope of bismuth is found in nature. Bismuth is most commonly found in sulphide minerals and is even more strictly chalcophilic than the other group VB elements, As and Sb. Bismuth is trivalent in all its minerals. Variations in Bi content in sulphide minerals range over several orders of magnitude. In general, bismuth is concentrated in galena to a greater extent than other sulphide minerals. Bismuth tends to be concentrated in residual melts during differentiation and is frequently found in granitic pegmatites and hydrothermal deposits. The element may be slightly more abundant in sedimentary rocks than igneous rocks but data are sparse. Concentrations of Bi in igneous rocks range from 0.02 to 2 p.p.m. Bismuth appears to be incorporated into manganese nodules although it varies widely in concentration—a common feature of several of the elements found in manganese nodules.

SULPHUR (S)

Sulphur commonly occurs in three types of minerals: simple sulphides; complex sulphides, containing combinations of sulphur and cations of the group V elements such as As, Sb, or Bi with Cu, Pb, Ag, Zn, or Hg; and sulphates which contain the fully oxygenated SO_4^{2-} ion. Sulphide and silicate melts are only partly miscible, causing the sulphide phase to separate early in the crystallization sequence. Most sulphide minerals formed in normal igneous rocks contain Fe as a major constituent. A variety of sulphur minerals are also formed during late-stage hydrothermal mineralization. In sediments under reducing conditions, such as the bottoms of some estuaries, fiords, and enclosed basins, sulphur can form sulphides. Bacteria act to reduce sulphates to native sulphur. Marine evaporite deposits include anhydrite and gypsum beds. Continental evaporites contain a variety of sulphate minerals.

SELENIUM (Se)

Six stable isotopes of selenium occur naturally. Selenium generally occurs in the −2 valency state but can also be found as native selenium and in the +6 oxidation state. Selenium is primarily chalcophilic and does not form silicates. It is frequently a trace element in sulphide minerals. In hydrothermal deposits, Se is concentrated in pyrite, chalcopyrite, arsenopyrite, galena and sphalerite. The element is also found in volcanic gases. Sindeeva (1964) gave an average of 0.14 p.p.m. Se for igneous rocks. In sedimentary rocks, selenium enters fine-grained, clastic sediments and is not found to any extent in evaporite deposits containing sulphur. Selenium is also readily adsorbed by iron hydroxides.

TELLURIUM (Te)

Tellurium has eight stable isotopes. It is strongly chalcophilic in nature and is not found in rock-forming silicate minerals. Tellurium forms a variety of independent minerals. Due to the great differences in ionic radii of Te, Se and S, Te does not substitute for S very readily, whereas Se will do so to a greater extent. Minerals of Te include gold, silver, bismuth, copper and mercury ores. According to Turekian and Wedepohl (1961), the average tellurium content in the upper continental crust is of the order of 0.002 p.p.m. It is found in trace amounts in volcanic gases. The Te concentrations in greywackes and shales are generally less than 0.1 to 1 p.p.m. Concentrations averaging about 48 p.p.m. Te have been reported in deep-sea manganese nodules (Lakin et al., 1963).

FLUORINE (F)

Minerals with fluorine as a major constituent are relatively few in number, the major ones being fluorite, topaz, cryolite, fluor-apatite and villiaumite. Fluorine is present in many other minerals, generally occupying OH^- sites. Most fluorine-bearing minerals are found in or associated with pegmatites. Hydrothermal deposits are another source of fluorine-bearing minerals. In general, alkalic and silicic rocks have higher F contents, averaging 800–1,000 p.p.m., than ultramafic rocks which average 100 p.p.m. (Koritnig, 1972). Fluorine is frequently found in micas, such as phlogopite, in ultramafic rocks and in Li micas, apatite, amphiboles, topaz, phosphates, carbonates, and fluorite in pegmatites. In metamorphic rocks the main F-bearing minerals are micas, chlorites and amphiboles. In clastic sedimentary rocks most fluorine is found in micas, clay minerals, or apatite. The average F concentration in shales is about half that in igneous rocks. Averages in limestones are even lower. Fluorine is found in evaporite deposits where it may be incorporated into the NaCl structure. The F is also found associated with organogenic phosphates.

CHLORINE (Cl)

Chlorine is present as Cl^- and is found in most igneous rocks. Because of its larger size, Cl^- does not replace F^- to a great extent except in apatite where Cl^-, F^-, OH^-, and CO_3^{2-} co-substitute. Chlorine has a lesser tendency to become concentrated in late stage crystallates than F. Due to the solubility of chlorides, essentially no precipitates form in sediments. Chlorine may become incorporated into marine sediments in interstitial waters. Marine evaporite deposits contain large amounts of Cl^- in salts such as NaCl. Continental evaporites also contain chlorine in various salts.

BROMINE AND IODINE (Br, I)

Bromine and iodine are dispersed elements and are never concentrated in igneous rocks. Due to the differences in ionic radii, Br^- and I^- do not co-substitute or replace the other halogens to any extent. In sedimentary rocks both Br and I are evidently derived from organic material and are found in brines associated with petroleum deposits and in marine phosphates. Some evaporite deposits are especially rich in Br and I. Iodine minerals are most abundant in nitrate beds.

NOBLE GASES

The geochemistry of the noble gases is reflected in the stability of their complete electron shells which makes them highly inert. The noble gases are helium, neon, argon, krypton, xenon and radon. Since these elements do not form compounds, their occurrence in the rocks of the earth's crust is rare. Both He and [40]Ar are end products of radioactive decay and have been useful in age-dating techniques. Helium is a basic decay product of several radioisotopes whereas [40]Ar is produced from the decay of [40]K. The main source of helium in crustal rocks is in natural gas fields, although it has also been found in volcanic gases, mineral springs and coal mines. Neon and argon are found in the upper lithosphere in minerals and rocks, mineral waters, mine gases, volcanic emanations, and natural gas. Krypton and xenon have been identified in gases from mineral springs and volcanic vents.

Conclusions

The geochemical composition of the Earth's crust is well studied and as more data, with increased accuracy and precision, become available, the level of knowledge will further improve. In the same way, analytical data on element concentration of individual mineral phases in geological materials will permit both an augmented knowledge of element partitioning and provide the basis for further developing concepts on the factors that influence and/or control elemental distributions in earth materials. Such advances will be due in part to technological achievements that lower the sensitivity and limitations of analytical methods, in part to continued careful sampling and sample handling to improve the precision and accuracy of geochemical data, and to physical, chemical and biological principles which permit the development of hypotheses that can evolve into principles that can be used to evaluate element distributions more completely.

James E. Mielke

References

ALBUQUERQUE, C. A. R. de; SHAW, D. M. 1972. Thallium. In: K. H. Wedepohl (ed.), *Handbook of Geochemistry*, Vol. II, No. 3. Berlin, Springer-Verlag.

CLARKE, F. W. 1889. The Relative Abundance of the Chemical Elements. *Bull. Phil. Soc. Washington*, Vol. 11. 131 p.

CLARKE, F. W.; WASHINGTON, H. S. 1924. *The Composition of the Earth's Crust.* (United States Geological Survey Professional Paper 127.)

CROCKETT, J. H. 1969. Platinum Metals. In: K. H. Wedepohl (ed.), *Handbook of Geochemistry*, Vol. II, No. 1. Berlin, Springer-Verlag.

DALY, R. A. 1910. Average Chemical Composition of Igneous Rock Types. *Proc. Amer. Acad. Arts & Sci.*, Vol. 45. 211 p.

DEGENS, E. T.; WILLIAMS, E. G.; KEITH, M. L. 1957. Environmental Studies of Carboniferous Sediments, Part I: Geochemical Criteria for Differentiating Marine from Fresh Water Shales. *Bull. Amer. Assoc. Petroleum Geologists*, Vol. 41, p. 2427.

—— 1958. Environmental Studies of Carboniferous Sediments, Part II: Application of Geochemical Criteria. *Bull. Amer. Assoc. Petroleum Geologists*, Vol. 42, p. 981.

EHMANN, W. D.; LOVERING, J. F. 1967. The Abundance of Mercury in Meteorites and Rocks by Neutron Activation Analysis. *Geochim. et Cosmochim. Acta*, Vol. 31, p. 357–76.

GOLDSCHMIDT, V. M. 1933. Grundlagen der Quantitativen Geochemie. *Fortschr. Mineral. Krist. Petrog.*, Vol. 17, p. 112.

GREEN, J. 1959. Geochemical Table of the Elements for 1959. *Bull. Geol. Soc. Am.*, Vol. 70, p. 1127–84.

——. 1972. Elements: Planetary Abundances and Distribution. In: R. W. Fairbridge (ed.), *The Encyclopedia of Geochemistry and Environmental Sciences*, p. 268–99. New York, Van Nostrand Reinhold.

HAMAGUCHI, H; KURODA, R.; ONUMA, N.; KAWABUCHI, K.; MITSUBAYASHI, T.; HOSOHARA, K. 1964. The Geochemistry of Tin. *Geochim. et Cosmochim. Acta.* Vol. 28, p. 1039.

HASKIN, L. A.; FREY, F. A.; SCHMITT, R. A.; SMITH, R. H. 1966. Meteoritic, Solar, and Terrestrial Rare-earth Distributions. *Physics and Chemistry of the Earth*, Vol. 7, p. 167–321. New York, McGraw-Hill.

HEIER, K. S.; ADAMS, J. A. S. 1964. The Geochemistry of the Alkali Metals. *Physics and Chemistry of the Earth*, Vol. 5, p. 255–380. New York, McGraw-Hill.

HEIER, K. S.; BILLINGS, G. K. 1970. Lithium. Sodium. Potassium. Rubidium. Cesium. In: K. H. Wedepohl (ed.), *Handbook of Geochemistry*, Vol. II, No. 2. Berlin, Springer-Verlag.

HERRMANN, A. G. 1970. Yttrium and Lanthanides. In: K. H. Wedepohl (ed.), *Handbook of Geochemistry*, Vol. II, No. 2. Berlin, Springer-Verlag.

HOEFS, J. 1969. Carbon. In: K. H. Wedepohl (ed.), *Handbook of Geochemistry*, Vol. II, No. 1. Berlin, Springer-Verlag.

HÖRMANN, P. K. 1963. Zur Geochemie des Germaniums. *Geochim. et Cosmochim. Acta*, Vol. 27, p. 861–76.

——. 1969. Beryllium. In: K. H. Wedepohl (ed.), *Handbook of Geochemistry*, Vol. II, No. 1. Berlin, Springer-Verlag.

HORN, M. K.; ADAMS, J. A. 1966. Computer-derived Geochemical Balances and Element Abundances. *Geochim. et Cosmochim. Acta*, Vol. 30, p. 279–7.

HORSTMAN, E. L. 1967. The Distribution of Li, Rb, and Cs in Igneous and Sedimentary Rocks. *Geochim. et Cosmochim. Acta*, Vol. 12, p. 1–28.

HURLEY, P. M.; HUGHES, H.; FAURE, G.; FAIRBAIN, H. W.;

PINSON, W. H. 1962. Radiogenic Sr^{87} Model of Continent Formation. *J. Geophys. Res.*, Vol. 67, p. 5315–34.

KNOPF, A. 1916. The Composition of the Average Igneous Rock. *J. Geol.*, Vol. 24, p. 620.

KORITNIQ, S. 1972. Fluorine. In: K. H. Wedepohl (ed.), *Handbook of Geochemistry*, Vol. II, No. 3. Berlin, Springer-Verlag.

KRAUSKOPF, K. B. 1967. *Introduction to Geochemistry.* New York, McGraw-Hill. 721 p.

——. 1970. Tungsten. In: K. H. Wedepohl (ed.), *Handbook of Geochemistry*, Vol. II, No. 2. Berlin, Springer-Verlag.

LAKIN, H. W.; THOMPSON, C. E.; DAVIDSON, D. F. 1963. Tellurium Content of Marine Manganese Oxides and Other Manganese Oxides. *Science*, Vol. 142, p. 1568.

LANDERGREN, S. 1948. On the Geochemistry of Swedish Iron Ores and Associated Rocks. A Study on Iron Ore Formation. *Sveriges Geol. Underskon.*, Ser. C, *Avhandl. och Uppsat.*, No. 496; Arsbok 42, No. 5.

LINN, T. A. Jr.; SCHMITT, R. A. 1972. Indium. In: K. H. Wedepohl (ed.), *Handbook of Geochemistry*, Vol. II, No. 3. Berlin, Springer-Verlag.

MAGAK'YAN, I. G.; PIDZHYAN, G. O.; FARAMAZYAN, A. S. 1963. Rhenium in Armenium Copper-molybdenum Deposits. *Doklady Akad. Nauk Armyans. S.S.R.*, Vol. 37, No. 2, p. 77–80.

MASON, B. 1958. *Principles of Geochemistry*, 2nd ed. New York, Wiley. 310 p.

MIESCH, A. T. 1969. The Constant Sum Problem in Geochemistry. In: *Computer Applications in the Earth Sciences*, p. 161–76. New York. Plenum.

MORRIS, D. F. C.; SHORT, E. L. 1969. Phenium. In: K. H. Wedepohl (ed.), *Handbook of Geochemistry*, Vol. II, No. 1. Berlin, Springer-Verlag.

NOCKOLDS, S. R. 1954. Average Chemical Compositions of Some Igneous Rocks. *Bull. Geol. Soc. America*, Vol. 65, p. 1007–32.

NORMAN, J. C.; HASKIN, L. A. 1968. The Geochemistry of Sc: A Comparison to the Rare Earths and Fe. *Geochim. et Cosmochim. Acta*, Vol. 32, p. 93.

ONISHI, H. 1969. Arsenic. Antimony. In: K. H. Wedepohl (ed.), *Handbook of Geochemistry.* Vol. II, No. 1. Berlin, Springer-Verlag.

POLDERVAART, A. 1955. Chemistry of the Earth's Crust. *Geol. Soc. America. Spec. Paper 62*, p. 119–44.

PUCHELT, H. 1967. Zur Geochemie des Bariums in Exogenen Zyklus. *Sitzungsber. Heidelb. Akad. Wiss. Math-nat. Kl.* 4 Abh.

——. 1972. Barium. In: K. H. Wedepohl (ed.), *Handbook of Geochemistry.* Vol. II, No. 3. Berlin, Springer-Verlag.

RANKAMA, K. 1944. On the Geochemistry of Tantalum. *Bull. Comm. Geol. Finlande*, p. 133.

RANKAMA, K.; SAHAMA, T. G. 1950. *Geochemistry.* Chicago, Ill., University of Chicago Press. 912 p.

ROGERS, J. I.; ADAMS, J. A. S. 1969. Thorium. Uranium. In: K. H. Wedepohl (ed.), *Handbook of Geochemistry*, Vol. II, No. 1. Berlin, Springer-Verlag.

RONOV, A. B.; YAROSHEVSKY, A. A. 1969. Chemical Composition of the Earth's Crust. *Amer. Geophys. Union Monograph 13-D.*

——. 1972. Earth's Crust Geochemistry. In: R. W. Fairbridge (ed.), *The Encyclopedia of Geochemistry and Environmental Sciences*, p. 243–54. New York, Van Nostrand Reinhold.

SAHÀMA, T. G. 1945. Spurenelemente der Gesteine im Sudlichen Finnisch-Lappland. *Null. Comm. Geol. Finlande 135.*

SANDELL, E. B.; GOLDICH, S. S. 1943. The Rarer Metallic Constituents of Some American Igneous Rocks, Part. I. *J. Geol.,* Vol. 51, p. 99. Part II, p. 167.

SINDEEVA, N. D. 1964. *Mineralogy and Types of Deposits of Selenium and Tellurium.* New York, N.Y., Interscience. (English translation: E. Ingerson (ed.).)

TATEL, H. E.; TUVE, M. A. 1955. Seismic Exploration of the Crust. In: A. Poldervaart (ed.), *Crust of the Earth,* p. 35–50 (Geological Society of America Special Paper 62.)

TAYLOR, S. R. 1964. Abundance of Chemical Elements in the Continental Crust: A New Table. *Geochim. et Cosmochim. Acta,* Vol. 28, p. 1273–85.

TUREKIAN, K. K. 1968. The Composition of the Crust. In: L. H. Ahrens (ed.), *Origin and Distribution of the Elements,* p. 549–57. Oxford, Pergamon Press.

TUREKIAN, K. K.; WEDEPOHL, K. H. 1961. Distribution of the Elements in Some Major Units of the Earth's Crust. *Bull. Geol. Soc. America,* Vol. 72, p. 175–92.

VINOGRADOV, A. P. 1962. Average Content of Chemical Elements in Main Types of Igneous Rocks of the Earth's Crust. *Geokhimiya,* No. 7, p. 555–71.

WAKITA, H.; SCHMITT, R. A. 1970. Cadmium. In: K. H. Wedepohl (ed.), *Handbook of Geochimistry,* Vol. II, No. 2. Berlin, Springer-Verlag.

WEDEPOHL, K. H. (ed.). 1969, 1970, 1972. *Handbook of Geochemistry,* Vol. I, II, No. 1, 1969; Vol. II, No. 2, 1970; Vol. II, No. 3, 1972. Berlin, Springer-Verlag.

——. 1971. *Geochemistry.* New York, N.Y., Holt, Rinehart & Winston. 231 p. (English translation.)

——. 1972. Zinc. In: K.H. Wedepohl (ed.), *Handbook of Geochemistry,* Vol. II, No. 3. Berlin, Springer-Verlag.

WLOTZKA, F. 1972. Nitrogen. In: K. H. Wedepohl (ed.), *Handbook of Geochemistry,* Vol. II, No. 3. Berlin, Springer-Verlag.

Weathering of rocks and formation of soils

Georges Pedro

Department of Soil Science (INRA),
Centre National de Recherches
Agronomiques,
Versailles (France)

Gaston Sieffermann

Department of Pedology,
Office de la Recherche Scientifique
et Technique d'Outre-Mer,
Bondy (France)

General framework of weathering phenomena at the Earth's surface

The rocks which emerge at the surface of the Earth are, in most cases, not at equilibrium with the conditions existing in the present surface environment. They are thus inevitably subject to change, and it is this change, in the form of a greater or lesser loss of solidity in the material, which is ultimately responsible for the development of soils or weathering complexes over the continental land masses.

In actual fact, the surface becomes friable in this way as the result of a chemical degradation of the 'vulnerable' minerals present in the original rocks. The consequent change in mass gives rise to new structures called 'secondary constituents' which are characteristic of the soils. These minerals, which then remain perfectly stable under surface conditions, have, generally speaking, the following characteristics: (a) microscopic size ($\phi < 2\,\mu$m), (b) a cryptocrystalline nature, and (c) a lamellar structure.

This chapter studies the formation of these various minerals according to the conditions of the surface environment and also studies their distribution within the principal types of soil.

DEFINITION OF WEATHERING

The most general equation representing weathering phenomena at the Earth's surface is schematically represented in Figure 1.

An examination of Figure 1 shows that the nature of the minerals formed and the type of geochemical process involved depend on three sets of factors:

The nature of the mineral subject to weathering; this is a passive factor but determines at the outset the rules that the geochemical process will follow.

The nature of the attacking reagents: composition, state of ionic dissociation, pH, redox potential, concentration.

The value of the thermodynamic parameters which are the active factors in the reaction: open or closed system, pressure, temperature, solution-mineral contact time.

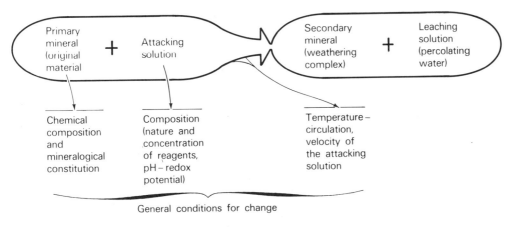

FIG. 1. Weathering: a schematic representation.

These three sets of factors interact with the three types of weathering conditions which will be hereafter briefly described; they are the lithological, the thermohydric and the physico-chemical conditions.

Lithological conditions

These determine the process which will take place in so far as the chemical composition of the secondary minerals formed depends on the composition of the primary minerals. Thus if the weathered product is to be a clay neoformation, the process must start off with endogenetic silicates. These may be:

Aluminosilicates (feldspars, micas, etc.) the composition of which is distinguished by a preponderance of silica and alumina.

Ferromagnesian silicates (of the peridot or pyroxene type), distinguished by an abundance of SiO_2 and (Mg, Fe)O.

These two major groups of minerals thus form two distinct groups from the weathering point of view:

The sialferric group $[SiO_2 - Al_2O_3(Fe_2O_3)]$ composed of the crystallized aluminosilicate rocks.[1]

The sifemic group $[SiO_2 - MgO(FeO)]$ which takes in all the ultrabasic rocks.

The first group is by far the biggest, accounting for more than 90 per cent of the lithosphere. It is thus this group which will be the major subject of the present article.

Thermohydric conditions

As geochemical changes in rocks require the presence of water in liquid form, chemical weathering phenomena will be negligible at the Earth's surface in a certain number of regions, particularly those distinguished by:

Sub-zero temperatures, i.e. ice-bound polar regions.

A rainfall of practically zero, i.e. absolute deserts.

Evapotranspiration so high that rain water returns immediately to the atmosphere, i.e. very arid zones.

In all other cases which thus possess humid environments, the phenomena of surface change will develop in normal ways. We should nevertheless note that although pressure remains more or less constant (about one atmosphere), other conditions may vary considerably from one region of the world to another, e.g. temperature that can range roughly from 0° to 40 °C and the mobility of the liquid phase in contact with the minerals, which is more or less marked depending on the efficiency of the drainage at each point of observation.

For this reason, there exists a fairly large number of different thermohydric situations at the Earth's surface.

Physico-chemical conditions

These depend on the nature of the reagents present in the attacking solutions, which consist of rain water with solubilized compounds, which are of atmospheric, biospheric or lithospheric origin.

Atmospheric origin

CO_2 and oxygen from the air are the main substances involved (the other gases, inert in practice, need not be taken into consideration).

Biospheric origin

In this group are the reagents formed by the physical, chemical and microbiological breakdown of vegetable litter and animal excrement. The number of compounds formed may be relatively large since it depends on: (a) the nature and composition of the initial raw products (carbon, nitrogen, sulphur, etc., compounds); (b) the general conditions under which the changes are taking place (aerobic or anaerobic, fungal or bacterial attack, temperature, humidity, etc.), and (c) the stage of biochemical change or, in other words, on whether this change stops at intermediate decomposition stages or continues until the final reaction products are obtained.

Nevertheless, of the group of compounds formed by biochemical synthesis (and thus polymerized to a greater or lesser extent), only water-soluble products are active as factors in weathering.[2] In the most common situation, which is that of the carbon compounds, two broad possibilities then occur from the physical and chemical points of view:

Under normal conditions, CO_2 is given off in abundance, hence the greater the degree of mineralization, i.e. the warmer and more humid the environment, the more CO_2 is liberated.

In cold and humid conditions, there is a characteristic formation of organic acids of low molecular weight, e.g. formic, oxalic, succinic, tartric, citric, parahydroxybenzoic, etc.

Lithosphere origin

Compounds of this type are generally formed by the dissolving of certain soluble minerals, in this case salts: limestone, gypsum, rock salt. They may sometimes, however, be the result of the decomposition of minor constituents, like the oxidation of pyrites which results, for example, in the release of sulphuric acid.

PHYSICO–CHEMICAL CONTEXT OF WEATHERING

If we try to group together all the various reagents we have so far considered by their physico-chemical characteristics, (especially pK and concentration in the attacking solutions), we have to draw a distinction between several types of environment, each permitting the development of a well-defined physico-chemical

1. Changes in vitreous volcanic rocks are a case apart (andosolization) and will not be mentioned here.
2. The other products must not be overlooked but can be considered as constituents of the soils rather than factors in soil formation.

TABLE 1. Major mechanisms of aluminosilicate weathering according to the physico-chemical characteristics of the weathering solutions

	pH		
Concentrations	pH < 5	5 < pH < 9.6	pH ⩾ 9.6
Dilute solution (around N/1,000) (I < 0.005)	Acidolysis	Hydrolysis	—
More concentrated ionic solutions (I > 0.005)	—	Salinolysis	Alkalinolysis

mechanism. The problem is then how to select suitable upper and lower values for each of the factors under consideration:

pH. The values selected naturally depend on the chemical characteristics, particularly the pK, of the main elements in the original aluminosilicates, namely Al and SiO_2. As the pK of aluminium is 5.0 and that of silicic acid, $Si(OH)_4$, 9.6, these are the two values which will serve as limits.

Saline concentrations. We shall take a concentration of N/1,000 as the limit, seeing that solutions with a lower content may be considered as dilute by the Debye-Huckel theory (ionic strength, I, lower than 0.005). We should mention, so as to indicate an order of magnitude, that an N/1,000 solution of NaCl, for example, corresponds to a concentration of 53 mg l^{-1}, or such a solution of $CaSO_4$ to one of 68 mg l^{-1}.

Using these parameters, we thus arrive at the framework set ou in Table 1 which shows the four major mechanisms by which aluminosilicates are weathered according to the physico-chemical characteristics of the weathering solutions.

PHYSICO–CHEMICAL CHARACTERISTICS OF THE FOUR BASIC WEATHERING MECHANISMS

The various mechanisms considered above all lead to the dislocation of the primary materials and to the release of the constituent elements but, as we shall see, what differentiates them is the form in which the various elements appear in the surface environment after degradation.

Hydrolysis

Hydrolysis as a weathering mechanism manifests itself essentially in the extraction and individualization of aluminium in the form of insoluble and weakly ionized compounds. These may be either free hydroxides (total hydrolysis) or basic salts of the argillaceous hydroxy-silicate type (partial hydrolysis). The other two types of element (SiO_2 and basic cations) remain in solution in the state of non-dissociated $Si(OH)_4$ acid, or of a highly ionized base (K^+, OH^-).[1]

In this case, the most complete reaction is written:

$$(Si_3Al)O_8K \xrightarrow{H_2O} 3\,Si(OH)_4 + Al(OH)_3 + K^+, OH^-. \text{(a)}$$

This is the mechanism which operates under the influence of pure water or of dilute solutions containing reagents with pK falling between 5 and 9.6.

Salinolysis

When there are more concentrated saline solutions containing acid salts and strong base salts, the type of change which the primary minerals undergo is the same for SiO_2 and Al_2O_3. On the other hand, strongly basic cations play a part in the double decomposition phenomenon and thus appear in the environment as soluble (strong acid) salts:

$$(Si_3Al)O_8K \xrightarrow[H_2O-Na(Cl)]{} 3\,Si(OH)_4 + Al(OH)_3 + K^+, Cl^-. \text{(b)}$$

Acidolysis

When the attacking solutions contain acid reagents with a pK lower than 5, these may salify not only the basic cations of the mineral but also the aluminium, which then appears in the form of soluble salts, dissociated to a greater or lesser extent. Neutralization may be partial or total. If it is complete, it corresponds to the reaction set out in schematic form below:

$$(Si_3Al)O_8K \xrightarrow[(HCl)]{H^+} 3\,Si(OH)_4 + Al^{3+}, \\ 3\,Cl^- + K^+, Cl^-. \text{[2]} \quad \text{(c)}$$

Acidolysis thus represents a type of mechanism the principal characteristic of which is the release of aluminium in the form of soluble aluminium salts, with silica remaining as $Si(OH)_4$.

Alkalinolysis

When there are basic reagents with a pK higher than 9.6, weathering leads to the individualization of the silica and then of the aluminium, released in the form of anions (silicate and aluminate) which are furthermore

1. In case of strong bases like KOH or NaOH, these in fact undergo a carbonation reaction.
2. HCl does not occur in the natural environment but has been taken here as the prototype of strong acids.

neutralized by the basic (external) cations of the alkaline attacking solution. We can thus write:

$$2(Si_3Al)O_8K \xrightarrow[(Na_2CO_3)]{OH^-} 6\,SiO_3^{2-},$$

$$2\,Na^+ + 2\,AlO_2^-, Na^+ + 2K^+CO_3^{2-}. \qquad (d)$$

The essential feature of this change is thus the appearance of silicates and alkaline aluminates in the weathering solutions.

GEOGRAPHICAL RANGE OF OCCURRENCE AT THE EARTH'S SURFACE

Figure 2 shows schematically the areas where the principal weathering mechanisms are active at the Earth's surface:

The typical 'acidolytic' environment, where the dominant weathering mechanism is acidolysis or acido-complexolysis. This environment corresponds very closely to the cold and humid, far northern area of coniferous forest-land (taiga). In these harsh climates, the vegetable litter decomposes very slowly and in all cases leads to the formation of a certain quantity of water-soluble organic acids responsible for the acidolytic changes.

The 'arid' environment—the boundaries of which as shown here are those established by Meigs (1952) on the basis of the data given by Thornthwaite (1948). Most of the salinolytic and alkálinolytic weathering found in the world is concentrated in the endorheic basins of this region.[1] This is therefore not a continuous area like those areas characterized by the two major mechanisms, acidolysis and hydrolysis.

Finally, the 'hydrolytic' environment made up of the remainder of the continental land masses. As this is by far the largest area, it is obvious that hydrolysis represents the main weathering mechanism at the Earth's surface. According to the areas occupied by the various environments under consideration, the order of importance of the mechanisms in question is thus as follows: hydrolysis, acido-complexolysis, salinolysis and alkalinolysis.

Crystallochemical weathering and soil formation processes: the formation of secondary minerals

INTRODUCTION—THE THREE WEATHERING PROCESSES

Before embarking on the crystallochemical aspect of weathering problems proper, it is necessary to describe the three main types of weathering which can be distinguished according to the nature and structure of the original minerals and to the environmental conditions.

(A) When the various elements released by the weathering process do not react with each other and remain whole in solution, this is said to be a phenomenon of solubilization.

For this reason, the composition of the solution in equilibrium is identical to the composition of the original solid or, in other words, the solubilization is of a congruent type (reaction A),

$$\begin{array}{c} \text{Primary} \\ \text{mineral} \\ \text{(lattice } X) \end{array} + \begin{array}{c} \text{Reagent} \\ \text{in} \\ \text{solution} \end{array} \underset{\longleftarrow}{\longrightarrow} \begin{array}{c} \text{Elements in} \\ \text{weathering} \\ \text{solution} \end{array} \quad (A)$$

(B) On the other hand, when weathering leads to the formation of secondary minerals which are, for their part, in perfect equilibrium with the conditions of the environment, the composition of the solution becomes different from that of the original solid. Solubilization is then of an incongruent type.

From the crystallochemical point of view, two kinds of mechanisms are possible:

(B1) In the first, all the elements are extracted simultaneously so that the structure of the primary mineral is completely lost. This is what happens to minerals with a closed structure, which prevents attacking solutions from penetrating (e.g. feldspars). In this case, therefore, it is impossible to extract some elements without acting on the rest. This leads to the complete disintegration of the primary lattice as the weathering proceeds.

At a later stage, the elements thus released react among themselves to produce a new solid phase with a well-determined lattice which, for its part, is stable in the weathering environment (reaction B1):

$$\begin{array}{c} \text{Primary} \\ \text{mineral} \\ \text{(lattice } X) \end{array} + \text{Reagent} \longrightarrow \begin{array}{c} \text{Newly formed} \\ \text{secondary} \\ \text{mineral} \\ \text{(lattice } Y) \end{array} + \text{Solution} \quad (B)$$

The process of change is thus characterized in this instance by a complete weathering accompanied by a phenomenon of neoformation.

(B2) In the second mechanism, certain elements only of the original mineral are extracted, so that its structural fabric remains (type of network X). This is what occurs with rocks with an open (or easily opened) structure like micas (phyllosilicates) or feldspathoids (tectosilicates). In fact, as it is possible under certain conditions to cause the compensating elements to be freed without disturbing the structure of the primary mineral, change appears as a gradual transformation

---

1. The other zones are either of the hyperarid and desert type, i.e. without chemical weathering, or give rise to very limited hydrolysis.

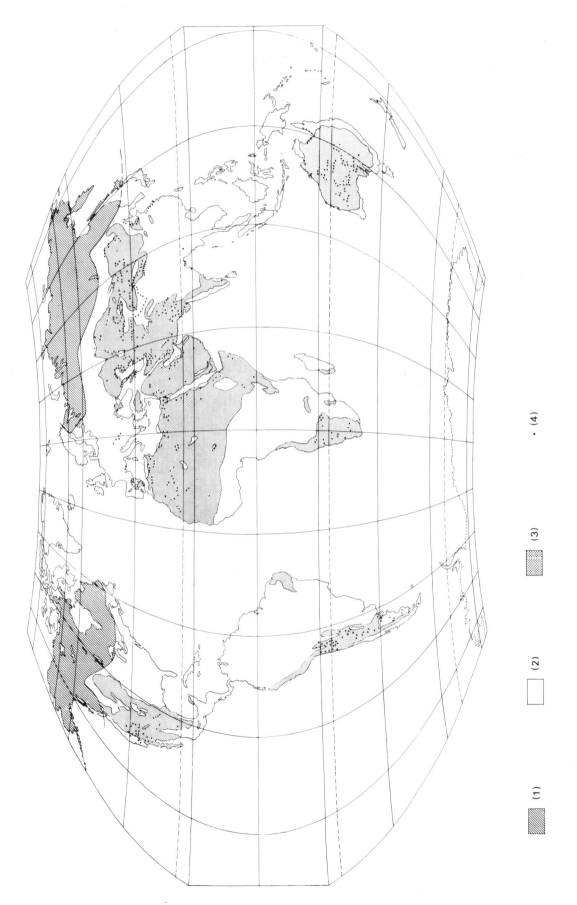

Fig. 2. Areas affected by the principal weathering mechanisms: (1) typically acidolytic environment; (2) hydrolytic environment; (3) arid environment; (4) saline soils (salinolysis and alkalinolysis).

(1)

(2)

(3)

· (4)

which brings about only discrete modifications (reaction B2),

$$\text{Primary mineral (lattice } X) + \text{Reagent} \longrightarrow \text{Transformed mineral (lattice } X) + \text{Solution} \qquad (B2)$$

All three transformation processes occur during surface weathering and soil formation phenomena, but only the latter two, which give rise to secondary minerals, play any real part in the constitution of soils and surface formation. There are therefore the processes which we shall examine in greater detail, and to do so we shall examine successively the weathering of feldspars and the neoformation of secondary minerals and the weathering of micas and 2/1 phyllosilicates by transformation.

THE WEATHERING OF FELDSPATHIC ROCKS AND THE NEOFORMATION OF SECONDARY MINERALS

We shall begin our study of the weathering of closed structure minerals like feldspars by analysing how they behave under the influence of the three principal mechanisms mentioned above—hydrolysis, acidolysis and alkalinolysis.

Hydrolytic weathering

Description of the crystallochemical weathering processes
We gave earlier a very general description of the mechanism of hydrolysis. In fact, since this mechanism develops with varying degrees of completeness (according to the conditions of the environment), we must now give consideration to these various degrees of the phenomenon.

Total hydrolysis. Hydrolysis is total when all the elements in the minerals appear in the reaction in the form of hydroxides. These may or may not be dissociated and may be basic or acidic in nature (reaction 1).

$$\underset{\text{Orthose}}{(Si_3Al)O_8K} \xrightarrow{H_2O} \underset{\text{Gibbsite}}{3\,Al(OH)_3} + \underset{\text{Solution}}{3\,Si(OH)_4 + K^+,\,OH^-}$$

$$Ki = \frac{SiO_2}{Al_2O_3} = 6 \qquad Ki = \frac{SiO_2}{Al_2O_3} = 0$$

$$R = \frac{SiO_2}{K_2O} = 6 \qquad L = \frac{SiO_2}{K_2O} = 6 \qquad (1)$$

The products are, in addition, soluble to varying extents. Thus, KOH and $Si(OH)_4$, which are soluble, will stay in solution while $Al(OH)_3$, which is insoluble ($pS = 33$), will precipitate *in situ* as crystallized hydroxides containing hexaco-ordinate aluminium. In natural weathering, it is usually gibbsite with a foliated structure which appears.

The crystallochemical process involved corresponds, using the term introduced by Harrassowitz (1926), to an allitization characteristic of the phenomena of lateritization and bauxitization.

Partial hydrolysis. When a portion of the silica extracted in weathering reacts with the whole of the alumina released and produces new silica compounds of the basic salt type, hydrolysis is said to be partial.

This mechanism thus leads to the individualization of aluminous hydroxysilicates of the argillaceous phyllosilicate group, such that the crystallochemical process corresponds to siallitization (Harrassowitz, 1926), with the development of an argillaceous weathering.

In fact, as this partial hydrolysis may occur with varying degrees of intensity, we are obliged to make a further distinction here between two broad categories.

In the first case, the weathering leads to the formation of a 1:1 clay of a kaolinite or halloysite type, i.e. distinguished by folia having one layer of silica tetrahedrons to one octahedral layer of aluminium (thickness = 7 Å) and with a zero electrical charge (reaction 2):

$$\underset{\text{Orthose}}{2(Si_3Al)O_8K} \xrightarrow{H_2O} \underset{\text{Kaolinite}}{Si_2O_5\,Al_2(OH)} +$$

$$+ \underset{\text{Solution}}{4\,Si(OH)_4 + 2K^+,\,OH^-}$$

$$Ki = \frac{SiO_2}{Al_2O_3} = 6 \qquad Ki = \frac{SiO_2}{Al_2O_3} = 2$$

$$R = \frac{SiO_2}{K_2O} = 6 \qquad L = \frac{SiO_2}{K_2O} = 4 \qquad (2)$$

This may be termed monosiallitization (Pedro, 1966).

In the second case, partial hydrolysis leads to the individualization of a 1:1 clay of the smectite type (montmorillonite, beidellite, etc.) whose sheets, made up by the superimposition of two layers of silica on one layer of gibbsite (thickness 10 Å), have a certain electrical charge which must therefore be compensated for by the retention of a certain number of basic cations in interfoliate positions (reaction 3):

$$\underset{\text{Orthose}}{2.3(Si_3Al)O_8K} \xrightarrow{H_2O} \underset{\text{Beidellite}}{(Si_{3.7}Al_{0.3})O_{10}Al_2(OH)_2K_{0.3}} +$$

$$+ \underset{\text{Solution}}{3.25(OH)_4 + 2(K^+,\,OH^-)}$$

$$Ki = \frac{SiO_2}{Al_2O_3} = 6 \qquad Ki = \frac{SiO_2}{Al_2O_3} = 3.2$$

$$R = \frac{SiO_2}{K_2O} = 6 \qquad L = \frac{SiO_2}{K_2O} = 3.2 \qquad (3)[1]$$

---

1. This is in order to balance a simple reaction completely. In fact, the presence of magnesium, for example, is vital for the formation of a typical montmorillonite.

The process occurring in this case corresponds to true bisiallitization (Pedro, 1966).

Geochemical characteristics and genetic consequences
The geochemical characteristics of the three hydrolytic processes can be seen by examining the composition of the percolating water, as it appears from an examination of reactions 1, 2 and 3 above. In all three cases, as the alumina is completely maintained within the newly formed structures, the geochemical differences can be shown only by a comparative study of the behaviour of the silica and the basic cations. For this purpose, it is possible to use parameters like: $R = SiO_2$ / Bases characteristic of the original materials, and $L = SiO_2$ / Bases corresponding to the weathering solutions (Pedro, 1964a).

Thus, first, in allitization, all the silica and all the cations released by the hydrolytic dislocation of the lattice, remain in solution and are eliminated with the leachate. The extreme geochemical condition under which such a process can take place can thus be written $L = R$, which means that in relative values, the silica has to be carried off at least as quickly as the basic cations and hence means that $(SiO_2) \geqslant$ (Bases).[1]

From the geochemical standpoint, allitization thus manifests itself as a desilication and complete dealkalinization of the material undergoing weathering.

Second, we saw that during monosiallitization, part of the silica removed by hydrolysis was retained in the form of kaolinite while the bases were completely leached. For this reason, we find ourselves unavoidably in conditions where $L < R$, or $(SiO_2) <$ (Bases). Accordingly, it becomes insufficient from the geochemical point of view to take only the parameter R into consideration. What is, in fact, needed is a comparison of L with a value broadly characteristic of the original material after its complete transformation into kaolinite or, in other words, with the ratio of non-kaolinizable SiO_2 to basic cations. We have called this new parameter R_k and have shown that R_k is equal to the ratio: $SiO_2 - 2\,Al_2O_3$/Bases in the original mineral (Pedro, 1966). Thus the extreme geochemical condition for monosiallitization corresponds to $L = R_k$, which means that if the silica is removed here less quickly (in comparison with the basic cations) than in allitization (R), its dynamics in solution, expressed by L, must nevertheless satisfy the inequality:

$$R_k < L < R.$$

Overall, monosiallitization thus manifests itself from the geochemical point of view by a total dealkalinization and a partial (66 per cent) desilication of the original mineral undergoing hydrolysis.[2]

Third, bisiallitization, for its part, is distinguished by partial dealkalinization (87 per cent for reaction 3) and by an even lesser desilication (46 per cent).[2] Under

1. The parentheses used in the expressions like (SiO_2), (Bases), (Al_2O_3) indicate relative speeds of elimination.
2. For orthose.

TABLE 2. Principal geochemical and crystallochemical data characteristic of hydrolytic weathering processes

Degree of hydrolysis		Total hydrolysis		Partial hydrolysis
Geochemistry of desilicification	Overall balance	Total desilicification		Partial desilicification
	Relative dynamics	$L \geqslant R$		$L < R$
		$(SiO_2) \geqslant$ (Bases)		$(SiO_2) <$ (Bases)
	SiO_2[1] concentration	$SiO_2 < 10^{-4.7}\,M$		$SiO_2 > 10^{-4.7}\,M$
Crystallochemical nature of constituents formed	General process	Allitization		Siallitization
	Co-ordination of aluminium	VI		VI
	Constituents formed	Aluminium hydroxide	Phyllites 1/1	Phyllites 2/1
	Example	Gibbsite	Kaolinite	Montmorillonite
	Crystallochemical process	Allitization	Monosiallitization	Bisiallitization
Geochemistry of the dealkalinization	Overall balance	Total dealkalinization		Partial dealkalinization
	Relative dynamics	$L > R_K$		$L < R_K$
	Alkaline cations and H^+ concentration[1]	$\dfrac{[K^+]}{[H^+]} < 10^{6.5}$		$\dfrac{[K^+]}{[H^+]} > 10^{6.5}$

1. Only for orthoclase (from Garrels, 1965) as an example.

these conditions—and in accordance with the above—L must not only be less than R but also less than R_k: $L < R_k$.

The principal crystallochemical and geochemical data characteristic of the three hydrolytic weathering processes are finally set out in a table that clearly shows the relationships that may exist between the geochemistry of the percolating water and the nature of the hydrolysis phenomena (Table 2). Here thus, in embryonic form, are all the essentials for a productive method of study since it is possible to determine rapidly the kind of process that is occurring simply by examining the composition of the leachates.

These, however, are geochemical consequences. We still have to clarify, very briefly, the nature of the parameters which underlie the development of any given type of hydrolysis.

Conditions for development
The geochemical characteristics can easily be discovered, as we have seen, by a qualitative study of the relative behaviour of the elements in solution, i.e. SiO_2 and basic cations. In fact, what governs the development of any hydrolysis phenomenon are the quantitative parameters, such as the actual concentration of SiO_2, basic cations and H^+ ions in the hydrolytic weathering water; Garrels (1965) and others have clearly demonstrated this on the basis of theoretical considerations of a thermodynamic nature. Thus we note from Figure 3—which gives an overall view of the results of the hydrolysis of orthose:

That allitization can develop only if the concentration of silica in solution is below a very precise value. This, which is fixed according to the kaolinite-gibbsite equilibrium constant, is $10^{-4.7} M$.

That, on the other hand, as soon as the SiO_2 concentration rises above $10^{-4.7} M$, hydrolysis leads necessarily to siallitization. In this case, it is the environment's basic cation concentration (and also H^+ ion concentration) which determines the formation of either kaolinite or bisiallites (Table 2).

Since we have an open system most of the time in the natural environment, the equilibrium concentration in solution is a function of the temperature and, especially, of the speed of circulation of the water in contact with the minerals. It is the latter which is really the active factor in hydrolysis phenomena (Lagache, 1965; Delmas, 1972).

In fact, when the time of contact between solution and mineral is short due to rapid drainage, the silica in the weathering environment is so dilute (<1 mg l^{-1}) that hydrolysis is total with a precipitation of $Al(OH)_3$. If the speed of leaching is less, the SiO_2 concentration increases but as the base content is still too small, the product finally reached is that of the solubility of kaolinite.

Finally, with slow drainage, the weathering solutions are more concentrated not only in SiO_2 but also in basic cations, so that it is a bisiallite which is produced.

In short, it is the hydrodynamic and thermal factors which effectively determine the degree of hydrolysis of an alumino-silicate and thus determine the crystallochemical process involved.

Further note on salinolytic weathering. We have already mentioned this mechanism (reaction b) on page 41. From the foregoing, it is easy to show that salinolysis can lead only to one crystallochemical process, namely, bisiallitization. In fact, the parameter L in the percolating water, corresponding here to [extracted SiO_2]/[extracted basic cations plus basic cations of the salt], is always very small, so that it is not only less than R but also less than R_k.

Acidolytic weathering

Description of the principal acidolytic processes
Acidolysis is the mechanism that develops whenever the attacking solutions contain reagents with a pK < 5. Nevertheless, just as we have described total and partial

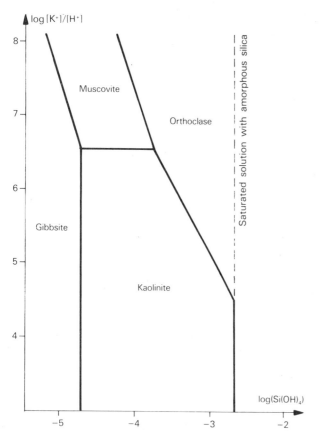

FIG. 3. Diagram of stability in the system.
$SiO_2 - Al_2O_3 - K_2O - H_2O$ at 25 °C and at atmospheric pressure (Garrels, 1965).

hydrolysis above, so we can now consider complete acidocomplexolysis and limited acidocomplexolysis:

Total acidocomplexolysis. Acidolysis is total when the various elements of a metallic nature in the primary minerals, i.e. not only the basic cations but also acid cations of the Al kind, all appear in the form of soluble salts, either dissociated or non-dissociated (complex), in the weathering solutions. The reaction is expressed in true acidolysis:

$$(Si_3Al)O_8K \xrightarrow[(HL)]{H^+}$$
$$\underbrace{3\,Si(OH)_4 + Al^{3+} + K^+ + \;K^+ + 4\,L^-}_{Solution} \quad (4)$$

(Si₃Al)O₈K — Orthose

and in acidocomplexolysis:

$$(Si_3Al)O_8K \xrightarrow[(HL)]{H^+} \underbrace{3\,Si(OH)_4 + AlL_3 + (K^+, L^-)}_{Solution}$$

Orthose

$$A_R = \frac{Al_2O_3}{K_2O} = 1 \qquad A_L = \frac{Al_2O_3}{K_2O} = 1 \quad (5)$$

The reaction which takes place is thus a straightforward solubilization of the aluminosilicates as occurs during podzolization.[1]

Limited acidocomplexolysis. Acidolysis is limited when the attack is not sufficiently powerful for the various cations—and all the aluminium in particular—to be present as soluble salts. To a certain extent this is again a partial hydrolysis, since a portion of the silica may then react with the non-salified aluminium which is present in the form of hydroxy cations, to give rise to clays of a bisiallitic type. The only difference lies in the fact that because of the acid environment and the low concentration of basic cations, compensation for loss of charge can only occur through the retention of hydroxy-aluminous cations in interfoliate positions (cf. Trichet, 1970; Hetier and Tardy, 1969).

The limited acidolysis reaction can be expressed as follows:

$$(Si_3Al)O_8K \xrightarrow[(LH)]{H^+} (Si_{3.5}Al_{0.5})O_{10}Al_2(OH)_2 \cdot Al(OH)_{2\ 0.5} +$$

Orthose — Aluminous intergrade Beidellite Al

$$+ Al^{3+} + \underbrace{4K^+ + 4L^- + 8.5\,Si(OH)_4}_{Solution}$$

$$A_R = \frac{Al_2O_3}{K_2O} = 1 \qquad A_L = \frac{Al_2O_3}{K_2O} = \frac{1}{4} \quad (6)$$

This is then a process which may be called aluminosiallitization. From the crystallochemical point of view this process corresponds to a bisiallitization but because of the presence of aluminium in the interfoliate

spaces, the composition of the secondary constituents comes close to that of kaolinite ($SiO_2/Al_2O_3 = 2$). For this reason, it is sometimes said to be a kind of disguised monosiallitization (Tardy, 1970).

Geochemical characteristics

In an acidolytic environment, all elements go into solution but it is the silica which then has the lowest solubility. Thus the geochemical characteristics of the processes which have just been described in reactions 4 and 5 can only be arrived at by considering the relative behaviour of the aluminium and the bases in the percolating water. For this purpose, one can refer to the parameters:

$A_R = Al_2O_3/Bases$ (characteristic of the original minerals); and

$A_L = Al_2O_3/Bases$ (in the weathering solutions).

TABLE 3. Geochemical and crystallochemical data characteristic of acidolytic processes

Intensity of acidolysis		Total acidolysis	Partial acidolysis
Geo-chemistry of de-alu-minization	Overall balance	Total de-alu-minization	Partial de-aluminization
	Relative dynamics	$A_L \geqslant A_R$ (Al_2O_3) $\geqslant$ (bases)	$A_L < A_R$ (Al_2O_3) < (bases)
	Co-ordination of aluminium	VI	VI
	Physico-chemical conditions	pH < 4 or very complexing agent	4 < pH < 5
Charac-teristic secondary minerals	Consti-tuents formed	—	Phyllites 2/1 − Al
	Nature of inter-foliated cations	—	$Al(OH)^{2+} - Al(OH)_2^+$ $[Al_x(OH)_y]\,n^+$
	Examples	—	Montmorillonite Al
Geochemical and crys-tallochemical process		Podzolization (complete)	Aluminosiallitization

1. In the confined environments of coastal regions where, as a result of the oxidation of pyrites, changes are conditioned by the presence of sulphuric acid (acid sulphate soils), crystallized basic sulphates may then appear either of aluminium (e.g. Alunite $KAl_3(SO_4)_2(OH)_6$), or of iron (e.g. Jarosite $K\,Fe_3(SO_4)_2(OH)_6$). cf. Viellefon (1974).

Thus, first, during complete acidocomplexolysis, all the aluminium released during weathering remains in solution and is carried away with the leachate. The extreme geochemical condition under which this process can develop is thus expressed as $A_L = A_R$, which means that, relatively speaking, the aluminium must be carried away at least as rapidly as the silica, i.e. $(Al_2O_2) \geqslant (Bases)$.

From the geochemical point of view, 'podzolization' thus leads first and foremost to a complete dealuminization of the mineral under attack.

Second, in the case of aluminosiallitization, we saw that part of the aluminium removed during partial acidolysis became fixed again in both the octahedral and interfoliate layers within the secondary phyllites. This explains why we have $A_L < A_R$, and hence why the basic cations are carried off much more easily than the aluminium: $(Al_2O_3) < (Bases)$. Table 3 recapitulates the various data concerning the crystallochemical and geochemical analysis of the problem.

Conditions for the start of the process

From the data given above which enabled us to define the two kinds of acidolysis, we can see that hydrodynamic conditions play practically no part in the development of either process. What determines the appearance of the phenomena are the physical and chemical conditions—particularly the greater or lesser acidity (H^+ concentration) and the more or less chelating nature of the attacking solutions.

Thus, when the environment is either very highly acid or markedly chelating, all conditions are present for the aluminium to be caught and maintained in solution as soluble salts. Owing to the great variation in the pK of organic acids in the natural environment and their varying chelating ability, it is difficult to establish the pH value below which the phenomenon of podzolization clearly develops. We can nevertheless say that this occurs as a general rule with pH about or less than 4. These values also must be smaller in proportion as the anion is less chelating.

On the other hand, when the environment is less acid ($4 < pH < 5$, approximately) and not highly chelating, most of the free aluminium exists not in the form of simple cations linked to anions but rather in the form of hydroxy cations: $Al(OH)^{2+}$, $Al(OH)_2^+$, $Al(OH)_4^{2+}$, etc., which are also more hydroxylated and more polymerized in proportion as the pH increases and approaches the precipitation pH of the hydroxide. In this case, it is these aluminous hydroxy cations which, not being very mobile, react with the silica to produce constituents of the aluminosiallite type.

Thus the pH of the attacking solutions and the reagents' chelating ability are the determining factors in acidolytic changes.

Alkalinolytic weathering

Determination and conditions for the development of alkalinolytic processes

Alkalinolysis is the mechanism set in motion when the reagents in the attacking solutions are markedly basic in character (e.g. sodium carbonate solution). Here, too, however, a distinction can be made between total alkalinolysis and limited alkalinolysis:

Total alkalinolysis. Alkalinolysis is complete:

1. When the environment is sufficiently basic[1] for all the silica and all the aluminium to appear during the changes as soluble alkaline aluminates and silicates (aluminium is then present in a tetracoordinate state).

2. When the leaching environment is suitably renewed so that the weathering solutions remain very dilute and the substances produced during the change are constantly carried away with the percolating water. In this case, it is just as if a purely alkaline dissolution of the primary mineral were taking place (reaction 7).

$$(Si_3Al)O_8K \xrightarrow[(Na_2CO_3)]{OH}$$

Orthose

$$\underline{3\,SiO_3^{2-},\ Na_2^+ + Al^VO_2,\ Na^+ + K^+,\ OH^-}$$

Solution

$$Ki = \frac{SiO_2}{Al_2O_3} = 6 \qquad P_L = \frac{SiO_2}{Al_2O_3} = 6$$

$$R = \frac{SiO_2}{K_2O} = 6 \qquad L = \frac{SiO_2}{K_2O + Na_2O} = \frac{3}{4} \qquad (7)$$

This is what occurs at the Earth's surface during the dealkalinization by water of alkaline soils which thus leads to the individualization of soil profiles with a bleached and quartzose A2 eluvial horizon, as in solods, hence the name solodization, which may be used to designate the geochemical process characteristic of total alkalinolysis.

Limited alkalinolysis. This second mechanism occurs in confined alkaline zones of the endorheic area when the concentration of the weathering solutions becomes sufficiently high for there to be a possibility of reaction between the tetrahedral silicate and aluminate anions of the environment and the *in situ* formation of three-dimensional aluminosilicates of the zeolite or feldspathoid type. At the same time, various alkaline carbonates may also be deposited (Maglione, 1974).

---

1. An environment of this kind calls for the presence of very alkaline cations such as sodium (pK = 14.7).

The weathering reaction is written thus (taking no account of the carbonates deposited) (reaction 8):

$$2(Si_3Al)O_8K \xrightarrow[(Na_2CO_3)]{OH^-} (Si_2Al^{IV})O_6Na, H_2O +$$

Orthose   Analcine

$$+ 4SiO_3^{2-} + AlO_2^- + 9Na^+ + K^+, OH$$

Solution

$$Ki = \frac{SiO_2}{Al_2O_3} = 6 \qquad Ki = \frac{SiO_2}{Al_2O_3} = 1$$

$$R = \frac{SiO_2}{K_2O} = 6$$

$$pL = \frac{SiO_2}{Al_2O_3} = 8$$

$$L = \frac{SiO_2}{K_2O + Na_2O} = \frac{4}{5} = 0.8 \qquad (8)$$

This shows the existence of a very specific crystallochemical process which can be given the name of sialicatization.

Geochemical characteristics

In an alkalinolytic environment, due to the presence of basic cations in the attacking solutions, the ratio $L = SiO_2/(Bases)$ is always very low in the water in the environment where the reaction is taking place (<1 in the examples taken here). Everything then revolves around the simultaneous behaviour of the silicate and the aluminium. In this connection, reference to the parameters $Ki = SiO_2/Al_2O_3$ in the original minerals and $pL = SiO_2/Al_2O_3$ in the weathering water, makes it possible to shed light on the geochemical problem:

1. Total alkalinolysis thus occurs when the elimination of the silica and the aluminium is complete, so that the aluminium is removed at the same speed as the silica ($pL = Ki$).
2. On the other hand, the process of sialicatization is accompanied by a preferential elimination of silica corresponding to $pL > Ki$.

Table 4 gives a summary of the situation in alkalinolytic regions.

GENERAL GEOCHEMICAL CHARACTERISTICS OF THE SURFACE WEATHERING OF FELDSPATHIC ROCKS

When one simultaneously examines the conditions affecting the three main weathering mechanisms we have just been studying, namely, acidolysis, hydrolysis and alkalinolysis, two main points emerge:

Each of these mechanisms is distinguished by the fact that two of the three basic elements in the feldspars are carried away first: Al_2O_3 and basic cations

TABLE 4. Geochemical and crystallochemical data characteristic of alkalinolytic processes

Intensity of alkalinolysis		Total alkalinolysis	Partial alkalinolysis
Geochemistry of aluminium and silica	Relative dynamics	$pL \leqslant Ki$ $(SiO_2) \leqslant (Al_2O_3)$	$pL > Ki$ $(SiO_2) > (Al_2O_3)$
	Co-ordination of aluminium	IV	IV
	Basic environment	Opened medium	Closed medium

Secondary mineral characteristics

Silicate constituents	Structural types	—	Zeolites Feldspathoides
	Examples	—	Analcime Sodalite Kenyaite–magadite Kanemite
Carbonate constituents	Structural types	—	Sodium carbonates
	Examples	—	Trona-natron-thermonatrite

Geochemical and crystallochemical process

		Solodization	Sialicatization

in the case of acidolysis, SiO_2 and basic cations in that of hydrolysis, and SiO_2 and Al_2O_3 in that of alkalinolysis.

The intensity of the phenomenon (total or limited) is directly related to the relative speed of elimination of the two most easily removed elements.

Systematizing the changes at geochemical level, as in Table 5, shows in the end that six broad geochemical and crystallochemical weathering processes in feldspathic rocks are possible.

THE WEATHERING OF MICAS AND 2:1 PHYLLOSILICATES BY TRANSFORMATION

General framework for the weathering of micas

This is a very special case of weathering since the primary minerals (micas) are already, to begin with, 2:1 phyllosilicates of the 'bisiallite' type. The problem thus does not arise in quite the same terms as for feldspars, although we are still dealing with a structure (but of a foliated kind) made up principally of SiO_2 and Al_2O_3,

TABLE 5. Broad geochemical conditions for the evolution of feldspathic rocks

Alteration mechanism	Acidolysis		Hydrolysis		Alkalinolysis	
Least mobile or most concentrated constituent under alteration conditions	SiO_2		Al_2O_3		Basic cations	
Most mobile constituents	Al_2O_3 — Bases		SiO_2 — Bases		$SiO_2 - Al_2O_3$	
Relative rate of elimination	$(Al_2O_3 \geqslant (Bases)$ $A_L \geqslant A_R$	$(Al_2O_3) < (Bases)$ $A_L < A_R$	$(SiO_2) \geqslant (Bases)$ $L \geqslant R$	$(SiO_2) < (Bases)$ $L < R$	$(Al_2O_3) < (SiO_2)$ $PL > Ki$	$(Al_2O_3) \geqslant (SiO_2)$ $pL \leqslant Ki$
Intensity of phenomenon	Total acidolysis	Limited acidolysis	Total hydrolysis	Partial hydrolysis	Partial alkalinolysis	Total alkalinolysis
Process	Podzolization	Aluminosiallitization	Allitization	Siallitization	Sialicatization	Solodization

and neutralized through the intermediary of compensating basic cations (K_2O).

The changes brought about at the surface may, in this case, follow one of two paths:

Either the weathering mechanisms permit the complete dislocation of the 2:1 sheets and we are brought right back to the previous problem with the development, depending on the environmental conditions, of the phenomena of solubilization or neoformation; or

The mechanisms involved affect the sheets very little and are essentially active in the interfoliar spaces, in which case these are transformation phenomena proper.

In these circumstances, the problem is one of determining in what way the 2:1 sheets will be stable or unstable according to the characteristics of the environment and the nature of the mechanisms involved.

Conditions leading to the instability of 2:1 sheets

The mica sheets become unstable when it is primarily the constituents of the sheet (F): SiO_2 and Al_2O_3 which are carried away rather than the interfoliar (Int) elements (Robert, 1971; Robert and Pedro, 1972).

From the geochemical point of view, dislocation therefore appears as the major phenomenon when it is possible to write:

$$(Int) < (F).$$

From this, it is possible to deduce quite easily the types of environment responsible for such weathering. In fact, this general relationship is verified: (a) When $L > R$, or $(SiO_2) > (Bases)$; this is what occurs in a very dilute environment with the development of total hydrolysis (allitization); or (b) When $A_L > A_R$, which corre-

sponds to $(Al_2O_3) > (Bases)$; this is then the result of an acid and highly chelating attack (total acidolysis) as in straightforward podzolization.

For all other eventualities, we pass into the field of transformations, even if in some cases these are possibly entirely transitory.

Conditions for and type of mica transformation

By analogy, it may be considered that the 2:1 sheets are stable whenever the extraction of interfoliar cations takes precedence over the release of the elements from the folia. Hence, from the point of view of the geochemical dynamic, the transformation is expressed by the relationship:

$$(Int) > (F).$$

This inequality is verified in the various following instances: (a) If $A_L < A_R$, hence when $(K_2O) > (Al_2O_3)$, as in the mechanism of limited acidolysis; (b) If $L < R$, which corresponds to $(K_2O) > (SiO_2)$ with the development of partial hydrolysis; (c) Finally, when the concentration of basic cations (e.g. Na, Ca, etc.) in the environment under attack is sufficiently high for this concentration easily to bring about a kind of spontaneous exchange with the interfoliar cations, e.g. in salinolysis and during alkalinolysis.

If we now turn to the crystallochemical aspect, it is possible to show that these various mechanisms lead in fact to two kinds of transformation depending on whether the interfoliar ions in the micas (K) are replaced by cations of a very basic nature (Na, Ca, Mg, etc.) or by cations which are already acid in nature (Al, for example).

The first type occurs in all mechanisms where there is a certain concentration of basic cations in the weathering solutions, namely, limited hydrolysis, salinolysis and alkalinolysis. The change then principally consists of the dry K^+ ions from the micas being replaced by other basic cations which are always hydrated in the surface environment: $Ca(H_2O)_x$, $Na(H_2O)_y$, etc., this causing an expansion of the lattice in a direction perpendicular to the sheets with individualization of phyllites which behave like vermiculite.[1]

$$[(Si_3Al)O_{10}Al_2(OH)_2]\ K \xrightarrow[(Ca)]{H_2O} [(Si_3Al)O_{10}Al_2(OH)_2]\cdot$$
Muscovite
$$Ca_{0.5}(H_2O)_x + K^+ \quad (9)$$
Vermiculite

This change may, in any case, continue until the smectite stage when changes occur in the tetrahedral layers as was envisaged by Millot et al. (1965) and Tardy (1970).

The transformation process set in motion thus leads, as in bisiallitization, to the formation of expandable 2:1 clays. Here, however, there has been no sheet neoformation, which is why Paquet (1970) described this as an apparent bisiallitization (vermiculitization).

The second type of transformation occurs when the acid cations (Al) in the weathering solutions take precedence over the basic cations. This is what happens, for example, in the case of limited acidolysis:

$$[(Si_3Al)O_{10}Al_2(OH)_2]\ K \xrightarrow[(Al)]{H^+} [(Si_3Al)O_{10}Al_2(OH)_2]\cdot$$
Muscovite
$$Al(OH)_2 + K^+ \quad (10)$$
Aluminous intergrade

The development of aluminous bisiallites then occurs, the behaviour and nature of which depend essentially

on the rate at which the interfoliar spaces are filled. This is the natural result of the initial charge of the folium but is also the result of the pH conditions since the hydroxylation and polymerization rate of aluminium cations increases with the pH. Depending on the physical and chemical characteristics of the environment, it is possible to imagine the individualization of aluminous intergrades (of the Al vermiculite kind) or of pseudochlorites when the layers of interfoliar hydroxides are discontinuous, or even of genuine secondary chlorites when the gibbsite layers are continuous.

The process involved here is essentially a kind of aluminization, the extreme form of which could be given the name of aluminochloritization.

In Table 6, the various possibilities occurring during the weathering of micas in a surface environment are summarized by way of conclusion.

Summing up and study of the geographical distribution of the main soils and pedological processes

All the data which have been set out in the previous section concerning the various crystallochemical processes which may take place during weathering at the

1. We are considering here only the extreme limits of the transformation. In actual fact, this kind of weathering is gradual and occurs sheet by sheet so all intermediate states may exist, represented in nature by interstratified argillaceous structures.

TABLE 6. Geochemical evolution of micas and micaceous lattices

Physico-chemical mechanism	Behaviour of basic interfoliate cations with regard to total layer (F)	Behaviour of basic interfoliate cations with regard to the component predominantly removed	Type of evolution of micas		Process
Acydolysis					
Total	$Int \leqslant F$	$(K_2O) = (Al_2O_3)$	Acidolytic destruction of layers (octahedral sheet)		Podzolization
Partial	$Int > F$	$(K_2O) > (Al_2O_3)$	Opening and aluminization (acid cations hydroxylated)	Conservation of the 2:1 layer	Aluminization (aluminochloritization)
Hydrolysis					
Partial hydrolysis and salinolysis	$Int > F$	$(K_2O) > (SiO_2)$	Opening and vermiculitization (basic cations hydrated)	Transformation	Apparent bisiallitization
Total hydrolysis	$Int \leqslant F$	$(K_2O) = (SiO_2)$	Hydrolytic destruction of layers (tetrahedral sheets)		Monosiallization Allitization

Characteristic secondary minerals		O	Clay silicates			Hydroxides Al	Tectosilicates (zeolites)
			2/1 type		1/1 type		
Mode of alteration		Solubilization	Transformation		Neoformation		
Acidolysis	Total	Podzolization	–		–	–	–
	Partial	–	Aluminization and aluminosiallitization (intergrades of Al and secondary chlorites		–	–	–
Hydrolysis	Partial and salinolysis	–	Bisiallitization Apparent (vermiculite-smectites)	Real (smectite)	–	–	–
	Intense/total desalcalinization	–	–	–	Monosiallitization (kaolinite-halloysite)	Allitization (gibbsite-boehmite)	–
Alkalinolysis	Partial	–	–	–	–	–	Sialicatization (analcime-sodalite)
	Total	Solodization	–	–	–	–	–

FIG. 4. Basic processes of the superficial evolution of sialferric rocks.

surface of the Earth, are presented in Figure 4 in which are shown simultaneously:

The physical and chemical mechanisms involved (hydrolysis, salinolysis, acidolysis, alkalinolysis).

The type of weathering (solubilization–neoformation–transformation).

The nature of the individualized secondary constituents (hydroxides–phyllosilicates–tectosilicates).

Figure 4 shows that four main geochemical types can, in theory, be identified:

The first is characterized by acidolytic solubilization phenomena, without there being any real formation of specific secondary minerals, and corresponds to the straightforward podzolization developed in podzols and podzolic soils (Rode, 1937; Souchier, 1971).

The second type is characterized by stability in silicated sheets of the 2:1 type, due to a neoformation or to a retention (transformation), irrespective of the nature of the interfoliar elements (Na, Ca, Al, etc.). This is then a case of bisiallitization in the broad sense of the term, characteristic of the 'brown earths' in general and of the 'vertisols'.

The third type involves intense hydrolysis phenomena leading to the complete dealkalinization of the primary minerals. The formation of 2:1 phyllites of the kaolinite type (monosiallitization) may then occur, possibly accompanied by the formation of aluminium hydroxides (allitization). This type corresponds to weathering phenomena which can be grouped together under the general name of lateritization and which are found in 'laterite' or, again, 'ferrallitic' soils (Lacroix, 1913; Harrassowitz, 1926; Harrison, 1933; Sieffermann, 1973; Chatelin, 1974).

Finally, the fourth type is distinguished in extreme instances by the neogenesis of alkaline tectosilicates and is evidence of the phenomena of widespread alkalinization (alkaline soils and alkaline zones of the endorheic area) (Maglione, 1974).

GEOGRAPHICAL DISTRIBUTION

The latter type, which is of great theoretical importance, is found in fact in relatively limited areas of the Earth's surface and these, as we have seen (Fig. 2), are

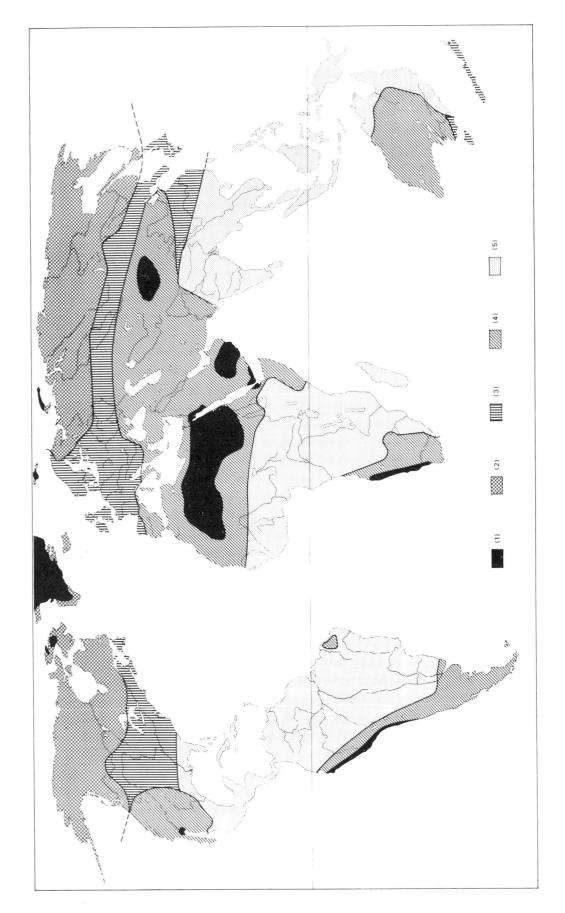

Fig. 5. The principal areas of surface weathering and soil formation: (1) area with no chemical weathering (ice-bound zones and absolute deserts); (2) northern podzolization area; (3) area of aluminization; (4) area of bisiallitization (in the strict sense); (5) area of lateritization.

scattered among the main endorheic basins. Hence outside the icebound zones and absolute deserts where geochemical changes are nil, the surface of the continents is divided among the first three types.

In this connection, it is possible to visualize these phenomena from the geographical sketch which is provided (Fig. 5). The main demarcation lines used are taken from the first study carried out in 1967 (Pedro, 1968).

In the northern evergreen forest zone (taiga),[1] podzolization is dominant, resulting from acidolytic solubilization and distinguished in practice by the low level of neoformation and transformation phenomena. It is thus spread over about 16 per cent of the continents.

Lateritization, which includes the processes of kaolinization and allitization, is concentrated in the warm and sufficiently humid tropical zones. Using the demarcation lines originally chosen (based on both an annual rainfall of over 500 mm and an average temperature of above 15 °C), this type, essentially associated with neoformation phenomena, covers about 31 per cent of the earth's surface.

The remainder, which corresponds to bisiallitization in the broad sense, thus takes up 39 per cent of the continents and extends around the temperate zone. It is thus the best represented type. In actual fact, it does not form a homogeneous unit.

First of all, 2:1 phyllites may be the result, as we have seen (Fig. 4), either of the phenomenon of neoformation or of the phenomenon of transformation. We should, of course, note that outside certain well-demarcated zones, which correspond to the dry tropical marginal areas and where the tendency is for vertisols to form, transformation is almost always more frequent than neoformation.

This, however, is not the main point, since the stable phyllite structure is in any case of the same type.

What is more important is to draw a distinction between the nature and composition of the interfoliar spaces which have a real physical and chemical, and hence geochemical, significance. In this connexion, we must consider:

An area of bisiallitization in the strict sense of the term in which the sheets are compensated by basic hydrated cations (vermiculitization, montmorillonitization, etc.). This is a characteristic of eutrophic brown earths, fersiallitic soils and vertisols (Paquet, 1970; Lamouroux, 1971; Blot and Leprun, 1973; Boulet, 1974).

A second area where the interfoliar ions are acid cations (aluminium), to a greater or lesser extent hydroxylated and polymerized. This can be called by the general name of aluminization since it in fact covers the processes of aluminosiallitization, intergradation and aluminous chloritization. This is the area of forest brown earths, brown podzolic soils and cryptopodzolic soils (Hetier and Tardy, 1969; Brethes, 1973).

The first of these two areas is concentrated in the warmer parts of the temperate and tropical zones; the second extends more particularly through the cool temperate zones. De Martonne's choice of a value higher than 20 for the index of aridity in the temperate zone seems to us perfectly justified for drawing the zonal limits for the area of aluminization and distinguishing it in this way from the area of bisiallitization.[2]

In conclusion, Table 7 brings together the various data that have been given about the types of geochemical change and soils, to which have been added

---

1. The other processes set in action in this zone are mainly of a hydromorphic nature: mineral hydromorphic soils or organic soils (peat bogs).
2. We should recall that the index of aridity $I = P/T + 10$, where P is the mean annual rainfall in mm and T the mean annual temperature in °C.

TABLE 7. The principal types of pedogeochemical evolution and the areas affected in the world

Type of pedogeochemical evolution	Area affected (percentage of earth surface)	Characteristic secondary minerals	Principal types of soils
Podzolization	16	None (solubilization) and (Clay 2/1)H	Podzols and podzolic soils
Aluminization	12.5	(Clay 2/1)Al	Acid-brown soils and brown podzolic soils
Bisiallitization s.s	26.5	(Clay 2/1)·Ca	Vertisols Brown tropical soils Fersiallitic soils
Lateritization	31	Kaolinite–Gibbsite	Ferrallitic soils
Alkalization	—	(Clay 2/1)·Na and tectosilicates	Alkaline soils (Solonetz-solod-natron deposits)

data about the areas respectively affected. At the Earth's surface there thus exist four main pedogeochemical categories, namely:
1. Podzolization, characteristic of the cold northern zones (16 per cent).
2. Aluminization in cool and humid temperate zones (12.5 per cent).

3. Bisiallitization, either by transformation or by neoformation, corresponding to the warm temperate and dry tropical zones (26.5 per cent).
4. Finally, lateritization in humid tropical zones (31 per cent).

References

BLOT, A.; LEPRUN, J. C. 1973. Influence de Deux-Roches-mères de Composition Voisine sur les Altérations et les Sols (Sénégal Oriental). *Cahiers ORSTOM. Géol.* Vol. V, No. 1, p. 45-7.

BOULET, R. 1974. Toposéquences de Sols Tropicaux en Haute-Volta. Equilibres Dynamiques et Bioclimats. Strasbourg, ORSTOM, 330 p. (Memoire No. 85, 1978.)

BRETHES, A. 1973. Mode d'Altération et Différenciation Pédogénétique sur Leucogranites du Massif du Morvan. Nancy. 97 p.

CHATELIN, Y. 1974. Les Sols Ferrallitiques: L'Altération. *Doc. Tech. ORSTOM.* No. 214. 144 p.

DELMAS, A. B. 1972. Sur les Rôles Respectifs de la Température et du Débit au Cours de l'Altération Expérimentale de l'Olivine par Lessivage à l'Eau. *C. R. Ac. Sc., Paris,* 274, p. 2413-15.

GARRELS, R. M. 1965. *Solutions, Minerals and Equilibria.* New York, N.Y., Harper & Row. 450 p.

HARRASSOWITZ, H. 1926. Laterit. *Forschr. Geol. Pal.,* Vol. 4, p. 253-565.

HARRISON, J. B. 1933. The Kalamorphism of Igneous Rocks und Humid Tropical Conditions. Harpenden, *Imper. Bur. Soil. Sc.,* 79 p.

HETIER, J. M.; TARDY, Y. 1969. Présence de Vermiculite Al, Montmorillonite Al, et Chlorite Al et leur Répartition dans quelques Sols des Vosges. *C. R. Ac. Sc., Paris,* Vol. 268, p. 259-61.

LACROIX, A. 1913. Les Latérites de Guinée et les Produits d'Altération qui leur sont Associés. *Nouv. Arch. Museum Hist. Nat.,* Vol. 5, p. 255-356.

LAGACHE, M. 1965. Contribution à l'Étude de l'Altération des Feldspaths dans l'Eau entre 100° et 200 °C. *Bull. Soc. Fr. Min. Crist.,* Vol. 88, p. 223-53.

LAMOUROUX, M. 1971. Étude de Sols Formés sur Roches Carbonatées Pédogenèse Fersiallitique au Liban. Strasbourg, *Mem. ORSTOM,* No. 56. 258 p.

MAGLIONE, G. 1974. Géochimie des Évaporites et Silicates Néoformés en Milieu Continental Confiné (Tchad). *Travaux et Documents* (Paris), No. 50, 331 p.

MEIGS, P. 1952. *La Répartition Mondiale des Zones Climatiques Arides et Semi-arides,* 2 maps. Paris, Unesco.

MILLOT, G.; LUCAS, J.; PAQUET, H. 1965. Evolution Géochimique par Dégradation et Agradation des Minéraux Argileux dans l'Hydrosphère. *Geol. Rundsch.,* Vol. 55, p. 1-20.

PAQUET, H. 1970. Evolution Géochimique des Minéraux Argileux dans les Altérations et les Sols des Climats Méditerranéens et Tropicaux à Saisons Contrastées. *Mémoire Serv. Géol. Als. Lorr.,* No. 30. 212 p.

PEDRO, G. 1964a. Contribution à l'Étude Expérimentale de l'Altération Géochimique des Roches Cristallines. Paris, *Ann. Agron.,* Vol. 15, p. 85-191, p. 243-333, p. 339-456.

——. 1964b. Principes Géochimiques de la Pédogenèse—Incidences Minéralogiques. *Trans. VIII Cong. Int. Soil Sc., Bucarest III,* p. 1087-94.

——. 1966. Essai sur la Caractérisation Géochimique des Différents Processus Zonaux Résultant de l'Altération Superficielle. *C. R. Ac. Sc., Paris,* Vol. 262D, p. 1828-31.

——. 1968. Distribution des Principaux Types d'Altération Chimiques à la Surface du Globe—Présentation d'une Esquisse Géographique. *Rev. Géol. Phys. Géol. Dyn.,* Vol. X, p. 457-70.

ROBERT, M. 1971. Étude Expérimentale de la Désagrégation du Granite et de l'Évolution des Micas. *Ann. Agron.,* Vol. 21, p. 777-817; Vol. 22, p. 43-93, p. 155-81.

ROBERT, M.; PEDRO, G. 1972. Établissement d'un Schéma de l'Évolution Expérimentale des Micas Bioctaédrique en Fonction des Conditions du Milieu. *4th Int. Clay Conf., Madrid,* p. 433-47.

RODE, A. A. 1937. Le Processus de Podzolisation. *Počvovedenie,* Vol. 10. p. 849-62. (In Russian.)

SIEFFERMANN, G. 1973. Les Sols des Régions Volcaniques du Cameroun. Strasbourg. *Mem. ORSTOM,* No. 66. 183 p. (Thesis.)

SOUCHIER, B. 1971. Évolution des Sols sur Roches Cristallines à l'Étage Montagnard (Vosges). *Mem. Serv. Géol. Als. Lorr.* Nancy, No. 33. 134 p. (Thesis.)

TARDY, Y. 1970. Géochemie des Altérations—Étude des Arènes et des Eaux des Quelques Massifs Cristallins d'Europe et d'Afrique. *Mem. Serv. Géol. Als. Lorr.,* Strasbourg. No. 31. 199 p.

THORNTHWAITE, C. W. 1948. An Approach Towards a Rational Classification of Climate. *Geogr. Rev.,* p. 55-94.

TRICHET, J. 1970. Contribution à l'Étude de l'Altération Expérimentale des Verres Volcaniques. *Trav. Labo. Géol. Ecole Norm. Sup.,* Paris, No. 4. 152 p.

VIELLEFON, J. 1974. Contribution à l'Étude de la Pédogenèse dans le Domaine Fluvio-marin en Climat Tropical d'Afrique de l'Ouest. Paris. 360 p. (Memoire No. 83.)

The role of major and minor elements in the nutrition of plants, animals and man

J. A. C. Fortescue

Department of Geological Sciences,
Brock University, St Catharines, Ontario (Canada)

Introduction

Geochemistry is holistic for two major reasons. First, because it is concerned with complex natural systems —such as landscapes—as 'totalities' involving interactions between the lithosphere, hydrosphere, atmosphere and biosphere; and second, because it is concerned with the circulation of all chemical elements in nature and not only those which happen to be nutrients.

Like ecology, geochemistry also relates to time. Some geochemical processes take place deep in the Earth during geological time while others occur at, or near the Earth's surface under one or more sets of climatic constraints. The ecological time-scale may involve short-term effects (e.g. diurnal, or seasonal changes) or longer time intervals involving the evolution of soils or the development of a climax type of plant cover within a particular area of country.

Many geochemical processes are now being, or have in the past been modified by man's activities. For example, man may have upset the delicate balance of supply and demand for nutrient elements within a particular area of country. One role of the geochemist interested in nutrition is to learn to recognize these effects in relation to the health of plants, animals or man. Another aspect of the geochemist's role is to predict short-term, as well as long-term effects on the geochemistry of landscapes of the addition of chemicals before they are added to the landscape in order to avoid the creation of new problems. Consequently, the role of the geochemist is of increasing importance to environmental teams of scientists and others who plan the optimum use of areas of country from a holistic point of view.

This chapter includes an overview of the role of chemical elements in the nutrition of plants, animals and man with particular reference to examples of imbalances in the nutrition and health of organisms caused by particular elements or groups of elements. In later sections of the chapter an outline is given of the relationship of geochemistry—particularly landscape geochemistry and ecology—to nutrition. The final section of the chapter is a discussion of ways in which research projects might study the interface between geochemistry and ecology in order to solve problems in the growing field of geoepidemiology.

Because of the broad scope of the chapter it has been necessary to impose certain constraints upon the subject-matter. For example, micro-organisms are only mentioned incidentally even though they play a major role in the circulation of nutrients in various kinds of landscapes. More information on micro-organisms in geochemistry may be obtained from Kuznetsov et al. (1963), Zajic (1969) and Alexander (1961, 1971). With respect to ecology, books by Brock (1966), Richards (1974) and the symposium edited by Gray and Parkinson (1968) should be consulted. Another constraint on the subject-matter concerns organic geochemistry. General information on this important aspect of geochemistry may be obtained from Eglington and Murphy (1969) and, in the more specialized field of non-marine deposits, from the book by Swaine (1970).

Some useful references on specific topics include Mason (1966), Krauskopf (1967), Wedepohl (1970), Levinson (1974) and Siegel (1974) for geochemistry, Bowen (1966) and Datta and Ottaway (1965) for biochemistry; Steward (1963), Treshow (1970), Epstein (1972) and Peel (1974) for plant nutrition; Russell and Duncan (1956), Mallette et al. (1960), Underwood (1966, 1971) and Schütte (1964) for animal nutrition; and Harper (1971) and Underwood (1971) for human nutrition and health. Geochemists interested in ecology and landscape geochemistry should refer to Stafelt (1972) and Odum (1971) before reading more advanced texts such as Sukachev and Dylis (1964), Perelman (1961), Collier et al. (1973) and Reichle (1970). A general introduction to the modern study of landscapes is given by Mitchell (1973) and the principles of landscape geochemistry have been described by articles by Fortescue (1973a, 1973b, 1974a,

1974*b*); for a more detailed study, see Kozlovskiy (1972).

Nutrient elements

Bowen (1966) noted that an element is a nutrient for a particular organism if the organism can neither grow nor complete its life cycle in the absence of the element and if the element could not be replaced by any other in the nutrition process; consequently, all nutrients are essential for the healthy growth of organisms. However, not all essential elements are nutrients; for example, in mammals the essential element fluorine is not a nutrient but if it is absent in drinking water, mammals will develop dental caries. This review includes reference to both nutrients and essential elements for plants, animals and man.

Nutrient elements may be divided into three groups: gaseous elements, macronutrients and micronutrients (Table 1). The list of gaseous elements and macronutrients has been known for several decades but some of the micronutrients have been proved essential only recently. As more sophisticated methods of chemical analysis become available it is likely that the list of essential elements in Table 1 will increase.

The distinction between macronutrients and micronutrients is an empirical one and is based on different criteria for plants and animals. In plants, macronutrients are nutrient elements which are required at a concentration of above 1 p.p.m. for individuals grown in nutrient solution, micronutrients being required at low concentrations (see Table 2). In animals, the presence of an element at a concentration less than 1 p.p.m. in oven-dried food may identify it as a micronutrient.

If the gaseous elements are excluded, there are 15 nutrients required by plants (Table 2). At least 9 micronutrients (iron, manganese, zinc, cobalt, copper, iodine, molybdenum, selenium and chromium) have been identified for animals and there is evidence that tin, nickel, silicon and vanadium are also micronutrients for animals and some species of plants. It is likely that lithium is an element which, although not a nutrient, may be essential in small quantities for the well-being of man and other animals (Jenner *et al.*, 1968).

Geochemists usually express the content of a nutrient in organic matter as a percentage (1 per cent = 10,000 p.p.m.) of the ash and over-dry weight or fresh weight of the individual sample concerned. For specialized purposes other bases may be used; for example, in the fat-free body tissues of animals (Scott, 1972). These estimates of element content are generally based on the assumption that the total amount of the specified element in the sample is determined. Such information is of particular importance in the case of certain micronutrients such as iodine or cobalt in animals, but is of less value for macronutrients such as magnesium

TABLE 1. Essential elements for the healthy growth of plants and animals

	Plants	Animals
Gaseous elements	Hydrogen	Hydrogen
	Nitrogen	Nitrogen
	Oxygen	Oxygen
	Chlorine	Chlorine
Macronutrients	Carbon	Carbon
	Potassium	Potassium
	Calcium	Calcium
	Magnesium	Magnesium
	Phosphorus	Phosphorus
	Sulphur	Sulphur
Micronutrients	Iron	Iron
	Copper	Copper
	Zinc	Zinc
	Manganese	Manganese
	Boron	Cobalt
	Molybdenum	Iodine
	Silcon (?)	Molybdenum
		Selenium
		Fluorine
		Chromium
		Tin (?)
		Nickel (?)
		Vanadium (?)
Essential elements	Silicon (grasses)	Fluorine
		Lithium (?)

Source: plants: Fortescue and Marten (1970); animals: Hoekstra (1972).

which play a number of roles in the metabolism of an organism. Consequently, biochemists may divide up a diet of man into component parts and make chemical analyses of each part separately. Harper (1971) listed six components of such a diet; carbohydrates, fats and proteins which yield energy, provide for growth and maintain tissues subject to wear; and vitamins, minerals and water, which do not yield energy but which are essential for the syntheses of metabolites including hormones and enzymes. In this chapter, total contents of elements in over-dry material are assumed unless other bases for measurement are mentioned.

The distribution of nutrients in plant and animal tissues is indicated in Tables 2 and 3. A typical observation which must be made on data of this type collected for specialized purposes is evident from Table 2, i.e. the estimates did not cover all nutrients but only those which were of particular interest. One contribu which the holism of geochemistry can make to the collection of data of this type is to stress the need to look at all nutrient elements simultaneously in the same material.

One aspect of nutrition which is constantly in the mind of geochemists is the widely varying concentra-

TABLE 2. The concentration of essential plant nutrients in natural materials (in parts per million)

Element	Rocks [a]				Soil[a]	Fresh water[a]	Nutrient solution[d]	Plant matter (oven dry basis)		Relative number of atoms with respect to molybdenum (plant)
	Igneous[1]	Shale[1]	Sandstone[1]	Limestone[1]				All plants[b]	Conifers[c]	
Micronutrients										
Molybdenum	1.5	2.6	0.2	0.4	2[1]	0.00035[1]	0.05	0.1	1	1
Copper	55	45	5	4	20	0.01	0.064	6	4	100
Zinc	70	95	16	20	50	0.01	0.065	20	30	300
Manganese	950	850	50	1,100	850	0.012	0.55	50	200	1,000
Iron	56,300	47,200	9,800	3,800	38,000	0.67	5.6	100	200	2,000
Boron	10	100	35	20	10	0.13	0.50	20	20	2,000
Macronutrients										
Chlorine	130	180	10	150	100	7.8	3.5	100	1,000	3,000
Sulphur	260	2,400	240	1,200	700	3.7	48	1,000	800	30,000
Phosphorus	1,050	700	170	400	650	0.00541	41	2,000	1,300	60,000
Magnesium	23,300	15,000	10,700	2,700	5,000	4.1	36	2,000	1,000	80,000
Calcium	41,500	22,100	39,100	302,000	13,700	15.0	134–300	5,000	2,500	125,000
Potassium	21,000	26,700	11,000	2,700	14,000	2.3	130–295	10,000	8,000	250,000
Nitrogen	20	—	—	—	1,000	0.23	140–284	15,000	15,000	1,000,000
Oxygen	464,000	483,000	492,000	497,000	490,000	889,000		450,000		30,000,000
Carbon	200	15,000	14,000	11,400	20,000	11,000		450,000		35,000,000
Hydrogen	1,400	5,600	1,800	860	15,000	111,000		60,000		60,000,000

1. All values for concentration of elements are total content in samples.
Sources: Fortescue and Marten, 1970; (a) data from Bowen (1966); (b) data from Epstein (1965); (c) data from Keay (1964) for conifers; (d) data for 'Long Ashton Solution' (Hewitt, 1966).
Reprinted with permission of Springer-Verlag New York Inc.

tions of elements required by organisms. For example, listed in the right-hand column of Table 2 the relative proportions of molybdenum to hydrogen in plant material is of the order of 1:60 million atoms and yet it is almost routine for agriculturalists and horticulturalists to establish symptoms of molybdenum deficiency in crop plants grown under controlled conditions (Hewitt, 1966).

The data in Table 3 indicate that, provided a suitable basis for comparison is selected, the nutrient content of many kinds of animals is very similar to that of man himself. It is also interesting to note that the chemical composition of animals is not constant during the life cycle but varies from birth to maturity. Table 3 is of limited value from the geochemistry viewpoint because it does not include all nutrient elements and data on other elements, which may not be nutrients but which participate in the development of animals. An example of the latter group of elements is aluminium, which may be found in significant quantities in both plants and animals (Underwood, 1971). This is also evident in relation to data listed in Table 4, which is con-

cerned with the nutrient content of common feedstuffs. One role geochemists can play in nutrition during the next two decades is to compile like sets of data on the distribution and amount of all elements in the periodic table in tissues and foods for plants, animals and man. Perhaps the most interesting element in Tables 2, 3 and 4 is iron, which is a trace element in plant and animal tissues but is a major constituent of most soils.

Hoekstra (1972) has taken the description of nutrition requirements for animals a stage further by calculating (Table 5) examples of minimal requirements for mineral elements by animals of different kinds. These data have been listed per unit of dry diet. One notable feature of this approach was that it indicated that the relative amounts of nutrients required by different kinds of animals was remarkably similar, with certain understandable exceptions (e.g. the calcium requirement of chickens).

In summary, we note that all plants and animals require a wide range of nutrient elements for their well-being. Each element performs one or more functions in the organism and is usually present at given concentra-

TABLE 3. The nutrient content of different animals

Constituent	Chicken	Man	Pig	Cat	Rabbit	Rat	Steer
Adult							
Body weight, kg	2.0	65	125	4.0	2.6	0.35	500
Composition :[1]							
Water, g	760	720	750	740	730	720	550
Total N, g	31.0	34.0	31.5	33.6	37.0	35.0	27.2
Mg, mg	400	470	450	450	500	400	540
Fe, mg	40	74	90	60	60	60	168
Cu, mg	1.3	1.7	2.5	1.5	1.5	2.0	
Zn, mg	35	28	25	23	50	30	
I, mg	0.4	0.7					0.1
Se, mg	0.25	0.2	0.2				0.1
New-born							
Body weight, g	40	3,560	1,260	118	54	5.9	
Composition :[1]							
Water, g	830	823	820	822	865	862	
Total N, g	20.8	22.6	18.0	24.4	18.1	15.6	
Mg, mg	220	260	320	260	230	250	
Fe, mg	40	94	29	55	135	59	
Cu, mg	1.3	4.7	3.2	2.9	4.0	4.3	
Zn, mg		19.2	10.1	28.7	22.5	24.4	
Se, mg	0.20		0.15				

1. Results expressed per kg of fat-free body tissue.
Source: Scott, 1972.
Reprinted from *Micronutrients in Agriculture*, 1972, p. 525–54 (Geographic Distribution of Trace Element Problems) by permission of Soil Society of America.

tion levels which, as is indicated in the case of cereal crops in Table 4, are characteristic of the species or group. It has also been noted that the similarities between the chemical composition of plants and animals tends to outweigh the differences, which is an important consideration when populations of plants or animals complete for available food within a given environment.

Determination of the nutrient content of biological material

Without going into exhaustive details, it should be noted that major advances have been made in techniques of chemical analysis of biological materials during the past decade and when these are combined with modern computerized data processing, the geochemist has a tool that is equal to the holistic tasks of biogeochemistry. For example, it is now possible to determine quickly the levels of 30 different elements in a $25 \mu l$ sample of

fluid with a precision better than 5 per cent for each, using a combination of a radio-frequency plasma source and direct reading emission spectrograph (Fassel and Kniseley, 1974). Morrison (1972) described how spark-source mass spectrometry could be used for the simultaneous determination of 32 elements in human hair samples and Losee et al. (1974) used the same approach to determine the content of 68 elements in human teeth. As these analytical techniques and others like them become available to geochemists, it will be possible to examine the geochemistry of environments in a way which would have been considered most improbable only a few years ago. This advance in technique must be accompanied by similar advanced thinking on the part of scientists who work at the interface between geochemistry, ecology and nutrition and who have the responsibility of perfecting the science of geoepidemiology. More orthodox information on the chemical analysis of geochemical samples may be obtained from books by Wainerdi and Unen (1971) and Energlyn and Brealey (1971) and the review paper by Lisk (1974). The remainder of this chapter bears in mind these advances in data collection which may be reasonably expected over the next few years.

TABLE 4. Nutrient contents of common feedstuffs

Feedstuffs	Mg (percentage)	Mn	Zn	Fe	Cu (parts per million)	Co	I	Se
Alfalfa meal, 17 per cent protein	0.3	43	35	200	10	0.18	0.5	0.05–0.45
Barley	0.12	16	17	50	7.5	0.1	0.05	0.1–0.3
Beet pulp	0.25	35	1	300	12.5	0.1	—	—
Blood meal	0.2	5	—	3,800	9	0.1	—	0.07
Bone meal, steamed	0.6	30	425	800	16	0.1	—	—
Citrus pulp, dried	0.16	7	14	200	6	—	—	—
Coconut oil meal	0.25	55	—	680	19	2.0	—	—
Corn, dent, yellow, No. 2	0.12	5	10	35	4.5	0.1	0.05	0.03–0.38
Corn gluten feed	0.3	25	—	500	50	0.1	—	0.2
Corn gluten meal	0.05	7	—	400	30	0.1	—	1.15
Cotton-seed meal	0.55	20	—	100	20	0.1	0.12	0.06
Distillers' dried corn grains	0.06	19	—	200	45	0.1	—	—
Distillers' dried corn solubles	0.64	74	85	600	80	—	—	0.50
Fish meal, menhaden	—	36	150	270	8	—	—	1.7
Fish meal, herring	0.1	10	110	300	20	—	1.0	1.5–2.45
Fish solubles, 30 per cent solubles	0.2	25	38	300	48	—	—	1.0
Hominy feed, white or yellow	0.24	14	—	10	2	0.1	—	0.1
Linseed oil meal	0.6	37	—	300	25	0.2	0.07	1.1
Meat and bone scrap, 50 per cent protein	1.13	19	100	500	12	0.2	1.3	0.1–0.8
Milk, cow's	0.01	0.06	4	2	0.3	—	0.04	0.04
Milo	0.13	13	17	50	14	0.1	0.02	—
Molasses, beet	0.23	5	—	100	18	0.4	1.6	—
Molasses, cane	0.35	42	—	100	60	0.9	1.6	—
Oats	0.17	38	—	70	6	0.06	0.06	0.05–0.22
Oystershell, ground	0.3	130	—	2,900	—	—	—	0.01
Peanut oil meal	0.24	24	20	20	30	—	—	0.28
Rice bran	0.95	200	30	190	13	—	—	—
Rye grain	0.12	35	35	45	6	—	0.05	0.2
Sesame meal	0.75	48	100	—	3	—	—	0.08–0.15
Skim milk	0.11	2	40	30	3	—	—	0.05–1.0
Soy-bean meal, 44 per cent protein	0.27	35	27	150	20	0.1	0.13	0.05–1.0
Soy-bean meal, dehulled	0.25	40	45	150	20	0.1	0.1	0.05–1.0
Wheat bran	0.55	115	80	150	12	0.1	0.07	0.6
Wheat grain	0.16	20	5	50	7	0.08	0.04	0.05–0.8
Wheat standard middling	0.37	118	150	100	22	0.1	0.1	0.28–0.88
Whey, dried	0.13	5	3	7	45	0.1	—	0.08
Yeast, dried, brewers'	0.23	6	40	50	30	0.2	0.01	0.11–1.1
Yeast, dried, torula	0.13	13	100	90	13	0.04	—	0.03–0.05

Source : Scott, 1972.
Reprinted from *Micronutrients in Agriculture,* 1972, p. 524-54 (Geographic Distribution of Trace Element Problems) by permission of Soil Science Society of America.

The principal function of nutrient elements in plants, animals and man

Some nutrients such as cobalt in animals are relatively easy to describe from the functional point of view whereas others such as carbon or oxygen are so complex in function that many books are required to outline their detailed behaviour in organisms. What is attempted here is a mention of the important roles particular nutrients play in organisms—with special attention being given to functions which are of current geochemical interest. (In the interests of space, the reviews of each nutrient have been kept of approximately the same length regardless of the complexity of the biochemistry of the nutrient concerned. The objective of this section is to stimulate holistic thinking by readers regarding the nutrition process in relation to geochemistry.)

J. A. C. Fortescue

TABLE 5. Basic minimum requirements for nutrients by animals of different kinds

	Species					
	Chickens (0–8 weeks)	Swine (20–35 kg)	Beef cattle (growing)	Sheep	Rats (growing)	Typical or average requirement
Reference No.	23	24	25	26	27	
	(percentage)					
Ca	1.0	0.65	0.18–0.60	0.21–0.52	0.6	0.6
P	0.7	0.5	0.18–0.43	0.16–0.37	0.5	0.5
K	0.2	0.26	0.6–0.8	0.50	0.18	0.3
Na	0.15	0.1	0.1	0.04	0.05	0.1
Cl	—	0.13	—	—	0.05	0.1
S	—	—	0.1	0.80–0.16	—	0.1
Mg	0.05	0.04	0.04–0.1	0.04–0.08	0.04	0.06
	(parts per million)					
Fe	80	80	10	30–50	25	50
Zn	50	50	10–30	35–50	12	50
Mn	55	20	1–10	20–40	50	30
Cu	4	6	4	5	5	5
I	0.35	0.2	0.1 (?)	0.1–0.8	0.15	0.2
Mo	—	—	—	> 0.5 (?)	< 0.2[1]	0.2 (?)
Co	—	—	0.05–0.1	0.1	—	0.1
Se	0.1	0.1	0.05–0.1	0.1	0.04	0.1

1. Expressed per unit of dry diet.
Source: Hoekstra, 1972.
Reprinted with permission of the New York Acedemy of Sciences.

GASEOUS ELEMENTS

Hydrogen

Hydrogen is below only oxygen, nitrogen and carbon in biological importance. Most hydrogen in the biosphere is present as water, or as constituents of hydrocarbons. Elementary hydrogen is rare in the biosphere, although it may be liberated as the result of cellulose fermentation (Rankama and Sahama, 1950). In an interesting discussion, Epstein (1972) considered the problem of why water is the matrix and medium of living things. He focused attention on the following points as the pivot of his discussion:

1. Water is abundant at the surface of the Earth: if spread evenly, it would provide a layer 2.5 km thick over the whole surface of the globe;
2. Water is a liquid at temperatures prevailing at much of the Earth's surface: this provides a suitable physical medium for the generation of organized units;
3. Water has a relatively large heat capacity: this acts as a buffer allowing for the development of organisms within a relatively narrow range of temperature;
4. Water has an extremely high surface tension: this functions to maintain discrete boundaries where moist cells abut on air and where aqueous phases border on limpid membranes.

5. Water is the best solvent known: most substances, even resistant minerals, are slightly soluble in water, which may be important in the context of geological time;
6. Water viscosity is low: this permits rapid diffusion of solutes—including inorganic ions—through it and makes possible the rapid flow of water through narrow conduits of vascular systems.

Many of these properties of water are related to the hydrogen bond in which hydrogen atoms form a 'bridge' between two other atoms and 'share' it. As the result of this bonding, molecules of water are loosely knit together. Further information on relationships between water and life have been discussed by Blum (1968) and Needham (1965).

Another role hydrogen plays in nutrition is as a constituent of carbohydrates. Carbohydrates (or sugars and starches) are widely distributed in both plants and animals. In plants they are formed as a result of photosynthesis and form structural components as well as being constituents of cells. In animals, carbohydrates serve as important sources of energy for vital processes and may perform highly specific functions. Carbohydrates are classified into four groups: monosaccharides, or simple sugars, which have a general formula $C_nH_{2n}O_n$ and cannot be hydrolized into a simpler form;

62

and disaccharides, oligosaccharides, and polysaccharides which are named according to the number of monosaccharide units they would yield on hydrolizing. An introduction to carbohydrate chemistry in plants is given by Stace (1963) and in animals and man by Harper (1971).

Water uptake by plants is controlled by physical and osmotic forces in solids and a plant with insufficient water available to it will wilt. Several years ago the water available to plants from soils was related to a static model which involved the determination of 'constants' (e.g. wilting coefficient) for particular soils. More recently this approach has been shown to have serious theoretical limitations and a dynamic model has been proposed which allows for the exact measurement of interrelationships between potential, conductivity, water content and flux in both plant and soil components of the system. Using this model it has been realized that relationships vary incessantly between the different components of the system and static situations are exceedingly rare. Further information on this aspect of the geochemistry of water which is of vital importance in soil science and plant nutrition may be obtained from Hillel (1971).

Relationships between the water uptake and utilization in animals and man are relatively simple compared with those for plants growing in soil. As an example Harper (1971) noted that 40–60 per cent of the body weight of man is water-distributed in two main compartments, the extra-cellular compartment and the intracellular compartment. When the intake of water is restricted, or the loss from the skin is excessive, man suffers from a severe thirst and nausea and vomiting may result. When correction for dehydration is carried out it may be important to add chloride, sodium and potassium to the water taken. This is a very simple example of a principle which is common to many nutritional processes, namely, that the intake of one element, or elements, relates directly to the intake of others. In other words, it is often realistic to adopt a holistic approach to element requirements of plants, animals or man. A general review of the importance of water in the nutrition of plants and animals, including man, in arid areas is given in the book by Dregne (1970).

Oxygen

Oxygen plays a major part in metabolism quite aside from its role as a constituent of water. For example, it is found as a constituent of the OH^- and HCO_3^- ions and it is an important constituent of protoplasm, enzymes and many other kinds of organic compounds. In fact only a few organic compounds, such as β-carotene (Fig. 1) do not contain oxygen.

The energy input to the biosphere is almost entirely derived from the sun. Light energy from·this source is trapped by photosynthesis and is eventually converted

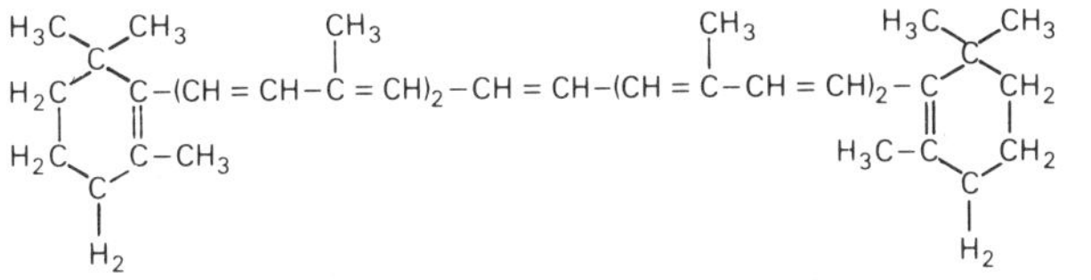

FIG. 1. The structure of β-carotine (from Harper, 1971).

into the chemical energy contained in stable, reduced carbon compounds which are later oxidized with the release of this energy. During photosynthesis, gaseous oxygen is given off into the atmosphere—or may be dissolved in the waters of the hydrosphere in which certain animals live. Oxygen from the atmosphere sink or from the hydrosphere, is required by organisms for respiration, during which it is exchanged for carbon dioxide which, in turn, is required for photosynthesis. Although the supply of oxygen from the air is unlimited this is not always the case in the aquatic environment where aerobic and occasionally anaerobic conditions may obtain. A limited supply of available oxygen, for example in lakes, may lead to the onset of the process of eutrophication which may seriously affect the geochemistry of the lake. For further information on this important topic see Rohlich (1969).

Nitrogen

Nitrogen occurs in all living organisms as a major constituent of proteins and amino-acids from which proteins are synthesized. Odum (1971) summarized the basic steps in nitrogen metabolism as shown in Figure 2. This series of syntheses and decompositions is at the heart of the 'nitrogen cycle', about which more information is given later on. Further details of the biochemistry of the cycle may be obtained from Gauch (1972) and Harper (1971). Although both oxygen and nitrogen occur in the atmosphere they differ in their biochemistry because whereas oxygen is readily available to plants and animals from the air or from water, nitrogen must first be 'fixed' from the atmosphere before it is suitable for uptake by plants. Under natural conditions, fixation of nitrogen occurs as a result of certain micro-organisms which may act alone or in root nodules of certain higher plants (e.g. legumes).

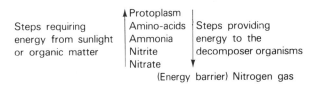

Chemical form of nitrogen

| Steps requiring energy from sunlight or organic matter | ▲ Protoplasm
Amino-acids
Ammonia
Nitrite
Nitrate | Steps providing energy to the decomposer organisms ▼ |

(Energy barrier) Nitrogen gas

FIG. 2. The basic steps in the synthesis and decomposition of organic compounds involving nitrogen (from Odum, 1971). Reproduced from *Fundamentals of Ecology* (3rd ed.) W. B. Saunders Co.

Due to this problem of nitrogen fixation, there is often a shortage of this element in agricultural and forest soils and chemicals including nitrogen in available form are commonly added to crops. Cooke (1972) noted that there are three common types of nitrogen fertilizers, including nitrates (which supply NO_3^-), ammonium salts (which supply NH_4^+) and simple amides which are not ionized but contain nitrogen as $-NH_2$. Urea, $CO(NH_2)_2$, is the most common type of amide fertilizer and contains about 46 per cent N by weight. For further information, see Black (1968).

Man's addition of nitrogen to the environment, either as fertilizers or as farmyard manure, may on occasion have adverse effects. For example, Turk et al. (1973) reported as much as 3,584 mg l^{-1} of nitrate in waters from wells in Runnels County, Texas, where cattle had died of nitrate poisoning and where the health of local residents was threatened.

Olsen (1972) noted that the application of nitrogen fertilizers may induce a zinc deficiency in citrus, and Underwood (1971) noted the relationship between molybdenum and the nitrogen-fixing capacity of the micro-organism Azotobacter.

Chlorine

Chlorine is found in all living organisms and is essential to their health, although proof of the essentiality of this element to plants was not generally accepted until the work of Broyer et al. (1954). Epstein (1972) noted that no enzymes, or other organic compounds, have been isolated from plants which contain chlorine as an essential constituent, although he pointed out that chloride is known to act in conjunction with an enzyme (or enzymes) of photosystem II (in the photosynthetic system) but its specific role was unknown. The chlorine requirement of animals is often considered together with the problem of dietary sodium (Underwood, 1966). According to Mallette et al. (1960), chloride is found throughout the tissues of animals and is absorbed to a considerable extent into proteins thus altering their zeta potentials. These writers also noted that chloride ions made up some 66 per cent of the anions in blood where the Cl$^-$ in serum is normally 0.36 per cent. Urine contains 0.06 per cent Cl$^-$ or less.

They also noted the difficulty of assessing the amount of chlorine required by animals and man. They noted that in animals fed on a diet not including chloride, none is excreted in the urine after a few days; thus the body effectively conserves its supply. Under normal conditions animals and humans require dietary chlorine because the content of this element in vegetables is relatively low. Deficiencies of chloride lead to abdominal cramp and loss of weight in animals. Animals seldom suffer from excess chlorine except when they drink sea water. Under normal conditions some 90 per cent of the chlorine ingested as chloride by animals is excreted in the urine.

The role of the macronutrient elements in the nutrition process

CARBON

Organic chemistry is essentially the chemistry of carbon and for this reason carbon can be considered the most important element in the biosphere. The element may be absorbed by organisms from the air, from water, from solid food or even from rock material. As Day (1965) noted, naturally occurring compounds are numerous, abundant and widespread. Plants obtain carbon as carbon dioxide from the air and photosynthetically fix the element in sugars and other substances. Animals obtain their carbon largely from the food which is absorbed by the gut. Both plants and animals return the element to the air by respiration.

The amount of carbon 'fixed' by a type of organism or a plant community during a given time is of particular importance since it can be used as a parameter of the growth rate of organisms. Bowen (1966) discussed this role of carbon (or more correctly, carbon dioxide) circulation in some detail and defined 'net productivity' as the amount fixed less that lost due to respiration. Even for green plants growing in strong sunlight, the difference between these two figures can be of the order of 10–50 per cent (Bowen, 1966). Bowen also pointed

TABLE 6. Estimates for the net productivities and net production of major communities

Community	Productivity kg m^{-2} yr^{-1}		Production; kg $\times 10^{-14}$ yr^{-1}	
	Carbon	Dry matter	Carbon	Dry matter
Coniferous forest	1.4	2.8	0.14	0.28
Deciduous forest	0.54	1.2	0.025	0.06
Tropical forest	2.3	5.0	0.34	0.73
Other forest	0.54	1.2	0.078	0.17
Arable land	0.91	2.0	0.21	0.46
Grassland	0.91	2.0	0.245	0.54
Desert	0.045	0.1	0.015	0.033
Tundra	0.045	0.1	0.005	0.011
Mean for land	0.71	1.53	1.058	2.284
Ocean	0.072	0.32	0.24	1.070
Continental shelf	0.16	0.73	0.043	0.195
Brown algal zone	1.0	2.9	0.009	0.026
Mean for sea	0.081	0.36	0.292	1.291
Mean for whole earth	0.265	0.70	1.35	3.575

Source: Bowen, 1966.
From Trace Elements in Biochemistry. © Academic Press Inc. (London) Ltd. Reprinted with permission.

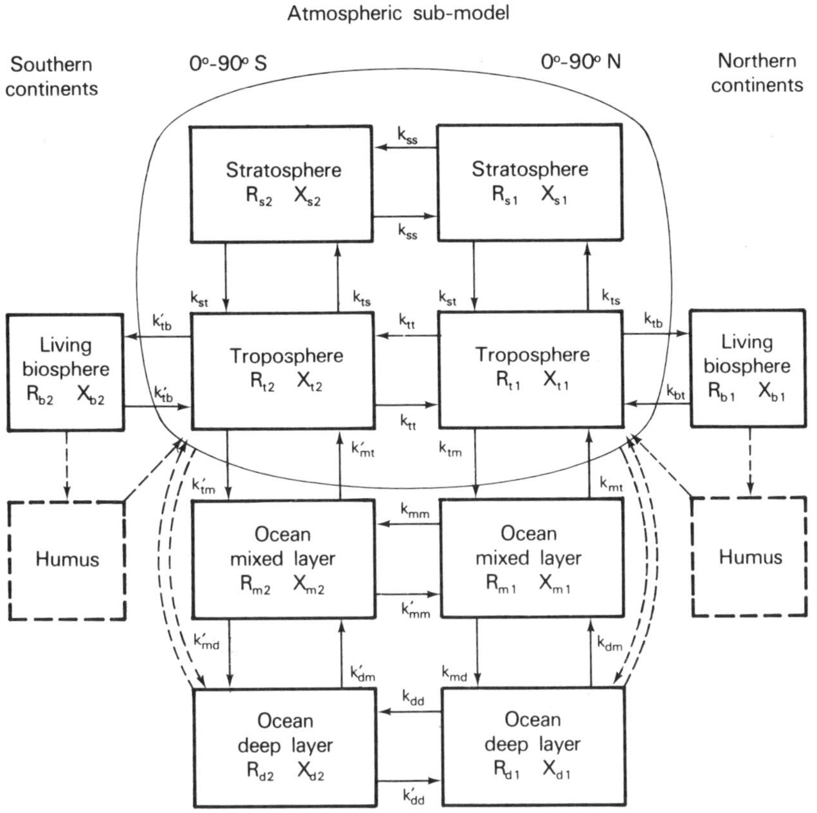

Atmospheric sub-model

FIG. 3. A model for the exchange of carbon and radio-carbon, generated in the atmosphere and redistributed to living and dead organic matter on continents and to mixed and deep layers of the oceans (from Olsen, 1970; after Nydal, 1968).

out that the productivity (i.e. amount of carbon dioxide fixed per year) per kilogram of unicellular organisms such as marine planktonic algae was several orders of magnitude greater than that of land plants. Closely related to the idea of productivity in the biosphere is the concept of 'biomass' which was defined by Bowen (1966) as the total mass of any organism, or organisms, which is found within a given area of country. Information on the net productivities of the major communities which together comprise the biosphere is given in Table 6. More comprehensive attempts have been made to describe the circulation of carbon in nature by means of systems analysis. For example, Olsen (1970) provided a flow diagram conceptual model of the circulation of carbon and radio-carbon on a global basis (Fig. 3).

. Radioactive tracers using ^{14}C (half-life 5,568 years) may be used in certain ecological experiments, for example, to test out the toxicity of particular chemicals as described by Metcalf (1974). Further information regarding the role of carbon in nutrition may be obtained from references listed at the commencement of this section.

POTASSIUM

Potassium is the only monovalent cation essential for the nutrition of all higher plants (Epstein, 1972), and this element is also essential for animals. Its principal role in these organisms is as an activator of enzymes. Evans and Sorger (1966) listed 66 enzymes derived from animals, higher plants and micro-organisms which require monovalent cations for maximum activity, of which potassium was the most effective, followed by the ammonium ion. Harper (1971) noted that while in man this element is the principal cation of the intercellular fluid, it is also a very important constituent of the extracellular fluid since it regulates all muscle activity, including that of the cardiac muscle. Within the cell, it influences the acid-base balance and cellular osmotic pressure.

Some animals require more potassium than sodium. For example, Mallette et al. (1960) noted that this was the case for rats, pigs and chickens. Potassium deficiencies in animals are unknown in nature but may be produced experimentally. According to Mallette et al.

(1960), potassium toxicity occurs in the terminal phases of Addison's disease. According to Harper (1971), high levels of potassium are found in the following foods: veal, chicken, beef liver, beef, pork, dried apricots, dried peaches and bananas.

CALCIUM

The calcium content of healthy plant tissues ranges from 0.2 per cent (oven-dry basis) to several per cent, and it is likely that these values are well in excess of the minimal metabolic requirements (Epstein, 1972). Wallace et al. (1966) showed experimentally that if other divalent metal ions were kept at low concentrations in the nutrient solutions, tobacco and corn plants could grow in nutrient solutions containing only 2 p.p.m. calcium. Calcium deficiency in plants leads to a disorganization of cells and tissues and this seems to indicate that this element is involved in maintaining membranes in a functional state (Epstein, 1972). Direct evidence for this has been obtained from the observation of the shoot apex in barley as affected by calcium deficiency by Marinos (1962). Stocking and Ongun (1962) found that the chloroplasts of tobacco and bean leaves contained some 60 per cent of total leaf calcium, and they concluded that chloroplasts were the site of calcium accumulation. Reviews of the role of calcium in plants have been published by Gauch (1972), Jones and Lunt (1967) and Burström (1968).

Calcium is the most abundant inorganic element in the higher animals, where it is located predominantly in the sekeletal tissues. With the exception of bones and teeth, calcium is distributed in small amounts and is precipitated in specialized functions such as the control of muscle contraction, nerve impulse or blood clotting. In man, a daily intake of 1.0 g calcium is advised for adults—with larger amounts necessary for children and pregnant women (Mallette et al., 1960). The effects of calcium deficiency in animals are of three types: (a) rickets, which involves poor bone formation and tooth development; (b) failure of the blood-clotting mechanism; and (c) tetany, which is a nervomuscular disorder. Abnormally heavy calcium intakes may lead to excessive calcification of joints and the formation of kidney stones. In man, good dietary sources of calcium include milk, cheese, egg yolk, lentils, nuts, figs, cabbage, turnip greens, cauliflower and asparagus (Harper, 1971).

MAGNESIUM

Epstein (1972) remarked, that chlorophylls were the only major stable compounds of plants which contained an atom of magnesium as a fixed (non-dissociable) constituent. Magnesium represents 2.7 per cent of the molecular weight of chlorophyll, but the chlorophyll represents only about 10 per cent of the total magnesium in the leaf. Plastids contain more magnesium than is found in chloroplasts since Mg takes part in the energy-

conversion process in the leaf. Magnesium is an activator of more enzymes than any other element. The importance of magnesium in plant nutrition was stressed in the classic text by Jacob (1958) which focused attention on the role played by this element in plant nutrition and the need for chemical fertilizers containing this element for agriculture. More recently, this topic has been discussed by Cooke (1972) in relation to agriculture and by Baule and Fricker (1970) in relation to forestry.

Like calcium, magnesium occurs in the skeletal parts of animals, where about 75 per cent of the total in the body is found in the teeth and bones (Mallette et al., 1960). In animals, as in plants, magnesium acts as an activator of enzymes. Natural deficiency of magnesium in animals is unknown although reduced levels in the blood were found to be related to a reduced diet of magnesium in experimental animals (Mallette et al., 1960). Magnesium balance is related to levels of both calcium and phosphorus; consequently, a serious imbalance in the diet of either of these two elements will interfere with the magnesium metabolism of an animal. The adult human body contains about 21 g of magnesium and 350 mg/day is required to maintain good health (Harper, 1971). Of interest to geochemists was the observation by Harper that a sub-acute, or chronic, deficiency of magnesium in man was not readily detectable but could be the cause—or a factor in the cause—of chronic disease in those systems (cardiovascular, renal and neuromuscular) which are particularly susceptible to magnesium.

PHOSPHORUS

Phosphorus is absorbed by plants as the dihydrogen phosphate ion, $H_2PO_4^-$, and in this respect it is like nitrogen (NO_3^-) and sulphur (SO_4^{2-}). However, unlike the other two elements, phosphorus is not reduced in the cell to a lower oxidation state. Phosphorus plays a key role in the metabolism of all living things due to the part it plays in the energy-transfer systems, which involve adenosine triphosphates. Phosphorus also regulates many enzyme processes in plants and activates certain enzymes (Epstein, 1972). According to Olsen (1972), many interactions occur in plants between phosphorus and the trace elements such as zinc, iron, copper and molybdenum; for example, Stout et al. (1951) showed that phosphorus enhanced the absorption and translocation of molybdenum by tomato plants.

Many soils are deficient in phosphorus and the addition of chemical fertilizers containing this element has long been an established agricultural practice. Further information on relations between phosphorus added as a chemical fertilizer and soils and plants may be obtained from Black (1968) and Cooke (1972).

Next to calcium, phosphorus is the next most abundant mineral element in animals. Most phosphorus is associated with calcium in bones and teeth, although there are many other biochemical roles of this element in other tissues (Mallette *et al.*, 1960). Humans require about 1 g of phosphorus per day in their diet.

Phosphorus deficiency in livestock is the most widespread of any on a world-wide basis and has been reported from more than twenty states in the United States of America (Mallette *et al.*, 1960). Symptoms of phosphorus deficiency include loss of appetite and a craving for wood and bones. Underwood (1966) noted that phosphorus deficiency was usually more severe in grazing cattle than in grazing sheep and also noted that the ability of animals to utilize phosphorus was dependent also on its vitamin D supplies.

In man, phosphate retention is a prominent cause of acidosis in severe renal disease and the close relationship that exists between calcium and phosphorus metabolism ensures that if either element is given in excess, the excretion of the other is inreased (Harper, 1971).

SULPHUR

Sulphur is absorbed by plants largely as the sulphate ion which is reduced intracellularly; the sulphur ions are then incorporated in proteins and other organic compounds. Volatile compounds of sulphur contribute to the characteristic odours given off by onion, mustards and other plants, although the function of these compounds is not clearly understood (Epstein, 1972). Sulphur activates some enzymes and may act in lieu of phosphate. Chemical fertilizers containing sulphur include ammonium sulphate (24 per cent sulphur) and superphosphate (12 per cent sulphur). Appreciable amounts of sulphur may occur in precipitation; for example, Cooke (1967) reported that 142 kg/ha sulphur year fell near an industrial area in Indiana in contrast with less than 3 kg/ha in rural areas of New Zealand. Sulphur is required by all living cells and, in animals, much of this element occurs in amino-acids, peptides and proteins. It is probable that the only sources of sulphur for man are the two sulphur-containing amino-acids, cystine and methionine (Harper, 1971). Keratin, the protein of hair, hoofs and nails, is rich in sulphur-containing amino-acids and, for this reason, the sulphur requirement of hairy animals is higher than that for human beings.

SUMMARY

Although these summaries of the roles of macronutrients in the nutrition of plants, animals and man have been extremely brief, they do emphasize the complexity of the role of each element in nutrition—as well as showing how each element has a unique and characteristic function within organisms of different kinds. It is of particular importance to geochemists that chemical elements in organisms interact with each other and so the study of a particular element may not provide information required in geoepidemiology. Another aspect

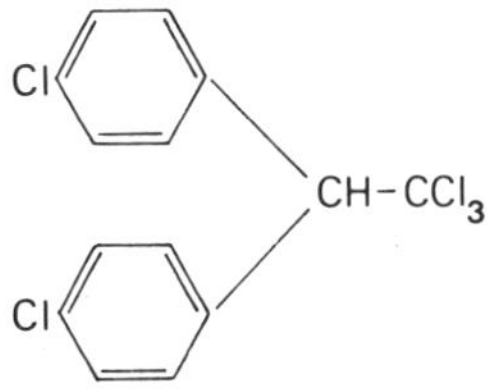

FIG. 4. The structural formula of DDT (dichlorodiphenyltrichloroethane).

is that the chemical form of an element may be of vital importance with respect to the health of an organism. For example, the residues of DDT contain just carbon, hydrogen and chlorine but when combined in the form shown in Figure 4, they constitute an environmental hazard to plants, animals, and man.

The role of the micronutrient elements in the nutrition process

Geochemists are often particularly interested in the circulation of the micronutrient elements in nature for a number of reasons. First, because—with the exception of iron, manganese and fluorine—these elements are found in the earth's crust at levels of less than 100 p.p.m. (Levinson, 1974) and are thus relatively rare in nature. The second aspect of the trace elements, which is of interest to scientists who study nutrition, is the variation in abundance of particular trace elements related to the genesis of particular rocks. For example, the element cobalt has an abundance in the Earth's crust of 25 p.p.m., in ultrabasic rocks of 150 p.p.m., in granites of 1 p.p.m. and in soils of 40 p.p.m. (Levinson, 1974). Consequently, when geochemists consider health problems, they often stress the trace elements when it might be more realistic to study the trace elements and the macronutrients together. As in the case of macronutrients, the outlines given for each element in this section are designed to stimulate the reader to further study rather than to describe in great detail the roles of the different elements.

IRON

In plants, there are many metabolites containing atoms of iron as fixed (not readily dissociable) constituents of the molecule. Iron can be either an integral part of a protein, or a part of a low molecular weight prosthetic group; the iron porphyrins (haems) are the most intensively studied iron-containing prosthetic groups (Epstein, 1972). Iron is essential for chlorophyll synthesis and, as Machold and Scholz (1969) showed in an elegant experiment, the radioactivity of ^{59}Fe supplied to the roots of a tomato plant suffering from acute iron deficiency was later found in an area of the leaves which coincided precisely with the green areas of the leaves.

(A photograph and an autoradiograph taken during the course of this experiment appeared in Epstein, 1972.) Price (1968) reviewed the effects of iron deficiency on plants. Lucus and Knezek (1972) listed conditions which induced iron deficiencies in plants, including low soil Fe, free $CaCO_3$, high HCO_3^-, extreme moisture, high levels of heavy metals, high soil P and excess soil acidity. Fruit trees and horticultural crops grown on calcareous soils often show an iron deficiency, which may be controlled by spraying with a 5 per cent ferrous sulphate solution (Cooke, 1972). Often it is difficult to get crops to absorb enough iron and the use of fertilizers incorporating chelated iron is often more effective (Cooke, 1972).

Underwood (1971) pointed out that the total iron content of the animal body varied with species, age, sex, nutrition and state of health. The normal adult human body contains 4–5 g of iron (or 60–70 p.p.m. for a 70 kg individual). This is twice the amount of zinc and twenty times that of body copper. Most of the iron is in complex forms bound to proteins, either as porphyrin or haem compounds (particularly haemoglobin and myoglobin), or as other protein compounds such as ferritin and transferrin. In the adult man the estimated distribution of iron is as shown in Table 7.

According to Underwood (1966), there is no convincing evidence that iron deficiency ever occurs in grazing stock under natural conditions—except possibly by heavy infestation with helminth intestinal parasites which may cause iron deficiency in lambs and calves. Earlier claims of iron deficiency in sheep and cattle grazing in particular localities in various parts of the world were invalidated by the discovery that the anaemia and muscular wasting (marasmus) characteristic of these areas was due to cobalt deficiency (Underwood, 1966). One problem, associated with the iron intake of humans, is that the amount of iron available from different foods is difficult to predict; another problem is the low iron content of breast milk, which cannot be relieved by providing the mother with more iron (Mallette et al., 1960).

COPPER

According to Epstein (1972), copper is a component of several plant enzymes, including ascorbic acid oxidase and the phenolases; hence, a deficiency of copper interferes with protein synthesis and will cause an increase in soluble nitrogen compounds. Visible effects of copper deficiency in plants include interveinal chloritic mottling in terminal leaves, and characteristic orange-brown striations may occur parallel to leaf margins. In peat soils, 'white tip' of cereals is caused by copper deficiency and results in young leaves becoming limp and chlorotic and remaining tightly rolled; in forest trees such as Sitka spruce (Picea sitchensis), copper deficiency also causes a spiral twisting of needles as well as ring-shaped necrotic lesions (Hewitt, 1966). Lucus and Knezek (1972) noted that the movement of copper applied to soils is usually minimal, so that a single application can supply the needs of plants for several years. The amount of the application should be corrected by making the applications of copper on a bulk weight basis rather than a weight basis. They also noted that the

TABLE 7. Distribution of iron-containing compounds in the normal adult human

	Total in body (g)	Iron content (g)	Percentage of total iron in body
Iron porphyrins (haem compounds)			
Haemoglobin	900	3.0	60–70
Myoglobin	40	0.13	3–5
Haemenzymes			
Cytochrome c	0.8	0.004	0.1
Catalase	5.0	0.004	0.1
Other cytochromes	—	—	—
Peroxidase	—	—	—
Nonporphyrin iron compounds			
Siderophilin (transferrin)	10	0.004	0.1
Ferritin	2–4	0.4–0.8	15.0
Haemosidena	—	—	—
Total available iron stores		1.2–1.5	
Total iron		4.0–5.0	

From H. A. Harper; V. W. Rodwell; P. A. Mayes. *Review of Physiological Chemistry,* 16th ed. ©1977 Lange Medical Publications, Los Altos, Calif. Reprinted with permission.

TABLE 8. Copper concentration in blood of different animal species

Species	Age and condition	Mean copper concentration (g/ml)
Human	Healthy, adult male	1.10±0.12[1]
Human	Healthy, adult female	1.23±0.16[1]
Human	Pregnant female, at delivery	2.69±0.49[1]
Human	Female, late pregnancy	2.80[1]
Human	Healthy, adult female	1.00[1]
Ovine	Healthy, mature	1.01±0.96[2]
Ovine	Healthy, mature	0.91[2]
Bovine	Healthy, mature	0.93[2]
Guinea pig	Healthy, mature	0.50±0.006[2]
Domestic fowl	Healthy, mature	0.23±0.008[2]
Domestic duck	Healthy, mature	0.35±0.007[2]

1. Serum.
2. Whole blood.
Source: Underwood, 1971.
From *Trace Elements in Human and Animal Nutrition* (3rd ed.). ©Academic Press Inc., New York. Reprinted with permission.

TABLE 9. Blood copper in various clinical conditions of man

Condition	No. of subjects	Whole blood Cu (μg/per cent)	Plasma Cu (μg/per cent)	Cell Cu (μg/per cent)	Volume of packed RBC (m/100 ml)	Plasma Fe (μg/per cent)
Normal	63	98 ± 13	109 ± 17	115 ± 22	47	115 ± 42
Pregnancy	30	169	222	130	37	91
Infection	37	141	167	116	41	57
Acute leukaemia	19	195	236	98	27	171
Chronic leukaemia	21	119	148	101	39	113
Hodgkin's disease	14	142	171	109	40	78
Pernicious anaemia	10	111	121	98	27	173
Aplastic anaemia	8	130	152	86	28	203
Iron deficiency anaemia:						
adults	9	114	132	109	30	26
infants	24	155	168	152	28	31
Haemochromatosis	14	103	134	—	—	234
Wilson's disease	3	79	55	110	42	64
Nephrosis	3	70	80	119	44	62

Source: Underwood, 1971.

From *Trace Elements in Human and Animal Nutrition* (3rd ed.) © Academic Press Inc., New York. Reprinted with permission.

availability (mobility) of copper in a given soil increases when the pH fell below 5.

Underwood (1971) noted that a healthy human body contains some 80 mg of copper. He also noted that the distribution of copper was relatively uniform throughout the body and tended to decrease with age in most organs with the exception of the brain—where it doubled with age. Exceptionally high copper levels exist in the eye. For example, in the iris of the trout eye, a level as high as 105 p.p.m. copper has been recorded (Underwood, 1971). Copper also occurs in human hair. Smith (1967) examined 277 samples of human hair of unspecified colour and found a mean level of 23.1 p.p.m. copper (range = 7.6–54.5 p.p.m.). Copper is also found in variable amounts in blood of different animals (Table 8) and varies according to whether the organism is healthy or diseased (Table 9). The data in Table 9 are of particular interest to geochemists since they indicate how the health of an individual may relate to the content of a trace element in the blood quite apart from environmental conditions. Unfortunately no data were readily available on the effect of genetic differences in man, although the uptake of copper by plants is known to be affected by considerations of this type (Brown et al., 1972). An adult man requires 2.5–5 mg of copper per day, which aids the absorption of iron from the gastrointestinal tract (Harper, 1971). Copper deficiency in animals may result in bone disorders, demyelination of the central nervous system, effects on the pigmentation of hair and wool and fibrosis of the myocardium (Scott, 1972).

ZINC

Zinc is the metal component of a number of metalloenzymes including several dehydrogenases, e.g. alcohol dehydrogenase and lactic dehydrogenase. There is some evidence that zinc may be related to the role of auxins in water absorption in higher plants (Gauch, 1972). Most disorders in the zinc metabolism of plants occur in calcareous soils where zinc is relatively insoluble (Lucus and Knezek, 1972). According to Hewitt (1966), changes in leaf morphology which occur as a result of zinc deficiency include 'little leaf' and 'rosette' in apple, peach and pecan, 'frenching' of citrus and 'bronzing' of tung trees.

In animals and man, zinc is a structural and functional component of the enzyme carboxypeptidase and participates directly in the enzyme's catalytic activity (Harper, 1971). Zinc is known to be a constituent of insulin and is involved in bone growth, wound healing, the reproductive process and in protein and carbohydrate metabolism. It seems clear that zinc is emerging as an element of increasing importance in the study of nutrition in plants, animals and man. According to Harper (1971) the recommended minimum intake of zinc by humans is 0.3 mg/kg body weight, an amount which is easily attained in a normal diet.

MANGANESE

Manganese is considered to be a catalyst in the oxygen-evolving reaction of photosynthesis (see Gauch, 1972). Epstein (1972) pointed out that manganese could substi-

69

tute for magnesium as an activator of phosphate-transferring mechanisms and that it was a prominent activator of enzymes involved in the Krebs cycle. Hewitt (1966) described deficiency diseases caused by a lack of manganese in plants—including the 'grey speck' of oats, 'marsh spot' of peas and 'speckled yellows' of beets. He also noted that an excess of manganese leads to an iron deficiency or a direct manganese toxicity in plants. Manganese deficiencies are commonly observed on well-drained soils with a neutral or slightly alkaline reaction; this is usually corrected by the addition of manganese sulphate in fertilizers (Murphy and Walsh, 1972).

In animals, manganese is required for general growth, proper bone formation and for the development and function of the reproductive system. Manganese deficiency is unknown in man, although deficiencies have been produced in domestic animals; these involve bone malformation and loss of reproductive capability (Mallette et al., 1960). Harper (1971) noted that the normal human intake of manganese was 4 mg/day, which is in excess of the minimum required. The total body content of manganese is about 10 mg, the kidney and liver being the chief storage organs. Most manganese is excreted in the bile, very little being found in the urine (Harper, 1971).

BORON

Boron is specifically a plant micronutrient. Epstein (1972) pointed out that no animal enzymes (or other essential organic compounds) had been isolated which contained boron as an active constituent, although it has been known for some time that boron plays a regulatory role in carbohydrate metabolism. In plants, the predominant pathway for the degradation of glucose is glycolysis but an alternative pathway, the 'pentose shunt pathway' does exist which involves boron-containing compounds and is significant in many tissues (Bonner and Varner, 1965). A full discussion of the role of boron in plant nutrition appears in Gauch (1972). Boron Boron deficiency produces a wide variety of visible symptoms in plants, including 'heart rot' of beet, 'stem crack' of celery and 'yellows' of alfalfa (Hewitt, 1966). Boron toxicity symptoms include chlorosis and marginal scorch, and Hewitt (1966) noted that 0.5–5 p.p.m. boron in irrigation water for citrus was sufficient to induce chlorosis. Lucus and Knezek (1972) noted that low total B in soil, moderate or heavy rainfall, neutral or alkaline soils, dry weather and high light intensity could all contribute to boron deficiency in plants (most boron-containing fertilizers are readily soluble in water, although colmanite ($Ca_2B_6O_{11} \cdot 5H_2O$) and boronated glass are only slightly soluble (Lehr, 1972)).

MOLYBDENUM

Epstein (1972) noted that molybdenum was known to occur in several metallo-enzymes, among which are those involved in nitrogen fixation and nitrate reduction in plants. Quite small amounts of molybdenum are required for the healthy growth of plants. For example, Gauch (1972) noted that molybdenum-sufficient tomato plants contained 10 p.p.b. Mo on a fresh-weight basis and that plants deficient in this element contained less than 0.1 p.p.m. of the element on a dry-weight basis. Symptoms of molybdenum deficiency in plants include bright yellow, green or pale orange interveinal mottling distributed fairly generally over the leaf and decreased (or suppressed) flower formation (Hewitt, 1966). Molybdenum deficiency and excess may be observed at levels of 0.00001 and 10 p.p.m., or over a 10^6 fold range (Hewitt, 1966). The discovery of the essentiality of molybdenum for plants was followed by the discovery of large molybdenum-deficient areas in Australia which were corrected by the addition of a few g/ha molybdenum in soluble salts to the soil (Kubota and Allaway, 1972).

The role of molybdenum in the nutrition of animals and man has been described by Underwood (1971). Briefly, the importance of molybdenum in their metabolic processes was first described in relation to a scouring disease of cattle in England called 'teart' in 1938. The need for molybdenum in animal nutrition was established in 1952, when it was discovered that xanthine oxidase, a molybdenum-containing metallo-enzyme, was found to depend for its activity on molybdenum. Since then the essentiality of this element for domestic animals has been demonstrated under controlled conditions—although no naturally occurring deficiency disease of animals owing to lack of molybdenum in the diet is known. There is little evidence that this element plays a significant role in the health or otherwise of man.

COBALT

Cobalt is not generally essential for the higher plants although certain legumes require it for their nitrogen-fixing bacteria and some plants have been found to benefit from its presence in their tissues (Gauch, 1972). Cases of natural cobalt deficiency have been reported from Australia but not from the United States (Kubota and Allaway, 1972).

Cobalt is required by certain micro-organisms and for animals, including man, where it is widely distributed throughout the body and not accumulated in any particular tissue or organ—with the exception of slightly higher levels in liver, kidney and bones (Underwood, 1971). As Underwood pointed out, careful study of the role of this element in animal metabolism resulted in the discovery that the anti-pernicious anaemia factor in liver was a compound containing 4 per cent cobalt (vitamin B_{12}) and that animals were dependent for the supply of this element entirely on the symbiotic activities of their own micro-organisms. Underwood also pointed out that the cobalt content of human blood was variable

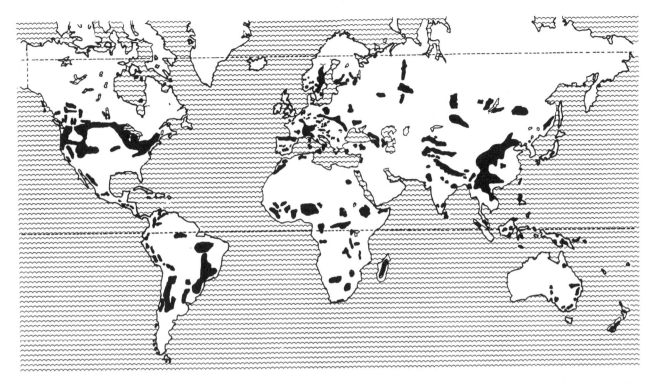

Fig. 5. World map showing occurrence of endemic goitre. Black areas indicate where the disease is located. (From Underwood, 1971. © Academic Press Inc., New York. Reproduced with permission.)

and that it could be very low in cow's milk, although supplementing the diet of animals with this element will correct this condition. Harper (1971) noted that cobalt affected blood formation in humans and that cobaltous chloride administration to human subjects produced an increase in red blood cells. He also noted that isotopic cobalt was quickly excreted via the kidneys.

IODINE

As Nason and McElroy (1963) pointed out, iodine has yet to be shown as essential to the higher plants—although certain algae require it and some brown algae, such as *Phaeophyta,* accumulate it (Epstein, 1972). Mallette *et al.* (1960) noted that this element was essential for animals, which ingest it as the iodide ion and store it in the thyroid gland. Most of the iodine is stored as thyroglobulin, which may act as a reservoir for thyroxin and perhaps 3,3',5-triiodothyronine. These last two compounds are hormones secreted by the thyroid gland and transported in the blood. A summary of the role of iodine in the metabolism of animals and man is given by Underwood (1971).

Probably the best documented relationship between the presence of a trace element in the environment and the health of animals and man concerns the relationship between iodine and endemic goitre. Endemic goitre is a relatively simple disease to treat, i.e. by the addition of

iodine to the diet, and a world map showing the areas where this disease has been reported is shown in Figure 5. Goitre has been said to be the most widespread of all mineral deficiencies in grazing stock (Underwood, 1971). Recent evidence from Colombia has indicated that endemic goitre can persist in spite of prolonged and adequate iodine supplementation; it has been concluded that the correlation of goitre prevalence and the source of drinking water indicate that organic microtoxins may be a major causative factor of the disease in that country. Consequently, the relationship between the prevalence of goitre and iodine in drinking water may not be as clear-cut as was once imagined.

According to Harper (1971) the iodine requirement for an adult human is 100–150 μg/day, which is amply provided for through the use of iodized salt.

SELENIUM

Like cobalt, selenium is an element which is essential as a nutrient for animals but not for plants. Unlike cobalt, selenium may be accumulated by pasture herbage to an extent at which it becomes toxic to livestock (Kubota and Allaway, 1972). In the northern plains and Rocky Mountain states, Kubota and Allaway also noted symptoms of selenium poisoning in animals eating plants with above 4 p.p.m. selenium; in other parts of the United States, a dietary deficiency of selenium resulting in 'white muscle disease' in lambs and calves has been observed. A relationship between vitamin E intake and the intake of selenium in chicks was clearly demonstrated by Thompson and Scott (1969). They showed that about 0.01 p.p.m. selenium was adequate for chicks if the diet contained 100 p.p.m. of vitamin E, but as the vitamin E was reduced to 10 p.p.m., the dietary selenium requirement increased to 0.1 p.p.m. It is likely that a similar relationship exists in man, although Harper (1971) and Underwood (1971) noted that evidence was lacking for both selenium toxicity and deficiency in humans.

FLUORINE

Underwood (1966) noted that fluorine has been recognized since 1805 as a constituent of bones and teeth, and for several decades as one of soft tissues and fluids of the body though no special function has yet been demonstrated. There is, however, abundant evidence that too little, or too much, fluoride in the diet of animals, including man, is related to the incidence of dental caries (Underwood, 1971). For example, Harper (1971) noted that mild mottling of teeth occurred in less than 2 per cent of children living in an area where the levels of fluorine in drinking water were between 0.6 and 1.3 p.p.m.; severity and incidence of mottling increased when the fluorine in the water exceeded this amount. Underwood (1971) noted that evidence had been produced that suggests that in areas of high fluoride there was decreased calcification of the aorta in men, compared with a similar group from a low-fluoride area. In the past, evidence of this type has occasionally led to the discovery of important relationships between the uptake of a particular trace element and the incidence of a known disease.

CHROMIUM

Gauch (1972) noted that there was no evidence for chromium being an essential element for plants and that, except in areas of serpentine soils, the chromium content of plants was usually less than 1 p.p.m. However, Coahran et al. (1973) showed that a continuous flow of chromium occurred into apple fruit during development and that this flow was similar to that for the micronutrients boron, zinc, iron and copper. Lyon et al. (1971) reported 2,470 p.p.m. chromium as an average content of 16 samples of a serpentine ecotype, Leptospermum scoparium, growing in New Zealand. The possibility of chromium being essential for plants has also been discussed recently by Huffman and Allaway (1973). Chromium became of interest in human health largely as the result of the observation by Schwartz and Mertz (1959) that in rats and mice, chromium was a cofactor of insulin for the regulation of normal glucose utilization and hence, for growth and longevity. Chromium, like fluorine, is an element which is still being in

tensively studied and its roles in plants and animals is unclear at the present time. For comprehensive review the occurrence and function of chromium in biological systems, see Mertz (1969).

TIN

It is by no means certain that tin is a micronutrient. Gauch (1972) noted that the content of tin in plant material was usually less than 2 p.p.m. and was not related to the concentration of tin in the soil. Underwood (1971) noted that little interest had so far been shown in the tin content of forage and pasture plants or in its uptake by domestic animals. However, Schwartz demonstrated a significant growth effect in rats related to the addition of small amounts of tin—comparable to that found in normal diets.

NICKEL

Treshow (1970) noted that nickel was normally toxic to plants when above 40 p.p.m. in the soil and that the nickel content of agricultural soils was usually lower than this except in areas of serpentine soils. Recently Tiffin (1971) presented evidence that nickel moved as an anion in the roots of five species of cultivated plants, although the significance of this observation is as yet unclear.

Underwood (1971) noted that no convincing evidence that nickel was essential for plants or animals had yet been presented. Some studies in which plastic 'isolator' systems were used have suggested that nickel affected metabolism in birds. For example, Underwood (1971) quoted some unpublished information of Neilson which indicated that chicks fed on a diet of less than 79 p.p.b. of nickel developed deficiency symptoms, including a bright orange leg colour, after a period of three weeks. The question of the essentiality of nickel for plants or animals remains unanswered.

VANADIUM

Vanadium is known to be essential as a micronutrient for certain of the lower plants, bacteria and algae, although evidence is scanty (Gauch, 1972). The vanadium uptake by barley roots has been discussed by Welch (1973) who concluded that this element was not absorbed by these roots and whose paper serves as a recent review of the status of this element. Underwood (1971) noted that this element should be considered as a micronutrient for chicks due to the increased growth rate of wing and tail feathers caused by the addition of 10 p.p.b. vanadium to chick diet under controlled conditions. It seems clear that proof of essentiality of micronutrients is often now required at the parts-per-billion level of concentration until the methodology for such studies becomes more readily available, the functions of elements such as vanadium will remain poorly defined.

SILICON

Gauch (1972) noted that silicon had no observable effects on the growth of tomato plants. Earlier, Hewitt (1968) noted that the ash of many plants contains silica and that this element might be required for special organs, such as the stinging hairs of nettle even though the proof of its essentiality was lacking.

Underwood (1971) noted that in the human foetus, the normal range for SiO_2 was 40–400 p.p.m. (dry tissue), compared with 50–1,000 p.p.m. for adult tissue. In the foetus, the lowest concentrations occur in the lungs and the highest in the muscles, whereas in adult humans the position is reversed; this may be due to dust inhalation. The same author also discussed the possibility that silicon was required at an early stage of bone calcification in animals, although this role for the element has not been confirmed.

SUMMARY

Altogether 15 elements have been included in this section, some of which are certainly required for the nutrition of plants, animals and man, and some of which are needed for one or the other. Still other elements have been included even though their essentiality is doubtful at the present time. From the geochemist's viewpoint, several observations may be made regarding the micronutrient elements described here. It is clear that some of these elements are found in particular geological formations and not in others; elements in this category include selenium, chromium and tin whereas others such as iron and copper are ubiquitous. It is interesting that some relatively common elements such as iron and silicon occur in quite small quantities in organisms.

It should be stressed that each element plays a role in an organism which often cannot be played by any other. Some, like iodine in animals and man, appear to have a relatively simple function, whereas others, like zinc, participate in numerous reactions in different parts of the body. Some elements concentrate in particular organs, e.g. iodine in the thyroid gland, whereas others are found in all tissues in roughly equal amounts. These and other characteristics of trace elements lead to complications in the study of the micronutrient elements as a group, even though they are of vital importance to the well-being of plants and animals including man.

Biochemistry and nutrition

It is one thing for a geochemist to consider the role played by each nutrient element in the metabolism of plants, animals and man; it is quite another to describe the pathways whereby these elements are incorporated into the structure of organisms. Korvalsky (1970) summarized, by means of a simple diagram, relationships between the inorganic and organic nutrition of plants, animals and man (Fig. 6). Here, interactions between the lithosphere, hydrosphere, atmosphere and biosphere are depicted in general terms, and for specific elements particular pathways are of considerable importance. For example, the element iodine may be obtained in sufficient quantity by animals and man from drinking water—plus a contribution from iodized salt if required. However, in coastal areas, another pathway via the atmosphere is of considerable importance and sufficient iodine reaches the drinking water from the atmosphere to fulfil requirements. Kubota and Allaway (1972) in support of this contention quoted the case of Japan, where goitre is generally absent. This example serves to focus attention on the fact that the sources for a particular element may differ in different parts of the world. In fact, the geochemical importance of the role of nutrient elements in the biosphere is largely due to their interactions with nutrient sources in the biosphere and the other three geospheres.

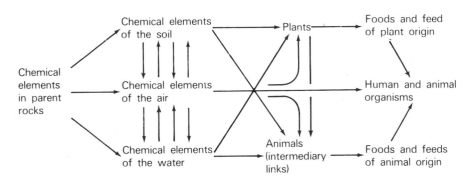

Fig. 6. Food chains and nutrition of plants, animals and man. (From Korvalsky, 1970. Reproduced with permission of Livingston, London.)

Ecology and biogeochemistry

Biogeochemistry studies the chemical composition of living organisms; ecology is concerned with fluxes of matter and energy which result in the growth of organisms and communities. Both disciplines are concerned with the biosphere which extends from the lower layers of the atmosphere to depths of 10,000 m below the ocean (Alexander, 1971). Ecologists today tend to adopt a holistic approach to the study of relationships between living and non-living matter. This approach is typified by the use of the term 'ecosystem'. Odum (1971) defined the term 'ecosystem' to include all living organisms in a given area (i.e. a community) which interact with the physical environment so that energy flows; continuously; this results in a clearly defined trophic structure, biotic diversity and material cycles (i.e. biogeochemical cycles for nutrient elements) within the system. He further noted that ecosystems have two trophic components: the autotrophic component, in which light energy, simple inorganic substances and the synthesis of complex organic substances predominate; and the heterotrophic component, in which the utilization, rearrangement and decomposition of complex materials predominate.

From the functional standpoint ecosystems are conveniently analysed in terms of the following aspects: energy circuits, food chains, diversity patterns in time and space, nutrient cycles (biogeochemical cycles), development and evolution, and control (cybernetics) (Odum, 1971). It is evident therefore, that when the role of chemical elements in nutrition is considered in the context of the ecosystem concept (as indicated in Fig. 6), one must consider the principles of geochemistry as they apply to important aspects of ecology—particularly in relation to food chains and nutrient cycles. This applies not only with respect to the nutrient requirements of particular organisms within specific communities (which vary with the degree of development of each organism through time), but also to the subtle interactions which occur continuously between organisms and their environment.

Another aspect of ecology which is pertinent to the nutrition process is the complexity of ecosystems. Odum (1971) provided a general list of ecosystem components as they are described from the ecological viewpoint as follows: (a) inorganic substances (C, N, CO_2, H_2O, etc.); (b) organic compounds (proteins, carbohydrates, lipids, humic substances, etc.); (c) climatic regime (temperature and other physical factors); (d) producers (autotrophic organisms, e.g. large green plants); (e) macroconsumers (heterotrophic organisms, chiefly animals which ingest other organisms or organic matter); and (f) microconsumers (heterotrophic organisms, chiefly bacteria and fungi, which break down the complex compounds of dead protoplasms, absorb some of the decomposition products and release inorganic nutrients.

These are usable by producers together with organic substances, and may provide energy sources or be inhibitory or stimulatory to other biotic components of the ecosystem).

It is clear that a detailed discussion of the role of each chemical element in organisms and organic matter within ecosystems is outside the scope of general geochemistry. It is therefore important to search for a basis for discussion between geochemistry, on the one hand, and the role of elements in the nutrition process, on the other. Fortunately such a basis exists which, although frequently of limited value in relation to some nutrient elements, is of considerable importance with respect to others, particularly the trace elements. This basis is called 'biomass' and, as Bowen (1966) pointed out, refers to the total mass of any organism, or group of organisms within an ecosystem, i.e. standing crop. The term biomass should always be qualified in two ways, the weight basis for the estimate should be given, i.e. either wet weight or oven-dry weight, and the extent of the estimate should be noted if it has been measured on a surface-weight basis. The distribution of particular elements within a given biomass is usually (a) expressed as a percentage or (b) in parts per million on an oven-dry-weight basis. This concept of biomass is of considerable importance in geochemistry because it enables data on the amount, distribution and circulation of chemical elements in biological materials derived from particular ecosystems to be expressed on a comparative basis. Since the process of oven-drying removes only water (and possibly, minor amounts of organic compounds) from organic material, the resulting material allows for the determination of the content of most of the nutrient elements, with the obvious exception of total hydrogen and oxygen. The use of carbon as an indicator of the gross productivity and net productivity of ecosystems has already been discussed, and data for dry matter and carbon have been presented (Table 6). The major element content of a number of different organisms was calculated by Bowen (1966) and is shown in Table 10.

Any consideration of biomass in relation to nutrition is incomplete without reference to effects of changing the amount of a nutrient (or nutrients) and the effects which this may have on biomass production. The analysis of response in crop and livestock production to the addition of chemicals in fertilizers or in animal diets has been summarized by Dillion (1969), although the effects of the addition of fertilizer treatments to crops can often be demonstrated on a simple cause-and-effect basis. For example, Cooke (1972) has described the effect of addition of nitrogen fertilizer to a particular field in different parts of which were grown grass, sugar-beet, cereals and potatoes. He noted that the addition of a certain amount of nitrogen would increase the yields for all four crops whereas further additions would benefit the grass, have no effect on the sugar-beet growth and actually decrease the growth rate of the cereals and pot-

TABLE 10. Mean elementary composition of organisms, percentage dry weight basis

Group	C	H	N	O	S	Other
Angiosperms	45.4	5.5	3.3	41.0	0.44	4.4
Brown algae	34.5	4.1	1.5	37.7	1.2	22.0
Marine plankton	22.5	4.6	3.8	44.0	(0.6)	24.5
Mean plant	45	5	3	40	0.5	6.5
Bacteria	48.5	7.4	10.7	27.0	0.61	(5.8)
Fungi	49.4	(5.5)	5.1	34.0	(0.4)	(5.7)
Mean decomposer	49	7	8	30	0.5	5.5
Coelenterata	43.6	(4.5)	6.3	27.1	1.9	16.6
Annelida	40.2	5.9	9.9	34.0	1.4	8.6
Nematoda	(40.0)	(6.0)	11.0	35.4	(1.3)	6.3
Mollusca (soft parts)	39.9	6.0	8.5	39.0	1.5	4.9
Echinodermata	(40.0)	(4.5)	4.4	45.6	0.5	5.0
Arthropoda (land)	44.6	7.3	12.3	32.3	0.44	3.1
Pisces	47.5	6.8	11.4	29.0	1.0	4.3
Mammalia	48.4	6.6	12.8	18.6	1.6	12.0
Mean animal	45	6.5	10	30	1.3	7.2

Source: Bowen, 1966.
From *Trace Elements in Biochemistry.* © Academic Press Inc. (London) Ltd. Reprinted with permission.

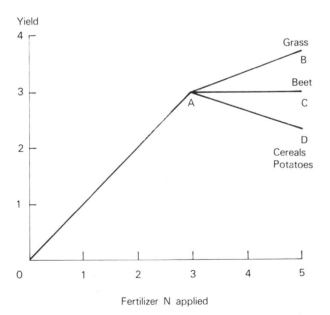

FIG. 7. Three possible relationships between the effects of nitrogen fertilizer on the yields of different kinds of crops. (From Cooke, 1972. Reproduced with permission of Crosby Lockwood & Son Ltd.)

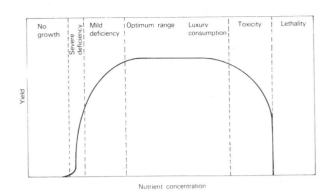

FIG. 8. Idealized diagrams of growth of an organism as a function of the concentration of an essential nutrient. (From Bowen, 1966. © Academic Press Inc. (London) Ltd. Reproduced with permission.)

atoes (Fig. 7). The data obtained from experience of this type indicates that additions of a single element to different crops may have different effects. When two or more elements are added to crops still less predictable results may be obtained. For example, Cook (1972) went on to describe experiments in the eastern United States which showed that neither potassium nor magnesium had any effect on the yield of peaches when either was applied alone; when both nutrients were added together, however the peach yield increased by 30 per cent. Information of this type indicates that idealized diagrams of the growth-response of organisms to increased nutrient concentration of the type shown in Figure 8 tell but a part of the whole story and that a partial or entirely holistic approach to the study of nutrition is desirable. It is here that the geochemist may be of considerable assistance to both ecologist and biologist because he may, on the basis of geological and climatic considerations, be able to predict which elements are likely to be in short supply within a particular area of country. In the remainder of this chapter, we shall take some of the implications of this statement a stage further in relation to the nutrient-cycling process as it affects the nutrition of plants, animals and man.

Landscape geochemistry

Relationships between ecology and biogeochemistry have been discussed in general terms in a previous section, where it was pointed out that the growth of the component organisms in the biosphere may depend on the amount of an element, or an assemblage of elements, which is available to the organism from the environment. In this section we shall discuss a geochemical approach to the description of these environments using the terminology of landscape geochemistry.

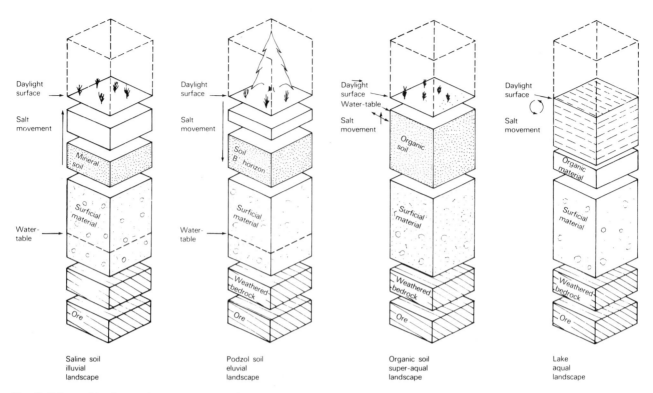

FIG. 9. Prisms of landscape drawn to represent four elementary landscape types (from Fortescue and Bradshaw, 1973).

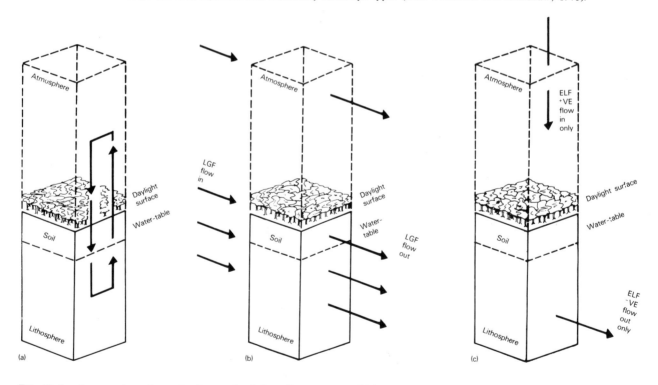

FIG. 10. Landscape prisms drawn to show each of three flow patterns which occur concurrently within a landscape type: (a) MMC: circulation of chemical substances within the landscape prism as a closed system; (b) LGF: flow of chemical substances through the landscape prism parallel to the daylight surface; (c) ELF: flow of chemical substances into the landscape prism where they accumulate (+ve) or out from the prism (−ve) (from Fortescue, 1974).

During the past thirty years in the Union of Soviet Socialist Republics, a branch of geochemistry has evolved largely as the result, of the ideas of the pioneer soil scientist and geochemist, B. B. Polynov: 'landscape geochemistry' is a holistic approach to the study of the behaviour of all elements in the periodic table at, or near the daylight surface of the Earth where the atmosphere interacts with the biosphere, hydrosphere and lithosphere. Although landscape geochemistry is still in its infancy and little known in the Western world, it provides an unique bridge between geochemistry and ecology which will surely become more important as time goes on.

The scope of landscape geochemistry has been outlined elsewhere (Fortescue, 1974a, 1974b) and more detailed information on the subject may be obtained from Perel'man (1961, 1972), Glazovskaya (1963) and Kozlovskiy (1972). Briefly, it is convenient to consider volume units, located to include the daylight surface of the earth extending from the atmosphere to the lithosphere. Although under natural conditions such units are irregular in shape, it is both practical and convenient to refer to isometric block diagrams in order to describe landscape features (see Fortescue, 1974b) and to focus attention on the interactions of landscape components. Polynov recognized three 'elementary landscape types', which are illustrated by means of conceptual models in Figure 9. These are: (a) eluvial landscapes, where the water-table is below the daylight surface and the precipitation is greater than the evaporation; (b) super aqual landscapes, where the daylight surface and the water-table coincide; and (c) aqual landscapes, where the hydrosphere lies above the solid matter of the Earth. In desert areas where evaporation exceeds precipitation, instead of an eluvial landscape, an illuvial landscape is found. From the geochemistry viewpoint, the concept of landscape types is of particular importance. This is because the relative mobility of elements and their behaviour during ecological and geological time may vary in each of the elementary landscape types and in the more common transition areas between them.

Kozlovsky (1972) provided a series of terms to describe the migration of chemical elements within landscapes and these may be simply combined with the prism concept shown in Figure 9. If the prism is considered to be a closed system and elements migrate from the soil to the plants and return to the soil again, i.e. nutrient cycling, this flow pattern is called the 'main migrational cycle' (MMC) (Fig. 10). If elements flow through a prism without the deposition or removal of matter from them, the flow is called a 'landscape geochemical flow' (LGF) whereas if material is deposited, or removed from the prism the flow is a positive, or negative, 'extra landscape flow' (+ELF or −ELF). Although these three types of flow have been separated out in Figure 10, they obviously occur together within a landscape prism and may be described for any element in the periodic table over any length of time. For

example, the movement of carbon dioxide in air may be measured in litres per minute, carbon in vegetable material in terms of years or the decomposition of carbonate rocks in terms of thousands or even millions of years. Consequently when one considers the mobility of elements in landscapes one must consider the chemical form of the element and the time-scale involved.

This brings us to a more general discussion of the circulation of chemical elements in natural systems. It seems clear that three hierarchies are always involved in discussions of this type: the hierarchy of *space* (for example, the volume units of landscape just described), which may include volumes of less than a litre or the whole earth's surface considered as a totality (e.g. in relation to the cycling of the element carbon; see Olsen, 1970); the hierarchy of *time* which, as mentioned above, may have a duration of less than an hour or more than a million years; and the hierarchy of *complexity,* which is concerned with the chemical form of an element when it circulates within a volume unit. From the practical viewpoint there is another hierarchy which is of considerable importance in geochemistry. This is the hierarchy of *effort*. We have noted above that it is relatively simple to make biomass estimates for an ecosystem at an instant in time but most difficult to estimate the growth and decay rates of different organisms during ecological time. Similarly, it is relatively simple to trace a radioactive isotope as it moves within a given ecosystem; it is much more difficult—or even impossible—to trace non-radioactive isotopes of the same element through the same system. Geochemists who are concerned with the description and prediction of the movement of elements through ecosystems or, more correctly through different landscapes, must always take into consideration these four hierarchies in order to collect information which not only explains the circumstances within a given area of country but may also be used as a basis for comparative studies elsewhere.

Landscape geochemistry and the nutrition of plants and animals

Elementary landscape type, climate and bedrock composition together with effects of the LGF and positive or negative ELF's all contribute to the availability of nutrient elements to organisms found in particular landscapes. To a lesser extent the effects of geological and ecological time, space and complexity of materials also affect the nutrition process.

With respect to the atmosphere, there is an LGF of the air through a given volume unit of landscape from minute to minute which supplies an excess of oxygen and nitrogen to terrestrial ecosystems at all times. In the case of super aqual and aqual landscapes, particularly bogs and lakes, the supply of dissolved oxygen in waters may limit the growth of plants or animals. The presence of reducing conditions in the lower layers of marshes and bogs is well known and in certain lakes eutrophication results largely from a lack of oxygen (Rohlich, 1971). Hence, the supply of nutrient elements may vary with landscape type within the same general area of country.

In addition to the major constituents of the atmosphere, there are certain minor constituents which are of considerable importance in nutrition. The most vital to plants is carbon dioxide, which is the source of carbon for photosynthesis and which is present in the atmosphere at a concentration of a few tens, or hundreds, of parts per million. Other nutrients such as nitrate, sulphate and many others may be obtained by plants from the atmosphere. For example, Spanish moss (*Tillandsia*

usneoides L.), an epiphyte which is common on the Atlantic and Gulf Coast plains of the United States, was found by Shacklette and Connor (1973) to contain significant amounts of over 30 elements, all of which had been obtained from the air. The trace element content of air has also been studied on a diurnal basis for as many as 26 elements concurrently, some of which were shown to vary in concentration by over one order of magnitude (Wesolowski *et al.*, 1973). Consequently, the air is emerging as a more and more important source of nutrient elements to plants, animals and man. Relationships between the intake of certain of the trace nutrients such as chromium and vanadium and the content of these elements in air may eventually be established. The MMC serves to fix nutrients within

TABLE 11. Threshold concentrations of chemical elements in soils and possible responses of organisms (endemic disease)

Chemical element	Number of determinations	Range of concentrations (percentage of air dry soil)		
		Deficiency (lower threshold concentrations)	Normal (range of normal regulatory functions)	Excess (higher threshold concentrations)
Co	2,400	$<2-7.10^{-4}$ Acobaltoses, anaemia, hypo- and avitaminosis B_2; aggravation of endemic goitre.	$7-30.10^{-4}$	$>30.10^{-4}$ Possible inhibition of vitamin B_{12} synthesis.
Cu	3,194	$<6-15.10^{-4}$ Anaemia, bone deformities, endemic ataxia, in case of excess of Mo and SO_4^{2-}. Lodging of cereals, low yields of grain, dry tops of fruit trees.	$15-60.10^{-4}$	$>60.10^{-4}$ Anaemia, endemic icterohaemoglobinuria, liver lesions. Chlorosis of plants.
Mn	1,629	$<4.10^{-2}$ Bone diseases, aggravation of goitre. Chlorosis and necrosis of plants, yellow specks on sugar-beet leaves.	$7-30.10^{-2}$	$>30.10^{-2}$ Bone diseases. Possible toxic effects on plants in acid soils.
Zn	1,927	$<3.10^{-3}$ Parakeratosis in swine. Chlorosis, 'little leaf' diseases in plants.	$3-7.10^{-3}$	$>7.10^{-3}$ Possible anaemia, inhibition of oxidative processes.
Mo	1,216	$<1 \cdot 5.10^{-4}$ Diseases of plants (clover).	$1 \cdot 5-7.10^{-4}$	$>7.10^{-4}$ Gout in humans. Mo toxicosis in animals.
B	879	$<3-6.10^{-4}$ Plant diseases, death of terminal buds of stem and roots. 'Heart rot' in sugar-beet, browning of the cabbage flowers, and of the core of the turnip.	$6-30.10^{-4}$	$>30.10^{-4}$ Boron enteritis in animals and humans. Plant abnormalities.
Sr	1,269		up to 6.10^{-2}	$6-10.10^{-2}$ Chondro- and osteo-dystrophy, Urov disease, rickets. Brittleness of bones of animals.
I	491	$<2-5.10^{-4}$ Endemic goitre; aggravation of goitre in case of imbalance of I with Co, Mn, Cu.	$5-40.10^{-4}$	$>40.10^{-4}$ Possible decrease of synthesis of I-compounds in the thyroid gland.

Source: Korvalsky, 1970.
Reprinted from *Trace Element Metabolism*, with permission of Livingston, London.

particular landscapes. This trend is most marked in mature forests, which have reached a more or less steady-state condition with respect to input and output of elements. Further information of the nutrient element balance of forests may be obtained from Rodin and Basilevich (1967) and Reichle (1970). From the landscape geochemistry viewpoint, the addition of fertilizers to farm crops is a positive ELF and results in a benefit to man. Unfortunately, the addition of nutrient elements to one kind of landscape type may affect conditions in adjacent landscapes. According to Gittings (1973) for example, the phosphorus content of Lake Erie increased threefold between 1942 and 1968 resulting in an increase in algal growth and a decrease in fish numbers.

These examples, and many others like them involving other nutrients, can be cited to illustrate the relationship between the simple basic concepts of landscape geochemistry and the availability of nutrients to plants and animals. In the case of man, these relationships tend to become more complicated owing to this varied diet.

Geochemical ecology

Korvalsky (1970) and others have carried the concept of the geochemistry of landscape a stage further. Basing their observations on general climatic and geological

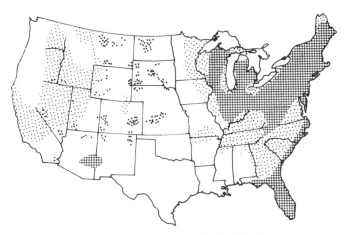

⊞ Low-approximately 80 per cent of all forage and grain contain < 0.05 p.p.m. of selenium.

⋰ Variable-approximately 50 per cent contains > 0.1 p.p.m.

☐ Adequate-80 per cent of all forages and grain contain > 0.1 p.p.m. of selenium.

● Local areas where selenium accumulator plants contain > 50 p.p.m.

FIG. 11. Geographic distribution of low variable, and adequate selenium areas in the United States. (From Kubota and Allaway, 1972. Reproduced by permission of Soil Science Society of America.)

data and the behaviour of elements during weathering and soil-information processes, they have divided up the U.S.S.R. into a series of biogeochemical zones and provinces within each of which the possibility of endemic disease of plants and animals, including man, has been considered. In Table 11, endemic diseases resulting from deficiencies or excesses of each of 8 micronutrient elements have been listed as an example of this approach to the geochemistry of disease in organisms.

It seems clear that more systematic investigations along these lines will be required in the future as the role of each nutrient in the environment becomes better known. A similar, but less systematic approach to the mapping of the distribution of nutrients elsewhere in the world has been adopted; for example, the distribution of selenium in the United States as described by Kubota and Allaway (1972) (Fig. 11).

Biogeochemical cycles

A dynamic approach to the study of the behaviour of nutrient elements in relation to the environment has also been developed and involves biogeochemical cycles. Biogeochemical cycles are now discussed in relation to several disciplines, including geography (Watts, 1971), ecology (Odum, 1971; Collier et al., 1973), forestry (Reichle, 1970), and geochemistry (Fairbridge, 1972). Biogeochemical cycles may be defined as the pathways which chemical elements or compounds (e.g. DDT) follow in ecosystems and they may involve abiotic (e.g. the atmosphere) as well as biotic components. Such cycles may be described for a given element (nutrient or non-nutrient) within a given ecosystem (e.g. Thomas (1969), who studied the calcium cycle in a forest) or may be generalized for the same element in all ecosystems. The systematic study of biogeochemical cycles in relation to landscape geochemistry has not been attempted on a world-wide scale, although data for biomass studies (e.g. Table 6) represent a first step in this direction.

Biogeochemical cycles are usually described by means of flow diagrams which may or may not include quantitative data (Figs. 12 and 13). Let us consider two diagrams, a qualitative diagram for the cycle of nitrogen (Fig. 12) and a quantitative one for the elements sulphur and carbon (Fig. 13). The nitrogen cycle is an example of a complex gaseous-type cycle which involves the decomposition of protoplasm by a series of steps (as shown in Fig. 2) carried out by bacteria to form nitrate, which is the form most readily taken in by green plants that synthesize protoplasm, thus completing the cycle. The atmosphere is a sink for nitrogen and acts as a safety valve for the system. If denitrifying bacteria participate in the decomposition process, gaseous nitrogen is returned to the atmosphere, but there are also nitrogen-fixing bacteria which, together with certain types of

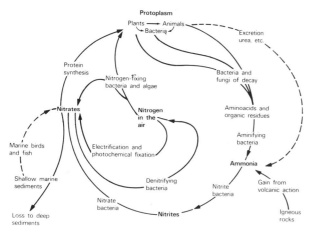

FIG. 12. The nitrogen cycle described by means of a generalized flow diagram showing the circulation of nitrogen between organisms and the environment with names of the key microorganisms involved. (From Odum, 1971. Reproduced with permission of W. B. Saunders Co.)

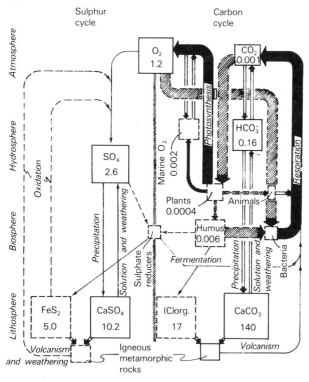

FIG. 13. Interaction between the biogeochemical cycles for sulphur and carbon indicating their mode of controlling the oxygen cycle. Line thickness denotes approximate magnitude of annual flow rates. Numerical values signify total oxygen contents in 10^{15} metric tonnes either as molecular oxygen, oxygen sink (FeS_2) or equivalent amount of oxygen released (C) org. (From Brooks et al., 1972. © 1972 Litton Educational Publishing Inc. Reproduced with permission of Van Nostrand Reinhold Company.)

algae, may synthesize nitrogen compounds from gaseous nitrogen. From man's viewpoint, there is a shortage of nitrates and consequently nitrate fertilizers are in constant demand by farmers all over the world. Geochemically, the cycle of nitrogen is unique because this element is present as a trace element in rocks, although it is a macronutrient and major constituent of both the atmosphere and the biosphere.

A more complex situation involving the biogeochemical cycles of carbon, sulphur and oxygen has been described in diagrammatic form by Brooks et al. (1972) (Fig. 13). Like nitrogen, sulphur has a number of valency states in the landscape, ranging from -2 in sulphides to $+6$ in sulphate. The reduced state is characteristic of anerobic conditions found in bogs or bottoms of lakes or ponds. Sulphides are also found in primary and secondary rocks and, in evaporite deposits, gypsum and anhydrite are often common. Weathering will convert sulphate to sulphide or, depending on environmental conditions, reverse this process. In either case, bacteria of different kinds (Zajic, 1969) facilitate these chemical reactions. General features of the carbon cycle involving the four geospheres are also indicated in Figure 3. In general, carbon occurs as organic matter or carbonates in rocks as a result of organic activity. In the biosphere, carbon is a major constituent of dead and living organic matter. The growth of plants involves the fixation of carbon from the atmosphere (as CO_2) and, as

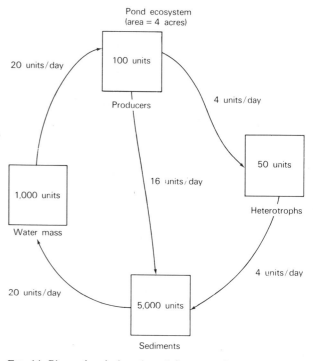

FIG. 14. Biogeochemical pools and flux rates for a hypothetical nutrient cycle. (From Collier et al., 1973. Reproduced with permission of Prentice-Hall Inc.)

a by-product of photosynthesis, the liberation of oxygen which has gradually accumulated there during geological time. In the hydrosphere, carbon is also present in organic matter as well as in solution, for example, as the bicarbonate ion. This example is included to illustrate the complex relationships which occur between the biogeochemical cycles of different elements and to show how the four geospheres participate in the cycling process.

Let us now consider biogeochemical cycles which occur within particular ecosystems. Collier *et al.* (1973) described in the simplest terms the cycling of a hypothetical nutrient within a pond ecosystem (Fig. 14). Such an arrow diagram affords a convenient method for the display of cycling data of this type because it allows for a description of relationships between possible flux rates and biogeochemical pools in the ecosystem considered as a whole. As these authors pointed out, the flow of matter in an ecosystem is a vehicle of energy flow. This is like a bus on a circular route where the bus repeatedly covers the same route and its fuel follows a one-way path only. These writers classify biogeochemical cycles

on a functional basis as perfect or imperfect, according to the degree of regulation of processes involved in the cycle or the extent to which the pools or flux rates in the cycle can re-establish normal conditions following a disturbance. If perfect biogeochemical cycles are disturbed, the ecosystem as a whole acts to restore the original balance; for example, the re-establishment of a mature forest after a forest fire. Imperfect biogeochemical cycles are less stable and as a result of disturbance —natural or anthropogenic—will not return automatically to their original pre-disturbance state.

Some idea of the amounts of a macronutrient which cycle in a forest ecosystem per year may be obtained from the data displayed in diagrammatic form in Figure 15. Briefly, the numbers in the diagram represent the average number of kg/hectare/year for inputs and outputs. Using the terminology of landscape geochemistry, described above, we note that positive ELFs of calcium are from the atmosphere (2.6 kg) and the rocks and soils (9.1 kg). Negative ELFs in waters removing calcium from the system amount to 12 kg so that the net loss from the system is 9.4 kg. During the

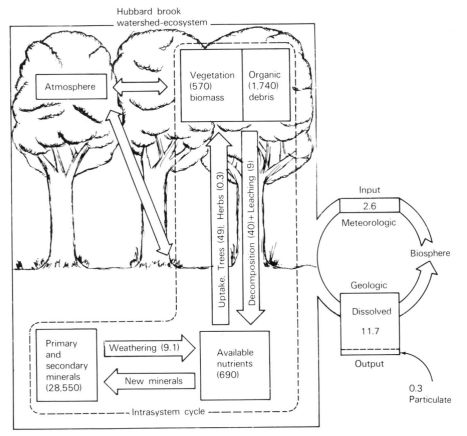

FIG. 15. The calcium cycle in an undisturbed ecosystem. Numerals represent the average number of kg/h per year. Thus the meteorological input to the ecosystem in precipitation and dust is 2.6 kg/h annually. A substantial amount of calcium is in the soil and rocks; 9.1 kg is released annually by weathering. Vegetation takes up 49.3 kg; 49 kg is returned to the soil by decomposition and leaching. Gross loss in stream drainage is 12 kg so that the net loss is 9.4 kg. (From Bormann and Likens, 1971. Reproduced with permission of Yale Scientific Publications Inc.)

MMC, 49.3 kg of calcium is taken up by the vegetation and 49.0 kg is returned to the soil again. This relatively simple example indicates how the terminology of landscape geochemistry may be conveniently applied to data of this type collected by biologists.

Nutrient cycling

So far, we have considered biogeochemical cycling in general terms with respect to the whole biosphere (Figs. 5 and 6) or with respect to a particular forest ecosystem (Figs. 7 and 8). Another approach is to study the role of a particular organism in the cycling process of a particular element through a given ecosystem. Studies of this type may be largely empirical or they may commence with an empirical approach to a problem which is later described by means of a model system. Studies of nutrient cycling are often carried out in forest plants which are relatively large and easy to observe using radioactive tracers—which may be added to plants at a definite time in a particular location (usually the stem) and which may be readily detected in the plants later on. A good example here is the study made by Thomas (1969) of the macronutrient calcium in the flowering dogwood tree (*Cornus florida*). Thomas introduced ^{45}Ca at the bases of the trunks in early May and sampled different components of the tree/soil system periodically during the growing season. Estimates of the total calcium and ^{45}Ca isotope were made on the sample material, and an arrow diagram was drawn to illustrate the percentage of ^{45}Ca transferred during a single growing season (Fig. 16). A large proportion (89 per cent) of the isotope was translocated to other parts of the tree during the growing season—almost three-quarters of it to the leaves (73 per cent). Smaller amounts (3 per cent)

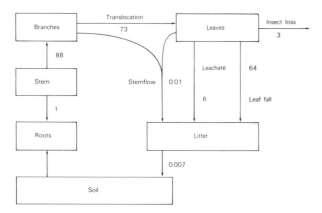

FIG. 16. Percentages of ^{45}Ca transferred, during one growing season, from inoculation site in stems of flowering dogwood trees to various locations in the dogwood-soil system. (From Thomas, 1969. Reproduced from *Stalfelt's Plant Ecology* (Jarvis) with permission of A.B. Svenska Bokforlaget.)

TABLE 12. Above-ground weight and calcium content of components of an 18-year-old loblolly pine stand in Tennessee (United States)

Component	Oven-dry weight (kg/ha)	Calcium (kg/ha)	Percentage of total	
			Over-dry weight	Calcium
Dogwood trees	1,030	12.5	0.2	1.8
Pine trees	382.940	525.0	89.3	74.1
Litter	45,020	170.6	10.5	24.1
TOTAL	428,990	708.1	100.0	100.0

Source: Thomas, 1969.

were lost to insect grazing, and 6 per cent was leached from the leaves by rainfall. Stem flow was found to be an insignificant pathway of calcium movement (less than 0.01 per cent), and the bulk of the calcium did not reach the soil until the leaf fall in the autumn. Direct measurements for total calcium loss by litter indicated that 79 per cent of the calcium was released during the first year of decomposition. Evidence for the importance of dogwood in the calcium cycle is presented in Table 12 where it is shown that although the dogwood trees comprised only 0.2 per cent of the total dry weight of the forest trees, they contained 1.8 per cent of the total calcium in the forest trees. Hence the dogwood plays an important role[1] in the MMC for calcium in the loblolly pine stands of east Tennessee in the United States.

Nutrient cycling is carried out in ecosystems of all types, and studies of this type are becoming more sophisticated year by year. A parallel development has been the application of the principles of systems analysis to nutrient cycling. Although a detailed discussion of this topic lies outside this chapter, it is important for geochemists to consider the implications of this approach to geochemistry. A good introduction to systems ecology, supported by a simple worked example for the behaviour of phosphorus in a hypothetical aquatic ecosystem, has been given by Smith (1970). Smith showed, using his example, how a systems model is developed from arrow diagrams describing flow patterns for phosphorus within the system. His model also illustrated the use of sensitivity analysis to determine the relative importance of the six different parameters included in the model and showed how the model could be used to simulate effects on the system of variations in the phosphorus input. For further information on this interesting approach, the reader is advised to refer to Smith (1970) or to books on systems ecology such as that by Watt (1966) or Patten (1972).

1. One cannot help commenting that it was a pity that the samples collected by Thomas could not have been examined for all nutrient elements instead of calcium alone.

The geochemistry of nutrient elements in plants and landscapes

It is evident from the brief summaries of the role that nutrient elements play in the growth of plants that each element has a unique contribution to make to the process. Similarly the function of each element is different from all others, and in many cases the pathway by which the element enters the plant is also unique. Because different elements have different functions within the plant, the content of nutrients varies from organ to organ. It is only when we consider landscapes as a totality, including solids, liquids and gases as components of the same system, that a holistic approach to plant growth at a particular instant in time can be obtained. Although geochemists tend to simplify the problem of the biochemical role of elements in plants by expressing amounts on an ash or oven-dry-weight basis, there still remains the problem of the chemical form in which an element is present in the soil and the way in which it is taken into the plant. Further complications result from the fact that plants, unlike most rocks, do not long remain the same size or chemical composition. Consequently, the flow rate of the elements through plants varies with many factors of the plant itself, e.g. the stage reached in the growth cycle, and of the environment, e.g. season. The general problem of environmental factors and their relation to plant growth is indicated in Figure 17 and, as has been shown in Figure 7, this approach is often discussed in relation to particular nutrient elements. Problems of plant growth in relation to environmental factors are systematically studied by plant physiologists, botanists, agriculturalists, foresters, ecologists and others, and the problem for geochemists is to discover where geochemistry can make its own contribution to the solution of such problems.

With the modern instrumentation now available it should be possible for geochemists to approach the problems of plant nutrition from one or more, of the following standpoints.

1. To describe carefully the amount of each nutrient element which is available for uptake by plants from a given area of country at a given instant in time;
2. To describe the reserves available for each element for plant growth within the landscape for a given length of time and to predict which element (or elements) will become in short supply within given lengths of time, providing that the landscape is not disturbed by man's activities or by some natural catastrophe;
3. To describe aspects of the landscape which result from its evolution and which may be expected to interfere with the growth of plants now or at some future time;
4. To predict problems involving the geochemistry of the environment which may result in long-term or short-term variations in the growth of plants in particular landscapes.

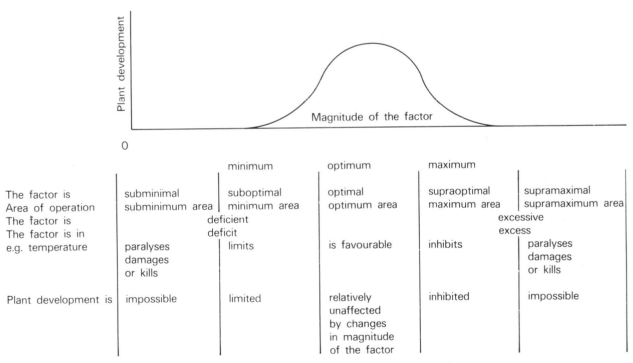

	minimum	optimum	maximum	
The factor is	subminimal	suboptimal	optimal	supraoptimal	supramaximal
Area of operation	subminimum area	minimum area	optimum area	maximum area	supramaximum area
The factor is	deficient			excessive	
The factor is in	deficit			excess	
e.g. temperature	paralyses damages or kills	limits	is favourable	inhibits	paralyses damages or kills
Plant development is	impossible	limited	relatively unaffected by changes in magnitude of the factor	inhibited	impossible

FIG. 17. Diagram showing relation between the development of a plant (or plant community) and the magnitude value of an environmental factor such as temperature (after Stafelt, 1972).

These tasks present a considerable challenge to geochemists who nowadays are often concerned with the effects of one or a small number, of elements in relation to a single or even a part of a single component of the landscape. It seems clear that if the problems above can be approached in a systematic way world-wide, then comparable sets of information could be collected and general principles synthesized. These general principles, which could be described empirically, conceptually or by means of mathematical models, would surely be of interest to scientists in developing countries faced with crop failures related to the geochemistry of the environment.

The geochemistry of nutrient elements in animals and landscapes

As in the case of plants, the elements essential for the nutrition of animals are numerous and varied. Animals, unlike plants, obtain the greater part of their food from plants or other animals, supplemented by elements dissolved in water or derived from the air. Because animals move about the landscape, their sources of food may not be so easily studied in detail by the geochemist unless kept in confined, controlled quarters. For this reason, animal nutrition studies often involve detailed studies of the chemistry of solids, liquids and gases ingested with respect to the health of the animal as a whole—or the condition of a particular organ such as the thyroid gland.

The role of geochemistry in animal nutrition may be outlined as follows:
1. The description of the distribution and amount of elements available as food in the landscapes within which the animals live;
2. The description of probable imbalances in the chemistry of the environment which may affect the health and growth of particular animal species;
3. The description of the effects that changes in environment geochemistry may have on the health of particular species at different stages in their life cycles.

As in the case of plants, systematic geochemical studies are required to study relationships between the chemistry of the environment and the intake of elements in available form for digestion or absorption by animals. It should not be forgotten that, in the case of animals, several additional elements, for example cobalt, are required for optimum health than in the case of plants.

The geochemistry of nutrient elements in man and landscapes

In certain parts of the world, where the food and water taken by humans is derived locally, relationships between nutrient intake and the environment may be firmly established. In cities, such relationships may not exist or may be very difficult to study. For example, the trace-element content of city water may vary due to several factors, one of which is the input of elements between source and tap owing to different kinds of plumbing. Another problem is that, in many countries, few people now drink significant amounts of water each day. But in spite of these problems, some relationships between human health and the environment have been described, one of the most striking examples being that between iodine and goitre already discussed.

Perhaps the most important role of the geochemist in relation to human nutrition is in the assessment of the trace element content of crops and livestock where they are produced. It is essential that foods have a sufficient content of elements, particularly micronutrient, for the healthy growth of humans. One of the problems which is now becoming of increasing importance is the addition of non-nutrient elements to farm land in sewage, which in turn could contaminate foods to a significant extent.

Summary

In this chapter the nutrient elements are defined and information of interest to the geochemist is provided for each element in relation to its role in plants and animals, including man. Although these summaries are very brief, they do provide the reader with information on the role each element plays in the nutritional processes. General relationships between the geochemistry of the environment and the growth of organisms are outlined, particular stress being placed on general principles of ecology and landscape geochemistry. After a discussion of geochemical ecology as it is practised in the U.S.S.R. and biogeochemical cycling in general, details are given of studies of nutrient cycling. Finally, reference is made to particular problems involved when the nutrition of plants, animals or man is studied by geochemists adopting a holistic approach.

Conclusion

It seems clear that geochemists will be called on in the future to play a much more active part in nutrition studies than they have done in the past. This will involve close co-operation between geochemists, ecologists and other scientists involved in the supply of elements to plants, animals and man. Perhaps the most important need at the moment is for all geochemists to agree on a series of simple standards and conceptual models which can be used for comparative purposes by workers in all parts of the world. This, and the later use of mathematical models, which will eventually be developed far enough to predict answers to the problems in nutrition of organisms before they occur, may allow future remedial measures to be taken in time.

References

ALEXANDER, M. 1961. *Introduction to Soil Microbiology.* New York, Wiley, 472 p.

——. 1971. *Microbial Ecology.* New York, Wiley. 511 p.

BAULE, H.; FRICKER, E. 1970. *The Fertilizer Treatment of Forest Trees.* München, BLV Verlagsgesellschaft mbH. 259 p.

BLACK, C. A. 1968. *Soil-Plant Relationships,* 2nd. ed. New York, Wiley. 792 p.

BONNER, J.; VARNER, J. E. (eds.). 1965. *Plant Biochemistry.* New York, Academic Press.

BORMANN, F. H.; LIKENS, G. E. 1971. The Ecosystem Concept and the Rational Management of Natural Resources. *Yale Scientific,* Vol. 45, p. 2–8.

BOWEN, H. J. M. 1966. *Trace Elements in Biochemistry.* London, Academic Press. 241 p.

BROCK, T. D. 1966. *The Principles of Microbial Ecology.* Englewood Cliffs, N.J., Prentice Hall. 306 p.

BROOKS, R.; RAND, I.; KAPLAN, R. 1972. Biochemistry. In: R. W. Fairbridge (ed.), *Encyclopedia of Geochemistry and Environmental Sciences,* Vol. IVA, p. 74–82. New York, Van Nostrand, Reinhold.

BROWN, J. C.; AMBLER, J. E.; CHANEY, R. L.; FOY, C. D. 1972. Differential Responses of Plant Genotypes to Micronutrients.

BROYER, T. C.; CARLTON, A. B.; JOHNSON, C. M.; STOUT, P. R. 1954. Chlorine, A Micronutrient Element for Higher Plants. *Plant Physiol.,* Vol. 29. 526 p.

BURSTRÖM, H. G. 1968. Calcium and Plant Growth. *Biol. Revs.,* Vol. 43, p. 287–316.

COAHRAN, D. R.; MAXWELL, R. C.; ZUCHER, M. 1973. Flow of Chromium into Apple Fruit During Development. *Plant Physiol.,* Vol. 52, p. 84–5.

COLLIER, B. D.; COX, G. W.; JOHNSON, A. W.; MILLER, P. C. 1973. *Dynamic Ecology.* Englewood Cliffs, N.J., Prentice Hall. 563 p.

COOKE, G. W. 1967. *The Control of Soil Fertility.* Darien, Conn., Hafner. 526 p.

——. 1972. *Fertilizing for Maximum Yield.* London, Crosby Lockwood. 296 p.

DATTA, S. P.; OTTAWAY, J. H. 1965. *Biochemistry.* London, Baillière Tindall & Cassell. 379 p.

DAY, F. H. 1965. *The Chemical Elements in Nature.* London, Harrap. 372 p.

DREGNE, H. E. 1970. *Arid Lands in Transition.* Baltimore. Md., American Association for the Advancement of Science, Horn Shafer. 542 p.

EGLINGTON, G.; MURPHY, M. T. J. (eds.). 1969. *Organic Geochemistry.* Berlin, Springer-Verlag. 828 p.

ENERGLYN, B.; BREALEY, L. 1971. *Analytical Geochemistry.* Amsterdam, Elsevier. 426 p.

EPSTEIN, E. 1972. *Mineral Nutrition of Plants: Principles and Perspectives.* New York, Wiley, 412 p.

EVANS, H. J.; SORGER, G. J. 1966. Role of Mineral Elements with Emphasis on the Univalent Cations. *Ann. Rev. Plant Physiol.,* Vol. 17, p. 47–76.

FAIRBRIDGE, R. W. 1972. *The Encyclopedia of Geochemistry and Environmental Sciences,* Vol. IVA. New York, Van Nostrand. 1321 p.

FASSEL, V. A.; KNISELEY, R. N. 1974. Inductively Coupled Plasma-Optical Emission Spectroscopy. *Anal. Chem.,* Vol. 46, No. 13, p. 1110A–20A.

FORTESCUE, J. A. C. 1972a. The Need for Conceptual Thinking in Geoepidemiological Research. In: *Trace Substances in Environmental Health—VI, A Symposium, 1973,* p. 333–9. Columbia, Mo., University of Missouri.

——. 1973b. *Relationship Between Landscape Geochemistry and Exploration Geochemistry.* St. Catharines, Ontario, Brock University, Department of Geological Sciences. (Report Ser. No. 17.)

——. 1974a. The Environment and Landscape Geochemistry. *Western Miner,* March 1974. 6 p.

——. 1974b. Exploration Geochemistry and Landscape. *Bull. Can. Inst. Min. and Metal,* November 1974. 9 p.

FORTESCUE, J. A. C.; MARTEN, G. G. 1970. Micronutrients: Forest Ecology and Systems Analysis. In: D. F. Reichle (ed.), *Analysis of Temperate Forest Ecosystems,* p. 173–98. New York, Springer Verlag.

GAUCH, H. G. 1972. *Inorganic Plant Nutrition.* Stroudsburg, Pa., Dowden, Hutchinson, & Ross. 488 p.

GITTINGS, J. C. 1973. *Chemistry, Man and Environmental Change.* San Francisco, Calif., Canfield Press. 472 p.

GLAZOVSKAYA, M. A. 1963. On Geochemical Principles of the Classification of Natural Landscapes. *Intl. Geol. Rev.,* Vol. 5, No. 11, p. 1403–31.

GRAY, T. R. G.; PARKINSON, D. 1968. *The Ecology of Soil Bacteria.* Toronto, University of Toronto Press. 681 p.

HARPER, H. A. 1971. *Review of Physiological Chemistry.* Los Altos, Calif., Lange Medical Publications. 529 p.

HEWITT, E. J. 1966. *Sand and Water Culture Methods Used in the Study of Plant Nutrition.* 2nd ed., East Malling, Maidstone, Kent, Comm. Bur. Hort. and Plant Crops. 547 p. (Commonwealth Agricultural Bureaux Technical Communication, No. 22.)

HILLEL, G. 1971. *Soil and Water.* New York, Academic Press. 288 p.

HOEKSTRA, W. G. 1972. Animal Requirements: Basic and Optimal, *Ann. N.Y. Acad. Sci.,* Vol. 199, p. 182–90.

HUFFMAN, E. W. D.; ALLAWAY, W. H. 1973. Growth of Plants in Solution Culture Containing Low Levels of Chromium. *Plant Physiol.,* Vol. 52, p. 72–5.

JACOB, A. 1958. *Magnesium, the Fifth Major Plant Nutrient.* London, Staples Press. 159 p.

JENNER, F. A.; GOODWIN, J. C.; SHERIDAN, M.; TAUBER, I. J.; LOBBAN, M. C. 1968. The Effect of an Altered Time Regime on Biological Rhythms in a 48-hour Periodic Psychosis. *Brit. J. Psychiat.,* Vol. 114, p. 213.

JONES, R. G. W.; LUNT, O. R. 1967. The Function of Calcium in Plants. *Bot. Rev.,* Vol. 33, p. 407–26.

KORVALSKY, V. V. 1970. The Chemical Ecology of Organisms Under Conditions of Varying Contents of Trace Elements in the Environment. In: Mills (ed.), *Trace Element Metabolism in Animals,* p. 385-97. London, Livingston.

KOZLOVSKIY, F. 1972. Structural Functional and Mathematical Model of Migrational Landscape Geochemical Processes. *Pochvovedeniye,* No. 4, p. 122–8. (Translation.)

KRAUSKOPF, K. B. 1967. *Introduction to Geochemistry.* New York, McGraw-Hill. 721 p.

KUBOTA, J.; ALLAWAY, W. H. 1972. Geographical Distribution of Trace Element Problems. In: J. J. Mortvedt *et al.* (eds.), *Micronutrients in Agriculture,* p. 521–54. Madison, Wis., Soil Science Society of America.

KUZNETSOV, S. I.; IVANOV, M. V.; LYALIKOVA, N. N. 1963. *Introduction to Geological Microbiology.* New York, McGraw-Hill. (Translated by Broneer.)

LEHR, J. E. 1972. Chemical Reactions of Micronutrients in Fertilizers. In: J. J. Mortvedt *et al.* (eds.) *Micronutrients in Agriculture.* Madison, Wis., Soil Science Society of America.

LEVINSON, A. A. 1974. *Introduction to Exploration Geochemistry.* Calgary, Alberta, Applied Publishers. 612 p.

LISK, D. J. 1974. Recent Developments in the Analysis of Toxic Elements. *Science,* Vol. 184, p. 1137–41.

LOSEE, F. T. W.; CUTRESS; BROWN, R. 1974. Trace Elements in Human Dental Enamel. In: *Trace Substances in Environmental Health—VII, A Symposium,* 1974, p. 19–24. Columbia, Mo., University of Missouri.

LUCUS, R. E.; KNEZEK, B. D. 1972. Climatic and Soil Conditions Promoting Micronutrient Deficiencies in Plants. In: J. J. Mortvedt *et el.* (eds.), *Micronutrients in Agriculture,* p. 265-88. Madison, Wis., Soil Science Society of America.

LYON, G. I.; PETERSON, P. J.; BROOKS, R. R.; BUTLER, G. W. 1971. *J. Ecology,* Vol. 59, p. 421.

MACHOLD, O.; SCHOLZ, G. 1969. Eisenhaushalt und Chlorophyllbildung bei Höheren Pflanzen. *Naturwiss.,* Vol. 56, p. 447–52.

MALLETTE, M. F.; ALTHOUSE, P. M.; CLAGETT, C. O. 1960. *Biochemistry of Plants and Animals.* New York, Wiley. 522 p.

MARINOS, N. G. 1962. Studies on Submicroscopic Aspects of Mineral Deficiencies: I, Calcium Deficiency in the Shoot Apex of Barley. *Am. J. Bot.,* Vol. 49, p. 834–41.

MASON, B. 1966. *Principles of Geochemistry,* 3rd ed. New York, Wiley. 329 p.

MERTZ, W. 1969. Chromium; Occurrence and Function in Biological Systems. *Physiol. Rev.,* Vol. 49, p. 163.

METCALF, R. L. 1974. Screening Compounds for Early Warnings About Environmental Pollution. In: *Trace Substances in Environmental Health—VIII, A Symposium 1974,* p. 213–17. Columbia, Mo., University of Missouri.

MITCHELL, C. W. 1973. *Terrain Evaluation.* London, Longman. 221 p.

MORRISON, G. H. 1972. Spark Source Mass Spectrometry for the Study of the Geochemical Environment. *Annals N.Y. Acad. Sci.,* Vol. 199, p. 162–73.

MURPHY, L. S.; WALSH, L. M. 1972. Correction of Micronutrient Deficiencies with Fertilizers. In: J. J. Mortvedt *et al.* (eds.), *Micronutrients in Agriculture.* Madison, Wis., Soil Science Society of America.

NASON, A.; McELROY, W. D. 1963. Inorganic Nutrient Nutrition and Microorganisms. *Plant Physiol.,* Vol. III, p. 451–536.

NYDAL, R. 1968. *Further Investigation on the Transfer of Radiocarbon in Nature. J. Geophys. Res.,* Vol. 75, p. 1617–35.

ODUM, E. P. 1971. *Fundamentals of Ecology,* 3rd ed. Philadelphia, Pa., Saunders. 574 p.

OLSEN, J. S. 1970. Carbon Cycles in Temperate Woodlands. In: D. E. Reichle, (ed.), Analysis of Temperate Forest Woodlands, p. 226-41. Berlin, Springer.

OLSEN, S. R. 1972. Micronutrient Interactions. In: J. J. Mortvedt *et al.* (eds.), *Micronutrients in Agriculture,* p. 243–64. Madison, Wis., Soil Science Society of America.

PATTEN, B. C. 1972. *Systems Analysis and Simulation in Ecology.* New York, Academic Press. 247 p.

PEEL, A. J. 1974. *Transport of Nutrients in Plants.* London, Butterworth. 258 p.

PEREL'MAN, A. I. 1961. *Landscape Geochemistry.* Geografgiz, Moscow.

——. 1972. *The Geochemistry of Elements in the Zone of Supergenesis.* Moscow, Nedre Publishing House. 273 p.

PRICE, C. A. 1968. Iron Compounds and Plant Nutrition. *Ann. Rev. Plant Physiol.,* Vol. 19, p. 239–48.

RANKAMA, K.; SAHAMA, T. 1950. *Geochemistry.* Chicago, Ill., University of Chicago Press. 912 p.

REICHLE, D. E. 1970. *Analysis of Temperate Forest Ecosystems.* Berlin, Springer. 304 p.

RICHARDS, 1974. *Introduction to the Soil Ecosystem.* New York, Longman. 266 p.

RODIN, L. E.; BASILEVICH, N. I. 1967. *Production and Mineral Cycling in Terrestrial Vegetation.* Edinburgh, Oliver & Boyd. 288 p. (Translated from the Russian by G. E. Fogg.)

ROHLICH, G. A. (ed.). 1969. *Eutrophication. Proceedings of a Symposium.* Washington, D.C., National Academy of Science. 661 p.

RUSSELL, F. C.; DUNCAN, D. L. 1956. *Minerals in Pasture: Deficiencies and Excesses in Relation to Animal Health.* Farnham Royal, Bucks, Commonwealth Agricultural Bureau. 170 p.

SCHÜTTE, K. H. 1964. *The Biology of the Trace Elements.* London, Crosby Lockwood. 228 p.

SCHWARTZ, K.; MERTZ, W. 1959. *Biochem. Biophys.,* Vol. 85, p. 292.

SCOTT, M. L. 1972. Trace Elements in Animal Nutrition. In: J. J. Mortvedt *et al.* (eds.), *Micronutrients in Agriculture,* p. 555–92. Madison, Wis., Soil Science Society of America.

SIEGEL, F. R. 1974. *Applied Geochemistry.* New York, Wiley. 353 p.

SMITH, F. E. 1970. Analysis of Ecosystems. In: D. E. Reichle (ed.), *Analysis of Temperate Forest Ecosystems,* p. 7-18. Berlin, Springer.

STACE, C. A. 1963. *A Guide to Cellular Botany.* London, Longman. 148 p.

STAFELT, M. H. 1972. *Plant Ecology.* New York, Wiley. 595 p.

STEWARD, F. C. (ed.). 1963. *Plant Physiology. Inorganic Nutrition of Plants,* Vol. III. 811 p.

STOCKING, C. R.; ONGUN, A. 1962. The Intracellular Distribution of some Metallic Elements in Leaves. *Am. J. Bot.,* Vol. 49, p. 284–9.

STOUT, P. R.; MEAGHER, W. R.; PEARSON, G. A.; JOHNSON, C. M. 1951. Molybdenum Nutrition of Crop Plants. I. The Influence of Phosphate and Sulphate on the Absorption of Molybdenum from Soils and Solution Cultures. *Plant Soil,* Vol. 3, p. 51–87.

SUKACHEV, V.; DYLIS, N. 1964. *Fundamentals of Forest Biogeocoenology.* London, Oliver & Boyd. 672 p.

SWAINE, F. M. 1970. *Non-marine Organic Geochemistry.* London, University of Cambridge Press. 445 p.

THOMAS, W. A. 1969. Accumulation and Cycling of Calcium by Dogwood Trees. *Ecol. Monographs.* Vol. 39, p. 101–20.

THOMSON, J. N.; SCOTT, M. I. 1969. The Role of Selenium in the Nutrition of the Chick. *J. Nut.,* Vol. 97, p. 335–42.

TIFFIN, L. O. 1971. Translocation of Nickel in Xylem Exudate in Plants. *Plant Physiol.,* Vol. 48, p. 273–7.

TRESHOW, M. 1970. *Environment and Plant Responses.* New York, McGraw-Hill. 422 p.

TURK, L. J.; HEIL, R. J.; KREITLER, C. W.; JONES, D. C. 1973. Nitrate Contamination of Ground Water in Runnels County, Texas. In: *Trace Substances in Environmental Health—V, A Symposoium 1972,* p. 153–66. Columbia, Mo., University of Missouri.

UNDERWOOD, E. J. 1966. *The Mineral Nutrition of Livestock.* FAO, Commonwealth Agricultural Bureaux. 237 p.

——. 1971. *Trace Elements in Human and Animal Nutrition,* 3rd ed. New York, Academic Press. 543 p.

WAINERDI, R. E., UNEN, E. A. (eds.). 1971. *Modern Methods of Geochemical Analysis.* London, Plenum. 396 p.

WALLACE, A. E.; FROLICH; LUNT, O. R. 1966. Calcium Requirements of Higher plants. *Nature,* Vol. 209, p. 634.

WATT, K. E. F. 1966. *Ecology and Resource Management*. New York, N.Y., McGraw-Hill. 450 p.

WATTS, p. 1971. *Principles of Biogeography*. New York, McGraw-Hill. 401 p.

WEDEPOHL, K. H. 1970. *Geochemistry*. New York, Holt, Rinehart & Winston. 231 p.

WELCH, R. M. 1973. Vanadium Uptake by Plants. *Plant Physiol.,* Vol. 51, p. 828–32.

WESOLOWSKI, J. J.; JOHN, W.; KAIFER, R. 1973. Lead Source Identification by Multi-element Analysis of Diurnal Samples of Ambient Air. In: Kothny (ed.), *Trace Elements in the Environment,* p. 1–17. American Chemical Society (Advances in Chemistry Series., No. 123)

ZAJIC, J. E. 1969. *Microbial Biogeochemistry*. New York, Academic Press. 345 p.

Excesses and deficiencies in rocks and soils as related to plant and animal nutrition

Hubert W. Lakin

United States Geological Survey,
Denver, Colorado (United States of America)

Introduction

The increased sensitivity of analytical methods and improved techniques of identification of biological factors in recent years have led to recognition of some roles played by trace elements in biological functions. It is reasonable to suppose that some elements now considered non-essential to the health of plants or animals (and even toxic to animals) will be found in the future to be essential in very minute quantities. The most pressing problem in geochemistry in its relation to health is a more precise knowledge of the elements required for good health and what constitutes toxic amounts of these elements. During the following discussion the reader must thus bear in mind that elements labelled non-essential or even poisonous are so identified by the present, imperfect understanding of their biological function.

Optimum concentrations of elements in the environment

The mere presence of an adequate amount of a given element in the soil in which our food plants are grown is insufficient to assure an optimum concentration for the maximum growth and health of the food plant. To illustrate, the average iron content of soils is 3.8 per cent—about 400 times the iron content of plants; however, in alkaline soils iron is relatively insoluble and iron deficiency in plants is common. Some of the factors that are operative in establishing optimum concentrations of elements for good health in either vegetable or animal nutrition or in establishing certain elements as poisonous are: (a) the particular species under consideration such as in plants that have adapted to abnormally high concentrations of sodium chloride; (b) the chemical combination in which the element occurs, as in organic mercurials versus inorganic mercury compounds; (c) the oxidation state of the element as it affects solubility of the element in aqueous solutions: (d) the oxidation state of the element as it affects utilization of the element by the animal ingesting the material; (d) the ratio of one element to another in the food for animals, as in the importance of the Cu:Mo dietary ratio in preventing hypocuprosis in cattle; and (e) the age and sex of the animal, as in iron deficiency in humans.

Food plants differ widely in their soil and climate requirements. This is seen in two major food grains—wheat and paddy rice—that are basic diet foods for large populations. Wheat is a plant of cool, temperate climate that yields grain of best quality (highest in protein) on neutral to alkaline semi-arid soils. The neutral to alkaline soils are likely to yield selenium and molybdenum to the wheat grain in good quantity, but iron, manganese, zinc and copper may be relatively low. Paddy rice thrives in warmer climate in submerged soils. The paddy soils for rice culture should tend to yield much iron, manganese, zinc and copper to the plant, but the plant may be low in selenium and molybdenum.

Although soils are mainly derived from weathered rocks, important quantities of materials are added to the soils by wind-transported dust, volcanic dust and volatiles, compounds formed by atmospheric reaction and organic volatiles from vegetation. These additions tend to decrease the differences in soils that might be expected as a result of differing underlying parent rocks.

Vast areas in China and the mid-continent of North America are overlain by loess (wind-deposited silt) to depths of tens of metres. In times of drought, these wind-deposited soils are subject to further wind erosion and deposition. During the drought of the mid-1930s, in the United States, fields in North Dakota were eroded by wind to plough depth of 15 cm. A brown snow that fell in New Hampshire, Vermont and northern New York State, on 24 and 25 February 1936, owed its colour to soil carried from west of the Missouri River; at Saint Johnsbury, Vermont, this single storm deposited 8 tonnes of dust per square mile (about 2.6 km^2) (Robinson, 1936). Dust storms rise from the Sahara desert to heights of more than 1,500 metres and travel as far as central Europe (Kürger, 1967). The hundreds of volcanic eruptions during historic times have contributed enormous amounts of particulate matter to the Earth's surface. The eruptions of Krakatoa in Indonesia in 1883, of Mount Katmai in the Alaskan peninsula in 1912, and of Mount Hekla in Iceland in 1947 contributed more particulate matter to the atmosphere than all of man's activity to date (Pecora, 1972). Fine dust carried in the upper atmosphere from the Mount Katmai eruption formed a haze which so reduced the intensity of sunshine as to cause the recorded cold summer of 1912 throughout the northern hemisphere (Griggs, 1922).

Volatile chemical compounds are also released by volcanism, are eventually precipitated by rainfall and become an important contribution in the distribution of elements in soils. Six days after the eruption of Mount Katmai, housewives in Vancouver, B.C., found that linen laundry hung out to dry had been ruined by sulphuric acid from gases of the Katmai volcano 2,400 km to the north (Griggs, 1922). Zies (1929) estimated that in 1919, seven years later, a total of 1.25 million tonnes of hydrochloric acid, 0.2 million tonnes of hydrofluoric acid, and 0.3 million tonnes of hydrogen sulphide were released to the atmosphere by fumaroles associated with the Katmai volcano. It has been estimated that in the Earth's history, degassing of the Earth by volcanism has been responsible for the deposition of 5,365 g of chlorine, 475 g of sulphur, 26 g of boron, 18 g of bromine, and 0.1 g of selenium per square centimetre of the Earth's surface (Rankama and Sahama, 1950).

The oceans also return matter to the continents by wind-blown evaporites from ocean spray. The rapid drop in concentration of chloride content of rain water towards the centre of the continent (Fig. 1) shows the marked effect of the oceans on chlorides deposited by rain (Junge and Werby, 1958). The curves for sodium in rain water show similar coastal concentrations to that of chloride, but sodium exceeds chloride stoichiometrically, presumably because further sodium is added by continental dust. Vinogradov (1959) stated,

The ocean is the reservoir from which all of the iodine of the atmosphere is drawn and transported far inside the continents ... Thus, precipitations introduce from 9 to 50 g of I each year on 1 hectare of soil; the average for the continent is probably about 10 g of I.

The atmosphere is the primary source of nitrogen in soils. Ammonia and nitrogen oxides are formed in the upper atmosphere by solar energy and in the lower atmosphere by the electrical discharge of lightning. It has been estimated that 90 million tonnes of fixed nitrogen are deposited by rain annually from the atmosphere. This natural 'pollution' of the Earth's surface is essential for plant growth and thus for man's existence. About 95 per cent of an estimated 8,000 million tonnes of the chemical compounds annually entering the Earth's atmosphere are derived from natural sources and less than 1 per cent of the 1,500 tonnes of hydrocarbons are derived from human activity (Pecora, 1972). Forest trees give off gaseous hydrocarbons that condense in the atmosphere to create a smog recognized in such names as the Blue Ridge Mountains and the Great Smoky Mountains in the United States. Curtin *et al.* (1972), and Curtin (1974) have determined the presence of Li, Be, Mn, Fe, Co, Ni, Cu, Zn, Ga, As, Sr, Y, Zr, Mo, Ag, Cd, Sn, Sb, Ba, and La in condensed volatile exudates of conifer trees.

Soils develop into definite categories due to climate from the parent material of weathered underlying rocks, or water-, ice, or wind-transported parent material. In arid to semi-arid climates, a zone of calcium carbonate accumulation is usually present with little or no downward movement of iron or aluminium; in humid, cool climates, usually forested, a bleached surface zone is underlain by a zone of iron and aluminium accumulation; in wet, tropical climates a surface zone of iron or aluminium oxide accumulation may extent to several metres depth with the leaching of silica from the parent material.

If one must give an optimum concentration for elements in soils, the average composition of temperate climate soils shown in Table 1 approaches an optimum value when the soil has a pH slightly below 7. Compositionally these soils are very similar to shales, which are richer in many trace elements than other sedimentary rocks. Soils developed in the humid tropics are likely to yield foodstuffs deficient in minor elements, as are temperate soils derived from sandstones or unconsolidated sands.

In summary, soils are definite morphological entities that owe their physical and chemical properties to

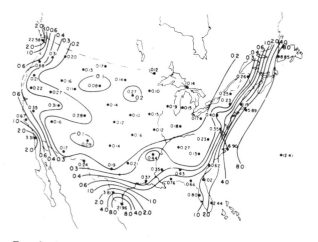

FIG. 1. Average Cl$^-$ in milligrams per litre (p.p.m.) in rain water in the United States from July 1955 to June 1956 (from Junge and Werby, Fig. 1, 1958).

the parent rock material from which they are formed, modified by the climate in which they are located and by rain-borne chemicals that are added to them.

Composition of soils related to parent rocks

Variation in the elemental composition of rocks from which soils are formed provides some insight into recognizing serious deviations from normal productive soils and provides a rough guide in predicting regions where nutritional problems may be encountered. In Table 1, 34 elements that are found in biological materials are listed in order of decreasing abundance in the continental crust. In addition, the abundance of each element in various rock types, in average soils, and in living organisms is given, together with brief comments on its function in biosystems.

In the data on the average composition of living matter (Perel'man, 1967), shown in Table 1, 98.8 per cent of living matter consists of oxygen, carbon, hydrogen and nitrogen that plants obtain directly or indirectly from the atmosphere—these elements are not included in the table. However, Gilbert (1957) estimated the mineral content of the body of mammals and plants at about 5 per cent (versus the 1.2 per cent used in Table 1). The low mineral content of biological systems is far more important in the metabolism of plants and animals than its weight percentage indicates. Twenty-one elements other than oxygen, carbon, hydrogen and nitrogen are required for the well-being of plants or animals.

The agronomist and soil scientist have worked assiduously for the past 100 years on various aspects of soil fertility and have found many variables that affect the supply of elements in the soil to the plant (Jacks, 1967; Sauchelli, 1969; Mortvedt, 1972; Kothny, 1973). Among these variables are the pH of the soil, the nature of the rock minerals from which the soil is derived and the extent of the leaching that has occurred in the development of the soil. Numerous quick tests have been developed for diagnosis of deficiency diseases of plants. The present use of large amounts of nitrogen, phosphorus and potassium in fertilizers has, in some instances, so increased the yields that a serious deficiency of available trace elements needed by the plants became evident. Other elements listed in Table 1 (and some that are not included) will no doubt be found to be essential.

Examination of Table 1 reveals that average soils of the temperate zone—for which all the soil data apply—contain adequate supplies of essential elements. The very real difficulty, however, lies in the amount of the element in the soil that is available to the plant in any given growing season. The soils of the conterminous United States contain ample total selenium but in many areas of the country cattle and sheep suffer from selenium deficiency because the selenium in the soil is unavailable to the forage crops on which these animals feed. The situation is similar for cobalt.

Soils developed from certain sedimentary rocks, particularly shales and some carbonates, contain abnormal concentrations of minor elements that are toxic to plants or animals. An example of this is the seleniferous (1–20 p.p.m. Se) soils developed on the Pierre Shale and the Niobrara Chalk in the mid-western United States; feeds grown on these soils are toxic to livestock because of their high selenium content. Such seleniferous foodstuffs have been reported all over the world.

Molybdenum toxicity has been associated with the Lower Lias shale in southern England. The shale contains up to 250 p.p.m. Mo. Vegetation grown on soils developed from the shale contains as much as 60 p.p.m. Mo dry weight. Granitic alluvium that forms neutral to alkaline poorly drained soils on both the east and west sides of the Sierra Nevada provides areas that produce vegetation toxic to animals owing to high molybdenum content. The granitic alluvium contains up to 15 p.p.m. Mo and plants contain up to 300 p.p.m. Mo (Kubota and Allaway, 1972).

In alkaline soils, the availability of cobalt, nickel, manganese and sometimes zinc is low, whereas the availability of molybdenum is enhanced. This is illustrated by data of Mitchell (1972) which showed that the liming of granitic soils, which raised the pH of the soil from 5.4 to 6.4, caused the cobalt, nickel and copper contents of red clover to fall to one-half of their previous values whereas the molybdenum content was increased sixfold.

Soils developed from basic rocks reflect the high magnesium, chromium and nickel and the very low potassium that are characteristic of these rocks; the soils are infertile and readily recognized by the sparse and distinctive native vegetation they support. The plant species growing in these soils have been termed 'serpentinophytes' to designate species tolerant of serpentine soils (Brown et al., 1972). Robinson (1935) attributed the infertility of serpentine soils in Maryland to toxicity of chromium and nickel, although he presented controversial evidence that a low lime–magnesia ratio is an added factor in the low productivity of these soils.

Soils developed in tropical forests are very porous and, as a result of intensive leaching, are usually low in available phosphorus, potassium and micronutrients.

Soils developed from many sandstones are low in available nutrients, as the sandstone is composed of minerals resistant to weathering. Truck farming on sandy soils in the eastern United States is frequently a modified system of hydroponics, using the soil primarily as a mechanical support for the plants.

Ore bodies are often surrounded by haloes of unusually high content of the ore metals. This fact is used in mineral exploration; the extent of this use is illustrated by the analysis of over 4 million soil and rock

TABLE 1. Elements commonly present in living organism listed in order of decreasing abundance in the continental crust with the abundance in various rocks, soils and living matter

Element [1]	Continental crustal abundances [a]	Ultra basic rocks (dunites, peridotites, pyroxenites)		Basic rocks [b] (basalts, gabbros, norites and andesites)	Intermediate rocks [b] (diorites and andesites)	Acid rocks [b] (granites, liparites and rhyolites)	Shales [c]	Sandstones [c]	Carbonates [c]	Soils [b]
				Weight percentage						
Si	28.15	20.5[b]	20.2[c]	22.80	26.00	32.30	7.3	36.8	2.4	33
Al	8.23	2.88	2	8.76	8.85	7.70	8	2.5	0.42	7.3
Fe	5.63	9.85	9.43	8.56	5.85	2.70	4.72	0.98	0.38	3.8
Ca	4.15	7.70	2.5	6.72	4.65	1.58	2.21	3.91	30.23	1.37
Na	2.36	0.57	0.42	1.94	3.0	2.77	0.96	0.33	0.04	0.63
Mg	2.33	14.10	20.4	4.50	2.18	0.56	1.5	0.7	4.7	0.63
K	2.09	0.5	0.004	0.83	2.31	3.34	2.66	1.07	0.27	1.36
P	0.105	0.12	0.022	0.14	0.16	0.07	0.07	0.017	0.4	0.08
Mn	0.095	0.13	0.162	0.22	0.12	0.06	0.085	0.00X	0.11	0.085
				Parts per million						
F	625	100	100	370	500	800	740	270	330	200
Ba	425	15	0.4	270	650	830	580	X0	10	500
Sr	375	27	1	440	800	300	300	20	610	300
S	260	3,000	300	2,000	1,000	400	2,400	240	1,200	850
V	135	140	40	200	100	40	130	20	20	100
Cl	130	200	85	200	200	240	180	10	150	100
Cr	100	2,000	1,600	300	56	25	90	35	11	200
Rb	90	2	0.2	45	70	400	140	60	3	100
Zn	70	50	50	130	72	60	95	16	20	50
Cu	55	80	10	140	35	30	45	X	4	20
Co	25	200	150	45	20	5	19	0.3	0.1	8
Sc	22	10	15	24	15	7	13	1	1	7
Li	20	2	0.X	15	20	70	66	15	5	30
Ga	15	4	1.5	18	20	30	19	12	4	30
Pb	12.5	—	1	8	15	2	20	7	9	10
B	10	4	3	1.5	—	5.5	100	35	20	10
U	2.7	0.03	0.001	0.8	1.8	3.5	3.7	0.45	2.2	1
Sn	2	—	0.5	6	—	45	6	X	X	10
As	1.8	2.8	1	2	2.4	1.5	13	1	1	5
Mo	1.5	0.4	0.3	1.4	0.9	1.9	2.6	0.2	0.4	2
I	0.5	0.8	0.5	0.5	0.3	0.4	2.2	1.7	1.2	5
Tl	0.45	0.06	0.06	0.2	0.15	2.4	1.4	0.82	0.0X	—
Cd	0.2	—	—	0.19	—	0.1	—	—	—	—
Hg	0.08	—	0.0X	0.09	—	0.04	0.4	0.03	0.04	0.01
Se	0.05	—	0.05	—	—	—	0.6	0.05	0.08	0.01

1. Elements known to be required by any organism are shown in bold type.
Sources: a. Taylor, 1964, Vol. 28, p. 145–6; b. Vinogradov, 1959; c. Turekian and Wedepohl, 1961, Vol. 72, p. 175–92; d. Shacklette *et al.* (1971a; 1971b; p. DI–71; 1973a; 1973b); e. Perel'man, 1967.

United States	Geometric mean of surface soils of conterminous United States[d]		Average composition of living matter[e]	Comments
	Western United States	Eastern United States		

Weight percentage

United States	Western United States	Eastern United States	Average composition of living matter[e]	Comments
—	—	—	0.2	Possibly essential to some plants; animals.
4.5	5.4	3.3	0.005	Always present in plants and animals. Use unknown
1.8	2.0	1.5	0.01	Essential to plants and animals
0.88	1.8	0.32	0.5	Essential to plants and animals
0.4	1.02	0.26	0.02	Essential to animals only
0.47	0.78	0.23	0.04	Essential to plants and animals
1.2	1.7	0.74	0.3	Essential to plants and animals
0.025	0.032	0.018	0.07	Essential to plants and animals
0.034	0.039	0.0285	0.0011	Essential to plants and animals

Parts per million

United States	Western United States	Eastern United States	Average composition of living matter[e]	Comments
180	250	115	5.1	Essential to animals
430	560	300	31	Probably not useful
120	210	51	21	Probably not useful
—	—	—	500	Essential to plants and animals
56	66	46	X	Micro-organism; animals
—	—	—	200	Essential to plants and animals
37	38	36	X	Essential to animals and plants
—	—	—	0.51	Probably not useful
44	51	36	5	Essential to plants and animals
18	21	14	2	Essential to plants and animals
7	8	7	0.2	Micro-organisms (*Rhizobium* sp.) Essential to animals
8	9	7	—	Probably not useful
20.4	23.3	26.1	0.11	Probably essential to animals
14	18	10	—	Possibly essential to plants
16	18	14	0.51	Present in plants and animals. Toxic
26	22	32	10	Essential to plants
—	—	—	<0.01	Probably not useful
—	—	—	0.51	Essential to animals
5.8	6.1	5.4	0.31	Possibly used by animals. Excess toxic
<3	<3	<3	0.11	Essential to plants and animals
—	—	—	—	Essential to animals
—	—	—	—	Promotes growth of plants; toxic to animals
—	—	—	—	Toxic
0.071	0.055	0.096	0.00X	Toxic. Fish contain unusual amounts
0.31	0.25	0.39	<0.01	Essential to animals and some plants. Excess toxic to animals

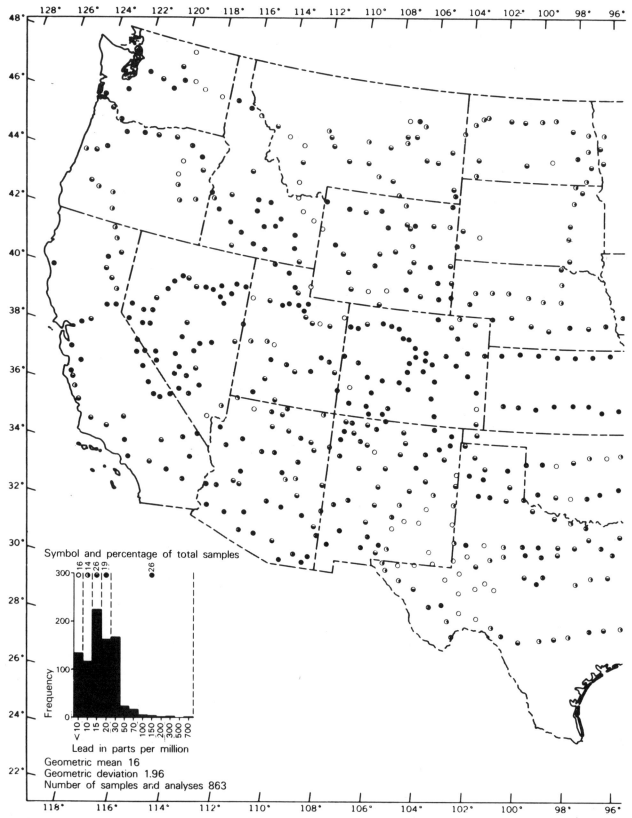

FIG. 2. Lead content of surficial materials in the conterminous United States.

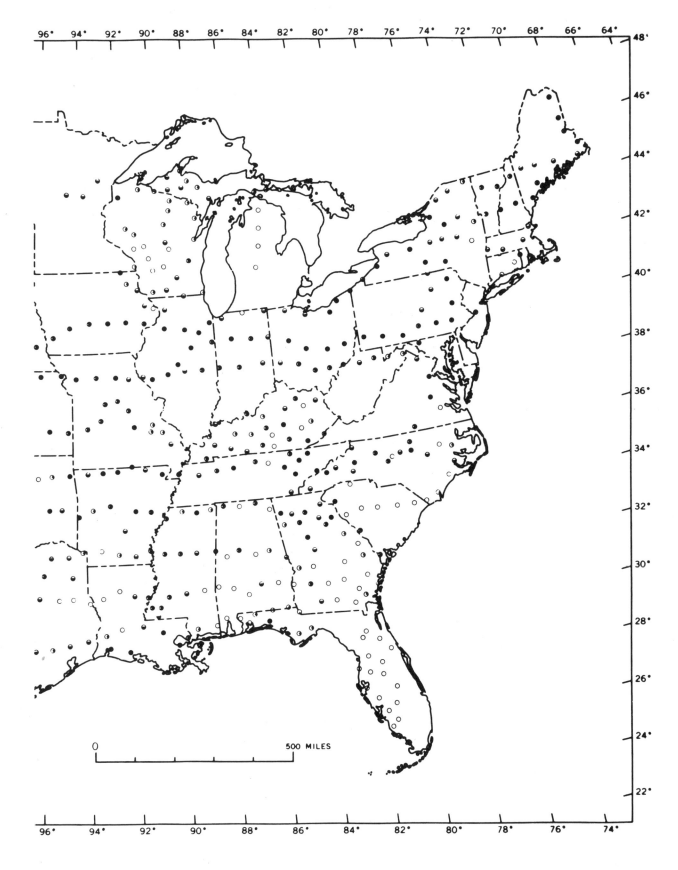

0   500 MILES

samples for mineral exploration in 26 countries during the year ending 31 May 1971. The enormous body of data on minor elements in soils and rocks of mineralized provinces of the world which is now accumulating in the files of mining and exploration companies, universities and government agencies could be collected in a single computer-storage-and-retrieval system. The constantly expanding data set would provide guidelines of great value to geochemists, environmentalists and health specialists.

Non-essential elements

LEAD

The average distribution of lead in some natural materials is summarized in Table 1. The misleading character of average values is apparent in the distribution of lead in 863 soils collected in conterminous United States by Shacklette *et al.* (1971*b*) shown in Figure 2. Mitchell (1964) gave a commonly found range of 3 to 200 p.p.m. Pb with unusual values as low as 0.3 p.p.m. and as high as 2,000 p.p.m. Pb. Lead content of soils in mineralized areas sometimes exceeds 10,000 p.p.m. (Huff, 1952) and the lead content of soils is now used widely in prospecting for lead ores (Kvalheim, 1967; Hawkes, 1972). A comprehensive report on lead in the environment is being prepared for publication by the United States Geological Survey (Lovering, 1976).

The data of Durum *et al.* (1971) on 717 surface waters as summarized by Fleischer (1973) shown in Table 2 further illustrate the wide fluctuation of lead in the environment. A study of 969 water-supply systems in the United States revealed that 14 of the water supplies contained lead in excess of $50 \, \text{ng} \, \text{l}^{-1}$, the upper limit set by drinking-water standards (United States Public Health Service, 1970).

The lead content of coal usually ranges from 2 to 50 p.p.m. and averages 15 p.p.m.. Some 50 tonnes of lead per year may enter the atmosphere as fly ash from a power-plant in the south-western United States burning

5 million tonnes of coal per year (Swanson, 1972). The mean lead content of vegetation growing near this power plant was found to be 70 per cent above the mean for the same species of plants collected prior to the installation of the power plant (Cannon and Anderson, 1971). In this study, vegetation collected near highways also reflected added lead from petrol burned by highway traffic.

The major source of lead contamination is the lead in petrol anti-knock additives. Deposition of lead on soils and plants adjacent to highways and in urban areas from petrol-powered vehicles has been amply demonstrated (Cannon and Bowles, 1962). Recent reduction in the use of leaded petrol will certainly change future figures. In 1969, 244,000 tonnes of lead were used in petrol in the United States (Morris *et al.*, 1973). It would take 5,000 power plants of the type and capacity discussed above to equal the automotive pollution. An excellent critical review of airborne lead was prepared by a committee of the National Academy of Sciences (1972).

Most human foods contain less than 1 p.p.m. lead; cow's milk contains 0.02–0.08 p.p.m. and muscle meats contain about 0.1 p.p.m. (Underwood, 1973). Human blood averages about 0.13 p.p.m. (Kubota *et al.*, 1968). In a summary discussion of the various sources of lead in the atmosphere, cigarette smoke and foods, Underwood (1973) concluded that there was no evidence that the small accumulation of lead in the human body (excess of intake over excretion of lead) was either harmful or otherwise.

CADMIUM

Cadmium is a rare and dispersed element seldom found in igneous rocks at levels above 0.2 p.p.m. (Table 3). It has been difficult analytically to determine trace quan-

TABLE 2. Dissolved lead in surface waters of the United States

$\mu g \, l^{-1}$ (p.p.b. Pb)	Percentage of 717 samples
< 1	35.8
1– 4	33.9
5– 9	15.8
10–19	9.8
20–29	2.9
30–79	1.5
Max. 80	0.3

Source: Fleischer, 1973.

TABLE 3. Concentrations of cadmium in some natural materials

Type of material	Range usually reported (p.p.m.)	Average (p.p.m.)
Continental crust	—	0.2
Basaltic igneous rocks	0.006–0.6	0.2
Silicic igneous rocks	0.003–0.18	0.15
Shales and clays	0–11	1.4
Black shales (high C)	0.5–8.4	2.0
Deep-sea clays	0.1–1	0.5
Limestones	—	0.05
Sandstones	—	0.05
Soils[1]	0.1–0.5	0.3
Phosphorites	0–170	30
Coals	—	2

1. Excluding soils from mineralized areas.
Source: Fleischer, 1973.

tities of cadmium in the presence of the other constituents of rocks, some of which seriously interfered with the analytical measurement, of cadmium. Recently, atomic absorption spectrometry has provided a moderately satisfactory analytical method, and data on cadmium are becoming available.

Cadmium is a chalcophilic element. It is found as a minor constituent in most sulphide minerals (Vlasov, 1966). It is particularly enriched in zinc sulphides, commonly in ranges from 1,000 to 5,000 p.p.m., and has been reported as high as 4.4 per cent in sphalerite (Fleischer, 1955). In magmatic rocks cadmium is reported to be concentrated in ferromagnesian minerals, but the lack of routine analytical methods with sufficient sensitivity has hindered systematic study of cadmium in the magmatic process (Vlasov, 1966).

The data in Table 3 compiled by Fleischer (1973) show cadmium to be enriched in clays, shales (especially black shales), coals and particularly phosphorites. Cadmium content of phosphate rocks ranges from less than 10 over 100 p.p.m. and in superphosphate from 1 to 170 p.p.m. (Swaine, 1962); use of these fertilizers makes cadmium available to agricultural crops.

Cadmium, like mercury, selenium and chromium, is accumulated in shales to an extent that far exceeds its content in magmatic rocks. Wedepohl (1968) suggested that much of the cadmium in shales came from the magmatic degassing of the lower crust and the mantle. Tourtelot et al. (1964) found an average of 1.4 p.p.m. and a median value of 0.8 p.p.m. Cd in samples of Pierre shale and some equivalent stratigraphic units; a rough correlation existed between cadmium and organic carbon content. Soils developed from shales—which account for about 75 per cent of all sedimentary rocks—should be expected to contain five times or more cadmium than soils developed from igneous rocks.

Shacklette et al. (1973a) reported that only 11 out of 912 soils collected in conterminous United States contained at least 1 p.p.m. Cd. The remaining 901 samples contained less than 1 p.p.m.—the limit of sensitivity of their analytical method (Nakagawa and Harms, 1968). Miesch and Huffman (1972) reported as much as 150 p.p.m. Cd in contaminated surface soils 0–4 in. (0–10 cm) deep near the smelter in Helena, Montana; cadmium contamination was not detected 8 km from the smelters. Because of its volatility, cadmium is enriched in the gases coming from the smokes-stack of smelters; several instances of cadmium contamination near smelters have been reported (Cannon and Anderson, 1971). Cadmium is recovered as a by-product during the smelting and refining of zinc and the recent improvement of zinc-refining techniques results in a more efficient recovery of cadmium (Wedow, 1973) and a corresponding reduction of cadmium losses to the atmosphere.

Data on cadmium in 727 surface waters of United States (Durum et al., 1971) are summarized in Table 4

TABLE 4. Cadmium in filtered surface waters of the United States

μg l^{-1} (p.p.b. Cd)	Percentage of 727 samples
< 1	54.2
1–4	32.3
5–4	32.3
5–9	7.7
10–20	4.1
21–40	0.9
41–130	0.8
	100.0

Source: Fleischer, 1973.

(Fleischer, 1973). Concentrations of cadmium in surface waters above 20μg l^{-1} are almost certainly due to industrial waste.

The principal use of cadmium is in electroplating, and wastes from electroplating introduce cadmium into sewage. Analyses of 42 sewage sludges from rural and industrial towns in England and Wales showed seven sludges to have cadmium levels above 100 p.p.m., presumably from towns with a diversity of light industry (Berrow and Webber, 1972). In the use of sewage sludge as a fertilizer, caution should be exercised with respect to the cadmium content as the cadmium is probably available to plants.

Cadmium is found naturally in plants, the concentration depending on which species and which tissue examined, as well as many soil factors. It can apparently be absorbed through the plant leaves from deposits of atmospheric dust (Shacklette, 1972). It has also been found in condensed volatiles from conifer trees (Curtin and others, 1973). Cannon and Anderson (1971) showed that plants are enriched in cadmium tenfold or more in vegetation near mining operations and zinc smelters.

MERCURY

Concern over the health hazards of mercury has prompted much research on the geochemistry of this element. Recent reviews (United States Geological Survey, 1970; Jonasson, 1970; Wallace et al., 1971; Jenne, 1972) adequately summarize the known geochemistry.

Mercury is concentrated in sulphide ores, especially those of zinc and, to a lesser extent, those of copper. It is concentrated in organic-rich shales (0.03–2.8 p.p.m.), phosphorites (0.001–0.95 p.p.m.) and coal (0.05–13.3 p.p.m.), and up to 2.9 p.p.m. Hg has been reported in petroleum (Fleischer, 1973). The enrichment of mercury in shales in excess of its abundance in igneous rocks is probably due to degassing of deep source rocks during volcanism (Wedepohl, 1968). The vapour pressure of mercury, even at surface temperatures, is so high that its direct volatilization to the

Hubert W. Lakin

TABLE 5. Dissolved mercury in filtered surface waters of the United States

μg l^{-1} (p.p.b. Hg)	Percentage of 706 samples
< 0.5	52.1
0.1–0.4	21.8
0.5–2.0	15.7
2.0–4.9	5.8
— EPA[1] upper limit —	
5.0–9.9	2.0
11.0–19.9	1.6
> 20 (max. 740)	1.0
	100.0

1. Environmental Protection Agency.
Source: Fleischer, 1973.

atmosphere is significant. This property has been used to prospect for ore deposits by sampling soil gases and even by measuring mercury collected from the air during flights over mineralized areas. The mercury content of air is highest over areas where the rocks are richest in mercury; mercury content of air at ground level over mercury ores may be as much as 20,000 ng/m^3, compared with a normal range of 3–9 ng/m^3 (McCarthy et al., 1970).

Data on the mercury content of filtered surface water of the United States (Durum et al., 1971)—as summarized by Fleischer (1973)—are shown in Table 5. A detailed account of mercury in waters of the United States was given by Jenne (1972). The mercury content of two river systems in Italy that radiate from the mercury district of mount Amiata is shown in Figure 3. Dall'Aglio stated that background values in the more than 300 samples of natural waters ranged from 0.01 to 0.05 p.p.b. (approximately 1–5 μg l^{-1}). The high values along the upper reaches of the Paglia River are probably due to mining and industrial wastes (Dall'Aglio, 1971). Mercury is rapidly adsorbed on stream sediments; this fact is illustrated by the high mercury content of stream sediments in the upper reaches of the Ohbrone and Paglia Rivers. Many of these samples contained in excess of 5 p.p.m. and reached a maximum of 50 p.p.m. Hg.

Mercury is present in volcanic emanations, fumarolic vapours, and hot springs associated with volcanism. Volcanism probably injects 10–100 times as much mercury into the planetary atmosphere as all of man's cumulated activities—all the mercury mined by man throughout history would total less than 0.001 per cent of that contained in sea water (Pecora, 1972).

Locally, man's activity has resulted in hazardous concentrations of mercury; as much as two-thirds of the daily input of mercury in Lake Ontario is from indus-

trial sources (Fleischer, 1973). In Japan, about 50 people out of more than 100 affected died after having eaten mercury-contaminated fish and shellfish from Minamata Bay, which had received methylmercury in waste effluents from a plastics factory (Underwood, 1973). Undeniably, care must be exercised in the industrial use of mercury.

Organic mercury compounds are used in agriculture as fungicides. Wheat grain coated with mercury compounds is intended only for seed grain; when in the past mistakenly consumed by people, it has been the cause of serious illness and death.

ARSENIC

The trace elements, lead, arsenic, cadmium and mercury, are frequently classified as toxic because their biological activity is evident mainly as toxic reactions. The toxicity of arsenic is a function of the valence state of the element; As^{3+} is more readily retained by the body than As^{5+} and thus smaller amounts of this form can cause toxic reactions.

Arsenic is especially concentrated in sulphides, predominantly as arsenopyrite (FeAsS), realgar (AsS) and orpiment (As$_2$S$_3$). Up to 1 per cent arsenic has been found in galena and sphalerite and 5 per cent in pyrite (Fleischer, 1955). Arsenic also occurs in the native state. It is enriched in sedimentary iron and manganese ores, attaining a concentration of 1.5 per cent in the manganese ores. The arsenic content of phosphate deposits ranges from 0,4 to 2,000 p.p.m. (Fleischer, 1973). Arsenic in superphosphate generally ranges from 100 to 200 p.p.m.; in rare instances it is as high as 500 p.p.m. (Goldschmidt, 1954); thus arsenic may be a contaminant in fertilizers and phosphate detergents.

Coal ash averages 86 p.p.m. As and ranges from less than 1 to 700 p.p.m. (Cannon, 1974). Goldschmidt (1954) stated that in certain coal ashes, the arsenic content ranged from 500 to 1,000 p.p.m. and in the ash of lignite was as much as 300 p.p.m.. In contrast, the coal ash of coals used in power plants in the south-western United States range from 1 to 5 p.p.m. As. The fly ash from these plants contains about 12 p.p.m. As. The contamination of the atmosphere with arsenic from the burning of coal is obviously variable, and in some areas may be a matter for concern.

Williams and Whetstone (1940) found from 1 to 40 p.p.m. As in soils from all the major soil groups of the United States. Their data showed a tendency for arsenic to accumulate in the B horizon of soils. The accumulation of arsenic in the iron-rich B zone is compatible with the known insolubility of basic ferric arsenates and probably reduces the availability of arsenic to plants. Heavy applications of lead arsenates to soils, e.g. 2,000 lb per acre (2,240 kg/ha), do not materially increase the arsenic content of vegetation but do retard growth and in some species prevent normal growth (string beans and onions).

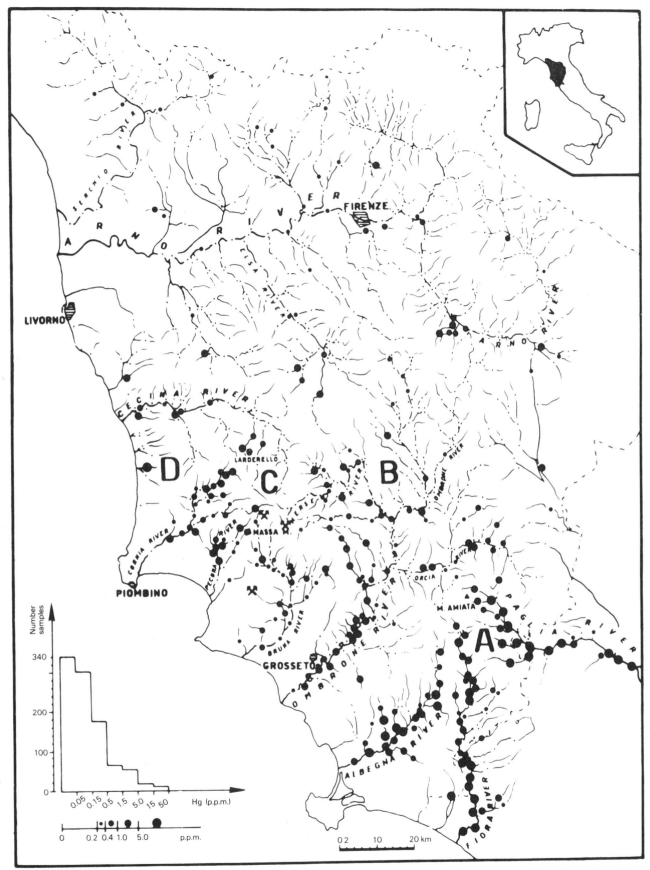

Fig. 3. Histogram and geographical distribution of mercury in approximately 1,000 stream-sediment samples collected over the whole of the Tuscany region, Italy, of 20,000 km². (From Dall-Aglio, 1971. Reproduced by permission of the Canadian Institute of Mining and Metallurgy.)

Cannon (1974) reported an average arsenic content of 0.3 p.p.m. in forage grasses, 5.4 p.p.m. in forage legumes and 1.0 p.p.m. in vegetables and fruits. According to Williams and Whetstone (1940), marine algae from Puget Sound, Washington contained 1–12 p.p.m. As; 60 samples of vegetables and grains contained less than 0.1 to 10 p.p.m. As and only 11 of the samples contained 1 p.p.m. or more arsenic.

Arsenic is often associated with valuable sulphide ores and has been used as an indicator element for mineralized volcanic pipes (Burbank *et al.*, 1972). Wild (1974c) reported that the arsenic content of plants growing on arsenical mine dumps ranged from 200 to 3,000 p.p.m.. He also observed characteristic arsenic-tolerant flora similar to the 'serpentine flora' growing on soils usually containing 300–5,000 p.p.m. As and as much as 20,000 p.p.m..

Essential elements

ZINC

Zinc tends to concentrate in igneous rocks containing dark-coloured ferromagnesian silicates that are present in basic rocks, such as basalts and gabbros; this concentration is probably due to the similarity in ionic radii of Mg^{2+} and Zn^{2+}, allowing zinc to substitute for Mg^{2+} in ferromagnesian silicate minerals. Minute grains of sulphide minerals containing zinc are also frequently present (Krauskopf, 1972). In acidic rocks (granites and rhyolites) rich in silica and relatively poor in iron, magnesium and sulphur, the zinc content averages less than half that of basic rocks (see Table 1). Shales are intermediate in their zinc content, approaching the zinc content of the continental crust. Sandstones (16 p.p.m. Zn) and limestones (20 p.p.m. Zn) are much lower in zinc content than the average soil (50 p.p.m.). The zinc content of coal varies in the range 50–150 p.p.m. and generally the zinc content of crude oil is only a few (usually less than 10) p.p.m. (Brehler and Wedepohl, 1970).

Samples of surface waters collected at over 130 sites in the United States have been analysed by direct-reading emission spectrography (Kopp, 1970). The results on 1,577 of these samples are shown in Table 6 with a comparison with the permissible concentration of these elements in public water supplies. Zinc equalled or exceeded 2 p.p.b. in 76.5 per cent of these samples but never approached the permissible level of 5,000 p.p.b. Zinc is readily adsorbed on clay minerals and manganese and iron oxides; consequently neither effluents of zinc-mining operations nor springs draining high-zinc areas greatly enhance the zinc content of major streams and rivers. Industrial waste to the Saale River (Germany) gave rise to several thousand micrograms of zinc per litre of river water. The zinc content decreased by a factor of 10 within 10–13 km (6–8 miles)

TABLE 6. Summary of dissolved trace elements in 1,577 surface-water samples (1 October 1962–30 September 1967) compared with permissible concentrations in public water supplies

Element	Frequency of detection (percentage)	Observed values $\mu g\,l^{-1}$			Permissible concentration $\mu g\,l^{-1}$
		Min.	Max.	Mean	
Lead	19.3	2	140	23	50
Cadmium	2.5	1	120	9.5	10
Arsenic	5.5	5	336	64	50
Zinc	76.5	2	1,183	64	5,000
Chromium	24.5	1	112	9.7	50
Copper	74.4	1	280	15	1,000
Molybdenum	32.7	2	1,500	68	—

Source: Kopp, 1970.

downstream. The scavenging ability of hydrous oxides of iron and manganese for zinc may be the principal agent in removing zinc from this river (Jenne, 1968). The zinc content of sea water ranges from 2 to 10 p.p.b. and varies with depth (Brehler and Wedepohl, 1970).

The usual range of zinc content of soils is 10–150 p.p.m., but values as low as 2 p.p.m. and in excess of 10,000 p.p.m. have been reported (Swaine, 1955). Soils that tend to be low in zinc are alluvials, regosols and soils under high rainfall. Humic gleys and organic soils may also be low in zinc content. Soils in which the zinc is relatively unavailable to plants are calcareous soils and those soils rich in phosphorus and nitrogen. Thus over-fertilization with phosphorus or nitrogen can induce zinc deficiency in soils for plants (Lucus and Knezek, 1972). In 1966, 43 states in the United States reported zinc-deficient areas; the known zinc-deficient areas aggregate 2.63×10^6 ha (Cunningham, 1972).

In mining areas a halo of the ore metal frequently surrounds a zinc ore deposit. In mineral provinces both soils and plants may be unusually high in zinc content. An abundant literature exists that presents many illustrations of naturally occurring residual and alluvial soil high in zinc content due to anomalous zinc content of a mineralized area. For further information, see Cameron (1967), Canney (1969), Boyle (1971) and Jones (1973) for sources of data on mineralized areas presented by representatives of many nations; the publications of the Royal School of Mines, London, the Geological Surveys of Canada, the United States and Norway on geochemical prospecting and the *Journal of Exploration Geochemistry* are also excellent sources of data.

Unusually high zinc content of soils is not necessarily related to ore deposits. Geochemical studies of zinc-bearing peats in western New York State show them related genetically to underlying mineralized beds of the Lockport Dolomite of Middle Silurian age. Although the mineralized beds of the dolomite contained only 600 p.p.m. zinc, the dried peat was found to contain as much as 160,000 p.p.m. zinc. Many of the peat bogs of the area were drained and the muck soil was put into cultivation. Cannon (1955) gave this account of the resulting experience of farmers in Orleans County near Manning, New York:

In some places the vegetable crops were poor over an area of an acre or more and smaller spots supported no vegetation (crops or other) whatsoever. Crops planted on these areas came up normally, but soon after the first leaves appeared the plants turned yellow, remained stunted and eventually died.

Hodgson (1970) expressed the relative importance of various environmental factors to plant composition by the statement:

Content of a given element in a plant, $P = f(R, S, A, P)$, where $R =$ parent rocks of soil, $S =$ soil-forming factors, $A =$ a measure of availability of the element in the soil, and $P =$ plant interaction with its environment. Although generalized, this statement emphasizes that all these factors function in controlling the metal content of a given plant in a given environment. Roughly stated, the zinc content of an oven-dry plant is 0.6 of the zinc content of the same weight of the oven-dry soil in which it was grown. These generalizations should not be allowed to exclude the fact that each soil-plant species pair is a special case which may or may not follow the general rule.

SELENIUM

Selenium is known to be essential to animals, and it may be essential to some species of plants. At levels below 0.04 p.p.m. in the diet of animals deficiency symptoms appear; at levels above 4 p.p.m. in the diet of animals it is toxic. Naturally toxic seleniferous vegetation has been reported in 19 countries (Lakin, 1972; Sindeeva, 1964). The occurrence of inadequate selenium in animal feedstuffs is far more common and involves greater areas than toxic seleniferous feedstuffs. Selenium deficiencies in livestock have been recognized in 13 countries (Allaway, 1969). Thus it seems especially appropriate to discuss deficiencies and excesses of selenium in rocks and soils.

Selenium is a rare and dispersed element (Table 1) whose chemistry closely resembles that of sulphur. The primary sources of selenium are volcanic emanations and metallic sulphides associated with igneous activity. Rankama and Sahama (1950) quoted Goldschmidt's estimate that selenium in gases emitted by volcanoes through the geologic ages amounted to 0.1 gram of selenium per square centimetre of the Earth's surface. A cartoon from *Punch* magazine (Fig. 4) seems appropriate. The selenium content of volcanic sulphur has been found to vary from less than 2 p.p.m. to 5.18 per cent (51,800 p.p.m.) (Lakin and Davidson, 1967). Selenium is equally variable in sulphide minerals, ranging from 1 or 2 p.p.m. to 5 per cent of the mineral. In sulphide deposits selenium minerals also occur, as clausthalite

Stanley by Murray Ball
Continuing the adventures of the Great Palaeolithic Hero

FIG. 4. Selenium from volcanic emanations through all of geologic time have an accumulative value of 0.1 g/cm² of Earth's surface (from *Punch* magazine).

(PbSe), tiemannite (HgSe), berzelianite ($Cu_{2-x}Se$), penroseite ($NiSe_2$) and many others (Sindeeva, 1964).

Only a small portion of the total selenium in the Earth's crust is accounted for by volcanism and sulphide ore deposits. Selenium is distributed throughout the igneous rocks that represent most of the crust and are eroded to form the sedimentary rocks so commonly the parent materials of soils. Goldschmidt and Strock (1935) analysed a large number of sulphide minerals from many different mineral assemblies and concluded that the ratio of sulphur to selenium was about 6,000. Using this ratio and an accepted crustal abundance of sulphur of 540 p.p.m., they estimated that the crustal abundance of selenium was 0.09 p.p.m. In 1961, Turekian and Wedepohl used 300 p.p.m. as the crustal abundance of sulphur and thus reduced their estimate of the crustal abundance of selenium to 0.05 p.p.m. Vlasov (1966) found 'only about 20 direct determinations of selenium in rocks, all for rocks of the U.S.S.R. (Sindeeva and Kurbanova, 1958)'. He concluded that the crustal abundance of selenium was uncertain. It is still uncertain but in recent years more sensitive methods than those available to Sindeeva have been developed, and Table 7 shows a collection of recent data.

We can draw some tentative observations from the data in Table 8. Selenium is known to be strongly associated with copper sulphides; the commercial source of selenium is from the slimes resulting from electrolytic refining of copper. Table 8 shows the copper content and selenium content of classes of igneous rocks; an approximate correlation is evident.

Hubert W. Lakin

TABLE 7. Selenium content of igneous rocks

Rock type	Number of samples	Range p.p.b.	Average p.p.b.
Intrusive rocks			
Ultrabasic rocks	11	34–104	72
Gabbro	31	< 20–230	108
Diorite	21	< 20–228	72
Granite	37	< 20–126	25
Finnish rapakivi granite	19	< 20–250	56
Basalts	2	170–380	220
Andesitic rock	2	470–920	690
Rhyolite	1	—	90
Extrusive rocks			
Basic andestitic volcanics	22	20–650	176
Acidic volcanics	18	< 20–280	86
Rhyolitic pumaceous ash	11	100–240	170
Andesitic ash	5	600–2,640	1,480
Basaltic ash	4	240–1,500	740

Source: Koljónen, 1973a; Wells, 1967.

TABLE 8. Copper and selenium content of some major igneous rock types

Rock	Copper, p.p.m.	Selenium, p.p.b.
Ultrabasic rocks	80	72
Gabbros	140	108
Diorites and andesites	35	72
Granites	30	25

Source: Vinogradov, 1959 (copper data); Koljónen, 1973a (selenium data).

It is also evident that extrusive rocks tend to contain more selenium than intrusive rocks (Table 7) as shown by comparing Iceland and New Zealand rocks with Finnish rocks. Koljonen (1973a) noted that vesicular obsidian contained more selenium (280 p.p.b.) than an adjacent dense glass from the margin of the same vein (less than 20 p.p.b.).

The selenium content of metamorphic rocks was found by Koljonen (1973b) to vary from less than 0.01 to 37 p.p.m., the high content being in black schist.

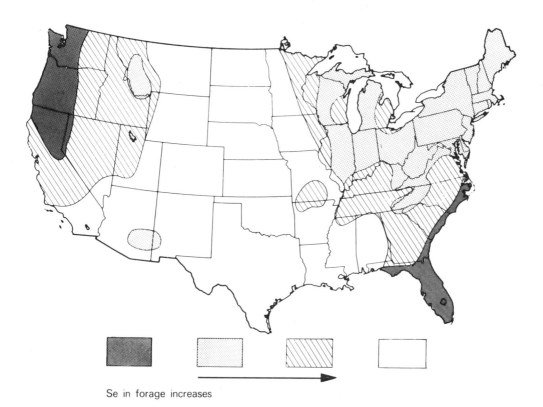

Se in forage increases

FIG. 5. Regional pattern of selenium in forage crops in the contiguous United States. (From Kubota *et al.,* 1967. © American Chemical Society. Reproduced with permission.)

In sedimentary rocks selenium is enriched over crustal abundance in black shales, coals and organic-rich carbonate rocks; thus, selenium is accumulated in biological sinks in the erosional cycle. Sandstones are usually very low in selenium content except for sandstones containing much organic debris, where selenium accumulates with uranium and vanadium. Koljonen (1973c) summarized published data on sedimentary rocks and added data on sedimentary rocks from Finland, Sweden and the U.S.S.R. Averages calculated from these new data are: sandstones less than 0.01 p.p.m.; shale 0.5 p.p.m.; and carbonates 0.03 p.p.m. Se. Sediments, mostly of Finland, contained less than 0.01 to 0.6 p.p.m. Se (Koljonen, 1974), with selenium lowest in sands (less than 0.01 p.p.m.) and highest in clays. Selenium content of humus, peat, and organic-rich, lake-bottom sediments ranged from less than 0.01 to 0.51 p.p.m..

The selenium content of soils varies from less than 0.01 p.p.m. to several p.p.m. The valency state of the selenium in soils is the major factor controlling the availability of selenium to plants. In alkaline soils of pH 8–8.5 in semi-arid areas, selenium is oxidized to selenate (SeO_4^{2-}) and is very soluble. In such areas, plants absorb sufficient selenium to be toxic to animals. In most farming areas of the world, a higher rainfall and resulting acid soil causes selenium to be present as selenite (Se_3^{2-}) or elemental (uncharged) selenium. Neither form is readily available to plants. The selenite is adsorbed or absorbed by ferric oxides and held in a most insoluble form (Lakin, 1961; Geering et al., 1968). Elemental selenium is not readily oxidized in soils and thus remains largely unavailable. A regional pattern of selenium concentration in plants (Fig. 5) shows a rough correlation of plant-available selenium with rainfall. The New England area of the United States produces forage deficient in selenium in spite of the enormous amount of coal, containing an average of 3 p.p.m. Se, that has been burnt in this area during the last 200 years. The selenium from the smoke stacks is most probably in elemental form and remains as an inert addition to the soils (Lakin, 1973). For an excellent critical review of the geochemistry and biochemistry of selenium the reader is referred to Allaway (1969).

CHROMIUM

Chromium is relatively abundant in the continental crust—twice as abundant as copper and four times as abundant as cobalt (Table 1). Chromium occurs almost exclusively in a valency of $3+$ in the Earth's crust; under strong oxidizing conditions it is oxidized to $6+$ (as in chromates from the Chilean nitrate deposits). Normally, the presence of CrO_4^{2-} indicates industrial pollution.

Chromium is a lithophilic element showing a strong affinity for oxygen and is enriched in the silicate crust. No sulphide minerals of chromium have been ob-

TABLE 9. World average content of magnesium, calcium, chromium, and nickel in the igneous rock series

Element	Ultramafic	Mafic	Intermediate	Felsic
Percentage				
Mg	25.9	4.5	2.2	0.6
Ca	0.7	6.7	4.7	1.6
Parts per million				
Cr	2,000	200	50	25
Ni	2,000	160	55	8

Source: Krauskopf, 1967, p. 588.

served in sulphide ore deposits. The mineral chromite ($FeCr_2O_4$)—with magnesium commonly substituting for ferrous iron and ferric iron and aluminium substituting for chromium—is the only commercial source of chromium (Thayer, 1973).

The chromium content of igneous rocks varies widely, being highest in ultrabasic rocks (dunites) and lowest in highly silicic rocks such as granites (Table 9). It is enriched in titaniferous iron ores up to 5,500 p.p.m. and in magnetite ores in gabbros up to 3,500 p.p.m. (Rankama and Sahama, 1950).

Chromite is one of the first minerals to form during the normal crystallization of a calc-alkalic magma and is found in stratified deposits in certain kinds of ultrabasic rocks.

Chromium is dispersed in many minerals, commonly substituting for Al^{3+}, whose ionic radius is close to that of Cr^{3+}. The chromium oxide content of diopsides and endiopsides (calcium–magnesium silicates) from olivine inclusions in basalts ranged from 0.8 to 2.12 and averaged 1.19 per cent (Deer et al., 1963, Table 6). Although chromium is often found enriched in diopsides, garnets, ilmenite and magnetite, these minerals are moderately stable and release chromium very slowly.

The ultrabasic igneous rocks are not abundant and only rarely are the parent materials of soils. More important to the soil scientists than the tightly bound chromium in chromite and related minerals is the occurrence of chromium in readily weathered minerals. An example is the mica, biotite, which according to much evidence almost universally contains chromium. Lovering (1969, 1972) analysed a suite of 200 biotite samples that were separated from 55 granites, 35 quartz monzonites, 41 granodiorites, 10 diorites, 9 pegmatites, 14 mafic intrusive rocks and 20 miscellaneous rocks. The chromium varied from 5 p.p.m. to 2,000 p.p.m. in these biotites, being highest in the mafic intrusive rocks and lowest in the quartz monzonite. The median value for biotite from the quartz monzonite is 200 p.p.m. Cr. Rimšaite (1964) reported a maximum of 3,700 p.p.m. Cr

103

in 18 biotite samples from igneous rock of Canada. Thus, it seems that chromium is enriched well above its average abundance in the continental crust in biotites, which thus form a ready source of chromium in soils.

In sedimentary rocks the average chromium content is 90 p.p.m. in shales, 35 p.p.m. in sandstones and 11 p.p.m. in carbonates (Table 1). The similarity of the ionic radii of Cr^{3+} and Al^{3+} allows Cr^{3+} to substitute for Al^{3+} in clay minerals and probably accounts for the chromium in shales. Chromium is especially high in phosphorites, ranging from 30 to 3,000 p.p.m. and averaging 300 p.p.m. The Phosphoria Formation of Idaho, Wyoming and Utah in the United States has an average chromium content of 1,000 p.p.m. with a positive correlation with phosphate in the shale member of the formation (Mertz et al., 1974). The chromium in sandstones is probably present as discrete chromium minerals which are resistant to weathering.

Metamorphic rocks generally have the same content of trace metals as the parent rocks. Of interest in a discussion of chromium are the serpentine rocks which are metamorphic derivatives of ultrabasic rocks. Soils developed from these rocks are often termed infertile. Chromium is closely associated with magnesium and nickel in peridotite, pyroxenite and serpentine. These rocks, low in calcium and high in chromium, nickel and magnesium (Table 9), yield upon weathering an infertile soil with a specialized vegetative cover. The infertility has been attributed variously to the unfavourrable Mg:Ca ratio, the toxic effects of nickel or the toxic effects of both nickel and chromium.

The chromium content of soil varies from a few p.p.m. to at least 3,000 p.p.m.. Shacklette et al. (1971b) collected 863 soil samples throughout the United States and found a geometric mean of 37 p.p.m. Cr; 25 per cent of the samples contained less than 30 p.p.m. and 12 per cent contained more than 90 p.p.m. Cr, ranging up to 1,600 p.p.m.. Mitchell (1964) gave the normal

range of chromium in soil at about 5–1,000 p.p.m., with the content in unusual soils reported as low as 0.5 p.p.m. and as high as 10,000 p.p.m. Cr. The chromium extractable from soils with acetic acid—used as a guide to available chromium to plants—was found by Swaine and Mitchell (1960) to vary much less than the total chromium. A soil developed from a serpentine rock containing 2,000 p.p.m. Cr in the B horizon contained only 0.24 p.p.m. acetic acid-extractable Cr, as compared with a soil developed from a sandstone containing 25 p.p.m. and 0.51 p.p.m. extractable Cr in the B horizon. Thus, predicting the chromium available to plants from the total chromium content of the soil is not possible.

Chromium in water is assumed to exist as the chromate ion and the presence of chromate in water usually indicates pollution by industrial wastes. Of 2,595 samples of 969 public water-supply systems serving 18.2 million people, only 0.2 per cent of the samples exceeded Cr values of 0.05 p.p.b. and the maximum value was only 0.079 p.p.b. (United States Department of Health, Education and Welfare, 1970). The literature on chromium in biological systems and natural waters has been reviewed by Mertz et al. (1969, 1974).

Chromium is present in measurable amounts in all plants. Content in diverse types of vegetation varied from 0.6 p.p.m. (dry weight) in conifer branch tips to 3.7 p.p.m. in leaves and twigs of shrubs (Table 10). Species of plants endemic to Rhodesian serpentines have been reported to contain from 9,000 to 48,000 p.p.m. Cr in the plant ash (Wilde, 1974a). The Phosphoria Formation produces soils high in chromium and 200 p.p.m. Cr (dry weight) has been reported in twigs of an Engelmann spruce growing in such soil (Lotspeich and Markward, 1963).

Mertz et al. (1974) assumed that growth response of plants to added chromium indicated that a chromium deficiency had existed. Soils low in chromium may be expected in sandy soils and soils developed from carbonate rocks.

Chromium is required by both plants and animals. Chromium deficiency—not chromium excess—is the major concern in human health (Mertz et al., 1974).

COPPER AND MOLYBDENUM

Copper and molybdenum are discussed together because high-molybdenum diet induces copper deficiency in cattle. Cattle suffer from diarrhoea on diets containing 20–100 p.p.m. Mo on a dry-weight basis. The disease (molybdenosis) can be controlled by massive oral doses of copper sulphate. However, as little as 5 p.p.m. of molybdenum in the diet of cattle having abnormally low copper intake will produce molybdenosis (Underwood, 1973). An unusual combination of events, triggered by an earthquake in New Zealand, led to the discovery that adequate molybdenum reduces the incidence of dental caries in children (Allaway, 1968). Both copper and molybdenum are essential to the growth of plants.

TABLE 10. Mean chromium content of various types of vegetation and soil in which they grew

Type of vegetation or soil collected more than 305 m from nearest road	Number of samples	Chromium (p.p.m.) mean in		Soil
		Vegetation		
		Ash	Dry	
Deciduous trees (leaves)	10	16	1.0	
Grass (part above ground)	15	22	1.8	
Shrubs (leaves and twigs)	22	46	3.7	
Conifers (branch tips)	20	16	0.6	
Average, all plants		27	1.5	
Soil 'B' horizon	39			80

Source: Cannon and Anderson, 1971.

Obviously, the geochemistry of these elements is important to man.

Copper is relatively abundant in the continental crust, which is estimated to contain 55 p.p.m. copper but only 1.5 p.p.m. molybdenum. In basalts, gabbros and other basic rocks, copper is enriched (140 p.p.m.) whereas molybdenum is not enriched; however, in acidic rocks (granites) copper is impoverished (30 p.p.m.) and molybdenum is slightly enriched (1.9 p.p.m.). Copper tends to be enriched in the ferromagnesium minerals as pyroxene and biotite. In 216 samples of biotite, Lovering (1969) found median values of copper for most rock types were about 300 p.p.m. and ranged from 70 p.p.m. in biotites from granites to 800 p.p.m. in biotites from mafic intrusive rocks. Molybdenum was detected in only 24 per cent of the samples (limit of detection = 3 p.p.m. Mo) but was found in quantities above 3 p.p.m. in all the biotites from pegmatites and 88 per cent of the pyroclastic rocks. From the data available on copper and molybdenum in igneous rocks, one would not expect soils developed from them to give rise to hypocuprosis.

The molybdenum content of sedimentary rocks and soils is higher relative to copper than that of most igneous rocks. Shales account for approximately 75 per cent of the sedimentary rocks, sandstones for about 15 per cent, and carbonates 8 per cent. The Earth's crust is assumed to consist of 44 per cent granites, 34 per cent granodiorites, 8 per cent quartz diorites and 13 per cent gabbros (Wedepohl, 1968).

Table 11 shows similar Cu:Mo ratios for acidic rocks (granites) and shales. These two rock types differ markedly in Cu:Mo ratio from other rocks of their class. The low ratio for granites in comparison with the markedly higher ratios of other igneous rocks may be due to (a) insufficiently accurate data on molybdenum content of all rock types; (b) molybdenum enrichment and copper impoverishment of late crystallization in magmatic differentiation; or (c) incorporation of molybdenum-rich sedimentary rocks into granite. Wedepohl (1968), discussing chemical fractionation in the sediments, stated:

Another group of elements is accumulated in shales to an extent exceeding a factor of two in relation to magmatic rocks: H, B, C, N, S, Cl, As, Se, Br, Mo, Cd, Sb, I and Hg. The respective amounts of these elements in sedimentary rocks (and in sea water) cannot be derived from decomposition of magmatic rocks. ... We know that these elements form readily volatile compounds under magmatic conditions.

In his study of the Valley of Ten Thousand Smokes in Alaska, Zies (1938) reported that molybdenum was present, visible as molybdenum blue (a hydrated oxide, ilsemannite) which covered several thousand square metres.

Barth (1968) was bold enough to state: 'All igneous material in the continents comes from sediments; and sedimentary differentiation is regarded as the main cause of the diversification of igneous rocks.' Geologists still debate this viewpoint, but it does allow us to suggest a sedimentary origin of at least some of the granites of the Earth.

Schroll (1968) suggested that conclusions regarding the abundance of trace elements were based on isolated analytical data and did not reflect complete analysis of the specific sample; furthermore, that they were subject to poor analytical data for the trace element. No one would deny the need for more refined analytical methods and careful control of the interpretation of the data.

The copper and molybdenum content of sea water is 0.9 and $10 \mu g \, l^{-1}$, respectively, reflecting the greater solubility of molybdates in strongly alkaline solutions. The content in stream waters is $7 \mu g \, l^{-1}$ Cu and $1 \mu g \, l^{-1}$ Mo (Turekian, 1969).

Coals contain about 15 p.p.m. Cu and 5 p.p.m. Mo. One extreme value of 6,000 p.p.m. Mo was found in Ruhr Valley coal. Molybdenum is also enriched in petroleum and phosphorites (Davis et al., 1974). Molybdenum content in phosphate rock ranges from 1 to 138 p.p.m., the high values being unusual; the content in superphosphate fertilizer is 0.1–15 p.p.m. Copper content varies from 1.5 to 394 p.p.m. in phosphate rock and 4 to 800 p.p.m. in superphosphate (Swaine, 1962).

The discrepancy between the copper-molybdenum ratios in all rocks and average soils and the occurrence of vegetation relatively rich in molybdenum is readily explained by the varying solubility of these elements with pH of the soils. In very alkaline soils of pH 8 and above, nearly all the molybdenum is water soluble; in soils of that pH copper is almost insoluble. Mitchell's (1971) data illustrated this inverse solubility and, therefore, availability to plants (Table 12).

TABLE 11. Comparison of copper–molybdenum ratios for various rock types

Rock type	Copper (p.p.m.)	Molybdenum (p.p.m.)	Ratio Cu : Mo
Continental crust	55	1.5	36
Ultrabasic rocks	10	0.3	33
Basic rocks	140	1.4	100
Intermediate rocks	35	0.9	39
Acidic rocks	30	1.9	16
Sandstone	2	0.2	10
Limestone	4	0.4	10
Shale	45	2.6	17
Black shales	70	10	7
Deep-sea clay	250	27	9
Soils	20	2	10

Sources: Copper and molybdenum data from Table 1 of this chapter; Davis et al., 1974 (sandstone and deep-sea clay); Vine and Tourtelot, 1970 (black shales).

TABLE 12. Effect of liming and soil treatment on the uptake of copper and molybdenum by red clover

	Copper (p.p.m. in dry material)	
	at pH 5.4	at pH 6.5
Soil	7.7	3.8
Copper sulphate added, 22 kg/ha	10.6	12.1
	Molybdenum (p.p.m. in dry material)	
	at pH 6.1	at pH 6.5
Soil	0.7	3.4
Sodium molybdate added, 1.1 kg/ha	18.6	35.6

Source: Mitchell, 1971.

Molybdenum is bound in organic material in a form available to plants. Kubota (1972) reported areas in the United States of molybdenum toxicity in peats and mucks in the California delta, the Klamath area in Oregon and the Everglades in Florida.

As noted in Table 11, granites are relatively high in molybdenum compared to copper (Cu:Mo = 16). Soils developed from granitic alluvium in California on the western side of the Sierra Nevadas and in Nevada on the eastern slope of the same mountains produce vegetation in moist fans that cause molybdenosis; adjacent, well-drained soils do not produce such vegetation, presumably because the water-soluble available molybdenum is flushed out of these soils.

Molybdenum toxicity has been observed in England and Wales on heavy textured soils derived from the Lower Lias clays and limestones, in Finland on soils derived from sediments, in Ireland on calcareous gleys and in Scotland on non-calcareous gleys and peaty gleys derived from argillaceous schists (Ryan *et al.*, 1967).

Molybdenum deficiency, most commonly associated with acidic soils, has been reported in nine countries in Europe.

Copper deficiency for plants occurs in the southeastern United States; it is associated with regions of high rainfall. It also occurs in 13 countries in Europe. Significantly, copper deficiency is frequently found in mucks and peaty soils.

An area in Australia (1,600 km) long and, in some places, as wide as 240 km is deficient in molybdenum. The addition of a molybdenum-superphosphate fertilizer is making this land which has more than 51 cm of annual rainfall into a useful pasture for stock (Anderson and Underwood, 1959).

Summary

Lead, zinc, copper, cadmium, mercury, arsenic, selenium, chromium and molybdenum occur in soils in amounts high enough to produce toxic effects either on the plants growing in the soils or on animals feeding on the plants produced from such soils. There is an abundant literature on the specialized species of plants that have adapted to high concentrations of specific elements (Antonovics *et al.*, 1971; Cannon, 1971; Trealease and Beath, 1949; Wild, 1974b; Rune, 1953). Evolutionary adaptation to unusual environments is apparently frequent in plants, and results in subspecies that are tolerant to unusual concentrations of specific elements in soils that are frequently lethal to non-adapted plants.

Zinc, copper and molybdenum are essential for plant growth. In areas in which one or more of these elements are either extremely low in the soil or exist in an unavailable form, plants suffer from a specific deficiency. Plant deficiencies have been observed for zinc, copper and molybdenum in many areas and in the past have caused significant loss to farmers and hence to world food supplies (Anderson and Underwood, 1959; Ryan *et al.*, 1967; Mortvedt, 1972).

Just 9 elements have been discussed—and they only briefly—in this chapter. Even the 25 trace elements listed in Table 1 are not all the elements that should be considered; we should be concerned with the naturally occurring radioactive elements—as well as those produced by modern technology.

Of considerable importance in evaluating the role of trace elements in food production are the interrelationships of these elements in the response of biosystems. The copper–molybdenum relation mentioned above is but one of many that are only vaguely understood. Much work is needed to increase the knowledge of the soil–plant–animal nutrition complex to a level of adequate understanding for maximum food production having maximum nutritional quality.

Finally, the geochemist and, to a lesser degree, the soil scientist have been inclined to consider only the inorganic solubilities and mobilities of the elements in the weathering and soil-forming processes. The action of the higher plant root system on solubilization merits study. For example, the wheat root exudes over 10 sugars, 19 amino-acids, 12 organic acids, 3 nucleotides, several flavonones, and 3 enzymes—a very complex system. Furthermore, the population of micro-organisms increases fivefold to tenfold in the nutrient-rich region surrounding the hair roots of plants (Rovira and Davey, 1974). All these biosystems contribute to the solubilization of relatively insoluble inorganic compounds by complex formation. Certainly much more knowledge of very intricate systems is needed to understand properly the chemistry of the growth of our food crops.

References

ALLAWAY, W. H. 1968. Agronomic Controls over the Environmental Cycling of Trace Elements. *Advan. Agron.*, Vol. 20, p. 235-74.

——. 1969. Control of the Environment Levels of Selenium. In: D. D. Hemphill (ed.), *Trace Substances in Environmental Health—II*, p. 181-206. Columbia, Mo., University of Missouri.

ANDERSON, A. J.; UNDERWOOD, E. J. 1959. Trace Element Deserts. *Scientific Amer.*, Vol. 200, No. 1, p. 97-106.

ANTONOVICS, J.; BRADSHAW, A. D.; TURNER, R. G. 1971. Heavy Metal Tolerance in Plants. *Advances in Ecological Research*, Vol. 7, p. 1-85. New York, Academic Press.

BARTH, T. F. W. 1968. The Geochemical Evolution of Continental Rocks, Model. In: L. H. Ahrens (ed.), *Origin and Distribution of the Elements*, p. 587-97. New York, Pergamon Press.

BERROW, M. L.; WEBBER, J. 1972. Trace Elements in Sewage Sludges. *J. Sci. Food Agric.*, Vol. 23, p. 93-100.

BOYLE, R. W. (ed.). 1971. *Proceedings 3rd International Geochemical Exploration Symposium, Toronto, 1970.* 594 p. Canadian Institution of Mining and Metallurgy (Special Vol. 11).

BREHLER, B.; WEDEPOHL, K. H. 1970. Zinc. In: K. H. Wedepohl (ed.), *Handbook of Geochemistry*, Vol. II, No. 2, p. 30-A-1 to 30-O-2. Berlin, Springer-Verlag.

BROWN, J. C.; AMBER, J. E.; CHANEY, R. L. FOY, C. D. 1972. Differential Responses of Plant Genotypes to Micronutrients. In: J. J. Mortvedt (ed.), *Micronutrients in Agriculture*, p. 389-418. Madison, Wis., Soil Science Society of America.

BURBANK, W. S.; LUEDKE, R. G.; WARD, F. N. 1972. Arsenic as an Indicator Element for Mineralized Volcanic Pipes in the Red Mountains Area, Western San Juan Mountains, Colorado. *U.S. Geol. Survey Bull.*, No. 1364, 31 p.

CAMERON, E. M. (ed.). 1966. *Proceedings, Symposium on Geochemical Prospecting, Ottawa 1967.* 282 p. (Geological Survey of Canada Paper 66-54.)

CANNEY, F. C. (ed.). 1968. Proceedings, International Geochemical Exploration Symposium. *Colorado School of Mines Quart.*, Vol. 64, No. 1, 1969, 520 p.

CANNON, H. L. 1955. Geochemical Relations of Zinc-bearing Peats to the Lockport Dolomite, Orleans County, New York, *U.S. Geol. Survey Bull. 1000-D*, p. 119-85.

——. 1971. The Use of Plant Indicators in Ground Water Surveys, Geologic Mapping, and Mineral Prospecting. *Taxon*, Vol. 20, Nos. 2 and 3, p. 227-56.

——. 1974. Natural Toxicants of Geologic Origin and their Availability to Man. In: White and Robbins (eds.), *Environmental Quality and Food Supply*, p. 143-64. Mount Kisco, N.Y., Futura Publishing Co.

CANNON, H. L.; ANDERSON, B. M. 1971. The Geochemist's Involvement with the Pollution Problem. In: H. L. Cannon and H. C. Hopps (eds.), *Environmental Geochemistry in Health and Disease*, p. 155-77. (Geological Society of America Memoir 123.)

——. 1972. Trace Element Content of the Soils and Vegetation in the Vicinity of the Four Corners Power Plant. *Report of Coal Resources Work Group, Southwest Energy Study*, Appendix J.III, p. 1-144. (United States Geological Survey Openfile Report.).

CANNON, H. L.; BOWLES, J. M. 1962. Contamination of Vegetation by Tetraethyl Lead. *Science*, Vol. 137, No. 3532, p. 765-6.

CUNNINGHAM, H. G. 1972. Trends in the Use of Micronutrients. In: J. J. Mortvedt (ed.), *Micronutrients in Agriculture*, p. 419-30. Madison, Wis., Soil Science Society of America.

CURTIN, G. C.; KING, H. D.; MOSIER, E. L. 1973. Trace Element Content of Volatile Exudates from Conifer Trees Growing in Mineralized Areas in Colorado and Idaho. *Mining Engineering*, Vol. 25, p. 47. (Abstract.)

——. 1974. Movement of Elements into the Atmosphere from Coniferous Trees in Subalpine Forests of Colorado and Idaho. *J. Geochem. Explor.*, Vol. 3, p. 245-63.

DALL'AGLIO, M. 1971. Comparison Between Hydrogeochemical and Stream-sediment Methods in Prospecting for Mercury. In: R. W. Boyle (ed.), *Proceedings, 3rd International Geochemical Exploration Symposium, Toronto, 1970.* p. 126-31. Canadian Institution of Mining and Metallurgy (Special Vol. 11).

DAVIS, G. K.; JORDEN, R.; KUBOTA, J.; LAITINEN, H. A.; MATRONE, G.; NEWBERNE, P. M.; O'DELL, B. L.; WEBB, J. S. 1974. Copper and Molybdenum. In: P. M. Newberne (ed.), *Geochemistry and the Environment*. Vol. 1, *The Relation of Selected Trace Elements to Health and Disease*, p. 68-79. Washington, D.C., National Academy of Science.

DEER, W. A.; HOWIE, R. A.; ZUSSMANN, J. 1963. *Rock-forming Minerals.* Vol. 2, *Chain Silicates*, London, Longman, 379 p.

DURUM, W. H.; HEM, J. D.; HEIDEL, S. G. 1971. *Reconnaissance of Selected Minor Elements in Surface Waters of the United States, October, 1970.* 49 p. (United States Geological Survey Circular 643.)

FLEISCHER, M. 1955, Minor Elements in Some Sulfide Minerals. *Econ. Geology*, p. 970-1024. (Fiftieth anniversary vol.)

——. 1973. Natural Sources of Some Trace Elements in the Environment. In: Curry and Gigliotti (comp.), *Cycling and Control of Metals, Proceedings of an Environmental Resources Conference, 1972*, p. 3-7. Cincinnati, Ohio, National Environmental Research Center.

GEERING, H. R.; GARY, E. E.; JONES, L. H. P.; ALLAWAY, W. H. 1968. Solubility and Redox Criteria for the Possible Forms of Selenium in Soils. *Soil Sci. Soc. Amer. Proc. 32*, p. 35-40.

GILBERT, F. A. 1957. *Mineral Nutrition and the Balance of Life.* Norman, Okla. University of Oklahoma Press, 350 p.

GOLDSCHMIDT, V. M. 1954. *Geochemistry.* London, Oxford University Press. 730 p.

GOLDSCHMIDT, V. M.; STROCK, L. W. 1935. Zur Geochemie des Selens II. *Nachr. Ges. Wiss. Gottingen Math.-Phys. Kl. IV*, Vol. 1, p. 123-43.

GRIGGS, R. F. 1922. *The Valley of Ten Thousand Smokes.* Washington, D.C. The National Geographic Society. 340 p.

HAWKES, H. E. 1972. *Exploration Geochemistry Bibliography, Jan. 1965–Dec. 1971.* Denver, Colo., Association of Exploration Geochemists. 118 p. (Special Vol. 1.)

HODGSON, J. F. 1970. Chemistry of Trace Elements in Soils with Reference to Trace Element Concentration in Plants. In: D. D. Hemphill (ed.), *Trace Substances in Environmental Health—III*, p. 46-58. Columbia, Mo., University of Missouri.

HUFF, L. C. 1952. Abnormal Copper, Lead, and Zinc Content of Soil Near Metalliferous Veins. *Econ. Geology*, Vol. 47, p. 517-41.

JACKS, G. V. (Ed.). 1967. *Soil Chemistry and Fertility.* International Society of Soil Science, Aberdeen (Scotland). The University Press. 415 p.

JENNE, E. A. 1968. Controls on Mn, Fe, Co, Ni, Cu, and Zn Concentrations in Soils and Water: The Significant Role of Hydrous Mn and Fe Oxides. In: Gould (ed.) *Trace Inorganics in Water*, p. 337–87. Washington, D.C., American Chemical Society. (Advances in Chemistry Series, No. 73.)

——. 1972. Mercury in Waters of the United States 1970–71. *U.S. Geol. Survey Open-file Rept.* 34 p.

JONASSON, I. R. 1970. *Mercury in the Natural Environment: A Review of Recent Work.* 39 p. (Geological Survey of Canada Paper 70–57.)

JONES, M. J. 1973. *Geochemical Exploration 1972. Proceedings 4th International Geochemical Exploration Symposium, London, 1972.* London, Institution of Mining and Metallurgy. 458 p.

JUNGE, C. E.; WERBY, R. T. 1958. The Concentration of Chloride, Sodium, Potassium, Calcium, and Sulphate in Rainwater over the United States. *J. of Meteorology*, Vol. 15, No. 4, p. 417–25.

KOLJONEN, T. 1973a. Selenium in Certain Igneous Rocks. *Bull. Geol. Soc. Finland*, Vol. 45, p. 9–22.

——. 1973b. Selenium in Certain Metamorphic Rocks. *Bull. Geol. Soc. Finland*, Vol. 45, p. 107–17.

——. 1973c. Selenium in Certain Sedimentary Rocks. *Bull. Geol. Soc. Finland*, Vol. 45, p. 119–23.

——. 1974. Selenium in Certain Finnish Sediments. *Bull. Geol. Soc. Finland*, Vol. 46, p. 15–21.

KOPP, J. F. 1970. The Occurrence of Trace Elements in Water. In: D. D. Hemphill (ed.), *Trace Substances in Environmental Health—III*, p. 59–73. Columbia, Mo., University of Missouri.

KOTHNY, E. L. (ed.). 1973. *Trace Elements in the Environment.* Washington, D.C., American Chemical Society. 149 p. (Advances in Chemistry Series, No. 123.)

KRAUSKOPF, K. B. 1967. *Introduction to Geochemistry.* New York, McGraw-Hill. 721 p.

——. 1972. Geochemistry of Micronutrients. In: J. J. Mortvedt (ed.) *Micronutrients in Agriculture*, Madison, Wis. Soil Science Society of America, p. 7–40.

KUBOTA, J. 1972. Sampling of Soils for Trace Element Studies. In: H. C. Hopps and H. L. Cannon (eds.), *Geochemical Environment in Relation to Health and Disease*, Annals of the New York Acad. Sci., Vol. 199, p. 105–17.

KUBOTA, J.; ALLAWAY, W. H. 1972. Geographic Distribution of Trace Element Problems. In: J. J. Mortvedt (ed.), *Micronutrients in Agriculture*, p. 525–54. Madison, Wis, Soil Science Society of America.

KUBOTA, J.; ALLAWAY, W. H.; CARTER, D. L.; CARY, E. E.; LAZAR, V. A. 1967. Selenium in Crops in the United States in Relation to Selenium-responsive Diseases of Animals. *J. Agr. Food Chem.*, Vol. 15, p. 448–53.

KUBOTA, J.; LAZAR, V. A.; LOSEE, F. 1968. Copper, Zinc, Cadmium, and Lead in Human Blood from 19 Locations in the United States. *Arch. Environ. Health.* Vol. 16, p. 788–93.

KÜRGER, C. 1967. *Sahara.* New York, Putnam. 183 p.

KVALHEIM, A. (ed.). 1967. Geochemical Prospecting in Fennoscandia. *Interscience Publ.*, New York, 350 p.

LAKIN, H. W. 1961. Geochemistry of Selenium in Relation to Agriculture. In: Anderson, Lakin, Beeson, Smith and Thacker (eds.), p. 3–12. *Selenium in Agriculture.* (United States Department of Agriculture Handbook No. 200.)

——. 1972. Selenium Accumulation in Soils and its Absorption by Plants and Animals. In: H. L. Cannon and H. C. Hopps (eds.), *Geochemical Environment in Relation to Health and Disease.* p. 45–53. (Geological Society of America Special Paper 140.)

——. 1973. Selenium in our Environment. In: *Trace Elements in the Environment*, p. 96–111. Washington, D.C., American Chemical Society. (Advances in Chemistry Series No. 123.)

LAKIN, H. W.; DAVIDSON, D. F. 1967. The Relation of the Geochemistry of Selenium to its Occurrence in Soils. In: Muth (ed.), *Selenium in Biomedicine*, p. 27–56. Westport, Conn., AVI Publishing Co.

LOTSPEICH, F. B.; MARKWARD, E. L. 1963. Minor Elements in Bedrock, Soil, and Vegetation at an Outcrop of the Phosphoria Formation on Snowdrift Mountain, S.E. Idaho. *U.S. Geol. Survey Bull. 1181–F*, p. F1–42.

LOVERING, T. G. 1969. Distribution of Minor Elements in Samples of Biotite from Igneous Rocks. In: *Geological Survey Research 1969*, p. B101–6. (U.S. Geol. Survey Prof. Paper 650–B.)

——. 1972. Distribution of Minor Elements in Biotite Samples from Felsic Intrusive Rocks as a Tool for Correlation. *U.S. Geol. Survey Bull. 1314–D*, p. D1–29.

LOVERING, T. G. (ed.). 1976. *Lead in the Environment.* (United States Geological Survey Professional Paper 957.)

LUCUS, R. E.; KNEZEK, B. D. 1972. Climatic and Soil Conditions Promoting Micronutrient Deficiencies in Plants. In: J. J. Mortvedt (ed.), *Micronutrients in Agriculture*, p. 265–88. Madison, Wis. Soil Science Society of America.

MCCARTHY, J. H.; MEUSCHKE, J. L.; FICKLIN, W. H.; LEARNED, R. E. 1970. Mercury in the Atmosphere. In: *Mercury in the Environment*, p. 37–9. (United States Geological Survey Professional Paper 713.)

MERTZ, W. 1969. Chromium Occurrence and Function in Biological Systems. *Physiological Reviews*, Vol. 49, p. 163–239.

MERTZ, W.; ANGINO, E. E.; CANNON, H. L.; HAMBIDGE, K. M.; VOORS, A. W. 1974. Chromium. In: P. M. Newberne (ed.), *Geochemistry and the Environment*, Vol. 1. *The Relation of Selected Trace Elements to Health and Disease*, p. 29–35. Washington, D.C. National Academy of Sciences.

MIESCH, A. T.; HUFFMAN, C. Jr. 1972. Abundance and Distribution of Lead, Zinc, Cadmium, and Arsenic in Soils. In: *Helena Valley, Montana, Area Environmental Pollution Study*, p. 65–80. U.S. Environmental Protection Agency A.P. 91, Office of Air Programs, Research Triangle Park, North Carolina 27711.

MITCHELL, R. L. 1964. Trace Elements in Soils. In: Bear (ed.), *Chemistry of the Soil*, p. 320–68. New York, Reinhold Publishing Corp.

——. 1971. Trace Elements in Soils. In: Trace Elements in Soils and Crops, *Tech. Bull. Min. Agric.*, Vol. 21, p. 8–20.

——. 1972. Trace Elements in Soils and Factors that Affect their Availability. *Geol. Soc. Amer. Bull.*, Vol. 83, p. 1069–76.

MORRIS, H. T.; HEYL, A. V.; HALL, R. B. 1973. Lead. In: D. A. Brobst and W. P. Pratt (eds.). *United States Mineral Resources*, p. 313–32. (United States Geological Survey Professional Paper 820.)

MORTVEDT, J. J. (ed.). 1972. *Micronutrients in Agriculture.* Madison, Wis., Soil Science Society of America. 666 p.

NAKAGAWA, H. M.; HARMS, T. F. 1968. Atomic Absorption Determination of Cadmium in Geologic Materials. In: *Geological Survey Research 1968.* p. D207–9. (United States Geological Survey Professional Paper 600–D.)

NATIONAL ACADEMY OF SCIENCE. 1972. *Biologic Effects of Atmospheric Pollutants: Lead, Airbone Lead in Perspective.* Washington, D.C., National Academy of Science. 330 p.

PECORA, W. T. 1972. Nature—An Environmental Yardstick. *U.S. Geol. Survey,* Washington D.C., United States Government Printing Office (Stock No. 2401-00214.)

PEREL'MAN, A. I. 1967. *Geochemistry of Epigenesis.* New York, Plenum Press. 266 p.

RANKAMA, K.; SAHAMA, T. G. 1950. *Geochemistry.* Chicago, Ill., The University of Chicago Press, 912 p.

RIMSAITE, J. 1964. On Micas from Magmatic and Metamorphic Rocks. *Beitrage zur Mineralogie und Petrographie,* Vol. 10, p. 152-83.

ROBINSON, W. O. 1935. Chemical Studies of Infertile Soils High in Magnesium and Generally High in Chromium and Nickel. *U.S. Dept. Agriculture Tech. Bull.,* No. 471. 29 p.

——. 1936. Composition and Origin of Dust in the Fall of Brown Snow, New Hampshire and Vermont, Feb. 24, 1936. *Monthly Weather Review,* Vol. 64, March, p. 86.

ROVIRA, A. D.; DAVEY, C. B. 1974. Biology of the Rhizosphere. In: Carson (ed.), *The Plant Root and Its Environment,* p. 153-204. Charlottesville, Va., University of Virginia Press.

RUNE, O. 1953. Plant Life on Serpentines and Related Rocks in the North of Sweden. *Acta Phytogeographica Suecica,* Vol. 31, Uppsala. 139 p.

RYAN, P.; LEE, J.; PEEBLES, T. F. 1967. Trace Element Problems in Relation to Soil Units in Europe. *World Soil Resources Rept. 31,* Rome. FAO. 55 p.

SAUCHELLI, V. 1969. *Trace Elements in Agriculture.* New York, Van Nostrand Reinhold Co., 248 p.

SCHROLL, E. 1968. Abundances of the Chemical Elements in the Main Rock Types of the Lithosphere in Relation to a System of Correlations. In: L. H. Ahrens (ed.). *Origin and Distribution of the Elements,* p. 599-617. New York, Pergamon Press.

SHACKLETTE, H. T. 1972. Cadmium in Plants. *U.S. Geol. Survey Bull. 1314-G.* p. G1-28.

SHACKLETTE, H. T.; BOERNGEN, J. G.; CAHILL, J. P.; RAHILL, R. L. 1973a. Lithium in Surficial Materials of the Conterminous United States and Partial Data on Cadmium. *U.S. Geol. Survey Circ. 673.*

SHACKLETTE, H. T.; BOERINGEN, J. G.; KEITH, J. R. 1973b. Selenium, Fluorine, and Arsenic in Surficial Materials of the Conterminous United States. *U.S. Geol. Survey Circ. 692.* 14 p.

SHACKLETTE, II. T.; BOERNGEN, J. G.; TURNER, R. L. 1971a. Mercury in the Environment-surficial Materials of the Conterminous United States. *U.S. Geol. Survey Circ. 644.* 5 p.

SHACKLETTE, H. T.; HAMILTON, J. C.; BOERNGEN, J. G.; BOWLES, J. M. 1971b. Elemental Composition of Surficial Materials in the Conterminous United States. *U.S. Geol. Survey Prof. Paper 574-D.* 71 p.

SINDEEVA, N. D. 1964. *Mineralogy and Types of Deposits of Selenium and Tellerium.* New York, Interscience Publishers. 363 p.

SINDEEVA, N. D.; KURBANOVA, N. Z. 1958. O Klarke Selena v Nekotorykh Gornykh Porodakh SSSR [The Clarke Value of Selenium in Certain Rocks of the U.S.S.R.]. *DANSSSR,* Vol. 120, No. 2, p. 353-5.

SWAINE, D. J. 1955. *The Trace-element Content of Soils.* Farnham Royal, Bucks (United Kingdom). Commonwealth Agricultural Bureaux. 157 p.

——. 1962. *The Trace-element Content of Fertilizers.* (Technical, Communication No. 52), Farnham Royal, Bucks (United Kingdom). Commonwealth Bureau of Soils, Harpenden, Commonwealth Agricultural Bureaux. 306 p.

SWAINE, D. J.; MITCHELL, R. L. 1960. Trace Element Distribution in Soil Profiles. *J. Soil Sci.,* Vol. 11, p. 347-68.

SWANSON, V. E. 1972. *Composition and Trace Element Content of Coal and Power Plant Ash,* Pt. 2. *Appendix J of Southwest Energy Study.* 61 p. (United States Geological Survey Open File Report.)

TAYLOR, S. R. 1964. The Abundance of Chemical Elements in the Continental Crust—A New Table. *Geochim. et Cosmochim. Acta,* Vol. 28, p. 1273-85.

THAYER, T. P. 1973. Chromium. In: D. A. Brobst and W. P. Pratt (eds.), *United States Mineral Resources.* p. 111-21. (United States Geological Survey Professional Paper 820.)

TOURTELOT, H. A.; HUFFMAN, C. Jr.; RADER, L. F. 1964. Cadmium in Samples of Pierre Shale and Some Equivalent Stratigraphic Units, Grate Plains Region. In: *Geological Survey Research, 1964,* p. D73-8. (United States Geological Survey Professional Paper 475-D.)

TREALEASE, S. F.; BEATH, O. A. 1949. *Selenium; Its Geological Occurrence and its Biological Effects in Relation to Botany, Chemistry, Agriculture, Nutrition, and Medicine.* Published by the authors, New York, 292 p.

TUREKIAN, K. K. 1969. The Oceans, Streams, and Atmosphere. In: K. H. Wedepohl (ed.), *Handbook of Geochemistry.* Vol. I, p. 297-323. Berlin, Springer-Verlag.

TUREKIAN, K. K.; WEDEPOHL, K. H. 1961. Distribution of the Elements in Some Major Units of the Earth's Crust. *Geol. Soc. America Bull.,* Vol. 72, p. 175-91.

UNDERWOOD, E. J. 1973. Trace Elements. In: *Toxicants Occurring Naturally in Foods* (2nd ed.), p. 43-87. Washington, D.C., Committee on Food Protection, Food and Nutrition Board, National Research Council, National Academy of Sciences.

UNITED STATES DEPARTMENT OF HEALTH, EDUCATION AND WELFARE. 1970. *Community Water Supply Study, Analysis of National Survey Finding.* Washington, D.C., Bureau of Water Hygiene, Environmental Health Service, United States Public Health Service, Department of Health, Education and Welfare. 111 p.

UNITED STATES GEOLOGICAL SURVEY. 1970. *Mercury in the Environment.* 67 p. (United States Geological Survey Professional Paper 713.)

——. 1972. Composition and Trace Element Content of Coal and Power Plant Ash, Pt. 2. In: *Appendix J. of Southwest Energy Study.* 61 p. (United States Geological Survey Open File Report.)

UNITED STATES PUBLIC HEALTH SERVICE. 1970. *Community Water Supply Study.* Washington, D.C., Bureau of Water Hygiene, Environmental Health, Department of Health, Education and Welfare.

VINE, J. D.; TOURTELOT, E. B. 1970. Geochemistry of Black Shale Deposits—A Summary Report. *Econ. Geol.,* Vol. 65, p. 253-72.

VINOGRADOV, A. P. 1959. *The Geochemistry of Rare and Dispersed Chemical Elements in Soil.* 2nd. ed. New York, Consultants Bureau. 208 p.

——. 1960. Regularity of Distribution of Chemicals in the Earth's Crust. *Geochemistry* (Ann Arbor, Mich.), p. 1-43. (Translation of *Geochimiya,* The Geochemical Society.).

VLASOV, K. A. (ed.). 1966. *Geochemistry and Mineralogy of Rare Elements and Genetic Types of their Deposits.* Vol. 1; *Geochemistry of Rare Elements.* Jerusalem. Israel Program for Scientific Translations Ltd., p. 688.

WALLACE, R. A.; FULKERSON, W.; SHULTS, W. D.; LYON, W. S. 1971. *Mercury in the Environment—The Human Ele-*

ment. Oak Ridge, Tenn., Oak Ridge National Library, 61 p. (ORNL-NSF-EP-1.)

WEDEPHOL, K. H. 1968. Chemical Fractionation in the Sedimentary Environment. In: L. H. Ahrens (ed.), *Origin and Distribution of the Elements,* p. 999–1016. New York, Pergamon Press.

WEDOW, H., Jr. 1973. Cadmium. In: D. A. Brobst and W. P. Pratt (eds.), *United States Mineral Resources,* p. 105–9. (United States Geological Survey Professional Paper 820.)

WELLS, N. 1967. Selenium Content of Soil-forming Rocks. *New Zeal. Geol. Geophys.,* Vol. 10, p. 198–208.

WILD, H. 1974a. Indigenous Plants and Chromium in Rhodesia. *Kirkia,* Vol. 9. Pt. 2, p. 233–41.

——. 1974b. Geobotanical Anomalies in Rhodesia, 4. The Vegetation of Arsenical Soils. *Kirkia,* Vol. 9, Pt. 2, p. 243–64.

——. 1974c. Arsenic Tolerant Plant Species Established on Arsenical Mine Dumps in Rhodesia. *Kirkia,* Vol. 9, Pt. 2, p. 265–78.

WILLIAMS, K. T.; WHETSTONE, R. R. 1940. Arsenic Distribution in Soils and its Presence in Certain Plants. *U.S. Dept. Agriculture Tech. Bull.* 732. 20 p.

ZIES, E. G. 1929. *The Valley of Ten Thousand Smokes.* I, *The Fumarolic Incrustations and their Bearing on Ore Deposition;* II, *The Acid Gases Contributed to the Sea During Volcanic Activity.* Vol. 1, No. 4, 79 p. (National Geographic Society, Contributed Papers, Katmai Series.)

——. 1938. The Concentration of the Less Familiar Elements Through Igneous and Related Activity. *Amer. J. Sci.,* Vol. 35A, p. 385–404.

Hydrogeochemistry: non-marine —unpolluted and polluted waters

Mary E. Thompson

Canada Centre for Inland Waters,
Burlington, Ontario (Canada)

With a section on 'Stable Isotope Geochemistry'
by Henry P. Schwarcz
McMaster University, Toronto, Ontario (Canada).

Introduction

The hydrologic cycle has been described in many publications and it will be described here only to develop a series of topic headings under which various recent developments in hydrogeochemistry can be described.

The term 'hydrogeochemistry' as used here, is very general. It refers to geochemical research on all aspects of the hydrologic cycle. In this chapter the discussions will mainly concern fresh waters.

In brief, the hydrologic cycle is as follows. Water evaporates, chiefly from the world's oceans but also from water bodies on land, and a considerable fraction of water falling as rain is returned to the atmosphere by plant evapotranspiration. Although this stage in the cycle is of prime importance in providing the means for constant replenishment of fresh-water bodies on land, the amount of water in the atmosphere at any one time is a very small portion of the total water content of Earth's hydrosphere. Estimates of the momentary atmospheric water vapour range from 0.0008 per cent (Garrels and Mackenzie, 1971) to 0.001 per cent (Nace, 1964).

In the atmosphere clouds form and water condenses and falls as rain or snow on land and sea. The amounts and patterns of these falls are influenced by climate and geography and may be affected by activities of man such as air pollution and cloud-seeding efforts. In recent years, concern has developed over the chemical composition of rain and atmospheric precipitation in general, with respect to its acid, heavy metal and nutrient content (Almer et al., 1974; Beamish, 1974; Beamish and Harvey, 1972; Bruce, 1972; Charlson et al., 1974; Drozodova and Makhon'ko, 1970; Fisher, 1968; Gorham, 1961; Granat, 1972; Likens, 1972; Likens and Bormann, 1974; Likens et al., 1972); Miller and de Perra, 1972; Peirson et al., 1973; Rodhe, 1972; Shiomi and Kuntz, 1973; Val'nikov, 1971a,b).

Water falling onto the land surface follows several pathways. Some falls as snow and may remain frozen (as in glaciers or on the ice-cap), for many years. In some parts of the world such as northern United States and southern Canada the annual springtime melting of the winter snows is a notable peak in the hydrologic cycle and must be allowed for in attempts to avoid excessive flooding.

Much water falls as rain, some of which runs off across the land surface and collects in lakes or streams and river channels. Some percolates into the soil, where some or all of it may be taken up by the roots of plants and returned to the atmosphere by evapotranspiration. Some may percolate on downwards through this unsaturated or phreatic zone to the water-table and be incorporated into a ground-water body. Ground water may re-emerge as springs or may move through aquifers and become available as well water.

Through all these varied pathways, the water acquires a chemical content and there is considerable interest among hydrogeochemists to provide explanations for the chemistry of individual lakes or rivers, and also to calculate the annual contributions of continental run-off to the seas. Very often the source and history of a particular water body can be learned by applying the techniques of measurement of stable isotope ratios for such elements as carbon, hydrogen, oxygen, sulphur and others naturally present in the water.

Another rapidly developing technology applicable to the study of fresh-water bodies and systems is that of remote sensing, whereby a single overflight can gather much information that on the ground in difficult terrain could be obtained only with great effort.

Concern for environmental pollution has been growing in recent years—not only among geochemists—and one of the new scientific techniques is that of multidisciplinary environmental research. These multidisciplinary approaches have brought much improved understanding of the natural pathways of pollutants.

Advances in desalination technology have made possible the provision of high-quality water supplies at competitive costs to sea-shore communities in arid regions, and to communities whose river- or groundwater supplies have become brackish.

Atmospheric precipitation

The atmospheric input to the hydrologic cycle comes to us in numerous forms (rain, mist, snow, sleet, freezing rain, hailstones, etc.) and with great range in intensity—from a light summer shower to sudden and violent thunderstorms, monsoons and blizzards). The most common type of atmospheric precipitation, however, in most parts of the world, is rain and many studies of its contributions to the chemistry of natural waters have been reported recently. Several studies of the chemistry of snow have also been published.

The recent intensified interest in air and environmental pollution has prompted a number of studies of the chemical composition of rain and atmospheric fallout. It is a normal characteristic of the hydrologic cycle for rain to be slightly acid since it dissolves carbon dioxide gas from the air to form a dilute solution of carbonic acid. From a consideration of the partial pressure of carbon dioxide in the atmosphere (generally taken as about $10^{-3.5}$ atmospheres, although it fluctuates considerably—Weiler, 1972, personal communication) one can predict a pH of about 5.7. This gives the falling water a capacity for chemical weathering of rocks and soils or, conversely, most rocks and soils (and statues and stone buildings) have a capability of neutralizing the acidity of the rain by releasing cations such as sodium, calcium and magnesium in exchange for protons. Not all rocks or soils, however, are able to neutralize acid rains and the problems of lakes in such areas are serious—not only does the water in such lakes become acidic, but the lower pH will tend also to increase the solubility (and availability) or heavy metals.

In recent years considerable attention has been focused on the increasingly acidic nature of the rain—a change caused by atmospheric pollution by sulphur dioxide (Almer et al., 1974; Beamish, 1974; Beamish and Harvey, 1972; Brosset, 1973; Bruce, 1972; Charlson et al., 1974), which oxidizes readily and dissolves in water to form sulphuric acid. The source of sulphur dioxide in the atmosphere is the greatly increased rate of burning of sulphurous fossil fuels. This is a problem for which no simple solution exists; however, at the present rate of consumption, such fuels will be quite soon used up (or, at least, difficult to obtain) and will, perforce, be replaced by other sources of energy.

CARBON DIOXIDE IN THE AIR

The carbon dioxide content of air has been increasing in recent years. Since 1957 the level of carbon dioxide has been monitored at Mauna Loa Observatory at

20° N. latitude. Ekdahl and Keeling (1973) reported annual variations of about 6 p.p.m. (low in summer, high in winter) and a net increase each year of nearly 1 p.p.m. (11 p.p.m. over 14 years). Seasonal changes in carbon dioxide levels are more prominent in the northern hemisphere than in the southern since most the land masses (and hence summer growth of vegetation) are in the northern hemisphere. Average values are systematically higher in the northern hemisphere, too because most of the industrial effluents of carbon dioxide are in this half of the globe (Wofsy et al., 1972). Local effluents can cause the carbon dioxide levels to rise considerably, Weiler (1972) observed levels as high as 370 p.p.m. on Lake Ontario.

Vinogradov (1972) pointed out that the optimum level of atmospheric carbon dioxide for plant growth was an order of magnitude higher than the present level. Increased plant growth in general is desirable but the possible effects on climate are uncertain. Increases of global temperature are postulated to accompany increases in carbon dioxide, which would tend to increase evaporation and perhaps glacial melting. On the other hand, increased evaporation could lead to increased rain and snow cover, the latter perhaps causing a net loss of heat because of the increased albedo of snow-covered areas (Kukla and Kukla, 1974).

ACID RAIN

Hydrogeochemists tend to interest themselves in studies of the chemical weathering of rocks and in estimations of the annual contributions of dissolved materials to the sea by land drainage. These estimates must be continuously updated, however, as industrial development and changes in fuel use bring changes in the regional characteristics of rainfall (Drozodova and Makhon'ko, 1970; Gorham, 1955, 1961; Granat, 1972). Likens and Bormann (1974), in a discussion of the chemistry of rain in the north-eastern United States, compiled data on changes in rain chemistry in the last 50 years. Interestingly, they found a lower sulphate content in the rain since 1950, concomitant with a change-over from the use of wood and coal as fuel for heating homes to natural gas. They suggested that the increased acidity of the rain was due to the mounting numbers of taller smokestacks (fitted with precipitators to remove the larger particles), that inject sulphur dioxide into the atmosphere at heights of 60–360 m thus causing long-range dispersion of sulphur dioxide and sulphuric acid.

In Sweden, concern over continuous acidification of lakes in the south-western part of the country and the severe damage to fish led to establishing a joint Scandinavian project led by Nordforsk (Scandinavian Council for Applied Research) in 1970 to study the problem (Brosset, 1973). The project was later expanded to include much of western Europe under the auspices of the Organization for Economic Co-operation and Development (OECD) because acid rain is a regional problem rather than local, and sulphate can be carried great distances from its source. Estimates of the distance range from 100 to more than 1,000 km.

A survey carried out between 1970 and 1972 in the Swedish west-coast region showed half the 3,000 lakes in the area had pH values below 6. Many have pH values less than 5 and the sulphate content of the water has risen. Undesirable changes are taking place in fish and plankton populations and some species are disappearing.

In Canada, in the neighbourhood of Sudbury, Ontario, where large nickel and copper sulphide mines and smelters are operated, the potential for acidic, metal-rich rain exists. Kramer (1973) compared the chemistry of atmospheric precipitation in northern Ontario, far northern Ontario snow, and continental United States, with respect to heavy metals, sulphate, phosphate, pH and several other parameters. He concluded that ambient values of cadmium, lead, zinc, sulphate and chloride in northern Ontario were 'inherited' from the United States, that is, they had similar concentrations. He found anomalously high concentrations of copper, iron, nickel, cadmium and sulphate near the smelting plants of Sudbury. He also pointed out the deleterious effects of acid rain in areas where the rocks did not have a neutralizing capability, e.g. pure orthoquartzites such as the Lorain quartzite south and west of Sudbury, and predicted that acid lakes would be found there.

Shiomi and Kuntz (1973) recently reported data from a two-year study of rain as bulk precipitation from stations around Lake Ontario. (The samplers were left open so that, to some extent, dust was collected as well; cumulative samples were taken each month.) They concluded that bulk precipitation was a significant source to the lake of nitrogen, phosphorus, lead and

zinc, while for other materials, e.g. major ions and other heavy metals, it was not. Lake Ontario is at the lower end of a chain of Great Lakes and its chemical character is quite well established, but rain with the chemical content described by Shiomi and Kuntz could have a significant effect on other lakes. Their volume-weighted mean concentrations for a mid-lake station (Main Duck Island) are shown in Table 1.

The concentrations of some of these parameters are affected by urbanization. Near built-up areas and highways, higher concentrations of calcium, sodium and chloride (used to de-ice roads) and of nitrate and lead were found.

An interesting study on the effects of atmospheric precipitation on stream chemistry was reported by Fisher (1968). Sufficient data were available for a mass balance to be computed of the chemical content of the rain over the area and the chemical content of the stream discharge from the area. He found, for example, that the sulphate and nitrate in the rain accounted for the total amount of these anions in the streams, suggesting a steady-state condition with respect to addition and loss of these ions in the river basin. Chloride carried by the streams was considerably more than that provided by rain and he presented an approximate calculation suggesting that human use of refined salt in the area might account for this higher chloride content of the streams. The cation load of the streams was *greater* than that of the rain; however, as rain is acidic and the streams were near neutral pH, it is likely that the higher content of calcium, magnesium and potassium in the streams reflected the end-product of the weathering processes that consumed the excess hydrogen ion content of the rain.

TABLE 1. Volume-weighted means using Main Duck Island data

Parameter	Volume-weighted mean concentration (mg l⁻¹)
Total phosphorus as P	0.144
Reactive orthophosphate as P	0.066
Nitrate + nitrite as N	0.770
Ammonia as N	0.712
Sodium	1.66
Potassium	0.61
Calcium	3.34
Magnesium	0.44
Chloride	0.63
Sulphate	5.74
Lead	0.007
Iron	0.028
Zinc	0.096
Copper	0.008
Cadmium	0.002
Nickel	0.004

Source: Shiomi and Kuntz, 1973.
© International Association of Great Lakes Research. Reprinted with permission.

SNOW CHEMISTRY

Relatively few analyses of snow have been made (Jonasson and Allan, 1973), probably because of the difficulty of obtaining representative and uncontaminated samples. Kramer, in an unpublished report to the Ontario Ministry of the Environment, compared the chemistry of far northern Ontario snow (49.4°–52.9° N. latitude, 84.6°–93.3° W. longitude) with that of snow in the area of Sudbury, Ontario, an intense mining and smelting centre. The northern snow was low in dissolved solids, specific conductance ranged from 4 to 30 μmho/cm and the common ions Na^+, K^+ and Ca^{2+} ranged from 10 to 30 p.p.b. with occasionally higher values; the heavy metal content was also low. The snow around Sudbury, on the other hand, was enriched in heavy metals, sometimes approaching the p.p.m. level. For purposes of fair estimation, Kramer rejected extreme values and presented this comparison of total metal concentrations (units = mg l⁻¹):[1]

1. Values were determined by atomic absorption spectrophotometry on samples that had been digested with nitric acid.

Species:	Cd	Cu	Fe	Pb	Ni	Zn
Northern Ontario snow:	0.0002	0.010	0.020	0.010	0.020	0.005
Sudbury region:	0.0010	0.020	0.200	0.025	0.100	0.050

CHEMICAL WEATHERING

Water is an important agent of chemical weathering and is also an important means of transport of dissolved and particulate matter from the continents to the sea. The rate and degree of chemical weathering is strongly influenced by rainfall, stream flow, climate, rock and soil type and by land use and land-management practice. The ground rock of a region may affect the kind and amount of soil and, in turn, the kind and amount of vegetative cover; equally, the characteristics of chemical weathering will be affected. Cleaves *et al.* (1974) studied weathering processes in an area underlain in part by serpentinite and in part by schist. The lack of certain minerals in the serpentinite precluded the formation of a saprolite cover and consequently the serpentinite watershed had increased flood-flow discharge, greater fluctuation in seasonal, instantaneous base-flow discharge and pronounced seasonal fluctuations in total discharge. The watershed underlain by schist, however, had a thick saprolite cover and hence a reduced rate of chemical weathering.

In urban areas where much of the land surface is paved and most of the rainfall will quickly run off (Lindh, 1972), chemical weathering will be minimal but chemicals will be added to the run-off waters from urban debris, such as oil and metal. In an experimental watershed-ecosystem study area in New Hampshire in the United States, the cutting down of trees and inhibition of vegetation regrowth was found to produce large changes in stream flow and in the chemical composition of streams (Likens *et al.*, 1970). Especially large increases in nitrate concentrations were seen (nitrate is undesirable in drinking waters but is an important nutrient that is conserved and re-used in a steady state ecosystem).

Factors controlling the chemistry of the Earth's surface waters have been discussed by Gibbs (1970). He compared data for the major rivers and lakes of the world and concluded that three factors determined the chemistry of a particular water body: atmospheric precipitation, rock dominance, and the evaporation–crystallization process. The tropical rivers of Africa and South America, having sources in thoroughly leached areas of low relief, obtain their chemical constituents chiefly from atmospheric precipitation. The Amazon River has its source in the Andes Mountains and therefore its water chemistry is dominated by water–rock interactions.

Stable isotope geochemistry

Many elements are mixtures of two or more isotopes (Table 2). The extra-nuclear structures of the isotopes of an element are identical, giving rise to nearly identical chemical behaviour. In reactions or physical changes where mass is important, however, 'isotopic fractionation' may occur. These mass difference are most pronounced among lighter elements.

Variations have been observed in the natural abundance ratio of the stable isotopes of hydrogen, helium, boron, carbon, nitrogen, oxygen, silicon and sulphur. In the hydrosphere, variations have been found in the $^{18}O/^{16}O$ and $^{2}H/^{1}H$ (or as is usually written, D/H) ratios of water as well as in the isotopic compositions of the various dissolved and gaseous species of C, O, and S.

ANALYSIS PROCEDURES

Gas-source mass spectrometers are used to measure $^{18}O/^{16}O$, $^{13}C/^{12}C$, $^{34}S/^{32}S$, and D/H ratios. Isotopic ratios are always given in terms of difference from a standard, by means of the 'delta' notation. For example,

$$\delta^{18}O = \left(\frac{(^{18}O/^{16}O)_{unk}}{(^{18}O/^{16}O)_{std}} - \right) 1 \times 1,000$$

where *unk* and *std* refer respectively to the unknown

TABLE 2. Data for certain stable isotopes

Atomic No.	Name	Symbol	Mass. No.	Abundance (in percentage)
1	Hydrogen	H	1	99.985
	(Deuterium)	D	2	0.015
2	Helium	He	3	0.00013
			4	~100
5	Boron	B	10	18.8
			11	81.2
6	Carbon	C	12	98.89
			13	1.11
7	Nitrogen	N	14	99.63
			15	0.37
8	Oxygen	O	16	99.759
			17	0.037
			18	0.204
14	Silicon	Si	28	92.17
			29	4.71
			30	3.12
16	Sulphur	S	32	95.0
			33	0.75
			34	4.2

and to a standard substance. The units of measurement are thus given in per mille deviation of the unknown from the standard; for example, a sample with an $^{18}O/^{16}O$ ratio 10 per mille lower than the $^{18}O/^{16}O$ ratio of the standard would be described as having $\delta O = -10$ per mille. The conventional standards are as follows: $^{18}O/^{16}O$ and D/H: Standard Mean Ocean Water (SMOW);[1] $^{34}S/^{32}S$: Canon Diablo Troilite (CDT);[2] $^{13}C/^{12}C$: Pee-Dee belemnite (PDB)[1] (This is also used as an $^{18}O/^{16}O$ standard for palaeotemperature studies); $^{15}N/^{14}N$: atmospheric nitrogen.

The same mass spectrometer can be used for analysis of isotopic variations in C and O (as CO_2) and in S (as SO_2); a modified instrument is needed for D/H ratio measurements. Readily obtainable precision for $\delta^{18}O$ is ± 0.1 per mille; for $\delta^{34}S$, ± 0.2 per mille; for δD, ± 0.5 per mille. Procedures for analysing water, and dissolved salts have been described in various references (Epstein and Mayeda, 1953; Hoefs, 1973; O'Neil and Epstein, 1966).

NATURAL NON-MARINE WATERS

Both oxygen and hydrogen are isotopically fractionated during evaporation, freezing and other physical or chemical processes. The temperature dependence of the fractionation factors for oxygen and hydrogen are similar, so that $\delta^{18}O$ and δD are linearly related for any meteoric water (since both are derived almost entirely from the same marine reservoir). This has been experimentally demonstrated by Dansgaard (1964) and by Craig and Gordon (1965).

Both ^{18}O and D contents of meteoric water change during the course of precipitation so that natural variations in isotope ratios can be used as isotopic tracers. During rain or snow storms, precipitation becomes progressively lighter, i.e. more deficient in the heavy isotopes of oxygen and hydrogen, producing changes of up to 20 per mille in $\delta^{18}O$ and 200 per mille in δD in meteoric waters in temperate zones. Thus, water masses may be identified by their positions on a $D-^{18}O$ correlation graph (Fig. 1). Both δD and $\delta^{18}O$ tend to increase with rising temperature so that seasonal fluctuations in the precipitation can occur and can be used to monitor the origin in time of a given water mass.

Progressive D and ^{18}O depletion of rain clouds as they migrate across the continents results in a characteristic geographic distribution of δD and $\delta^{18}O$ in precipitation. Rain and snow-melt water inputs into ground- and surface-water systems can thus be geographically identified in some cases. In general, $\delta^{18}O$ and δD tend to decrease away from the oceans, towards the poles, and with increase in elevation of the site of precipitation. Glacial ice is the most ^{18}O, D depleted water reservoir. In polar region $\delta^{18}O$ and δD decrease linearly with decreasing temperature as measured by ground temperature at the site of precipitation. Seasonal isotopic variations in glacial ice accumulation have been recognized in ice cores penetrating to mid-Wisconsin ice deposits.

Ground waters tend to exchange oxygen—but not hydrogen—with their aquifer, so that they are shifted slightly off the O–D correlation for meteoric water. Thus ground-water contributions to surface-water masses can be identified. Isotopic studies have shown that meteoric water is the principal source of water in hot springs, oil field brines and in hydrothermal waters responsible for ore deposition (Clayton et al., 1966). 'Juvenile' water, i.e. of deep mantle derivation, is found always to be diluted with large excesses of recycled meteoric water, though its isotopic effects can be recognized in hydrous minerals of some deep-seated ore deposits.

Waters which have undergone evaporation follow a different $\delta^{18}O - \delta D$ path from normal meteoric waters and thus lakes with long residence times can be recognized (Fontes et al., 1970; Gat, 1970). Both heavy isotopes are enriched in the residual liquid and therefore $\delta^{18}O$ and δD increase with increasing salinity (or conductivity).

DISSOLVED SPECIES

Dissolved ions and molecules involving H, C, O and S exhibit isotopic ratio variations attributable either to the isotopic composition of their source or to subsequent isotopic fractionation effects.

The various dissolved species of carbon may vary in $\delta^{13}C$ by as much as 25 per mille as a result of two types of processes: (a) inorganic isotopic fractionation between the dissolved species or with respect to atmospheric CO_2; (b) organic isotope effects enriching ^{12}C in organic matter with respect to dissolved HCO_3^- or atmospheric CO_2. The first set of effects are dependent on temperature and pH. The second effect provides a tracer of organic activity through the ^{12}C-labelling of HCO_3^- derived from oxidized organic matter. For example, in lake waters, equilibration of HCO_3^- ions with the atmosphere occurs near the surface, while deeper waters tend to contain 'light' (^{12}C-rich) carbon from oxidation of organic matter. Seasonal variations in $\delta^{13}C$ occur, allowing recognition of transfer of carbon by diffusive and convective processes (Deevey and Stuiver, 1964).

Carbon and oxygen isotope ratios in ground waters carried through aquifers in carbonate terrains exhibit isotopic effects indicative of exchange between carbonate host and dissolved C and O species (Sackett and Moore, 1966). These effects may help indicate residence times and pathways in ground waters. Other possible isotopic tracers of ground- and surface-water movement include D/H ratios and tritium (^3H) content (Brown, 1970) and $^{234}U/^{238}U$ ratios (Kronfeld, 1971; Osmond et al., 1968).

1. Craig, 1957; 1961a, b.
2. Thode et al., 1961.

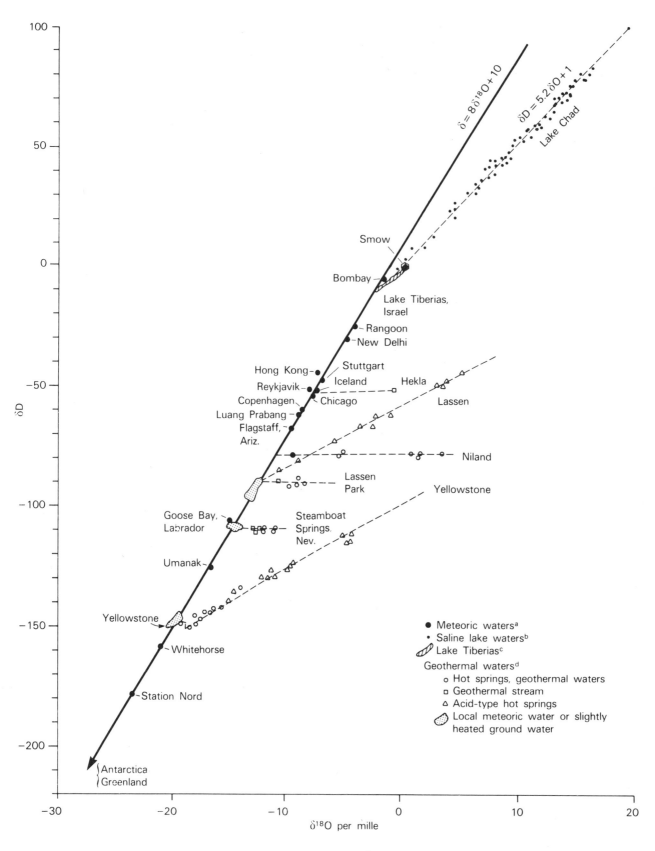

Sulphur and oxygen isotope ratios of sulphate dissolved in lake and river waters are indicators of its sources via: (a) natural oxidation of metal sulphides; (b) leaching of evaporites; (c) rain-borne sea-water sulphate; and (d) pollution by atmospheric SO_2 or anthropogenic waste sulphate solutions. Each source has a characteristic range of $\delta^{18}O$ and $\delta^{34}S$ values and its influence can thus be traced in the water. Sulphate may be reduced to HS^- by the action of *Desulphovibrio* and other bacteria, resulting in enrichment of ^{34}S in the residual sulphate. The aerobic oxidation of the HS^- results in contribution of a ^{32}S-rich sulphate which may locally alter the $\delta^{34}S$ of dissolved sulphate.

Helium isotopes can be used as a tracer of volcanogenic or other deep-seated sources of helium ground waters. The ^4He is derived from α-decay of U and Th. The ^3He is outgassed from the deep Earth's interior and is locally enriched relative to atmospheric $^3He/^4He$ ratios, indicating sites of outgassing of the Earth (Clarke and Kugler, 1973).

Remote sensing[1]

New developments in optical, radio, space and computer technologies have given mankind the ability to make large-scale studies of the radiation properties of Earth's surface. The term 'remote sensing' is acquiring a specific meaning, referring to instrumented observations and measurements of Earth's surface from aeroplanes, rockets and satellites. Remote sensing is a progressing science, with visible widespread applications appearing during the 1960s. At the present stage of development, this science remains strongly dependent on supportive field studies, so-called 'ground-truth' verifications.

The part of the electromagnetic spectrum used in remote sensing ranges from the ultraviolet (0.3 μm), through the visible and infra-red to the micro- and radio-wave region (>1 mm). Certain remote-sensing studies focus in the region of gamma radiation ($\sim 10^{-11}$ m).

FIG. 1. Synoptic graph summarizing some $\delta D - \delta^{18}O$ variations observed in non-marine waters. Meteoric waters lie on line $\delta D = 8\delta^{18}O + 10$, as do cold ground waters. ^{18}O and D are depleted with increasing latitude, and distance from sea. Hot, neutral-pH waters and geothermal steam are enriched in ^{18}O by exchange with rock (for most rock types, $5 < \delta^{18}O$ (rock) < 30 per mille). Acid, hot-spring waters are displaced in both δD and $\delta^{18}O$ from meteoric water source. Lake Chad is enriched in ^{18}O by evaporation. Lake Tiberias water reflects both seasonal (temperature dependent) and evaporative variations in δD and $\delta^{18}O$. (from: a. Dansgaard, 1964; b. Fontes, 1970; c. Gat, 1970; d. Craig, 1961a, b).

The chief source of light to the Earth is the sun, its radiation ranging from the ultraviolet region (0.3 μm) up into the infra-red region (3–4 μm), with the most intensive part around the green band (0.5 μm) of visible light. Earth's atmosphere, however, is opaque to much of this radiation. The air is transparent to radiation around the visible region from the ultraviolet (0.3 μm) up into the near infra-red (1.2 μm), in several bands in the infra-red, the most important being at 3–5.5 μm and 8–14 μm, and in the micro- and radio-wave region (>1 mm).

The radiations observed in these spectral regions are of two kinds: reflected solar radiation and thermal radiation from the target itself. All substances radiate thermal energy in the form of electromagnetic waves. The temperature range at Earth's surface is from, say, $-40°$ to $+50°C$, and the intensity of radiation from bodies at Earth's surface is therefore much less than that of the sun, which has a surface temperature of about 6,000 °K. Different substances, however, reflect and emit radiation in various amounts; this, in many cases, can be used to characterize the particular object.

The instruments used for remote sensing include ordinary photographic cameras, television cameras, multispectral scanners, various spectrometers, laser and radar devices and microwave detectors.

The launching of ERTS-I on 23 July 1972, opened a new era in the viewing of Earth from space (Maugh, 1973a, b). One of two planned for launching by the United States National Aeronautics and Space Administration, ERTS-I, has so far provided significantly greater capabilities for the inventory of land use, mapping water resources and water pollution, and prospecting for oil and minerals than had been anticipated. The satellite is equipped with a pair of imaging systems called the return beam vidicon (RBV) and the multispectral scanner (MSS), that scan the Earth in the visible and near infra-red, and a data-collection system (DCS) that relays telemetered information from ground stations in remote regions of North America. ERTS-I revolves around the Earth in a circular orbit 914 km above sea-level every 103 minutes. The orbit and coverage are designed to provide systematic and repetitive global coverage. The satellite crosses the equator at about 9.30 a.m. local time on the north-to-south leg of each orbit, a time of day when the shadow angles are most useful. During each orbit the satellite photographs strips of Earth's surface 185 km wide; 14 orbits each day cover the world. The next day's path overlaps the previous by 14 per cent, and every 18 days ERTS-I follows an identical orbit, at the same time of day.

Because of electronic difficulties for much of the time, the RBV has been shut down and only the MSS has been transmitting. The MSS is a line-scanning device that operates in three bands of the visible spectrum

1. An excellent, though brief, review of the basic principles of remote sensing by Dyring (1973) is the source of most of the information summarized below.

and one in the near infra-red, in the regions between 0.5 and 0.6 μm, 0.6 and 0.7 μm, 0.7 and 0.8 μm and 0.8 and 1.1 μm. Each image covers an area of about 185 km square. The black and white images are enhanced by assigning a different colour to each of the four MSS bands, typically blue and green for the lower wavelengths and red for the higher. Figure 2 shows such a false colour image of a region of southern Ontario–northern United States. Similar images are being used to show effluent plumes from industrial discharges into lakes and algal blooms in over-enriched lakes.

Additional uses of ERTS images are in mapping, observation of land uses such as irrigation, flooding, snow cover and glaciers (Link and Shindala, 1973; Pluhowski, 1972; Salomonson and Rango, 1974; Schmer *et al.*, 1972; Staelin *et al.*, 1973; Stoertz and Ericksen, 1972).

ERTS possesses a third instrument system that relays telemetered data from ground-based radios, known as data collection platforms (DCP), to NASA ground communication sites. The DCS can receive data from a swath that is more than 4,000 km wide as ERTS passes overhead, and can also provide data from the dark side of the Earth.

The United States Geological Survey (Paulson, 1974) is conducting an experimental study in the Delaware River basin. Twenty DCPs are being installed and interfaced with water resources stations, which include stream-gauging stations, ground-water observation wells and water quality monitors. The DCS provides data twice a day throughout the ERTS 17-day cycle and provides ground truth for image interpretation during imaging passes.

The Earth-resources data-acquisition systems on ERTS provide data with great potential for resources management. As it is possible under operational conditions for the data flow to be in real time, the value of the system when judgements must be made with respect to flood control, water diversions, etc., is obvious.

Water pollution

Pollution of water bodies due to industrial and municipal discharges of wastes (including heavy metals) has had tragic consequences in several parts of the world. The now well-documented mercury poisoning at Minamata in Japan, and the finding of high levels of mercury in fish in fresh-water lakes in Sweden, and in Lake Saint Clair, Lake Erie and in other fresh-water bodies in North America, has stimulated much research into the chemical and biological changes constituting the pathways in nature along which this metal moves. The research has been not only geochemical but has also included development by analytical chemists of sophisticated techniques to determine trace amounts of mercury in its various chemical forms. The elucidation of the bacteria-induced changes of chemical forms of mercury was done by microbiologists, and studies of the routes and pathways through the food chain were carried out by biologists.

It had been thought that metallic mercury was a relatively innocuous species when discharged into a water-way; being heavy it was supposed to settle into the bottom and be quickly buried in an inert form. But high contents of mercury in fish showed that mercury was somehow getting into the food chain.

Environmental pollution by mercury has recently been reviewed by Wood (1971, 1974), who has also done much research on the microbial transformations of mercury. It is now known that bacteria are capable of transforming metallic mercury and its higher valency states, Hg_2^{2+} and Hg^{2+}, to methyl mercury and dimethyl

FIG. 2. False-colour composite photograph compiled from ERTS bands 5, 6 and 7 for 20 March 1974, in the region between 42° 55′ and 43° N and 78° 43′ and 79° 43′ W, showing western Lake Ontario, the Niagara peninsula, and eastern Lake Erie. The assigned colours provide a clear contrast between water masses and land. Clouds appear as white and cast shadows. Snow cover appears as white, and urban areas as blue and brown. Sediment plumes around Long Point, Lake Erie, and off the Niagara River in Lake Ontario and other longshore sediment movements are clearly shown in shades of red. Ice piled up at the extreme east end of Lake Erie near Buffalo, New York, shows as a white mass, shaded with red showing surface melting. At the western end of Lake Ontario on the sand spit bounding Hamilton Harbor, a small square can be seen showing the man-made land that is the site of the Canada Centre for Inland Waters. The spacial resolution is 80 m. (*Photo*: Remote Sensing Section, CCIW.)

20MAR74 C N43-00/W079-48 N N42-55/W079-43 MSS-7 -D SUN EL40 A2141 194-8435-P-2-A-P-1L CCRS E-1605-15303-7

NOT PRECISION PROCESSED

POSITION ERROR 10.00KM

TRACK 9B   FC 1930 NTS 30M
IMAGE DATA CREATED 22MAR74

235-■567■

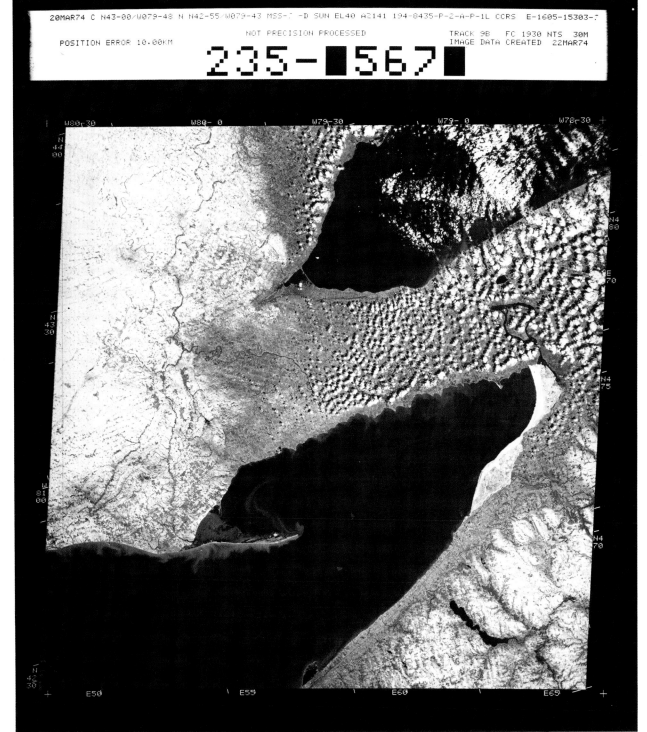

mercury under both aerobic and anaerobic conditions. Other micro-organisms can reduce methyl mercury to metallic mercury liberating methane. So no matter in which form mercury is discharged to a water body, highly toxic forms will soon be present; also, a certain proportion will enter the food chain and accumulate in fish, mostly as the highly toxic, methyl mercury.

In April 1970, a 'mercury alert' was called in Canada and the United States, when fish in many of the Laurentian Great Lakes were found to contain high levels of mercury. Fish containing more than 0.5 p.p.m. of mercury are considered unsafe for human consumption—especially if fish is a main part of the diet. The sediments too of the Saint Clair River (which drains Lake Huron) and Lake Saint Clair were found to contain high levels of mercury (Thomas, 1974). It has been estimated that as much as 90,700 kg of mercury have been deposited in the Saint Clair River system from industrial and municipal discharges in the last few decades. The present loading of mercury in the fish population is of the order of 45 kg. Obviously the potential exists for a very long-term contamination of fish in that area.

Downstream transportation of the mercury-polluted sediment is to be expected. Lake Saint Clair drains into the Detroit River which discharges into the western basin of Lake Erie. The commercial taking of certain species of fish from these waters has been banned since 1970 and since, as yet, there is little evidence of decreased concentrations of mercury in the fish, the ban may stay in effect for a considerable length of time.

An experimental project is now under way to see if the contaminated sediments can be removed by dredging and successfully confined behind a dyke out of the ecosystem. It will be several years before the results can be evaluated.

Wood has recently reviewed the problem of toxic elements in the environment, pointing out how both the inorganic chemistry and the microbiological synthesis of metal-organic compounds (as well as microbiological breakdowns of such compounds) must be considered. Also, metals that can be methylated, and that are employed in ways that aid their dispersion—such as platinum and palladium catalysts and other heavy metal alkyls that may be used in place of lead in gasoline—will have to be monitored carefully.

NITRILOTRIACETIC ACID

At about the same time that concern about environmental pollutants such as mercury arose, concern over various persistent pesticides and herbicides was beginning and because polyphosphates in laundry detergent were increasingly being indicated as pollutants in water bodies (causing over-production of algae and aquatic plants), the detergent industry proposed to substitute a chelating agent, nitrilotriacetic acid (NTA), for some or all of the polyphosphate in their products.

The industry had already carried out a number of tests of NTA's direct toxicity to animals and fish and believed it to be safe for general distribution. Geochemists and limnologists were not so convinced. The geochemists noted NTA's strong affinity for heavy metals, as compared with sodium tripolyphosphate (STP) and, as biodegradation of NTA had not yet been demonstrated, concern was expressed that NTA might persist in receiving water. This might facilitate the mobilization and transport of heavy metals with unknown but possibly deleterious consequences to the environment, and perhaps to human health.

These concerns, and the imminent introduction of widespread use of NTA, stimulated a great deal of research in Canada (Rudd and Hamilton, 1972), the United States and Sweden. A calculation by Childs (1971) showed that, such was the affinity of NTA for the cupric ion, virtually all of the NTA in a fresh-water body would be present as Cu-NTA until the molar concentration of NTA actually exceeded that of cupric ion. Natural waters have a certain chelating or complexing ability derived from naturally occurring organic compounds such as humic acids. A technique to measure this chelating capacity was developed by Chan (1971) and Chau and Lum-Shue-Chan (1974). This technique, when applied in 1972 at Hamilton Harbor showed that, on a molar basis, the complexing capacity of the water far exceeded that of the NTA in the water. It was decided to continue to monitor NTA in Hamilton Harbor, and to begin monitoring dissolved copper only when the complexing capacity of the NTA approached the base line complexing capacity of the harbour water. This has yet to occur.

A bacterial mutant that, in the absence of other sources of organic carbon degraded NTA rapidly was cultured and isolated by Wong et al. (1972). Apparently similar mutations must have occurred in nature because a continuing monitoring programme provides no evidence of any accumulations of NTA in lakes, harbours, rivers and ground waters.

PESTICIDES AND ORGANIC WASTE DISCHARGES

The publication, in 1963, of Rachel Carson's book, *Silent Spring,* brought to the general public for the first time good scientific information on the problems and limitations of over-dependence on chemical pesticides. DDT,[1] when its pesticidal properties were first discovered, was hailed as a 'wonder' chemical. At low levels it controlled undesirable insects and saved millions of lives by eradicating malaria.

In the last two decades, however, a series of incidents, in which birds and fish were killed as well as the target insects, have occurred. Many unwanted effects

---

1. *Dichlorodiphenyltrichloroethane.*

were found to be produced if water was DDT-contaminated. Fish are particularly susceptible to DDT poisoning since the immense volume of water that passes through their gills makes possible the cumulative extraction of large amounts of DDT, even though itself present in the water at very low concentrations. Also, the insects and other invertebrates that are the main food for some species of fish, may (a) accumulate the pesticide and die—so that fish starve, or (b) survive—and so contaminate the predator fish.

It has further been found that many insects have developed resistance to DDT; thus continued applications have less and less effect.

As it was becoming realized that DDT and similar chlorinated hydrocarbon pesticides could be concentrated in food chains, other industrial chemicals released in the environment, not necessarily used in pesticides, were also found to be accumulating in fish, birds and man (Lichtenstein, 1972); for example, the group of chemicals called polychlorinated biphenyls (PCBs). In a very recent study, Kaiser (1974) found such a chlorinated hydrocarbon (used as a flame retardant under the name of 'Dechlorane' or as an insecticide called 'Mirex') in fish from Lake Ontario. Because the 'Mirex' peak occurs at the same position on the gas chromatography chart as a prominent PCB peak, the possibility exists that this chemical may be found to be even more widespread than is now known.

Rivers that receive discharge from industrial plants and from which water is taken for municipal use downstream, may be found to be contaminated with so-called 'biorefractory' chemicals that are not removed by conventional water-treatment procedures. These are low-molecular weight chemicals of low volatility—chlorinated aliphatic and aromatic hydrocarbons that are not broken down by biological activity. Little is known at present of the long-term effects of drinking water contaminated with such chemicals although it is known that some of them create noxious tastes and odours (Miller, 1973).

It is clear that care will have to be taken in the future in the use and application of pesticides. The cleanliness of municipal water supplies should be considered when planning the location of industrial plants, and diversion from the natural drainage of heavily contaminated waste waters may be necessary.

A number of books have appeared in recent years on pesticides and on environmental contamination, and the reader is referred to them for further details (Graham, 1970; Mellanby, 1967; Muirhead-Thomson, 1971; Rudd, 1965).

Lake restoration

Four projects in lake restoration (Bjork, 1972)—three in Sweden and one in Tunisia—were undertaken by a team of limnologists at the University of Lund, Sweden. Their objectives were to develop methods of solving some anthropogenic environmental problems and, especially, to the restoration of certain lakes which they considered to be environmentally important (the project had the useful by-product of training young limnologists and ecologists in techniques of solving practical problems in lake management.) The approaches involved diverting or improving treatment of sewage, removal of over-enriched bottom sediments, artificial aeration of the bottom waters of a thermally stratified lake, mowing and harvesting of obnoxious, rooted vegetation. They stressed that each lake must be evaluated individually and given the most appropriate treatment. Their project, still under way, already shows marked improvement, in decreased nutrient content of the water, increased clarity and, at Lake Homborga, increased waterfowl populations.

Schindler (1974) has demonstrated that algal blooms can be induced in certain experimental lakes that are normally oligotrophic (poorly nourished) by additions of nitrogen, carbon *and phosphate,* and that the lake recovers promptly when additions of phosphorus alone are discontinued. He has concluded that the reduction of the phosphorus input to most lakes should produce a proportional abatement in phytoplankton blooms and other symptoms of eutrophication (over-nourishment).

LAKE ERIE—A LAKE
IN NEED OF RESTORATION

Erie, the smallest in volume and shallowest of the North American Laurentian Great Lakes, has been the one most subjected to municipal and industrial pollution pressures. Many articles in the press have described it as dead and unreclaimable, and while such alarmist claims are exaggerated, it must be admitted that over-enrichment of the supply of nutrient chemicals to that lake has had a number of undesirable consequences. Lake Erie is not 'dead', however; instead it might be said that it is too alive—the problem of depletion of dissolved oxygen in the bottom waters of the central basin in the summer is the result of over-abundant growth of algae in the surface waters. These algae bloom and, dying, sink to the bottom and decay. The process of decay uses up all of the dissolved oxygen in the bottom water.

For its location in the temperate climate of central North America, it might be said that Lake Erie (specifically its central basin) is either too shallow or too deep. Lake Erie may be divided into three basins; the westernmost is very shallow and extremely eutrophic. But, because it is so shallow, thermal stratification, if it forms, persists for only short periods of time, and although algal growth is lavish, dissolved oxygen is not used up before it is replenished from the atmosphere.

Thermal stratification. Lakes that are cold in the winter and warm in the summer may become thermally stratified because the surface waters warm much more rapidly than do the deeper waters. In time, the temperature difference may amount to 10–20 °C. The density of pure water varies with temperature as follows:

Temperature	Density (g/ml)
4°	1.00000
10°	0.99973
15°	0.99913
20°	0.99823
25°	0.99707

The temperature of the bottom waters in deep lakes tends to remain near 4 °C, the temperature of maximum density, and the surface water may warm up to 20 °C or higher. The warmer water is less dense and tends to float on the bottom water, depending on the density difference and degree of turbulence. A stable stratification exists in the central basin of Lake Erie when the bottom water temperature is about 10 °C and the surface-water temperature is about 20 °C.

A plot of the water temperature as a function of depth is shown in Figure 3.

Most of the bottom area of the central basin of Lake Erie is at a depth of 20–22 mm, and the thickness of the hypolimnion is only a few metres. If the basin were deeper the hypolimnion would be thicker and would begin the season of stratification with a larger oxygen reservoir; if it were shallower, stratification would probably be intermittent as in the western basin.

The eastern basin is sufficiently deep so that oxygen depletion does not penetrate low enough to become a problem.

The complete removal of dissolved oxygen from the hypolimnion of the central basin makes it impossible for cold-water fish species to survive; it also creates taste and odour problems in municipal water supplies, and facilitates regeneration of nutrients from the bottom sediments, exacerbating the over-enrichment problem.

In the summer of 1970, a joint project was carried out by scientists from the Canada Centre for Inland Waters and the United States Environmental Protection Agency, to make a quantitative study of phenomena as-

sociated with summer oxygen depletion in the central basin of Lake Erie (Burns and Ross, 1971). The approach was multidisciplinary; physicists, chemists, biologists and bacteriologists participated, and the survey used ships and launches, moored current metres, temperature sensors, wind gauges, and dissolved oxygen metres. An automatic underwater camera was set up to provide serial photographs of the bottom sediment. Some of the most interesting observations were made by scientists using scuba-diving techniques and underwater photographic equipment. They saw algae growing at the thermocline and falling through the hypolimnion onto the bottom sediments. They observed colour and texture changes in the sedimented algae as it matted down, and worm (*Oligochaeta*) tubes protruding through the mat; further afield the algae changed in colour from green to brown and the mat developed black patches. These changes were correlated with changes in the dissolved oxygen content of the water. The black patches represent reduced-oxygen regions, where as a result of reduction of ferric ion and sulphate, a precipitate of ferrous sulphide formed. Bacteriological studies confirmed the postulated changes as the dissolved oxygen was reduced to zero. Sharp decreases in the populations of nitrifying bacteria corresponded to the appearance of the black patches and of decreases in nitrate and increases in ammonia. Sulphur bacteria were so abundant at the sediment–water interface that they formed visible white clouds near the black patches.

The regeneration of nutrients from the sediments to the water was found to be much increased during anoxic conditions; therefore, the duration of anoxic conditions becomes of critical importance to the annual nutrient budget of the lake. The project leaders concluded that the input of phosphorus to Lake Erie must be reduced immediately (1972). In Canada, some reduction was achieved by legislation limiting the levels of phosphate allowable in laundry detergents, and by a programme of improving sewage treatment plants to include nutrient removal. In the United States, only a few local governments have limited phosphates in laundry detergents; the federal government's reliance has been on nutrient removal by improved sewage treatment.

The quantitative data and the public notice given to Project Hypo were both important factors in influencing the governments of the two countries to reach agreement on co-ordinated plans to improve the quality of the waters of the international Great Lakes (Canada–United States Agreement, 1972).

Desalination

Desalination offers hope to sea-shore communities in arid countries and to cities whose customary river- or ground-water supplies have become increasingly brackish in recent years. The technology has advanced consid-

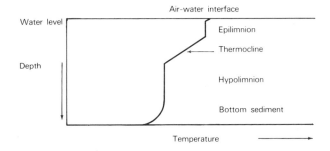

FIG. 3. Idealized plot of water temperature as a function of depth.

erably in the last few years, and in some areas high-quality water supplies are now becoming available from desalination plants at economically viable costs (Heckroth, 1974).

Desalting techniques may be considered under two general headings: those that utilize a phase-change—evaporation or freezing, and those that separate water and salt molecules within the liquid phase (Faber *et al.*, 1972). Among the phase-change procedures are multi-stage evaporation, vapour compression, vacuum freezing/vapour compression, and secondary-refrigerant freezing. These processes and their variants are best applied to sea water; the single phase systems are best applied to brackish waters.

The distillation process at Key West (Kennedy and Chillcott, 1972) uses the multistage flash technique. Sea water is pre-treated by pH control and two stages of degassing. The brine is circulated through forty-eight heat-recovery stages and two heat-reject stages. Each stage is at a progressively lower temperature and pressure, ranging from 21°C and pressures above atmospheric to 38°C and pressures somewhat below atmospheric. The major pumps are driven by steam turbines that exhaust to the brine heater to provide the evaporator heat input. The distillate is pumped out of the evaporator at 32°C with a total dissolved solids content of less than 100 mg l^{-1}. It is necessary to add about 15 mg l^{-1} of hardness to stabilize the product and chlorine is added before the water is pumped into the distribution system.

Freezing processes (Fraser and Gibson, 1972) are based on the principle that when a saline solution is frozen, the ice crystals that form are pure water. Once ice crystals are formed, the problem is to remove the adhering brine. Two types of processes have been developed; in one, the water vapour itself is the refrigerant a vacuum process; in the other, an immiscible refrigerant is mixed in direct contact with the feed water.

The more promising single phase processes are electrodialysis, ion exchange, and reverse osmosis. The electrodialysis process involves using electrical energy to drive ionized solids across semi-permeable membranes. The basic unit consists of an inlet water channel, two semi-permeable membranes and two electrodes. One electrode is negatively charged, allowing the passage of only positively charged ions, the other is positive and permits only negatively charged ions to pass. The inlet water must be nearly free of organic material, suspended solids, iron and manganese, and pre-treatment may be required. The recovery yield may be limited by the need to prevent precipitation of calcium sulphate but can approach 87 per cent of the brackish water supply.

The ion-exchange procedures make use of solid synthetic resins with chemically active exchange sites. In the softening process, calcium and magnesium ions are exchanged for sodium by a cation resin. In desalting processes, cations are exchanged for hydrogen ions by an acid resin; the effluent is passed over an anionic resin by which anions are exchanged for hydroxyl ions; the hydrogen and hydroxyl ions recombine to form water. Water recovery rates of about 90 per cent can be obtained when reducing the total dissolved solid content of brackish water from 1,250 mg l^{-1} to 500 mg l^{-1}. For saltier waters the recovery rate is less.

The tendency for water to move across a membrane (permeable only to water) from the less salty to the saltier solution is termed osmosis. The osmotic pressure is that pressure on the more concentrated solution that is sufficient to stop the diffusion of water into it. If an external force is used to apply a still greater pressure to the more concentrated solution, the direction of water diffusion is reversed, and the process is called reverse osmosis. Commercial development of reverse osmosis as a desalination process has depended on the development of membranes that could reject salts adequately and also produce suitable flux rates.

A simple reverse osmosis system (Lynch and Mintz, 1972) consists of the membrane, a structure to support the membrane, a vessel to contain the pressure and a pump to pressurize the saline water. The first reverse osmosis membrane, introduced about 1960, was a cellulose-acetate film, known as the Loeb type after its developer. This membrane has undergone extensive development but still retains the features of the original Loeb design, e.g. a dense layer or skin, a porous substrate, and cellulose-diacetate as the basic constituent. Other recent developments are the hollow fibre, the tubular and the spiral-wrap membrane.

Advantages of reverse osmosis processes are that corrosion is minimal because the systems are essentially non-metallic, operations are at low temperatures and capital and energy requirements are comparatively small.

As technology advances, the future looks bright for a number of desalination techniques.

Acknowledgements

I should like to thank Professor H. Schwarcz of McMaster University who wrote the section on 'Stable Isotope Geochemistry'. I also thank several colleagues at the Canada Centre for Inland Waters for assisting me in reviewing the literature by providing reprints and bibliographies in their areas of interest, notably M. T. Shiomi (rain chemistry), R. R. Weiler (CO$_2$ in air and lakes), K. Thomson (remote sensing) and K. Johnston (desalinization).

References

ALMER, B.; DICKSON, W.; EKSTROM, C.; HORNSTROM, E.; MILLER, U. 1974. Effects of Acidification on Swedish Lakes. *Ambio*, Vol. 3, No. 1, p. 30–6.

BEAMISH, R. J. 1974. Loss of Fish Populations from Unexploited Remote Lakes in Ontario, Canada, as a Consequence of Atmospheric Fallout of Acid. *Water Research*, Vol. 8, No. 1, p. 85–95.

BEAMISH, R. J.; HARVEY, H. H. 1972. Acidification of the La Cloche Mountain Lakes, Ontario and Resulting Fish Mortalities. *Jour. Fish Res. Board Can.*, Vol. 29, No. 8, p. 1131.

BJORK, S. 1972. Swedish Lake Restoration Program Gets Results. *Ambio*, Vol. 1, No. 5, p. 153–65.

BROSSET, C. 1973. Air-borne Acid. *Ambio*, Vol. 2, No. 2, p. 2–9.

BROWN, R. M. 1970. Distribution of Hydrogen Isotopes in Canadian Waters. *Isotope Hydrology*, p. 3–21. International Atomic Energy Agency, Vienna.

BRUCE, J. P. 1972. Meteorological Aspect of Great Lakes Pollution. In: *Selected Papers on Meteorology as Related to the Human Environment*, p. 127–32. (WMO Special Environmental Report, No. 2.)

BURNS, N. M.; ROSS, C. 1971. Project Hypo; A Description of an Intensive Study of the Lake Erie Central Basin Hypolimnion and Related Surface Water Phenomena. In: *Proceedings, 14th Conference on Great Lakes Research*, p. 740–2. Ann Arbor, International Association for Great Lakes Research.

Canada–United States Agreement on Water Quality in the Great Lakes. Signed by the President of the United States and the Canadian Prime Minister, Ottawa, 15 April 1972.

CARSON, R. 1963. *Silent Spring*. London, Hamish Hamilton.

CHARLSON, R. J.; VANDERPOL, A. H.; COVERT, D. S.; WAGGONER, A. P.; AHLQUIST, N. C. 1974. Sulphuric Acid—Ammonium Sulphate Aerosol: Optical Detection in the St. Louis Region. *Science*, Vol. 184, No. 4133, p. 156–8.

CHAU, Y. K. 1971. Determination of Trace Metals in Natural Waters. In: *Proceedings, International Symposium on Identification and Measurement of Environmental Pollutants*, Vol. 9, p. 354–7. Ottawa, National Research Council.

CHAU, Y. K.; LUM-SHUE-CHAN, K. 1974. Determination of Labile and Strongly Bound Metals in Lake Water. *Water Research*, Vol. 8, p. 383–8.

CHAU, Y. K.; SHIOMI, M. T. 1972. Complexing Properties of Nitrilotriacetic Acid in the Lake Environment. *Water, Air and Soil Pollution*, Vol. 1, p. 149–64.

CHILDS, C. W. 1971. Chemical Equilibrium Models for Lake Water which Contains Nitrilotriacetate and for 'normal' Lake Water. *Proceedings, 14th Conference on Great Lakes Research*, p. 198–210. Ann Arbor, Mich., International Association for Great Lakes Research.

CLARKE, W. B.; KUGLER, G. 1973. Dissolved Helium in Groundwater: A Possible Method for Uranium and Thorium Prospecting. *Econ. Geol.* Vol. 68, p. 243–51.

CLAYTON, R. N. 1971. Stable Isotope Geochemistry. *Trans. Amer. Geophys. Union.* IUGC, Vol. 52, p. 106.

CLAYTON, R. N.; FRIEDMAN, I.; GRAF, D. L.; MAYEDA, T. K.; MEENTS, W. F.; SHIMP, N. F. 1966. The Origin of Saline Formation Waters. 1. Isotopic Composition. *Jour. Geophys. Res.*, Vol. 71, p. 3869–82.

CLEAVES, E. T.; FISHER, D. W.; BRICKER, O. P. 1974. Chemical Weathering of Serpentinite in the Eastern Piedmont of Maryland. *Geol. Soc. America Bull.*, Vol. 85, p. 437–44.

CRAIG, H. C. 1957. Isotopic Standards for Carbon and Oxygen and Correction Factors for Mass Spectrometric Analysis of Carbon Dioxide. *Geochim. et Cosmochim. Acta*, Vol. 12, p. 133–49.

——. 1961a. Standard for Reporting Concentrations of Deuterium and Oxygen-18 in Natural Waters. *Science*, Vol. 133, p. 1833–4.

——. 1961b. Isotopic Variations in Meteoric Waters. *Science*, Vol. 133, p. 1702–3.

CRAIG, H. C.; GORDON, L. I. 1965. Isotopic Oceanography: Deuterium and Oxygen-18 Variations in the Ocean and Marine Atmosphere. *Proceedings Symposium Marine Geochemistry Univ. Rhode Island, Occ. Publ. 3*, p. 277–334.

DANSGAARD, W. 1964. Stable Isotopes in Precipitation. *Tellus*, Vol. 16, p. 436–69.

DEEVEY, E. S.; STUIVER, M. 1964. Distribution of Natural Isotopes of Carbon in Linsley Pond and Other New England Lakes. *Limnol and Oceanog.*, Vol. 9, p. 1–11.

DROZODOVA, V. M.; MAKHON'KO, E. P. 1970. Content of Trace Elements in Precipitation. *Jour. Geophys. Res.*, Vol. 75, p. 3610–12.

DYRING, E. 1973. The Principles of Remote Sensing. *Ambio*, Vol. 2, No. 3, p. 57–64.

EKDAHL, C. A.; KEELING, C. D. 1973. Atmospheric Carbon Dioxide and Radiocarbon in the Natural Carbon Cycle: 1. Quantitative Deductions from Records at Mauna Loa Observatory and at the South Pole. In: G. M. Woodwell and E. V. Pecan (eds.), *Carbon and the Biosphere*. Technical Information Center, Office of information Services, USAEC.

EPSTEIN, S.; MAYEDA, T. 1953. Variation in O^{18} Content of Waters from Natural Sources. *Geochim. et Cosmochim. Acta*, Vol. 4, p. 213–23.

FABER, H. A.; BRESLER, S. A.; WALTON, G. 1972. Improving Community Water Supplies with Desalting Technology. *Jour. Amer. Water Works Assoc.*, Vol. 64, No. 11, p. 705–10.

FISHER, D. W. 1968. Annual Variations in Chemical Composition of Atmospheric Precipitation, Eastern North Carolina and Southeastern Virginia. *U.S. Geol. Survey Water Supply Paper 1535 M*. Washington, D.C., United States Government Printing Office.

FONTES, J.-C.; GONFIANTINI, R.; ROCHE, M. A. 1970. Deuterium et Oxygène-18 dans les Eaux du Lac Tchad. In: *Isotope Hydrology*, p. 387–404.

FRASER, J. H.; GIBSON, W. E. 1972. Secondary-Refrigerant Desalination. *Jour. Amer. Water Works Assoc.*, Vol. 64, No. 11, p. 746–8.

GARRELS, R. M.; MACKENZIE, F. T. 1971. *Evolution of Sedimentary Rocks*. New York, W. W. Norton.

GAT, J. R. 1970. *Environmental Isotope Balance of Lake Tiberias*. p. 109–27, New York, W. W. Norton.

GIBBS, R. J. 1970. Mechanism Controlling World Water Chemistry. *Science*, Vol. 170, p. 1088–90.

GORHAM, E. 1955. On the Acidity and Salinity of Rain. *Geochim. et Cosmochim. Acta*, Vol. 7, p. 231–9.

——. 1961. Factors Influencing Supply of Major Ions to Inland Waters with Special Reference to the Atmosphere. *Geol. Soc. America Bull.*, Vol. 72, p. 795–840.

GRAHAM, F. Jr. 1970. *Since Silent Spring*. Boston, Houghton Mifflin.

GRANAT, L. 1972. On the Relation Between pH and the Chemical Composition in Atmospheric Precipitation. *Tellus*, Vol. 24, p. 550–60.

HECKROTH, C. W. 1974. Future of Desalting is Bright but Costs Must Drop. *Water and Wastes Engin.*, Vol. 11, No. 7, p. 20–3, 52–4.

HOEFS, J. 1973. *Stable Isotope Geochemistry*. Berlin, Springer-Verlag. 140 p.

JONASSON, I. R.; ALLAN, R. J. 1973. Snow: A Sampling Medium in Hydrogeochemical Prospecting in Temperate and Permafrost Regions. In: M. J. Jones (ed.), *Geochemical Exploration, 1972*. p. 161–76. London, Institution of Mining and Metallurgy.

KAISER, K. L. E. 1974. *Mirex: An Unrecognized Contaminant of Fishes from Lake Ontario.*

KENNEDY, J. R.; CHILLCOTT, L. H. 1972. Operating Expenses at Desalting Plants—Joint Discussion, Key West, Fla., Desalting Plant. *Jour. Amer. Water Works Assoc.*, Vol. 64, No. 11, p. 772–4.

KRAMER, J. R. 1973. Fate of Atmospheric Sulphur Dioxide and Related Substances as Indicated by Chemistry of Precipitation. *Final Report to Ministry of Environment, Canada.*

KRONFELD, J. 1971. Hydrologic Investigations and the Significance of $^{234}U/^{238}U$ Disequilibrium in the Ground Waters of Central Texas. In: *Development of Remote Methods for Obtaining Soil Information, Project THEMIS, Semi-annual Report to U.S. Army Engineers.* Houston, Texas, Department of Geology, Rice University.

KUKLA, G. J.; KUKLA, H. J. 1974. Increased Surface Albedo in the Northern Hemisphere. *Science,* Vol. 183, p. 709–14.

LICHTENSTEIN, E. P. 1972. Persistence and Fate of Pesticides in Soils, Water and Crops: Significance to Humans. In: *Fate of Pesticides in Environment.* London, Gordon & Breach Science Publishers.

LIKENS, G. E. 1972. The Chemistry of Precipitation in the Central Finger Lakes Region. *Cornell Univ. Water Resour. Mar. Sci. Cent. Tech. Rep. 50.*

LIKENS, G. E.; BORMANN, F. H. 1974. Acid Rain: A Serious Regional Environmental Problem. *Science,* Vol. 184, p. 1176–9.

LIKENS, G. E.; BORMANN, F. H.; JOHNSON, N. M. 1972. Acid Rain. *Environment,* Vol. 14, p. 33–40.

LIKENS, G. E.; BORMANN, F. H.; JOHNSON, N. M.; FISHER, D. W.; PIERCE, R. S. 1970. Effects of Forest Cutting and Herbicide Treatment on Nutrient Budgets in the Hubbard Brook Watershed Ecosystem. *Ecological Monographs,* Vol. 40, p. 23–47.

LINDH, G. 1972. Urbanization: A Hydrological Headache. *Ambio,* Vol. 1, No. 6, p. 185-201.

LINK, L. E.; SHINDALA, A. 1973. Utilization of Remote Sensing in River Basin Studies. *Water Resources Bull.*, Vol. 9, p. 907.

LYNCH, M. A. Jr.; MINTZ, M. S. 1972. Membrane and Ion-exchange Processes—A Review, *Jour. Amer. Water Works Assoc.*, Vol. 64, No. 11, p. 711–25.

MAUGH, T. H. 1973a. ERTS. Surveying Earth's Resources from Space. *Science,* Vol. 180, p. 49–51.

——. 1973b. ERTS (II): A New Way of Viewing the Earth. *Science,* Vol. 180, p. 171–3.

MELLANBY, K. 1967. Pesticides and Pollution. London, Collins.

Michigan Water Resources Commission Report. Heavy Metals in Surface Waters, Sediments, and Fish in Michigan, July 1972.

MILLER, S. S. 1973. Are you Drinking Biorefractories Too? *Envir. Sci. & Tech.*, Vol. 7, No. 1, p. 14–15.

MILLER, J. M.; DE PERRA, R. G. 1972. Contribution of Scavenged Sulphur Dioxide to the Sulfate Content of Rainwater. *Jour. Geophys. Res.*, Vol. 77, p. 5905–16.

MUIRHEAD-THOMSON, R. C. 1971. *Pesticides and Freshwater Fauna.* London and New York, Academic Press.

NACE, R. L. 1964. Water of the World. *Natural History,* Vol. 73, No. 1, p. 10–19.

O'NEIL, J. R.; EPSTEIN, S. 1966. A Method for Oxygen Isotope Analyses of Milligram Quantities of Water and Some of its Applications. *Jour. Geophys. Res.,* Vol. 71, p. 4955–61.

OSMOND, J. K.; RYDELL, H. S.; KAUFMAN, M. I. 1968. Uranium Disequilibrium in Groundwater: An Isotope Dilution Approach in Hydrologic Investigations. *Science,* Vol. 162, p. 997–9.

PAULSON, R. W. 1974. The Use of ERTS-1 for Relaying Hydrologic Data in the Delaware River Basin. *Jour. Amer. Water Works Assoc.,* Vol. 66, No. 5, p. 301–5.

PEIRSON, D. H.; CAWSE, P. A.; SALMON, L.; CAMBRAY, R. S. 1973. Trace Elements in the Atmospheric Environment. *Nature,* Vol. 241, p. 252–6.

PLUHOWSKI, E. J. 1972. Hydrologic Interpretations Based on Infrared Imagery of Long Island, New York. 200 p. United States Geological Survey Water Supply Paper 2009-B.)

RODHE, H. 1972. A Study of the Sulphur Budget over Northern Europe. *Tellus,* Vol. 24, p. 128–35.

RUDD, R. L. 1965. *Pesticides and the Living Landscape.* London, Faber & Faber.

RUDD, J. W. M.; HAMILTON, R. D. 1972. Biodegradation of Trisodium Nitrilotriacetate in a Model Aerated Sewage Lagoon. *Jour. of the Fisheries Research Board of Canada,* Vol. 29, No. 8, p. 1203–8.

SACKETT, W. M.; MOORE, W. S. 1966. Isotopic Variations of Dissolved Inorganic Carbon. *Chem. Geol.*, Vol. 1, p. 323–8.

SALOMONSON, V. V.; RANGO, A. 1974. ERTS-1. Applications in Hydrology and Water Ressources. *Jour. Amer. Water Works Assoc.,* Vol. 66, No. 3, p. 168–72.

SCHINDLER, D. W. 1974. Eutrophication and Recovery in Experimental Lakes: Implications for Lake Management. *Science,* Vol. 184, p. 897–9.

SCHMER, F. A.; TIPTON, M. J.; RYLAND, D. W.; HAYDEN, J.; BEAVER, G. 1972. Investigation of Lake Water Quality in Eastern South Dakota with Remote Sensing Techniques. In: *Proceedings of the 8th International Symposium on Remote Sensing of Environment* (Ann Arbor).

SHIOMI, M. T.; KUNTZ, K. W. 1973. Great Lakes Precipitation Chemistry: Part 1. Lake Ontario Basin. *Proceedings 16th Conference Great Lakes Research, International Association Great Lakes Research,* p. 581–602.

STAELIN, D. H.; BARRETT, A. H.; WATERS, J. W.; BARATH, F. T.; JOHNSTON, E. J.; ROSENKRANZ, P. W.; GANT, N. E.; LENOIR, W. W. 1973. Microwave Spectrometer on the Nimbus 5 Satellite: Meteorological and Geophysical Data. *Science,* Vol. 182, p. 1339–41.

STOERTZ, G. E.; ERICKSEN, G. E. 1972. Expected Results of Hydrologic and Geologic Studies Using ERTS Imagery on the Atacama Desert, Altiplano, and Puna de Atacama, South America (Southwest Bolivia, Northwest Argentina, and Northern Chile). In: *Proceedings of the 8th International Symposium on Remote Sensing of the Environment* (Ann Arbor).

THODE, H. G.; MONSTER, J.; DUNFORD, H. B. 1961. Sulphur-Isotope Geochemistry. *Geochim. et Cosmochim. Acta,* Vol. 25, p. 159–74.

THOMAS, R. L. 1974. The Distribution and Transport of Mercury in the Sediments of the Laurentian Great Lakes System. In: *Proceedings of International Conference on Transport of Persistent Chemicals in Aquatic Ecosystems* (Ottawa).

VAL'NIKOV, I. U. 1971a. Amount of Sulphur Depositing on the Earth's Surface with the Atmospheric Precipitation in Tatar S.S.R. *Biol. Nauki,* Vol. 14, p. 110–1.

——. 1971b. Sulphur Turnover and its Importance for Crop Production in Tatar A.S.S.R. *Pochvavedenie,* Vol. 3, p. 40–7.

VINOGRADOV, A. P. 1972. The Effects of Human Activity on the Atmosphere. *Geochemistry International,* Vol. 9, No. 11, p. 1–6.

WEILER, R. R. 1972. (Unpublished data.)

WOFSY, S. C.; MCCONNELL, J. C.; MCELROY, M. B. 1972. Atmospheric CH_4, CO and CO_2. *Jour. Geophys. Res.,* Vol. 77, No. 24, p. 4477–93.

WONG, P. T. S.; LIU, D.; DUTKA, B. J. 1972. Rapid Biodegradation of NTA by a Novel Bacterial Mutant. *Water Research,* Vol. 6, p. 1577–84.

WOOD, J. M. 1971. Environmental Pollution by Mercury. In: J. N. Pitts Jr. and R. L. Metcalf (eds.), *Advances in Environmental Science and Technology,* Vol. 2. New York, Wiley-Interscience.

——. 1974. Biological Cycles for Toxic Elements in the Environment. *Science,* Vol. 183, p. 1049–52.

Chemical aspects of coastal marine pollution

M. Grant Gross

Chesapeake Bay Institute,
The Johns Hopkins University,
Baltimore, Maryland (United States of America)

Michael L. Healy

Department of Oceanography,
University of Washington,
Seattle, Washington (United States of America)

Introduction

Pollutants from many sources flow into the ocean. Industrial waste and municipal sewage discharge reach the sea by stream and river. Other types of waste are transported in barges for offshore disposal or discharged directly into coastal water through outfalls. Rivers carry polluted materials from mines, farm operations, and land development activities to the ocean. Insecticides, toxic metals, radioactive and other polluting substances reach the ocean through runoff by air movements and by direct discharge from pipelines. Dredges remove wastes from harbours and barges dump them at sea. Severe coastal pollution problems are evident in many local areas and regional pollution problems have become obvious in recent years in the Mediterranean and Baltic Seas. Our purpose in this section is to examine the marine pollution likely to be experienced by developing and industrializing countries.

There is a general concern that build-up of certain materials in the oceans may be a future problem; petroleum hydrocarbons, radioactive waste, and synthetic organics have been identified as pollutants of worldwide significance. Goldberg (1976b) recently summarized our knowledge of metals and chlorinated hydrocarbons and their global distribution.

Because of the concern about radioactivity, several national and international organizations have reviewed the sources of radionuclides that reach the ocean, sketched their routes, major reservoirs and estimated their rates of movement through the world's oceans (National Academy of Sciences, 1971a). In fact, most of the summaries of marine pollution have dealt with pollution of the open ocean and paid much less attention to coastal problems.

Our attention will be focused primarily on those aspects of marine pollution that are most likely to be detected in coastal regions and to be problems in developing countries.

First we shall consider the characteristics of coastal and open ocean areas that can influence the behaviour of waste materials.

The open ocean and coastal ocean

The ocean can be conveniently divided into two major domains: the open ocean and the coastal ocean. The open ocean lies generally seaward of the continental shelf and is not significantly affected by its boundaries, the continents, or the ocean bottom. The coastal ocean includes estuaries and adjacent wetlands, lagoons, the waters over the continental shelves, and marginal seas (such as the North Sea, Sea of Japan, or Gulf of California). There is no clear-cut boundary between the open ocean and the coastal ocean; but for our purposes, water seaward of the 1,000-m depth contour will be considered part of the open ocean. Consideration of ocean conditions is necessary for making predictions of movements and the eventual fate of oil and other substances spilled, dumped or otherwise deposited in marine areas.

THE OPEN OCEAN

Beyond the continental shelves, the circulation of the open ocean is relatively unaffected by its boundaries. There are no lateral restrictions and the ocean bottom is about 4–7 km below the surface. Over most of the ocean, warm surface waters are separated from the cooler deep waters by the pycnocline (a layer of water exhibiting a marked increase in density with depth), which effectively separates the surface from deeper waters. The deeper waters move sluggishly after forming in polar regions (primarily the Antarctic) and return to the surface about 600–1,000 years later; they do not affect the fate or movement of materials spilled on the ocean surface. Therefore, for most pollution problems, we can ignore movements of deep ocean waters and concentrate on short-term, ocean processes such as surface currents.

Away from the continents that block surface-water movements, ocean currents are primarily directed east to west. Near the boundaries the currents are deflected to the north or south, forming major boundary currents. Open-ocean currents generally move surface waters at speeds of a few kilometres per day. In western boundary currents such as the Gulf Stream or the Kuroshio, surface waters move at speeds of 20–100 km per day.

Winds set ocean surface waters in motion. In the open ocean where tidal currents are relatively weak, these wind-drift currents account for about 40 per cent of the surface currents.

THE COASTAL OCEAN

The coastal ocean is the water adjacent to the coast, where fresh-water runoff from the land and local winds contribute significantly to the movement and mixing of near-shore waters. The width of the coastal ocean is variable and its outer boundary is not well defined. It may be quite narrow along coasts where the continental shelf is narrow and where oceanic conditions and 'permanent' currents come close to shore. Examples include the Peru–Chile coast or the Pacific coast of North America. Conversely, off a coast where the continental shelf is wide, such as the North Sea or the Atlantic coast of North America, the coastal ocean may be tens or even hundreds of kilometres wide. The coastal ocean is not, however, always coincident with the continental shelf. Where the shelf is narrow, the coastal ocean may extend beyond the edge of the shelf. Where the shelf is quite wide, it may extend out from shore for only a part of the shelf width.

Although coastal ocean waters make up only 12.5 per cent of the ocean surface, they are heavily used for water-borne commerce, fishing (recreational and commercial), boating—and waste disposal. Despite these heavy and often conflicting uses, coastal waters are still the most productive part of the world ocean; an estimated 90 per cent of the world's marine food resources are harvested there. These same coastal waters are most often affected by pollution resulting from waste disposal (Gross, 1972).

There is considerable variation in the movement of coastal waters, commonly making it difficult to predict movements of oil slicks and other pollutants near the shore. Tidal currents are often strong, normally the strongest currents being nearest the shore. They generally run parallel to the coast and may be either oscillatory or rotary in character depending on local geography.

Near the shore where surface waves break, a long-shore current usually develops and flows parallel to the shore line. The current direction depends upon the angle at which waves approach the shore. For example, along a coastline oriented in a north–south direction and the ocean to the east, waves approaching the shore line from the north-east will produce a long-shore current that flows southward. Long-shore currents are highly variable because of their dependence on local waves and can reverse direction and change speed in a few hours when the wave regime changes.

Outside the surf zone (where waves break before striking the beach) local winds and salinity gradients, caused by runoff from the land, combine to dominate non-tidal currents in near-shore areas. Since waves breaking on a shore line at any given time may have been generated for the most part by storm systems some distance away, the direction of wave-generated currents nearest the beach may be opposite to the direction of the currents seaward of the surf zone. Suspended sediment concentrations are usually high in these waters, providing ample surface area for chemical reactions. Particles apparently play major roles in the transport and removal of many pollutants in estuarine and coastal ocean systems.

Along coasts having relatively large fresh-water inflows to the ocean, such as the Atlantic coasts of North America and Europe, the salinity and generally the density of surface water increases with distance offshore. The average net (non-tidal) flow in the near-shore zone will, along such coasts in the northern hemisphere, be directed so that the shore is on the right of an observer looking down current. Such long-shore currents are particularly well developed along coasts lying to the right (looking seaward) of the mouths of major estuaries in the northern hemisphere—south of the entrance to Chesapeake Bay, for example, or to the left in the southern hemisphere, as with the Amazon River plume which flows northward along the coast of South America.

Currents in the coastal ocean may be dominated by local winds or by density-induced effects resulting from these winds. Hence, offshore winds will transport surface waters offshore, particularly if the wind is blowing at an angle to the shoreline so that an observer with his back to the wind has the shore to his left. Under these conditions warmer surface waters are transported offshore and cooler subsurface waters from offshore are brought to the surface near the shore, a process known as upwelling. In the northern hemisphere, the resulting density distribution produces a current flowing along the coastline with the shore to the left of the current (looking down current).

When such an offshore or long-shore wind ceases, the warmer surface layers move shoreward. In this way, oil from an offshore spill can be carried onto the beach following the cessation of a wind that has caused an offshore movement of the slick. In the presence of an onshore wind, the wind-induced currents run parallel to the shore in a zone just outside the breakers.

Materials introduced into the near-shore zone will be transported away from the point of discharge by currents and will be mixed into nearby coastal waters. Where there is continuous discharge of materials into near-shore waters, a diverging plume (like the plume from a smoke-stack) will extend down current with concentrations decreasing away from the source.

Because of changes in current patterns (particularly tidal currents), the plume will commonly change direction, sometimes folding back on itself. More often, when the current direction approximately reverses, the remnants of the older plume will be carried along a path somewhat offset from the path of the new plume, extending from the source in the new direction of the current. If we think in terms of a large volume of water with low oil concentrations, this will accumulate near the source over an extended period. Transient plumes of higher concentration plumes will also be superimposed over this water mass. The steady-state level of this general background concentration will depend upon the rate at which near-shore waters are exchanged with waters of adjacent coastal or open ocean areas.

We know little of the processes by which near-shore waters are renewed or mixed with open-ocean waters. Where the runoff of fresh water from the land into the coastal ocean is sufficient to produce measurable salinity gradients, and where the rates of fresh-water inflow can be determined, the mean residence time (or replacement time) for near-shore waters can be estimated. Such estimates have been made for the waters of the coastal ocean bordering the United States between Cape Hatteras and Cape Cod. Considering the salinity of the waters in this segment of the coastal ocean and the annual volume rate of flow from all rivers discharging from the adjacent coast, one can obtain a mean residence time of about one year. Typical residence times range from a few months for a small estuary to several years for large coastal ocean sectors. (Residence time may be considered as the length of time required for river discharge to cause the lowering of salinity observed in coastal waters.)

ESTUARINE CIRCULATION

Most of the world's major ports and harbours are located on estuaries. Despite their variety of shapes and sizes, estuaries commonly exhibit similar physical and biological processes. To understand the fate and effects of wastes released in these areas, we must consider estuarine processes.

An estuary is a semi-enclosed coastal body of water that has a free connection with the open sea; within an estuary, sea water is measurably diluted with fresh water which runs off from the land. Four types of estuaries can be distinguished: drowned river valleys, e.g. the Thames estuary, Chesapeake Bay; bar-built estuaries, e.g. Pamlico Sound; fjord-like estuaries, e.g. Strait of Georgia, Puget Sound; and tectonically formed estuaries, e.g. San Francisco Bay.

Drowned river valleys are the most familiar. They

occur along the Atlantic coasts of the United States and Europe. Because they are generally confined to coasts with relatively wide coastal plains, these water-ways have also been called coastal-plain estuaries.

The basic circulation pattern for estuarine waters is as follows. In a typical estuary, the salinity of the water increases with depth as well as in a seaward direction; in other words, the freshest waters occur near the river mouth, the saltiest near the sea. There is usually a surface layer in which vertical salinity changes are small, an intermediate layer in which salinity increases rapidly with depth, and a deep layer in which the rate of salinity increase with depth is small, as in the surface layer. Vertical mixing takes place between the surface and bottom waters.

Tidal currents usually dominate in estuaries. Superimposed on the tidal currents (usually oscillatory) is a net circulation pattern (called the estuarine circulation) in which there is a net seaward flow of near-surface waters and a net flow from the mouth towards the head of the estuary in subsurface waters (when the currents are averaged over many tidal cycles). There is also a small net flow from the deeper layers to the surface layers. The volume of water flowing towards the head of the estuary (per unit time) decreases from the mouth to the head of the estuary, since water is simultaneously being moved upwards (entrained) from the deeper layers to the surface layers. Consequently, the amount of seaward flow of surface waters increases from the head toward the mouth of the estuary.

Estuarine circulation prevails in most coastal ocean areas, where the input of the fresh water from river runoff and rain exceeds the loss by evaporation from the surface. Thus, over the continental shelf, surface waters generally have a net motion seaward, while near-bottom waters move generally towards the shore as well as moving along the coast.

Extensive mixing of surface and subsurface waters, combined with the estuarine circulation supply nutrients (phosphates and nitrogen compounds) to surface waters that would otherwise have been depleted from the water column. With a plentiful nutrient supply and abundant sunlight, the phytoplankton (minute floating plants) of coastal waters produce the food to support the abundant marine life including finfish and shellfish. Eggs and larvae of various organisms are also abundant. In the open ocean where there is less mixing, surface waters are depleted of nutrients and annual production of phytoplankton is much less than in coastal waters.

In coastal areas, winds from certain directions tend to move surface waters seaward and subsurface waters are brought to the surface near the coast in a process known as upwelling. With the abundance of nutrients, the productivity of phytoplankton is generally high and upwelling areas support some of the world's richest fisheries. One small area offshore from Peru to Chile supplied about one-fifth of the world's annual fish catch in the early 1970s. Broad continental shelves such as the Grand Banks of eastern North America can be very productive areas. The Bering Sea supports a fishery equal to one quarter of the world annual catch.

Ocean pollutants

Having briefly examined oceanic processes as they generally affect pollutants in the ocean, we can now turn to the consideration of some selected contaminants.

PETROLEUM HYDROCARBONS

Petroleum hydrocarbons are conspicuous pollutants in the ocean and increased amounts are likely to be discharged in the future. The rising demand for petroleum, particularly in industrialized countries, has caused further intense exploration and development of offshore petroleum resources and increased shipment of oil across the ocean. Thus hydrocarbons, which typify the behaviour expected of surface-associated pollutants, are found to be major pollutants in many ocean areas, including large parts of the open ocean.

Oil transport is unevenly distributed over the ocean; it is more heavily concentrated in coastal ocean

TABLE 1. Estimates of petroleum hydrocarbons (crude oils and refined products) entering the coastal and open oceans (in millions of tonnes per year)

Source	Coastal ocean	Open ocean	Coastal and open ocean	Percentage
Natural seep	0.6	—	0.6	9.8
Offshore production	0.08	—	0.08	1.3
Transportation				
Tanker operations (including bilges, bunkering and dry docking)	0.6	1.25	1.85	30.3
Accidents	0.2	0.1	0.3	4.9
Terminal operations	0.003	—	0.003	—
Rivers	1.6	—	1.6	26.2
Atmosphere	0.06	0.54	0.6	9.8
Urban and industrial areas				
Sewage	0.3	—	0.3	4.9
Runoff	0.3	—	0.3	4.9
Industrial wastes	0.3	—	0.3	4.9
Refineries	0.2	—	0.2	3.3
TOTAL	4.2	1.9	6.1	100
Percentages	68.9	31.1	100	

Source: Modified after National Academy of Sciences, 1975*a*.

areas and in some marginal seas such as the Mediterranean and Red Seas and Persian (Arabian) Gulf. Pollution of such areas may result in more ecological damage than if the oil were discharged primarily into open ocean waters (Table 1).

Sources

Hydrocarbons are derived from natural sources such as decaying organisms and from seeps of oil and gas from petroleum reservoir rocks. They are also introduced through the use and transport of fossil fuels. Marked compositional differences exist between hydrocarbons from living organisms on the one hand and spilled fuels or natural seepages on the other (National Academy of Sciences, 1975*a*). These differences can be detected by chemical analysis; this provides the basis for the distinction between naturally occurring hydrocarbons and pollutants. Natural seepage to the world's oceans is estimated to be about 600,000 tonnes per year (Table 1) though the amounts of hydrocarbons produced by phytoplankton are probably much larger.

Production, transport and use of hydrocarbons are the primary source of marine oil pollution. One estimate of the direct input of petroleum to the ocean was 6 million tonnes per year (Table 1). Transfer of hydrocarbons through the atmosphere to the ocean is also virtually unknown—one estimate of atmospheric transport was about 70 million tonnes per year (Goldberg, 1976*b*). About a half of this amount was due to marine transport, municipal wastes and offshore oil and gas operations. A quarter of the petroleum discharges to the oceans is accounted for by river runoff, which includes oil discharged from industrial operations, motor vehicles and recreational uses. The amount of hydrocarbons actually reaching the ocean surface via atmospheric transport was estimated at 0.6 million tonnes per year—roughly the same as the coming from seeps.

World petroleum production is expected to double between 1970 and 1980, from 2.2 to 4.0×10^9 tonnes per year. At the same time, offshore crude oil production is expected to triple (from 440 to $1,300 \times 10^6$ tonnes per year), while the amount of oil transported by tanker doubles. These data suggest that pollution problems caused by petroleum hydrocarbons will also increase. As Table 1 indicates, about two-thirds of the petroleum released to the ocean occurs in coastal ocean waters.

Behaviour of petroleum in the ocean

Oil is a complex mixture of hydrocarbons whose chemical composition can be altered significantly by evaporation, dissolution and microbial and chemical oxidation after a spill (Kolpach *et al.*, 1973). Because the varying constituents of oil are affected at different rates by these 'weathering' processes, the relative composition and, therefore, the biological effects of spilled oil also vary widely (Goldberg, 1976*b*).

Evaporation depletes most of the more volatile components but will cause little separation (fractionation) between hydrocarbons that have the same boiling-points (but often substantially different structures and effects on organisms). Hydrocarbons lost through evaporation go into the atmosphere where we have no information on their ultimate fate.

Dissolution also preferentially removes the lower molecular weight components from an oil slick. Some potentially toxic fractions such as the aromatic hydrocarbons are more soluble than other less-toxic compounds (paraffins) of the same boiling-point. Once dissolved in sea water, these soluble constituents may follow quite different pathways than the more conspicuous slick (Horn et al., 1970).

Microbial attack affects compounds with much wide boiling range than evaporation and dissolution. Hydrocarbons with the same general structures are attacked roughly at the same rates. Normally, paraffins are most readily degraded. Continued biochemical degradation causes gradual removal of the branched alkanes. Cycloalkanes and aromatic hydrocarbons are more resistant and disappear at much slower rates.

Chemical degradation processes of oil during weathering are not well understood. Oxidation most readily affects aromatic hydrocarbons of intermediate and higher molecular weight. The effect of these weathering processes is the rapid (within 48–96 hours) depletion of lower boiling fractions (boiling-point less than 250 °C) from a spilled slick by evaporation and dissolution and the slow degradation (in terms of years) of higher boiling fractions by microbial and chemical oxidation.

Oil incorporated into marine sediments apparently does not undergo the same changes as those observed in oils exposed to the atmosphere. Also, the lack of dissolved oxygen in most marine sediments and the shielding from sunlight cause the oil to remain unchanged for much longer periods (Blumer and Sass, 1972). Thus, oil originally incorporated in sediments may be later released to the environment essentially unchanged if the deposit is distributed by the actions of waves, tidal currents or dredging. After release, the oil can be moved into other areas where it can cause harmful effects.

When mixed with sea water, oil tends to form emulsions (Berridge et al., 1968). Depending on the chemical composition of the oil and presence of surface-active constituents (surfactants), the emulsions may be either oil in water (as is milk) or water in oil (as in butter). Emulsion formation changes the physical characteristics of the oil and its physical and chemical behaviour in the ocean and also modifies its effects on marine organisms.

Many oils when vigorously mixed with sea water form oil-in-water emulsions. The fine oil droplets are then dispersed through a large volume of water, often disappearing from the ocean surface. In a stable, dispersed form the oil does not 'wet' surfaces and this

phase also provides the maximum surface area for microbial and chemical degradation of the oil. The large surface area also permits soluble constituents in the oil to dissolve more readily in sea water.

Much oil discharged in coastal waters washes ashore forming crusts of weathered oil on rocks or sand. The outer crust is depleted of volatile constituents but according to a study by Levy (1972), petroleum in the interior of the crust apparently changed little even after a year's exposure.

Petroleum discharged in open ocean waters from sources such as tanker washings or bilges forms pelagic tar falls and lumps which are widely dispersed at sea (Butler et al., 1973).

Observations of the elimination of oil films from large lakes suggest that aerosol formation may remove oil from the sea surface and inject it into the atmosphere to be carried away by the winds. Oil films from recreational boating are eliminated from the water surface in 70–90 hours after boating stopped and there is no evidence that these oils are removed by evaporation or either dissolution or emulsification in the water. Instead, it appears that most of the removal occurs by formation of tiny droplets which are injected into the atmosphere as aerosols. Furthermore, the droplets so formed have been found to preferentially absorb salts and certain polar organic materials from the water (Baier, 1972).

TABLE 2. Major chemicals transported world-wide in 1970

Commodity	Amount (in thousands of tonnes)
Naphtha *	2,867
Liquid sulphur	1,865
Caustic soda (sodium hydroxide)	1,196
Molasses	825
Methanol	618
Xylene *	603
Styrene *	476
Toluene *	366
Phosphoric acid	299
Benzene *	280
Tallow	183
Ethylene dichloride *	241
Sulphuric acid	177
Ethylene glycol	170
Alcohols	160
Lube additives *	158
Cumene *	151
Ethyl benzene *	126
Carbon tetrachloride *	114
Fish oil	106

* Possible deleterious effects.
Source: After Inter-governmental Maritime Consultative Organization, 1973. Cited in GESAMP, 1976.

The combined action of bursting bubbles and ultraviolet irradiation was found in laboratory experiments to be most effective in removing oily films from the water surface. Exposure to the atmosphere and ultraviolet radiation apparently oxidized the hydrocarbons in these natural films. The film materials preferentially pick up other oil-soluble constituents in the water. Surfaces of droplets are also likely to be colonized by bacteria and other micro-organisms. While conclusive data are still lacking, these processes may be the cause of reported rapid disappearance of oil slicks in many marine areas. The behaviour and ultimate fate of petroleum in the atmosphere remains unknown.

INDUSTRIAL CHEMICALS

Many relatively unfamiliar industrial compounds can affect the ocean and marine life. These materials are transported in tankers in megaton quantities (see Table 2). Some of these are likely to be spilled in the ocean, especially in coastal or estuarine waters, due to mishaps in unloading or through other accidents.

Spills of many products such as molasses or tallow are not likely to cause problems but naphtha (a solvent), or xylene and ohter benzene-related compounds, may damage marine organisms if spilled in large quantities. More work is needed to study the chemical behaviour of these compounds in the ocean, especially in sediment-bearing coastal waters.

SYNTHETIC ORGANIC COMPOUNDS

Because of their widespread use in industry and agriculture, large quantities of chlorinated hydrocarbons such as DDT,[1] PCB[2] and hexachlorobenzene and volatile organic liquids and gases such as freon and dry-cleaning solvents reach the ocean. Some of these compounds are toxic, and many resist chemical and biological degradation to the point that they are considered persistent. They can readily escape into the atmosphere and large quantities are moved to the oceans by winds.

The insecticide DDT is one of the best-known synthetic hydrocarbon compounds and large quantities were produced and used in the United States before it was banned in 1971. DDT is relatively insoluble in water (1.2 p.p.b.) and tends to be sorbed on soil and silt particles. Therefore, river runoff is a less-important pathway for this pollutant than atmospheric transport. In soil and sediments, the half-life of DDT has been estimated to be a decade or more (see Goldberg, 1976b).

In the presence of oxygen, DDT degrades to DDE.[3] DDT can also be metabolized in the absence of oxygen to other forms such as DDD,[4] but DDE is the most persistent of the degraded forms of DDT. In fact, DDE may be the most abundant synthetic organic pollutant in the ocean and it has been estimated that DDE comprises at least 80 per cent of the DDT residual in marine organisms; DDE is essentially just as toxic as DDT. Moreover, there is no known organism normally found in the sea that metabolizes DDE, thus making it extremely persistent. DDT and its derivatives are known to accumulate in the fat tissue of marine organisms so that DDT concentrations tend to increase as it moves up through the food chain. Because of this step-by-step 'bio-concentration', fish-eating birds may contain DDT concentrations exceeding that found in the water by a factor of 1 million.

There is considerable variability among animals in their ability to tolerate chlorinated hydrocarbons. Birds seem especially vulnerable. One well-known effect of DDT is interference with the reproductive organs of female birds by preventing the mobilization of calcium in the oviduct—so that eggs may be laid with a thin shell or no shell at all. In the United States, salmon from Lake Michigan and mackerel from the Pacific Ocean were banned for human consumption because they contained DDT above approved levels.

PCBs are also wide-spread pollutants in the marine environment. The PCBs are stable, fire-resistant compounds that are widely used as an additive to paints, plastics, coating compounds and in transformers, capacitors and heat-transfer systems. As in the case of DDT and its residues, the observed distribution of PCBs in the oceans suggest that atmospheric transport and subsequent deposition on surface waters are important pathways.

Effects of PCBs in the marine environment and on marine organisms are poorly known. Biphenyls containing fewer chlorine atoms may be more toxic (Dexter and Pavlou, 1972). PCBs also appear to move through food chains in ways similar to DDT. The PCB compounds are toxic; five or more Japanese died in 1968 from accidental exposure, and hence much research is required to determine the amounts of PCB being released to the marine environment, its pathways and its potential effect both on marine life and human health.

TOXIC METALS

Toxic or potentially toxic metals are transported to the ocean in large quantites through sewers and atmospheric fall-out. Certain metals are highly toxic and highly persistent, and may retain their toxicity—against plants and animals—for an extended period. Many metals are accumulated in the bodies of organisms, remain there and function as cumulative poisons (Goldberg, 1976b).

About two dozen metals are known to be toxic to plants and animals. Of this group, mercury and lead are considered to be the most threatening pollutants.

The rate at which heavy metals are being released to the ocean environment is assumed to be related to world production of these metals. World production of

---

1. Dichlorodiphenyltrichloroethane.
2. Polychlorinated biphenyl.
3. 2,2'-(bis p-chlorophenyl)-dichloroethylene.
4. 2,2'-(bis p-chlorophenyl)-dichloroethane.

TABLE 3. Fluxes of mercury in the environment

Type of flow	Tonnes per year
Natural flows	
Continents to atmosphere (by degassing of the earth's crust):	
Based on precipitation with rain	8.4×10^4
Based on atmospheric content	15.0×10^4
Based on content in Greenland Glacier	2.5×10^4
River transport to oceans	$< 3.8 \times 10^3$
Flows involving man	
World production (1968)	8.8×10^3
Entry to atmosphere from fossil fuel combustion	1.6×10^3
Entry to atmosphere during cement manufacture	1.0×10^2
Losses in industrial and agricultural usage	4.0×10^3

Source: Weiss *et al.*, 1971.
© 1971 American Association for the Advancement of Science.
Reprinted with permission.

metals—with few exceptions—increasing. For example, lead production in 1970 was 3.4 million tonnes, an increase of 25 per cent over the amount produced in 1965 (Goldberg, 1976*b*).

Metals enter the ocean through river runoff, industrial waste discharge, municipal sewage systems, dumping and atmospheric transport. Most of the metal input occurs in coastal ocean waters and most evidence of elevated metal concentrations in waters and marine organisms comes from coastal areas such as the Irish Sea (Preston, 1973; Preston, *et al.*, 1972).

Mercury

World production of mercury is about 9,000 tonnes per year and about 5,000 tonnes per year is estimated as the input into the ocean from man's activities (Table 3). This input is unevenly distributed and has caused local and regional pollution problems, mostly in coastal or estuarine areas. Human input to the open ocean is probably significantly less than the natural fluxes from volcanic processes.

In the marine environment, most mercury compounds decompose to inorganic mercury and gradually methyl mercury is formed. Methyl mercury is more toxic to organisms (including humans) than elemental mercury. Most mercury remains in coastal waters.

Lead

The waters of the surface (wind-mixed) layer of the oceans of the northern hemisphere have average lead concentrations significantly greater than the concentrations in deeper waters, apparently an effect of pollution (Goldberg, 1976*b*). Effects of this increase in lead upon

the marine environment are unknown. Lead is known to interfere with some biochemical reactions in plant cells and lead concentrations have probably been elevated in higher organisms near the end of the marine food chains. Discharges of industrial lead have accumulated in the surface layers of near-shore sediments and have been found to cause, in such areas, hazardous amounts of lead in shellfish.

In 1940, world lead production was 1.9 million tonnes, with 36,000 tonnes being burned in leaded fuels. In 1966, world lead production was up to 3 million tonnes, with about 310,000 tonnes burned in leaded fuels (Goldberg, 1976*b*). Lead enters surface ocean waters through atmospheric transport of lead-containing aerosol particles. Thus, lead may constitute a global pollution problem in surface ocean waters.

Despite widespread concern about toxic metals and their effects, their concentrations are generally so low that accurate determinations are extremely difficult. Consequently, reported ranges of oceanic concentrations of heavy metals show considerable variations. The reported concentrations have tended to be lower as analytical and sampling methods improve. The bulk of the recent literature, however, brackets the concentrations of the commonly analysed metals in the sub-parts-per-billion range. An investigation of the distribution of copper, nickel and cadmium in the Sargasso Sea was conducted by Bender and Gagner (1976) using the co-precipitation method of Boyle and Edmond (1975). Sufficient analytical control was exercised to demonstrate that these elements were depleted in Sargasso Sea water compared with the North Atlantic deep water below. Their results favour the hypothesis that many metals are transported vertically from the surface waters. Previously, these removal processes have been difficult to

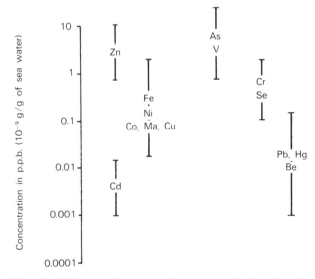

FIG. 1. Estimates of concentration ranges of selected toxic metals in average open ocean waters.

observe because of sample contamination and analytical error.

Considering the amount of metals introduced to the ocean from rivers and through the atmosphere and the data on the accumulation of metals in sediments (usually in the many-parts-per-million range), then removal processes must undoubtedly occur. If the plethora of analyses of oceanic trace metals are all high—and the actual concentrations are in the sub-parts-per-trillion range—then the removal processes must be enormously powerful and probably cannot be explained with our present knowledge of the chemical processes in the ocean. The approximate trace metal concentrations can be deduced from the recent data for general oceanic concentrations (Fig. 1).

Due to the extremely low concentrations of most metals in sea water, precautions must be taken to avoid contamination of samples during sampling or analytical procedures.

Since the concentration of several pollutants in the surface-water column appear to be of the order of 0.1 p.p.b., one can then ask what magnitude of fluxes would be required from common sources to produce that concentration in the surface waters of the world ocean. If a dissolved pollutant were distributed through surface waters (200 m thick) that had a residence time of 20 years (Broecker, 1974), the supply flux is:

$$0.1 \ \mu g^{-1} \times 200 \ m \times 10^2 \frac{cm}{m} \times 10^{-3} \frac{litre}{cm^3} \times \frac{1}{20} \ years^{-1},$$

or $\qquad 0.1 \ \mu g \ cm^{-2} \ yeas^{-1}.$

From a river source, this pollutant would be supplied in river water which flows (on the world average) into the sea at a rate of $10^{-2} \ l \ cm^{-2} \ year^{-1}$, so:

$$\frac{0.1 \ \mu g \ cm^{-2} \ year^{-1}}{10^{-2} \ l \ cm^{-2} \ year^{-1}} = 10 \ \mu g \ l^{-1}.$$

In other words, concentration of the metal would have to be only about 10 p.p.b. in river water to account for the observed concentrations in ocean water.

Contamination from an aerosol source may be similarly examined. The average bulk depositional velocity of marine aerosols is 3 cm sec^{-1} (Sehmel and Sutter, 1974). For a supply flux of 0.1 μg cm^{-2} year^{-1}, $10^{-9} \ \mu g$ cm^{-3} would be the atmospheric concentration required to supply this contaminant.

What would be the needed rate of supply from near-shore sediments? If 5 per cent of the sea floor provides the flux of 0.1 μg cm^{-2} year^{-1}, then the sediment must be leached of 2.0 μg cm^{-2} year^{-1}. To infer the rate of release of pollutant material from the sediments, a diffusion coefficient must be assumed. If an average diffusion coefficient for silica is representative, then a value can be picked from the treatment by Schink et al. (1975) that is modified for the properties of sea-floor sediments. Using $K = 3.5 \times 10^{-6}$ cm^2 sec^{-1}, the required concentration gradient of the pollutant in the sediments is 1,800 μg l^{-1} m^{-1}, or about 1.8 p.p.m. per metre.

The conclusion from these simple calculations is that any of the above supply processes could account for the persistence of substances in surface waters as long as only diffusion and advection were removing the metals (Bender et al., 1975).

WASTE LIQUIDS

Sewage effluents and urban runoff

Untreated sewage and treated sewage effluents reach the coastal ocean directly, through pipelines or, indirectly, through river and estuarine discharges. An increasing amount of the runoff from cities in developed countries enters the ocean through sewage disposal.

Sewage effluents are usually major sources of nutrients (phosphorus, nitrogen compounds) to coastal waters near cities. Such effluents can cause locally increased phytoplankton production. The organic matter from the sewage effluent, together with the phytoplankton, use large amounts of dissolved oxygen as they decompose. If currents fail to supply sufficient dissolved oxygen, near-bottom waters can be depleted in dissolved oxygen, thus killing bottom-dwelling organisms and fish which are unable to escape to oxygen-bearing waters.

TABLE 4. Composition of rainfall, untreated sewage, and runoff from urban and rural areas

Constituent (values in mg l^{-1})	Rainfall	Untreated sewage	Runoff Urban	Runoff Rural
Suspended solids	13	200	230	310
Chemical oxygen demand	16	350	110	—
Total nitrogen (as N)	1.3	40	3.1	9
Inorganic nitrogen (as N)	0.7	30	1.0	5
Total phosphate	0.08	10	0.4	0.6

Source: Weibel, 1969; Biggar and Corey, 1969.

TABLE 5. Characteristics of treated waste-water effluents

Constituent (values in mg l^{-1})	Level of sewage treatment None	Primary	Secondary	Advanced
Suspended solids	200	100	15	1
Biochemical oxygen demand	200	135	22	1
Phosphorus	10	10	7	< 1
Nitrogen (NH$_3$)	30	30	10	< 1
Nitrogen (organic)	15	15	6	< 5

Source: Metcalf and Eddy, Inc., 1972.
© 1972 McGraw-Hill Book Co. Reprinted with permission

Runoff of precipitation from urban areas also contributes pollutants to coastal waters. These discharges are similar in composition to untreated municipal waste waters and markedly different from rain water (Tables 4, 5).

WASTE SOLIDS

Long-ignored as a pollution problem, waste-solid disposal has emerged as a problem of significance in developing coastal ocean areas. Large volumes of wastes are produced and the impact of their disposal in the ocean may be long lasting due to changes in bottom topography and the physical and chemical characteristics of the bottom.

Ocean dumping involves transportation of large volumes of waste materials from coastal ports aboard barges or ships for disposal in ocean waters. Waste materials disposed of at sea include dredged materials, incinerator residues, garbage, sewage, sewage sludge, chemical wastes, discarded military equipment and munitions, excavation debris and other industrial, municipal and agricultural wastes (see Table 6).

In undisturbed areas, matter is transported to the oceans by rivers, winds or glaciers. Human activites in coastal urban and industrialized areas have added two additional pathways: ships and waste outfalls. Although we do not as yet have accurate figures on mass emission rates, there is little question that the volumes of waste matter dumped in coastal waters each year from developed countries constitutes a significant fraction of the total discharge of solids to the coastal ocean. For example, waste solids discharged from the New York metropolitan region exceeded the sediment discharge of all rivers flowing into the Atlantic Ocean between the United States–Canadian border and Chesapeake Bay (Gross, 1972).

TABLE 6. Types and amounts (in thousands of tonnes) of vessel-discharged wastes disposed of in United States coastal waters

Waste type	1973	1974	1975
Industrial waste[1]	5,059	4,592	3,446
Sewage sludge[1]	4,899	5,010	5,040
Construction and demolition debris	974	770	396
Solid waste (refuse, garbage)	0.2	0.2	—
Subtotal	10,933	10,373	8,882
Dredged materials	44,208	98,665	87,826
TOTAL	55,141	109,038	96,708

1. Liquid wastes, approximately 5 per cent solids by weight, containing approximately 496,000, 480,000 and 424,000 tonnes of dry solids per year, respectively.
Source: United States Environmental Protection Agency.

Dredged wastes

Dredging, whose wastes include sand, silt, clay, rocks and various other waste deposits, is usually designed to improve and maintain navigation channels. Spoils are generally disposed of in open coastal waters less than 30 m deep. Dredged wastes are the largest category of ocean-dumped waste material—84 per cent of the 1968 total (Smith and Brown, 1971)—and about one-third of the dredged material is considered to be polluted. Expanding needs of marine commerce will necessitate increased dredging and hence increased amounts of dredged materials requiring disposal.

Accumulations of wastes at the dump site and problems with turbidity are locally significant effects of ocean disposal of dredged spoils. Rapid build-up of waste deposits and movement over the bottom through gravity or bottom currents can suffocate bottom-dwelling organisms, reduce food supplies and vegetational cover, trap organic matter and thus induce anaerobic bottom conditions and absorb organic matter, including oil. Turbidity resulting from disposal of dredge spoils can have adverse effects on the marine ecosystem. Turbidity resulting from dredging and disposal operations can cause: (a) reduced growth and decreased survival of larval stages of fish and shellfish; (b) reduction in light penetration resulting in reduced photosynthesis; (c) reduction of visibility to some feeding organisms; and (d) flocculation and consequent settling of phytoplankton.

Effects of the disposal of dredged materials depend to a large extent on the location and characteristics of the disposal site. Materials discharged in shallow, normally turbid waters, for example, have less effect than the same materials discharged in deeper, less turbid waters. Also, dredged materials containing oil, sewage, chemicals and other wastes are likely to have more deleterious effects than the same quantity of clean sediment. Dredged waste disposal operations are usually highly localized and involve sudden releases of several thousand tons of material; this can produce a thick layer of sediment which may or may not be dispersed by current action.

No release of harmful chemical substances has been shown to result from dredging activities although it may well occur.

SEWAGE SLUDGE

Sewage sludges are semi-liquid residues of sewage treatment (Table 7). In 1960, almost 4 million tonnes (wet basis) of sewage sludge were dumped in the New York Bight alone and another 0.5 million tonnes were disposed of by the municipality of Philadelphia at a dump site off Cape May, New Jersey.

Undigested sewage sludge contains only 2–7 per cent of solid matter, which can be further treated by anaerobic digestion to decrease the sludge volume (by

TABLE 7. Typical characteristics of municipal waste-water treatment sludges

Constituent	Primary		Activated secondary	
	Range	Average	Range	Average
(percentage)				
Total solids	2–7	4	0.5–2	1
Volatile solids	60–80	70	60–80	75
Nitrogen	1.5–8.0	3.0	4.8–6.0	5.6
Phosphorus (P$_2$O$_5$)	0.8–2.8	1.6	3.0–7.4	5.7
(percentage in dry weight)				
Organic matter	60–80	70	62–75	69
Cellulose	—	3.8	—	7
Grease, fats	7–35	21	5–12	9
Protein	22–28	25	32–41	37
Humus	—	33	—	41

Source: Hecht *et al.*, 1975.

TABLE 8. Selected chemical components of digested sewage sludge

Constituent	Range	Median
Metals	(parts per million)	
Boron	15–11,000	100
Cadmium	5.0–2,000	15
Chromium	50–30,000	1,000
Copper	250–17,000	1,000
Lead	136–7,600	1,500
Mercury	1–100	10
Nickel	25–8,000	200
Zinc	500–50,000	2,000
Chlorinated hydrocarbons		
PCBs	1.2–105	3.2
Chlordane	3–30	—
Dieldrin	0.03–2.2	0.16
(percentage)		
Total solids	2–10	6
Volatile solids	40–60	50
Nitrogen	1.6–6.0	3.7
Phosphorus – P$_2$O$_5$	0.9–6.1	1.7

about a half) and reduce odours and pathogens. Although most of the sewage-sludge production in the United States is disposed of on land, much is not and, in 1960, sewage sludges comprised the third largest category of wastes disposed of at sea, amounting to 7 per cent of the total. Large quantities of sludge solids, for example, are discharged from the Los Angeles–San Diego area from submarine pipes 2–10 km offshore.

Sewage-sludge production in the United States is expected to increase dramatically in the future, possibly causing severe disposal problems for municipalities. In addition to normal population growth and industrial expansion, more stringent pollution laws require higher levels of treatment which, in turn, produce larger volumes of sludge. Industry, too, seeking to meet new control regulations, may discharge waste into municipal treatment facilities, making sludge disposal a still greater environmental problem.

Sludge contains some metals and other substances toxic to marine organisms (Table 8). Behaviour of these complex materials in sea water or in sediments is poorly known.

INDUSTRIAL WASTES

These—both liquid and solid—are the second largest category of waste dumped in the ocean. In 1968, industrial wastes in the United States accounted for 4.7 million tonnes or 8 per cent of the total. New York City alone disposed of 2.7 million tonnes at sea, of which 90 per cent was acid waste (Smith and Brown, 1971). Other industrial wastes include: acids (58 per cent), refinery (12 per cent), pesticides (7 per cent), paper-mill wastes (3 per cent), and a variety of other materials in relatively low quantities such as those from oil drilling, pharmaceutical manufacture and others. Relatively innocuous wastes are usually dumped in coastal waters. Highly toxic waste matter is usually packed in drums and dumped in depp water further out from the shore.

Some of the largest volumes of solid and liquid industrial waste are those extracted during production of titanium dioxide and alumina. These wastes cause local changes in pH and increased turbidity though their actual effects in the ocean appear to be slight and transitory.

WASTE MATERIALS AS ENVIRONMENTAL 'TRACERS'

Many of the waste discharges to the aquatic environment are little different from materials supplied by natural sources. Such waste disposal operations offer opportunities for study of processes not normally operative; they employ techniques easily available to marine scientists. One example is the study of transport processes using discharges of dredged wastes or extracted Al or Ti ores.

Other waste discharges, such as new synthetic chemicals and artificial radionuclides provide opportunities to study entirely new and often unsuspected transfer processes and interactions. The new materials may cause significant effects—often deleterious on organisms and ecosystems. This permits investigation of pathways that might otherwise be nearly impossible to identify.

One good example of a new area for geochemical work is in the assessment of the significance of potential pollutants (National Academy of Sciences, 1975b). During the process of screening synthetic chemicals as potential pollutants, several criteria have been identified. Some of these criteria also are useful in selecting substances for geochemical studies: (a) high production volume of wastes; (b) high probability of release to environment in significant quantities; (c) long life (or residence time) in the marine environment; (d) localized or point-source discharges. If biological effects are also to be included, two other criteria are pertinent: (e) potential bioaccumulation; (f) high toxicity.

Selection of marine pollutants and studies of their interactions has been done based primarily on global-scale effects; studies of radioactive fall-out or the atmospheric dispersal of chlorinated hydrocarbons are two well-known examples. Studies of regionally distributed discharges, however, can be quite useful. Sediment-associated wastes provide 'tags' for studying transport pathways and deposition of waste-associated sediments. Both radionuclides (NAS, 1971a) and metals (Gross, 1972) have been used for such studies. Waste discharges from mining and processing of ores may also provide useful tags.

Major questions are still unanswered regarding rates of petroleum input to the ocean and fates of hydrocarbons in the aqueous environment. For instance, river discharge of petroleum was estimated at 1.6 million tonnes per year (National Academy of Sciences, 1975a) but this estimate was based on the single unpublished analysis of Mississippi River sediment. Data on the atmospheric inputs are equally unreliable. Much research is needed to generate sufficiently reliable data to improve the estimates. Data on discharges of other waste materials are scanty and therefore many investigations concerned with the transfer and fate of pollutants in the ocean or atmosphere are severely limited.

On a much smaller scale, there are complex problems regarding transfer of materials to marine ecosystems—and possibly to man. While these may pose substantial public-health questions, they also constitute biogeochemical experiments of enormous importance (Siegel, 1974). In brief, some waste-disposal operations provide excellent opportunities to learn about ocean processes as part of carefully planned studies of waste-disposal operations. And in addition to scientific returns, the research can pay handsome dividends in devising better and more effective monitoring schemes (Goldberg, 1976a) and in programmes to protect and utilize the resources of the ocean.

International studies of marine pollution

Several international studies of marine pollution problems were organized during the 1970s. Some of the studies are globally oriented such as the Integrated Global Ocean Station system (IGOSS) jointly organized by the World Meteorological Organization (WMO) and Unesco's Intergovernmental Oceanographic Commission (IOC). The IGOSS programme began with the pilot project to monitor petroleum hydrocarbons, particularly in the open ocean.

Other investigations have taken more regional approaches. The IOC's Global Investigation of Pollution of the Marine Environment (GIPME) has stressed the regional approach to pollution studies. Much of the GIPME programme has built on the experience of the regional studies (North Sea, Baltic Sea) of the International Council for the Exploration of the Seas (ICES). The Health of the Ocean report by Goldberg (1976b) was an early product of GIPME activities; comparable reports on the world ocean and on selected regions are planned.

References

BAIER, R. S. 1972. Organic Films on Natural Waters: Their Retrieval, Identification and Modes of Elimination. *J. Geophys. Res.* Vol. 77, p. 5062–75.

BENDER, M. L.; GAGNER, C. 1976. Dissolved Copper, Nicker and Cadmium in Sargasso Sea. *J. Mar. Res.*

BENDER, M. L.; KLINKHAMMER, G. P.; SPENCER, D. W. 1975. Manganese in Sea Water and the Manganese Marine Balance. *Deep Sea Res.*

BERRIDGE, S. A.; THEU, M. T.; LORISTON-CLARKE, A.-G. 1968. The Formation and Stability of Emulsions of Water in Crude Petroleum and Similar Stocks. *J. Inst. Petrol.* Vol. 54, p. 333–57.

BIGGAR, J. W.; COREY, R. B. 1969. Agricultural Drainage and Eutrophication. *Eutrophication: Causes, Consequences, Correctives,* p. 404–45. Washington, D.C., National Academy of Sciences.

BLUMER, M.; SASS, J. 1972. Indigenous and Petroleum-derived Hydro-carbons in a Polluted Sediment. *Mar. Pollut. Bull.* Vol. 3, p. 92–4.

BOYLE, E.; EDMOND, J. M. 1975. Copper in Surface Waters South of New Zealand. *Nature,* Vol. 253, p. 107–9.

BROECKER, W. S. 1974. *Chemical Oceanography.* New York, Harcourt, Brace & Jovanovich. 214 p.

BUTLER, J. N.; MORRIS, B. F.; SASS, J. 1973. *Pelagic Tar from Bermuda and the Sargasso Sea.* St. George's, West Bermuda, Bermuda Biological Station. 346 p.

DEXTER, R. N.; PAVLOU, S. P. 1972. Chemical Inhibition of Phytoplankton Growth Dynamics by Synthetic Organic Compounds. *J. Etud. Poll.* Vol. 1972, p. 155–7.

GESAMP. 1976. *Review of Harmful Substances.* New York, 79 p. (GESAMP Reports and Studies, No. 2.)

GOLDBERG, E. D. (ed.). 1976a. *Strategies for Marine Pollution Monitoring*. New York, Wiley-Interscience. 390 p.

——. 1976b. *The Health of the Oceans*. Paris, Unesco. 172 p.

GROSS, M. G. 1972. Geological Aspects of Waste Solids and Marine Waste Deposits, New York Metropolitan Region. *Geol. Soc. Amer. Bull.* Vol. 83, p. 3163–76.

HECHT, N. L.; DUVAL, D. S.; PASHIDI, A. S. 1975. *Characterization and Utilization of Municipal and Utility Sludges and Ashes*. Vol. II: *Municipal Sludges*. Washington, D.C., United States Environmental Protection Agency. (NTIS PB 244 311.)

HORN, M. H.; TEAL, J. M.; BACKUS, R. H. 1970. Petroleum Lumps on the Surface of the Sea. *Science*, Vol. 68, p. 245–6.

IAEA. 1972. *Radioactive Contamination of the Marine Environment*. Vienna, International Atomic Energy Agency. 786 p.

KOLPACK, R. L.; MECHALAS, B. J.; MEYERS, T. J.; PLUTCHAK, N. B.; EATON, E. 1973. *Fate of Oil in a Water Environment*. American Petroleum Institute Publishers.

LEVY, E. M. 1972. The Identification of Petroleum Products in the Marine Environment by Absorption Spectrophotometry. *Water Res.* Vol. 6, p. 57–69.

METCALF AND EDDY, INC. 1972. *Wastewater Engineering*. New York, McGraw-Hill.

NATIONAL ACADEMY OF SCIENCES. 1971a. *Radioactivity in the Marine Environment*. Washington, D.C., National Academy of Sciences.

——. 1971b. *Marine Environmental Quality*. Washington, D.C., National Academy of Sciences.

——. 1975a. *Petroleum in the Marine Environment*. Washington, D.C., National Academy of Sciences. 107 p.

——. 1975b. *Assessing Potential Ocean Pollutants*. Washington, D.C., National Academy of Sciences, 438 p.

PRESTON, A. 1973. Heavy Metals in British Waters. *Nature*, Vol. 242, p. 95–7.

PRESTON, A.; JEFFRIES, D. F.; DUTTON, J. W. R.; HARVEY, B. R.; STEELE, A. K. 1972. British Isles Coastal Waters: the Concentrations of Selected Heavy Metals in Sea Water, Suspended Matter and Biological Indicators—A Pilot Survey. *Environ. Pollut.* Vol. 3, p. 69–82.

SCHINK, D. R.; GUMASSO, N. L.; FANNING, K. A. 1975. Processes Affecting the Sediment-Water Interface of the Atlantic Ocean. *J. Geophys. Res.* Vol. 80(21).

SEHMEL, G. A.; SUTTER, S. S. 1974. Particle Deposition Rates on a Water Surface as a Function of Particle Diameter and Air Velocity. *Res. Atmos.* VIII, p. 911–20.

SIEGEL, F. R. 1974. *Applied Geochemistry*. New York, Wiley. 353 p.

SMITH, D. D.; BROWN, R. P. 1971. *Ocean Disposal of Barge-delivered Liquid and Solid Wastes from U.S. Coastal Cities*. Washington, D.C., United States Environmental Protection Agency.

WEIBEL, S. R. 1969. *Urban Drainage as a Factor in Eutrophication. Eutrophication: Causes, Consequences, Correctives*, p. 383–403. Washington, D.C., National Academy of Sciences.

WEISS, H.; KOIDE, M.; GOLDBERG, E. D. 1971. Mercury in a Greenland Ice Sheet, Evidence of Recent Input by Man. *Science*, Vol. 174, p. 692–4.

Composition of the atmosphere

C. E. Junge and P. Warneck

Max Planck-Institut für Chemie (Otto Hahn-Institut),
Mainz (Federal Republic of Germany)

This article was written during the first half of 1974 so that by necessity it reflects the research situation existing at the end of that time. Since then, due to the long delay in bringing this article to press and owing to the rapid advances in the field of atmospheric chemistry, new facts and also new problems have emerged. Their inclusion into the discussion would require more than just minor revisions of several sections. Thus, the authors feel, obliged to caution the reader that he will have to supplement the information given here with material from recent scientific literature.

General introduction

According to a widely accepted model, the atmospheres of the terrestrial planets in the solar system are derived from the outgassing of virgin planetary matter, the cause being heating by radioactivity and gravitational energy. Since it appears reasonable to assume that the planets Venus, Earth and Mars have accreted from the same type of meteoritic material, one would expect the outgassing products, and hence the atmosphere of the three planets to feature rather similar compositions,

provided just a simple accumulation took place and the individual atmosphere did not undergo any further change. The principal constituents of the three atmospheres as observed today are listed in Table 1. It shows that the atmospheres of Venus and Mars are indeed fairly similar, the predominant constituent being carbon dioxide in both cases. In the Earth's atmosphere, by contrast, the abundance of carbon dioxide is much less than that of the major constituents nitrogen and oxygen. However, if all the CO_2 buried as carbonate in the oceans and the sediments were brought into the atmosphere, carbon dioxide would rank as a principal atmospheric constituent on Earth also (Rubey, 1951). These observations suggest that carbon dioxide was a major outgassing product on all three planets, but that on Earth it has been largely removed by sedimentation. The presence of the oceans on Earth but not on the neighbouring planets follows from the different distances from the sun and, to some degree, from the different thermal histories of these planets (Rasool and DeBergh, 1970).

Among all the planets, the Earth is unique also because its atmosphere contains oxygen as a major constituent. Neither on Venus nor on Mars is oxygen present in significant amounts. The geological record indi-

TABLE 1. Comparison of abundances of constituents in the atmosphere of terrestrial planets

| Planet | Surface | | Atmospheric abundances[1] | | | | | |
	Temperature (degrees K)	Pressure (kg/cm²)	CO_2	H_2O	CO	N_2	O_2	O_3
Venus[2]	750	~ 92	5.3×10^7	$> 7 \times 10^3$	5×10^3	—	$< 2 \times 10^3$	—
Mars[3]	200 ± 50	0.01	7.8×10^3	~ 4	7	—	10	$\sim 3 \times 10^{-4}$
Earth	300	1	2.5×10^2	2×10^3	0.08	6×10^5	1.6×10^5	~ 0.3

1. Given by the thickness (in cm) of a uniform layer of the gas if its pressure is one atmosphere and its temperature 273 °K.
2. Abundances are only approximate. The value for CO_2 was calculated from the data by Marov et al. (1973).
3. Abundances are reasonably well known (McElroy and Donahue, 1972).

cates that oxygen was essentially absent also on the primitive Earth (Rutten, 1971). Its accumulation in the Earth's atmosphere is assumed to have resulted mainly from biological activity. Certainly today, photosynthesis by plants and algae acts as the major generator of atmospheric oxygen.

The Earth's atmosphere is not at thermochemical equilibrium at the currently prevailing temperatures. If it were, oxygen and nitrogen, for example, would not coexist. Instead they would combine with each other and with water to form nitrate as the chemically more stable entity. Atmospheric nitrogen is indeed being oxidized continuously at a rate of a few 10^{11} grams/year as the result of the action of ionizing cosmic radiation to which the atmosphere is exposed (Warneck, 1972). The rate of this process is too slow to deplete the present amount of nitrogen within the lifetime of the planet Earth, but the existence of such a process demonstrates that the Earth's atmosphere cannot maintain its non-equilibrium status indefinitely unless counteracting forces are at work which prevent the atmosphere from reverting to equilibrium. Biological processes are again thought to provide the necessary driving forces. Ultimately, of course, it is solar radiation that provides the required energy.

The preceding remarks make evident two important aspects concerning the present atmosphere of the Earth: (a) it is inherently unstable and is kept in a steady state by various driving forces; and (b) it is coupled to other geochemical reservoirs, i.e. the lithosphere, the hydrosphere and the biosphere. Hence it cannot be treated as an isolated body. Some atmospheric gases have simply accumulated by exhalation from the lithosphere. Most of the rare gases belong to this category. Other gases participate more or less strongly in the general geochemical cycles of material transport. To gain an insight into the atmospheric subcycles, one must consider for each atmospheric constituent within this category the influx from and the outflux into other geochemical reservoirs, in addition to chemical conversion within the atmosphere itself. Accordingly, the abundance of any individual atmospheric constituent is not a fixed quantity. It is determined by the balance between the various external and internal sources and sinks, respectively, of the atmospheric reservoir. Among the sources, one must also include those that result from human activities. These anthropogenic sources have, in recent decades, reached proportions comparable in some cases to natural sources so that they may disturb the balance set up by nature not only on a local but even on a world-wide scale. An intimate knowledge of the cycle of atmospheric constituents would thus enable us to forecast the effect of anthropogenic emissions in the future. Unfortunately, our knowledge of many of the cycles is still insufficient to make reliable predictions.

Table 2 presents a list of the most important gaseous components of our atmosphere, their approximate

mixing ratios, and information on their cycles (Junge, 1972). Also given are estimates of the associated times of residence in the atmosphere. The inverse of the residence time is a measure of the average rate of production or removal from the atmosphere and thus constitutes an important parameter for any atmospheric constituent. Table 2 is introduced here mainly to provide the required background information and guidance for the more detailed discussion to follow. It also illustrates the division of the atmospheric gases into two categories: cumulative and cyclic constituents. It is noteworthy that among those gases undergoing a cycle, the biosphere plays an important role and is, therefore, a determining factor in the composition of our atmosphere.

As for any other planetary atmosphere, the density of the Earth's atmosphere decreases approximately exponentially with increasing distance from the surface of the planet. The atmospheric region adjacent to the surface is called the troposphere. It is characterized by a fairly linear decrease of mean temperature with altitude

TABLE 2. Atmospheric gases arranged according to their cycle

Gas	Average mixing ratio p.p.m.	Residence time[2]	Cycle[2]	Satuts
A	9,300	—	1. No. cycle	Accumulation during Earth's history
Ne	18	—		
Kr	1.1	—		
Xe	0.09	—		
N$_2$	78×10^4	~10^6 years	2. Cycle mainly biological and microbiological	?
O$_2$	21×10^4	~10^4 years		
CO$_2$	315	~15 years		
CH$_4$	1.4	~9 years		
H$_2$	0.55	~5 years		
N$_2$O	0.25	~10 years		
CO	0.1	~1 years		
H$_2$S/SO$_2$	Variable in the p.p.b. range	Days to weeks	3. Sources mainly microbiological Sinks mainly physicochemical	Quasi-steady state or equilibrium (H$_2$O, CO$_2$?)
NH$_3$				
NO/NO$_2$				
HC[1]				
H$_2$O	Variable	~10 days	4. Cycle physicochemical	
O$_3$	Variable	~0.3 days		
He	5.2	~10^6 years		
Rn	Variable	3 days		

1. Various hydrocarbons, natural and anthropogenic, except CH$_4$.
2. Primarily tropospheric, exchange with ocean is not included.

up to the tropopause at a level of 12 ± 4 km depending on latitude and meteorological conditions. The atmospheric region above the tropopause is called the stratosphere. In this region the temperature rises again up to an altitude of approximately 50 km. Still higher layers are the mesosphere and the thermosphere. The data given in Table 2 pertain to the troposphere. This atmospheric domain comprises about three quarters of the entire atmosphere by mass and is reasonably well mixed. It is also that portion of the atmosphere that is coupled directly to the other geochemical reservoirs. Accordingly, the troposphere will be the centre of our interest. The stratosphere and the mesosphere are only weakly coupled to the troposphere by exchange processes and may be considered as separate reservoirs. They are, nevertheless, important atmospheric regions because of their enhanced photochemical activity. The resulting photo-oxidation reactions affect the mixing ratios of several trace gases in these atmospheric domains—although those of the main constituents remain unchanged. To the troposphere, the stratosphere may appear as a source or sink region depending on the trace component considered and the chemical reactions involved. The best known example of the influence on the tropospheric trace gas by the stratosphere is ozone, which originates in the stratosphere. Ozone appears also to be the only atmospheric constituent not coupled to other geochemical reservoirs.

The various aspects and pieces of background information assembled above essentially determine the format of this article. First, it will be useful to provide an outline of the definitions, concepts and criteria employed in describing the behaviour of atmospheric gases. Subsequently, the gases listed in Table 2 will be discussed individually with respect to their tropospheric abundances, the importance of natural and anthropogenic sources and sinks as well as correlated parameters and the interrelationships with other geochemical reservoirs.

General concepts and criteria

In an attempt to understand the origin of the chemical composition and the behaviour of the Earth's atmosphere, it is helpful to consider some general concepts before the discussion of the individual constituents. Such concepts are needed to distinguish the different categories of atmospheric gases, and to help us understand the processes responsible for maintaining the observed abundances. It is surprising that only very recently has the question of why our atmosphere has its present composition been considered, although this composition has been known for a long time. Classical meteorology took the composition simply for granted and most of the recent research contributions to our knowledge about the origin of the atmosphere came from geochem-

ists, geologists, oceanographers, biologists and astrophysicists rather than from meteorologists.

A very helpful and widely used concept is that of the geochemical reservoir. It can be defined as a limited volume of material of fairly uniform character with reasonably well delineated boundaries. Ideally, the content of a reservoir should be well mixed, i.e. the rate of internal mixing should be higher than the rate of exchange with any adjoining reservoir. The interactions between different reservoirs are studied by so-called box models, where it is usually assumed that the exchange rate is a first-order process. The choice of the number of boxes, i.e. reservoirs in the model, depends to some extent on the problem at hand and requires considerations, among other parameters of the time of residence of a substance in each reservoir. For long-term global developments of geological time-scales, the atmosphere and the ocean must be considered as a combined reservoir. For shorter time-scales, they behave as separate reservoirs. For time-scales of the order of years or less, the atmosphere must be subdivided: vertically into the tropospheric reservoir and the domain above the tropopause; and horizontally, into the two hemispheres.

A second concept, which has evolved only recently, concerns the status of an atmospheric constituent. As indicated in Table 2 one can distinguish between two categories of atmospheric gases: those gases which during the Earth history have primarily accumulated in the atmosphere or within the atmosphere–ocean system; and the other gases, whose concentrations are determined on a much shorter time-scale by dynamic or cyclic processes. Off-hand, it is not always clear to which category an atmospheric constituent belongs because a cumulative gas may also undergo a geochemical or biological cycle. Due to their inert nature, the rare gases—with the exception of helium as will be explained later—belong to the first category. Nitrogen has probably accumulated, but at the same time it undergoes a cycle. Carbon dioxide in the atmosphere is controlled largely by its chemical equilibrium with the oceans, but again things are complicated by the existence of at least two cycles, a fast one and a slow one. For oxygen, the situation is similarly complex. Evidently, a set of criteria is required which will enable us to assign the gaseous constituents of the atmosphere to one of the two described categories. The following criteria provide an objective and quantitative basis for such an assignment.

The total amount, (M_t), of a volatile constituent exhaled during the history of the Earth may occur as a 'mobile' fraction, (M_g), in the atmosphere–ocean system or in bound form as a deposit, (M_s), in the sediments of the lithosphere:

$$M_t = M_g + M_s. \tag{1}$$

If most of the constituent mass is present in the atmosphere–ocean reservoir, one has:

$$M_g \gg M_s, \tag{2}$$

and one can term the gaseous constituent accumulative. If, on the other hand:

$$M_g \ll M_s \quad M_t, \qquad (3)$$

it must be assumed that the fraction remaining in the atmosphere is governed by processes of material exchange with the lithosphere and related reservoirs. It is clear that there may be intermediate cases but currently it appears that oxygen is the only candidate for which neither criterion fully applies. For all the other gases the distinction between these two categories is clear cut.

The discussion of the main gases will show that oxygen, carbon dioxide and perhaps nitrogen undergo more than one cycle. Accordingly, the question arises of whether the concentration of a gaseous constituent is determined mainly by accumulation or by its cycle, or, when two cycles exist, which of the two cycles is the controlling one. In these cases an additional criterion is required. Let M_r denote the amount of the constituent in the geochemical reservoir with which M_g exchanges ($M_r \leqslant M_s$) during a cycle. If the situation is such that:

$$M_r \ll M_g, \qquad (4)$$

the cycle will have little influence on M_g and $M_r + M_g \quad M_g$ must then be determined by other processes. If, on the other hand:

$$M_g \ll M_r, \qquad (5)$$

it is very likely that the cycle provides the controlling factor. In the subsequent discussion of the main constituents of the atmosphere, N_2, O_2 and CO_2, these criteria will prove important.

For all the gases listed in Table 2 below carbon dioxide, except H_2O, one finds that $M_g/M_r \ll 1$. Thus criterion (5) is satisfied so that the atmospheric abundances of these gases are controlled by dynamic processes leading to a steady-state situation. Helium is a special cases in that it escapes into interplanetary space; the system is thus open and M_r cannot be determined. The residence time of helium is also much greater than that of the other gases in this category. Carbon dioxide and water vapour are controlled primarily by thermodynamic equilibria and not by their cycles, but these are nevertheless important with respect to the time-scales with which the· equilibrium concentrations readjust themselves.

Next, we consider the behaviour with time of a constituent in a geochemical reservoir. The amount, M, in the reservoir varies with time, t, according to the equation:

$$\left(\frac{dM}{dt} = Q - S, \right) \qquad (6)$$

where Q and S denote the sums of all the internal and external sources and sinks, respectively. The units most frequently employed and used also here for M are grams; those for Q and S grams per year. The sources and sinks may be chemical reactions within the reser-

voir or they may be fluxes through its boundaries. Specifically for the tropospheric reservoir, the sources and sinks may be reactions within the atmosphere, or reactions at the Earth's surface, and the fluxes may occur through the tropopause, for example, or the ocean's surface. In equation (6), the time variation of M results, of course, from the temporal behaviour of Q and S. In their variation with time one can distinguish two components: a short-term component such as diurnal or seasonal variations and a component varying on a much larger time-scale. If one averages equation (6) over a time period sufficiently long compared with the short-term variations, and if the long-term variation is small over the same time period, then $d\bar{M}/dt \approx O$, i.e. $\bar{M}$ will be essentially constant. For these conditions, equation (6) assumes the steady state approximation:

$$\bar{Q} - \bar{S} = O, \qquad (7)$$

where the bars indicate the time averages. For the atmospheric reservoir the averaging period should be a few years in order that the seasonal variations are averaged out.

In practically all cases only the sinks but not the sources are functions of M so that one can write:

$$S(\bar{M}) = Q. \qquad (8)$$

The time τ which, on the average, a molecule of a gaseous constituent spends in the atmospheric reservoir is given by:

$$\tau = \bar{M}/\bar{Q}. \qquad (9)$$

This quantity is here called the residence time of the constituent. In the literature one frequently finds also the term 'life-time' of the constituent. However, the expression residence time is preferable, because lifetime implies elimination by chemical reactions within the reservoir whereas in many cases the constituent is merely transported through the boundaries of the reservoir. The definition (9) is independent of the mathematical function of $S(M)$. A typical and simple case is a first order removal mechanism, either as a chemical reaction or as a transport through the boundaries. Then, $\bar{S} = s_o \times \bar{M}$, where s_o is a constant. It follows that $\tau = s_o^{-1}$. First-order chemical processes are rare in the atmosphere, because they require both a first-order reaction with another constituent and a uniform concentration (not mixing ratio!) of this constituent. The latter condition would require a vertical increase of the mixing ratio inverse to the decrease of air density with altitude.

The abundance of a constituent in a parcel of atmospheric air mass can be expressed as a concentration or as a mixing ratio. Concentrations, C, are given in units such as grams/m³ of air or number densities of molecules. Concentrations are required in considerations concerning transport, chemical reactions, etc. They change as the pressure and the temperature of the air parcel changes. Mixing ratios, m, are given in units such as grams per gram of air (mixing ratio by mass) or number of molecules per number of molecules of air (mixing ratio by volume). The mixing ratios are conservative properties in that they do not vary with temperature and pressure. It is obvious that in a well-mixed reservoir, m is constant. Transport by mixing occurs only along gradients of m, not of C; m and C are related by the air density, ρ, in the appropriate units:

$$m \cdot \rho = C. \qquad (10)$$

Usually, the sources and sinks of any atmospheric constituent are quite variable with space and time. In conjunction with the complex atmospheric mixing processes, any uneven distribution results in variations of m with time and space. But the variability of m depends also on the residence time. If m is the mixing ratio of a constituent averaged over the whole troposphere (and over a period of at least one year to eliminate daily and seasonal variations), the local values $M(x_i, t)$ can be expressed by:

$$m(x_i, t) = m + m'(x_i, t) \qquad (11)$$

where x_i ($i = 1, 2, 3$) are the three space co-ordinates, t is the time, and m' is the space and time deviation from m. A suitable measure for the variability of $m(x_i, t)$ is the relative standard deviation:

$$\sigma = \sigma^*(m')/\overline{m}, \qquad (12)$$

where $\sigma(m')$ is the absolute standard deviation. In Figure 1, σ is plotted against the residence time (both related to the whole troposphere) for all gases for which data are available. We see that in general there is an inverse relationship between σ and τ. The apparent exception of helium is quite certainly due to the inadequate accuracy of the measuring techniques needed to detect

the expected small variations. For oxygen, the accuracy is apparently just on the borderline. The sizes of the boxes indicate the estimated limits of uncertainty for σ and τ. Since the residence time, τ, is only one factor which has an influence on σ, some scatter of the data around the broken line must be expected as real. The solid line was obtained from simplified model calculations. Figure 1 suggests the approximate relationship $\sigma \times \tau = 0.14$. Although this relationship probably holds only within a factor of two or three uncertainty for any individual case, it is useful in air chemistry in that it offers a simple way to obtain an estimate for τ. The residence time, in turn, gives directly the global source or sink strength from equation (9). It is usually very difficult to obtain reliable data on strengths of sources and sinks for atmospheric gases on a global basis, particularly when the details of their cycles are still obscure. This fact will become evident in the subsequent discussion of the individual gases. If the residence time is unknown, a suitable set of data of m within the troposphere provides $\overline{m}$ and σ, and hence estimates of τ and $\overline{Q}$ or $\overline{S}$, so long as it can be assumed that for the gas, steady-state conditions hold. In addition to the possibility of obtaining estimates of τ and Q, the data on m also furnish information on the distribution of important sources and sinks, which in turn provide a clue for their nature. These considerations clearly indicate the importance of a solid documentation of the behaviour of m in the atmosphere before meaningful conclusion about the cycle of the gas constituent can be reached.

Another aspect of Figure 1 is the following. If it has been ascertained that a constituent has a residence time of more than 0.5 year, one can expect that σ is of the order of 0.3 or less. As a consequence, the constituent is fairly uniformly distributed throughout the troposphere from pole to pole and one does not have to pay too much attention to the complex atmospheric transport and mixing processes unless, of course, one is interested in greater details. On the other hand, as τ drops below one month, the variations of m will become so large that considerations of the regional and large-scale transport process become mandatory. These aspects are important in dealing with the questions of the global spread of pollutants which affect the biosphere (e.g. pesticides and polychlorinated biphenyls) or the climate (e.g. aerosols).

We mentioned above that, for a well-defined geochemical reservoir, the internal mixing time should be small compared to the time of exchange within the adjoining reservoirs. Since the residence times of atmospheric gases cover a wide range over several orders of magnitude, it is perhaps useful to convey some information on the internal mixing and exchange times of tropospheric reservoirs. The characteristic mixing time within a reservoir can be estimated by the well-known Einstein equation:

$$\overline{X^2} = 2 D \cdot t \qquad (13)$$

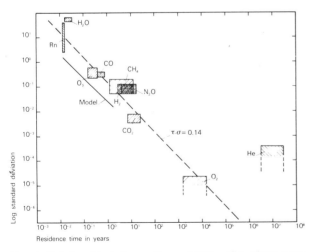

FIG. 1. Relationship between the variability of a minor atmospheric constituent gas and its residence time in the troposphere. The empirical relationship is $\sigma \cdot \tau$ 0.14, the solid line from a simple model predicts a similar but numerically slightly different result.

where x is a characteristic dimension of the reservoir and D is the *eddy* diffusion coefficient. For lateral mixing within both tropospheric hemispheres, $D \approx 3 \times 10^{10}$ cm^2 sec^{-1} (Bolin and Keeling, 1963) and x can be taken as two-thirds of the distance between each pole and the equator. The resulting time is $t \approx 5$ months. The application of (13), which was derived for molecular diffusion is, however, only justified if the speed of the mixing processes characterized by the magnitude of D is sufficiently uniform in the troposphere. There is strong evidence that at the equator the effective value of D drops considerably compared with that at higher latitudes, resulting in an exchange time between two hemispheres of about a year (Czeplak and Junge, 1974). These facts provide a sound basis for approximating the troposphere by a two-box model each box representing respectively the southern and the northern hemispheres, but any further subdivision of the troposphere is not justified. The inter-hemispheric exchange time of about a year indicates that there will be little global net transport across the equator for any constituent with residence times of the order of about one month or shorter.

For vertical mixing within the troposphere an average value for the eddy diffusion coefficient of D 2×10^5 cm^2 sec^{-1} can be assumed, resulting in a mixing time of about a month. It should be emphasized that these mixing times apply only to average meteorological conditions and that high-reaching convection processes and large-scale vertical motions in active weather systems can achieve a vertical exchange time of a few days, even though only over a restricted area.

For several gases the ocean acts as a source or as a sink (it is a source for N_2O, CO and H_2 and a sink for SO_2 and CO_2). Hence it is necessary to consider the exchange between the ocean and the atmosphere. The exchange rate for gases with low solubility such as CO_2, CO, H_2, N_2O and CH_4 is controlled essentially by molecular diffusion in a thin laminar layer at the water surface. The vertical flux within the layer is given by:

$$F = (C_i - C_w)D/z \qquad (14)$$

(Tsung-Hung Peng, 1973) where C_i is the concentration of dissolved gas at the air–water interface, in equilibrium with the air concentration according to Henry's law, and C_w is the respective concentration in the main body of water, both concentrations given in mole cm^{-3}: D, the molecular diffusion coefficient of the gas in ocean water, is given in cm^2 sec^{-1} and z (in cm) is the thickness of the stationary liquid film. This simple model has shown to be applicable with fair accuracy for a variety of gases. The parameter, z, accounts for the hydrodynamic properties of the system and depends on the state of agitation of the ocean surface. On the average this parameter was found to vary between 50 and 80 μm for different regions of the oceans and for different seasons.

The ocean itself is by no means a uniform reservoir. A layer at the ocean's surface with a depth of about 100 m is mixed fairly rapidly by wave action and other atmosphere–ocean interactions. This mixed layer can be considered the ocean reservoir with which the atmosphere exchanges fairly rapidly. Its lower boundary is poorly defined, however, and may not even exist everywhere, depending on the ocean structure and stratification. Hence it is doubtful that the exchange with the deeper waters of the ocean is nearly a first-order process. The turnover times associated with the larger body of the deep ocean are long, of the order of 500 to 1,000 years. The apparent exchange times between the mixed layer and the deep ocean are shorter, i.e. several decades (see, for example, Keeling 1973).

The land surface can act as another important source or sink for trace gases. Contrary to the gas exchange mechanism at the ocean surface, the details of the processes occurring at the land surface are not well understood. If the soils acts as a source, the production of the gas occurs in a layer of certain depth, as in the cases of CO_2, N_2O, or radon. The exchange with the atmosphere is controlled by diffusion through the soil capillaries. This process depends in a complex manner on the soil structure, the moisture content and meteorological conditions (Israel, 1962). If the earth surface acts as a sink, the destruction occurs at all surfaces including vegetation, etc. In some cases microbiological activites apparently play an important role (CO, H_2). Above such surfaces the concentration of the gas in the air decreases by turbulent diffusion towards the surface. The turbulent diffusion flux will equal the chemical or other destruction rate. For rough estimates, the magnitude of the flux towards the surface can be represented by the deposition velocity at which the gas column in the surface air must move through an area surface element to equal the measured flux (see, e.g. Slade, 1968). Velocities for different gases do not differ greatly under normal conditions and are of the order of 10^{-1} cm sec^{-1}.

At the upper boundary of the troposphere, the tropopause, exchange takes place with the stratospheric reservoir. Studies of the behaviour of radioactive debris from bomb tests injected into the stratosphere have indicated residence times within the lower stratosphere; these increase from less than about 0.5 year in the lowest region, a few km above the tropopause, to about 2 years for the region at about 25 km altitude (Martell, 1970). It is important to note that, due to thermal stability, the stratosphere is anything but a well-mixed reservoir, so the concept of residence times should be applied only with great caution. The vertical eddy diffusion coefficient, D, decreases rapidly from tropospheric values of about 2×10^5, to average values around 3×10^3 cm^2 sec^{-1} in the lower stratosphere. According to (13), this value gives vertical mixing times of about 2.5 years for $X = 5$ km and 0.4 year for $x = 2$ km—in reasonable agreement with experimentally measured values. It should be noted that these D-values are effective values derived from observations on a variety of different exchange processes, such as slow organized circulations, injections from the stratosphere into the troposphere through the tropopause breaks, small-scale turbulence, penetration of the tropopause by high-reaching cumulonimbus clouds and the seasonal changes of tropopause height. If we take an average exchange time of 2 years as representative of the whole mass of the stratosphere, and a mass ratio troposphere to stratosphere of 4:1 the total exchange between the troposphere and the stratosphere would require 8 years. This demonstrates that, chemically, both reservoirs are fairly well decoupled so that only long-lived constituents will be subject to exchange. For constituents with tropospheric residence times of a few years or less, the stratosphere cannot provide significant sink.

Formation of the Earth's atmosphere by exhalation and the presence of the noble gases in the atmosphere

It is now fairly well accepted that the atmosphere of the Earth, including the water of the oceans, the CO_2 contained in the carbonates and most of the other volatile substances in the sedimentary shell, has accumulated in the course of the Earth's history at the earth surface by thermal outgassing processes and volcanic activity. Brown (1949) and Suess (1949) first noted that, compared with nuclear abundances in the solar system, the abundance of the rare gases on the surface of the Earth is exceedingly rare. For xenon, for example, the deficiency factor is of the order of 10^{-7}. Similar deficiencies have been obtained for primordial rare gases in certain meteorites (Signer and Suess, 1963). These observations are generally interpreted to indicate that the chemically inert gases had been separated from the condensed matter before the final agglomeration of the Earth took place and that only a small fraction was incorporated into the solid matter. A similar separation is assumed to have occurred for other gases, so that in the very early stages of its history the Earth had practically no atmosphere.

On the basis of considerations of the thermal history of the Earth (Birch, 1965), it can be assumed that the exhalation rate due to thermal processes was initially much faster than it is today, so that most of the volatile material had already accumulated on the Earth's surface during the first 1.5×10^9 years after its formation (Li, 1972). The present rate of exhalation by volcanic activity is only a very small fraction of the early one. Geochemical data show that the present composition of the sediments, the ocean and the atmosphere combined, can be quantitatively explained by the exhalation of such volatile components as CO_2, H_2O, H_2S,

HCl, etc., their subsequent reaction with the igneous rocks then present and the formation of clay minerals. It is, therefore, reasonable to expect that the small fraction of primordial noble gases retained in the solid matter prior to the formation of the Earth was exhaled along with the other volatile constituents and accumulated slowly at the Earth's surface. The pertinent gases are neon, argon 36/38, krypton and xenon. The possibility cannot be excluded, however, that a portion of the neon may have been derived from solar wind. Because of their chemically inert character, probably only a very minor fraction of these noble gases was re-incorporated into the sediments and system. The greater part had no choice other than to remain in the atmosphere–ocean system. Thus there is good reason to regard these rare gases as accumulated gases and they are listed as such in Table 2.

The atmospheric components helium and argon 40 (^{40}Ar), are of special interest, because neither are primordial but rather derived from radioactive decay. Argon is heavy enough to be retained by the Earth's gravitational field, but helium can escape slowly to interplanetary space; thus it is not accumulated to the same degree as argon. ^{40}Ar is derived from ^{40}K, whereas helium results from the decay of uranium and thorium. If the Earth's content of these elements (concentrated in the crust plus upper mantle) were sufficiently well known, the present production rate and the total amount of argon and helium produced since the formation of the Earth would be directly calculable. Specifically for argon, one would then be able to deduce the partitioning between argon released to the atmosphere and that retained by the solid Earth. At present, only an order of magnitude estimate is available for the average potassium content of the Earth. Using the potassium abundance value recently given by Ganapathy and Anders (1974) 170 p.p.m. by mass, coupled with the abundance of ^{40}K in potassium, 1.2×10^{-4}, a total argon production over the last 4.5×10^9 years may be calculated which is about twice the amount existing at present in the atmosphere and dissolved in the oceans. If the Earth had the composition of chondritic meteorites (an incorrect assumption), the total production of argon would be 5 times higher (Macdonald, 1964) and this is certainly the upper limit. These estimates demonstrate that even though a major fraction of the radiogenic argon produced has been liberated to the atmosphere, a comparable fraction has not reached the surface. Presumably this is true also for the other non-radiogenic rare gases and helium.

Similar considerations for helium result in a total production over the period of 4.5×10^9 years of more than 1,000 times the amount existing at present in the atmosphere. It is unlikely that the solid Earth has retained significantly greater amounts of helium than of argon. For example, Zartman et al. (1961) have shown that most natural gases, including samples from geothermal areas, have radiogenic ^4He:^{40}Ar ratios which, on the average are close to their production ratio in normal rocks: ^{4}He:^{40}Ar 10. The helium deficiency in the atmosphere is generally attributed to escape from the outer boundary of the atmosphere. The escape rate is difficult to estimate quantitatively. In addition to thermal escape of neutral particles, there exists an ionic escape mechanism which may even be more efficient (Lemaire and Scherer, 1973). The combined escape rate is diminished by the accretion of helium from the solar wind (Reasoner, 1973). Rather detailed model considerations are required to treat the complexities of these processes. Thus it is not definitely established that a steady-state situation has developed for helium. If steady state is assumed, the residence time of helium in the atmosphere can be estimated from the current radiogenic production rate. Wasserburg *et al.* (1963) have combined the most frequently observed ratio of radiogenic helium to argon in natural gases with the known content of argon in the atmosphere to derive a radiogenic helium flux of about 3×10^9 g/year; this leads to a residence time for helium in the atmosphere of about 10^6 years. Clearly, these values are beset with considerable uncertainties, but at least they do not depend on any assumption concerning the uranium and thorium contents of the Earth. In fact, however, a similar helium flux can be obtained from the approximately known abundances of these elements.

The main constituents of the Earth's atmosphere

This section is devoted to those constituents which, from the quantitative viewpoint as well as that of importance, have to be considered the major gases of our atmosphere. All of these gases play a vital role on Earth. With the exception of nitrogen they determine to a large extent the radiation budget and thus the thermal structure of our atmosphere: oxygen and its reaction product ozone are principal absorbers in the ultraviolet and visible wavelengths; carbon dioxide and water vapour absorb and emit radiation primarily in the infrared spectral region. Accordingly, the abundances of these gases and their possible variations are of special concern to atmospheric science. An additional point of interest is that these constituents differ so widely among the atmospheres of the inner planets of our solar system a fact which raises the interesting question of why these differences developed.

NITROGEN

In the preceding section it was shown that, with the exception of helium, the abundances of all the noble gases are determined by accumulation. Nitrogen and possibly oxygen are the only other atmospheric constituents

whose abundances are also determined mainly by accumulation—as shown in Table 2. Geochemical data indicate that about 80 per cent of the total nitrogen present on the earth surface resides in the atmosphere, 20 per cent is bound in sediments or rocks or occurs as nitrate of biogenic origin in the soil and in the oceans (Wlotzka, 1972). Nitrate is a vital nutrient for all plants and micro-organisms and is formed and decomposed within the biological cycle on a quasi-steady-state basis. Atmospheric N_2 is part of this cycle in that it serves as a supplier of bio-nitrogen. From the amount of nitrogen in living organic material and the rate of decay, one can estimate that the biological cycle of nitrogen results in an atmospheric residence time of N_2 within this cycle of roughly 10^6 years. In the presence of free oxygen, nitrate is more stable on the thermochemical energy scale than nitrogen. If the entire amount of oxygen in the atmosphere (21 per cent by volume) were used to oxidize nitrogen to nitrate by the net reaction:

$$N_2 + 2\tfrac{1}{2}O_2 + 2\,OH^- \longrightarrow H_2O + 2NO_3^-,$$

the amount of nitrogen in the atmosphere would be reduced by about 11 per cent. The oxidation of nitrogen to nitrate is prevented by the biological cycle which keeps the reservoir of biogenic nitrate at a relatively low level. Nitrate is demineralized primarily by microbiological activites producing N_2, N_2O and perhaps some NO. No satisfactory data exist on the relative proportions of the individual production rates but it appears that N_2 is the major product (Matsubara and Mori, 1968). Because the nitrate–nitrogen reservoir is small compared with the free nitrogen reservoir in the atmosphere, it is clear from criterion 4 in Table 2 that the biological cycle of N_2 cannot determine its atmospheric abundance. It is quite likely, on the other hand, that the nitrate reservoir is determined by the N_2 reservoir. Consequently we conclude that the present level of N_2 in the atmosphere is primarily controlled by accumulation during the earth history and only to a very small degree by the biological cycle.

(It should be pointed out that N_2 may also participate in a very slow inorganic cycle similar to those discussed later for oxygen and CO_2, based on the turnover of the sediments, but its magnitude is entirely unknown. However, such a cycle would also be unimportant in determining the abundance of atmospheric nitrogen for the same reasons as the fast organic N_2 cycle is unimportant.)

OXYGEN

Oxygen is the second most abundant atmospheric constituent. As for nitrogen, the amount of oxygen dissolved in the ocean is only about 1 per cent of the total, so that the total free oxygen approximates to the levels in the atmosphere- As indicated in Table 2, it is not clear whether the present level of atmospheric oxygen is primarily determined by slow accumulation of biologically produced O_2 or whether it is controlled by a steady state due to a cycle involving production and loss. The following discussion about oxygen (and also, to some extent about CO_2) concerned with geochemical aspects, is largely based on a critical review by Schidlowski et al. (1974). Figures of reservoirs, fluxes, residence times for the cycles of these gases and other data in the subsequent section are taken from this review with minor modifications and simplifications and will not be referenced individually.

It is now generally accepted that the primordial atmosphere was highly reducing and did not contain free O_2. Considerable geological evidence recently summarized by Rutten (1971) showed that free oxygen did not appear to any degree in the atmosphere until after about 2×10^9 years ago. On the other hand, it is certain that photosynthetic production of O_2 was already in progress more than 3,000 million years ago. Geochemical data show that only 5 per cent of all the oxygen ever produced is present as free O_2 in the atmosphere, whereas 95 per cent, i.e. the dominant part, was used to oxidize iron and sulphur in the sedimentary shell. The data show further that the total oxygen in the atmosphere is approximately equivalent to that of organic carbon found in sediments. Only 1 per cent of the organic carbon is present in the form of coal or oil deposits which can be commercially exploited. These facts strongly support the idea that atmospheric oxygen derived from CO_2 by photosynthesis. An alternate source of O_2 would be the dissociation of H_2O by short-wave solar radiation in the high atmosphere followed by the escape of the resulting hydrogen into space from the Earth's exosphere. Estimates of the present rate of H_2 escape, however, if correct, indicate that it is too low by several orders of magnitude to explain the total O_2 reservoir in the Earth's crust (Hunten, 1973); it is possible that the exosphere may, in the past, have featured higher temperatures than today, a factor that would have increased the hydrogen escape flux. Even though the role of O_2 production via H_2O dissociation is still an open question, available evidence today tends to indicate that O_2 production from CO_2 by the process of photosynthesis was the dominant factor.

It is not impossible that the 5 per cent O_2 in the atmosphere accumulated only after most of the sulphur and bivalent iron had been oxidized and that this would not have occurred if a greater amount of sulphur and/or bivalent iron had been available at the Earth's surface. However, there are still considerable amounts of oxidizable sulphur and bivalent iron available in the present sedimentary shell. Possibly, the oxidation process depends on the partial pressure of oxygen in the atmosphere. As the amount of oxidizable material decreased, the O_2 partial pressure would then have increased to the same extent as the further oxidation of sulphur and bivalent iron became increasingly difficult.

While the processes of this kind can be considered accumulative, oxygen is also involved in at least two well-distinguishable cycles which, in principle, might be responsible for the present O_2 partial pressure since a steady state may be set up (Schidlowski et al., 1974). These two cycles differ considerably in the resulting residence times for O_2 in the atmosphere–ocean system and we shall distinguish between them accordingly.

The fast cycle is due to the production of O_2 by photosynthesis, on one hand, and the consumption of O_2 by respiration and the microbiological decomposition of organic matter on the other. The turnover time of atmospheric O_2 due to this cycle would amount to 2×10^4 years. The time constant can be estimated fairly well from the corresponding CO_2 cycle which is well known and will be treated subsequently.

In terms of geological time-scales, photosynthesis results in a very short and active cycle. The O_2 reservoir with which the atmospheric oxygen exchanges is that of the free CO_2 in the atmosphere–ocean system and this reservoir is only about 10 per cent of that of the free oxygen in the atmosphere. It has been well established that the actual O_2 molecule liberated by the photosynthetic process originates from the water molecules involved, but the size of the exchangeable reservoir is determined by the available CO_2. According to criterion (4), therefore, this fast, biological cycle cannot determine the atmospheric O_2 reservoir. This conclusion is quite similar to that in the case of N_2.

The second, slow O_2 cycle seems to be more important as a potential stabilizer for the atmospheric partial pressure of O_2. The slow cycle has its origin in the fact that the fast cycle is not exactly balanced. Each year between 0.1 and 1 per cent of the biomass present at the Earth's surface is buried as organic carbon in the normal process of sedimentation together with other reduced material (e.g. sulphur and bivalent iron) and a corresponding surplus of O_2 remains in the atmosphere. Oxygen from this source is in turn used up in the process of oxidizing those amounts of organic carbon, sulphide and bivalent iron which become exposed due to weathering during the slow turnover of the sediments. The latter process occurs on a time-scale of about 2×10^8 years. The resulting residence time of O_2 in the atmosphere–ocean system would be about 3×10^6 years, i.e. still short with respect to the geological time-scales.

It can be assumed that this cycle is fairly well balanced and may control, at least to some extent, the O_2 partial pressure by the steady-state conditions.

The O_2 exchange reservoir in this case comprises practically the total amount of O_2 contained in sedimentary SO_4^{2-} and Fe_2O_3, an amount about 20 times greater than that of atmospheric O_2. Thus on the basis of criterion (4), this cycle would have the capacity for controlling the atmospheric partial pressure. Clearly, the slow O_2 cycle depends among other things on certain geological parameters such as the average burial rate of organic carbon, sulphides, etc., or the exposure of these materials to the atmosphere through the turnover of the sediments. The parameters may show long-term fluctuations so that in equation (7), both $S(M)$ and Q and, as a consequence, also M may vary to some extent.

In conclusion, it appears that the O_2 abundance in the atmosphere may be determined either by a rather late accumulation process, or by the slow and variable geochemical cycle discussed above—or by both; it is not possible at the present time to provide a definitive answer to the question of which of the two mechanisms is the controlling factor.

CARBON DIOXIDE

Carbon dioxide in the atmosphere is of particular importance for life on Earth as well as for the long-wave radiation budget of the atmosphere at altitudes up to 100 km. This gas has received considerable attention as a climatic factor, particularly since it has been demonstrated that its partial pressure is at present continuously rising due to the burning of fossil fuels. In general, we have now a fairly good understanding of the processes determining the behaviour of CO_2 in the atmosphere, even if there still remain some open questions, especially with respect to a quantitative interpretation of CO_2 observations. The following specific points will be discussed: (a) The chemical fixation of CO_2 by the ocean; (b) The slow CO_2 cycle; (c) The fast CO_2 cycle; (d) The CO_2 increase due to man's activities; and (c) The prediction of future CO_2 levels.

The chemical fixation of CO_2 by the ocean

The atmosphere–ocean system is much more complex for CO_2 than for other gases since CO_2 in air stays in equilibrium with three different chemical species in water: H_2CO_3, HCO_3^- (bicarbonate) and CO_3^{2-} (carbonate). The most important cation for CO_2 in ocean water is Ca^{2+}. The relative abundance of the various carbonate species depends strongly on the pH value and the temperature of ocean water, as well as on the partial pressure of CO_2 in air. Only the concentration of H_2CO_3 in water is directly proportional to the CO_2 partial pressure in air—according to Henry's law. In contrast to practically all other atmospheric gases, the total amount of CO_2 dissolved in the ocean is much higher than that present as CO_2 in air. About 60 times more CO_2 resides in the oceans, so that the total CO_2 reservoir of the atmosphere–ocean system is practically equivalent to that of the ocean. There are several papers which discuss this sytem in detail, e.g. by Keeling (1973).

It is very likely that the CO_2 partial pressure in the atmosphere has been fixed at about the present levels since very early pre-Cambrian times. According to recently developed concepts of the origin of the oceans and the sedimentary shell, the accumulation of volatiles at the Earth's surface by volcanic exhalation during the history of our planet can be compared to a giant titration process (Sillen, 1963) in which acids such as H_2SO_4, HCl and H_2CO_3 were responsible for the wea-

thering of the igneous rocks and other material exposed at the Earth's surface. Together with the process of reversed weathering, the oceans were formed together with the salts dissolved in them, as well as the sediments composed of quartz, carbonates, clay minerals, trivalent iron deposits, etc. (Li, 1972). This concept suggests that the relative composition of the sediments and of the ocean water did not change very much throughout the ages. In particular, it can be assumed that both the Ca^{2+} and the H^+ ion concentrations in the oceans were fairly well fixed by the ion-exchange capacity of the clay minerals. Both these ions control the concentrations of bicarbonate and carbonate ions, which in turn, determine the CO_2 partial pressure in the atmosphere. It follows that the CO_2 partial pressure has remained at more or less the present level during most of the Earth's history. This conclusion is supported by the fact that the oldest known carbonate sediments—more than 3×10^9 years old—were formed by the activity of photosynthetic micro-organisms related to the blue algae species still living in our oceans today. Apparently, the basic environmental conditions, excepting the absence of oxygen, were rather similar.

The oceans are practically saturated with respect to carbonates and any CO_2 exhalations by volcanic activities must result in an increase in total carbonate (and organic carbon as shall be discussed later) in the sedimentary shell. If the oceans had not been present on earth no capability of carbonate formation and corresponding fixation of the CO_2 partial pressure would have existed. The exhaled CO_2 now buried in the sediments would have remained within the atmosphere, resulting in conditions similar to those on Venus and Mars, where CO_2 is the dominant constituent of their atmospheres.

According to this concept, the CO_2 level of our atmosphere is fixed by a chemical equilibrium, a fixation process apparently unique among all the atmospheric gases of lesser concentrations, most of which are in a steady state determined by cycles of production and loss processes. However, as indicated in the beginning of this section, cyclic transport occurs also for CO_2. Depending on the relaxation time of the CO_2 fixation process by the oceans, deviations from the chemically fixed CO_2 level are expected to occur by perturbations in these CO_2 cycles; for example, by variations in the injection rates of CO_2 by volcanoes or by the burning of fossil fuels. This question will be discussed in more detail subsequently.

The slow CO_2 cycle

Similar to O_2 and possibly N_2, carbon dioxide participates in two cycles, a fast one caused by biological activities and a slow one, caused by weathering. Accordingly, the question of whether these cycles can be of importance in controlling the atmospheric CO_2 level or what other roles they may assume requires discussion.

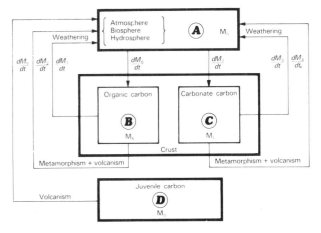

FIG. 2. Box model of the terrestrial carbon cycle (Holland, 1974). Juvenile carbon constituents (mostly CO_2) released to the atmosphere–ocean system (flux dM_5/dt) leave the latter as either organic or carbonate carbon (fluxes dM_6/dt and dM_7/dt, respectively). C_{org} and C_{carb} of the sedimentary reservoir are being gradually recycled by weathering (dM_1/dt, dM_3/dt) or metamorphism (dM_2/dt, dM_3/dt). There is reason to believe that the juvenile supply is at present almost exhausted, with reservoirs A, B and C being in a steady state.

Carbonates exposed to the atmosphere are dissolved as bicarbonates according to the weathering equation:

$$CaCO_3 + H_2O + CO_2 \longrightarrow Ca(HCO_3)_2 .$$

The soluble bicarbonate is carried into the oceans by the rivers of the world at a rate of about 2×10^{14} g carbon per year. Due to the near saturation of the oceans with carbonate, the identical amount of carbonate is deposited annually in the sediments.

As a consequence of tectonics, these carbonates will eventually become exposed again at the Earth's surface and be subjected again to weathering whereupon the cycle is closed. Figure 2 shows a box model for these processes where some CO_2 reflux by metamorphosis of carbonates is also included. Further, a corresponding cycle for the organic carbon in the sediments is shown (Holland, 1974). The model implies that the total amount of sediment remains unchanged by weathering and metamorphosis (the latter assumption may be questionable) and increases only as a result of juvenile exhalations. The carbonate cycle is accompanied by a cycle of organic carbon, which occurs in the sediments at a fairly constant proportion of 25 per cent of carbonate carbon (Ronov, 1968). After exposure, this organic carbon is either oxidized to CO_2, or is carried by the rivers as detritus into the ocean. The proportions depend on the atmospheric O_2 partial pressure and other environmental factors. Under steady-state conditions this loss is balanced by an equivalent amount of organic carbon being buried in the sediments. Since the organic carbon is produced by photosynthesis, this part of the slow carbon cycle is closely connected with the slow oxygen cycle as previously discussed. It is generally assumed that the weathering rate of organic carbon, which is primarily contained in the more resistant shales, is slower by a factor of about 2.

The sizes of the various carbon reservoirs and the resulting residence times within these reservoirs based on the weathering rates are indicated in Figure 2. It is seen that the turnover time for the ocean–atmosphere system due to carbonate weathering is about 2×10^5 years, and that for the carbonates in the sediments is 2×10^8 years. If we assume complete oxidation of organic carbon during its cycle, the corresponding residence times are about 8 times higher in the first case and 2 times higher in the second.

The slow carbonate cycle is of great importance in geology but does not have any net effect on the average CO_2 level of the atmosphere–ocean system. However, the residence time of 2×10^5 years sets the time-scale at which any perturbations of CO_2 in the atmosphere–ocean system will be eliminated. Such perturbation may arise by long-term fluctuations of the volcanic exhalation rate, by slow changes in the weathering rate due to geological changes in the exposure of carbonates, or by present human activities. Since the CO_2 level is controlled by the Ca^{2+} and H^+ concentrations in sea water, at least one cycle of weathering is needed to restore these concentrations due to the slow processes of reversed weathering, ion exchange and other processes. Although 2×10^5 years is a short period with respect to geological time-scales, it may be of interest with respect to climate variations and related phenomena.

The cycle of organic carbon alone is not as rigidly controlled as the carbonate cycle and its variation may result in perturbations of the CO_2 level. But any perturbations of this cycle affecting the CO_2 level will be smoothed out by the carbonate cycle as discussed above, so that again the slow organic-carbon cycle can have no influence on the long-term average CO_2 level in the atmosphere–ocean system. In conclusion, we can therefore state that the slow carbon cycle does not exert any control on the CO_2 level but will be of importance for the time-scales at which any perturbations of this level will be eliminated.

The fast CO_2 cycle

The fast cycle of CO_2 is determined by photosynthesis and is closely connected to the fast cycle of O_2. The associated residence time of CO_2 in the atmosphere alone, i.e. excluding the oceans, is extremely short, namely about 25 years. The corresponding residence time for the whole atmosphere–ocean system is about 1,500 years. Since the turnover time of the deep ocean is estimated to be between 500 and 1,500 years, it is clear that the internal mixing time of this reservoir and the residence time of CO_2 are of the same order of magnitude. This situation requires that the cycle must be delicately balanced or otherwise considerable deviations may occur. The result of this fast cycle is a global seasonal variation of the CO_2 content in the northern hemisphere (where most plants are growing) of about 7 p.p.m. out of 320 p.p.m. (Bolin and Keeling, 1963). It can be shown that this amplitude is in close agreement with independent calculations of the total carbon uptake of the plant cover (Junge and Czeplak, 1968). About 20 per cent of the organic matter produced by photosynthesis is directly converted. The rest is liberated as CO_2 from dead organic matter by microbiological activity on a time-scale ranging from less than 1 year to more than 5 years. Only a small fraction (of about 0.1 per cent) is buried as organic carbon in the sediments as discussed in connection with the slow cycles of O_2 and CO_2.

The fast CO_2 cycle cannot control the level of CO_2 in the atmosphere–ocean system for two reasons. First, the total biomass at the Earth surface and within the ocean is only about one-tenth of the total CO_2 content of the atmosphere–ocean reservoir, so that criterion (4) applies and, secondly, the CO_2 level is chemically fixed by the oceans anyway. The extremely short turnover time of this cycle may nevertheless result in small perturbations over periods from a few years to several hundred years due to imbalances between the annual production and decomposition of the biomass as a result of seasonal climatological deviations over large regions with intense plant growth and its slow and poor mixing with the deep ocean. The question of the effects of such deviations from steady state on the atmosphere–ocean CO_2 system leads us to the next point.

The CO_2 increase due to man's activities

The burning of fossil fuel by man is an unintended large-scale geochemical experiment in which large amounts of organic carbon stored over ages in the sediments are put back into the atmosphere over a period of 100–200 years, a time-scale very short compared with geological time-scales. What can be expected to happen as a result of such an injection into our atmosphere?

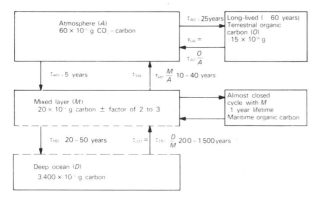

Fig. 3. Box model for the exchange of CO_2 in the atmosphere–ocean–biosphere system. Reservoir contents and exchange times are indicated.

The exchange time of atmospheric CO_2 with the mixed layer of the ocean as indicated in the box model shown in Figure 3 is about 5 years, i.e. fairly short. The resulting decrease of the CO_2 partial pressure in the atmosphere will be small due to the comparatively poor buffering capacity of the ocean water. Its pH will decrease correspondingly. Due to the small buffering capacity of the oceans, only about 10 per cent of the CO_2 input into the air is absorbed by the mixed layer, i.e. by a CO_2 reservoir of approximately the same size as the atmosphere. Equilibrium with the whole ocean including the deep sea would limit the atmospheric CO_2 increase to about 15 per cent of the input, but to achieve the equilibrium would require about 1,000 years, i.e. the average turnover time of the deep sea. The remaining 15 per cent of the increase will be reduced slowly as the ocean is flushed by the slow weathering cycle whereby the Ca^{2+} and H^+ concentrations are restored. As discussed above, this process occurs on a time-scale of about 2×10^5 years. It is quite clear from these qualitative considerations that the reduction or even elimination of man-made perturbations will be extremely slow compared with the time-scales of human activities.

It is now empirically well established that due to the use of fossil fuel, the atmospheric CO_2 level has risen on a global scale from about 290 p.p.m. at the turn of the century to the present value of 320 p.p.m.—and is rising at a rate of almost 1 p.p.m. per year (Smic Report, 1971). During this process, close to 50 per cent of the total CO_2 production has remained in the atmosphere. The other half has been absorbed by the upper layers of the oceans and the land biosphere. Parallel with the increase of atmospheric CO_2 the specific ^{14}C activity of atmospheric CO_2 has decreased by about 2 per cent (Suess effect).

Because of the possibility of changes in climate as a result of the increased CO_2 level, several attempts have been made in recent years to simulate these observations quantitatively in model calculations. The basis for such calculations were box models with 3, 4 or 5 separate reservoir of which Figure 3 gives an example. Included in this figure are the sizes of the various reservoirs and typical values of the exchange times between these reservoirs, some of which are still rather uncertain (e.g. those between the mixed layer and the deep sea). It is generally assumed that an exchange between the reservoirs is first order. If the parameters of these models are adjusted within the estimated limits of uncertainty to fit the data, most models can approximate the observations fairly well. However, this result does not necessarily indicate that the models are correct. It remains to be shown that at least the assumptions of well-mixed reservoirs and first-order exchange rates between the reservoirs are applicable. Probably the assumptions are rather crude for the ocean and the biosphere. For details, the reader is referred to the excellent review of these models by Keeling (1973).

Prediction of future trend

The main purpose of the models is to understand the atmosphere–ocean–biosphere CO_2 system quantitatively and to make an attempt to predict the future trend. Besides the difficulties inherent in the various box models, the most uncertain quantity which enters into these predictions is the future trend of fossil fuel consumption. This uncertainty may be the reason that practically none of the predictions go beyond the year 2000. For the near future it is probably reasonable to assume that the world consumption of fossil fuel will continue to grow at a rate of about 4 per cent a year until 1980 and then drop to 3.5 per cent due to the expansion of nuclear sources of energy (Smic Report, 1971). With such assumptions, the models predict increases of atmospheric CO_2 up to the year 2000 ranging between 20 and 50 p.p.m. with values around 30 to 40 p.p.m. being most likely.

However, the year 2000 is not far away and it is obvious that the major increase is to be expected in the next century—unless energy requirements and global economic concepts are drastically modified. On the basis of a fairly simple three-box model and careful consideration of the available fuel resources and the expected consumption rates, Ziemen and Altenheim (1973) predicted a doubling of the CO_2 level in the atmosphere by the middle of the next century. Even though our present capability of calculating the resulting rise of world-wide surface temperatures due to a CO_2 increase of this size are still rather limited, there seems little doubt that a doubling of CO_2 partial pressure would create not only a climatological problem but perhaps also an ecological one due to acidification of the ocean surface waters. The future development of atmospheric CO_2, therefore, is a matter for serious concern.

WATER VAPOUR

Next after CO_2, water vapour is of vital importance to meteorology and to life. Disregarding the upper atmosphere, the H_2O cycle in the atmosphere involves no chemical reactions; it is thus much simpler in most respects than the cycle of other atmospheric gases. Basically, the abundance of atmospheric H_2O is determined by the temperature distribution at the Earth's surface and within the troposphere. Water vapour is released into the atmosphere by evaporation from the ocean and land surfaces and by transpiration from plants. For all practical purposes the sink is the fall of precipitation. The small amounts of water vapour absorbed at dry or cold land surfaces is of little importance. At temperate latitudes, the average water-vapour content in the atmosphere at the Earth's surface is about 1.5 g/cm², corresponding to an average mixing ratio of about 2,500 p.p.m. A global average precipitation rate of about 100 g/cm² results in residence times of 5 to 10 days. Steady-state conditions are therefore always closely fulfilled.

For no other gas is the sink and its distribution with time and space so well known as for water vapour. The basic facts about the distribution of water on the Earth are given in several textbooks or monographs (see, for example, Sellers, 1965). Its great variability is clearly indicated in world precipitation charts and may remind us that we should expect a similar situation for most other gases and for their sources as well. The reservoir with which atmospheric water vapour is exchanged is provided by the oceans and the fresh-water lakes on land. Since this reservoir is much larger than the atmospheric reservoir, criterion (4) implies that the atmospheric cycle of H_2O controls its atmospheric abundance and that the steady state condition applies. The levels are limited by the temperature distribution as mentioned above, although the average relative humidity in the troposphere is only about 50 per cent.

A most important part of the water cycle as far as the biosphere is concerned is that occurring on and below the continents. Water is the only substance whose cycle is the subject of a whole scientific field, namely, hydrology. Important climatological consequences derive from the fact that sources and sinks, though globally balanced, differ widely on a local basis. If P and V denote the local rates of precipitation and evaporation respectively the balance equation reads:

$$P - V = A,$$

where A is the local runoff by surface and ground water. In this equation small changes in the ground-water reservoir are neglected. The expression holds for any region. On a global basis it is:

$$P_L - V_L = A_R = V_m - P_m,$$

where L and m refer to the total land and ocean surfaces respectively, and A_R is the total runoff into the ocean of all rivers combined. Detailed studies show that the subtropical oceans, where P is very small, are the major net source areas for atmospheric water vapour. Over most tropical and temperate zones, the re-evaporation of precipitated water is also of some importance, even over the land areas.

Heavy water, D_2O, has a slightly lower vapour pressure than H_2O. As the water content of the atmosphere is depleted by successive precipitation, the D:H ratio decreases also. Thus, the D:H ratio can drop by 50 per mille in polar snow compared with ocean water. Accordingly, measurements of deuterium can be of special help in hydrological studies.

Minor atmospheric constituents

In addition to the main constituents and the rare gases, the Earth's atmosphere contains a large number of trace gases of various, predominantly biogenic, origins. The trace gases may be classified conveniently into two groups: one comprising gases occurring with concentrations in the p.p.m. concentration range, and a second group comprising gases whose concentrations fall into the p.p.b. or less. As Table 2 shows, the first group includes methane, hydrogen, carbon monoxide and nitrous oxide. (The tropospheric concentrations of ozone are actually intermediate between the two ranges but if the stratospheric ozone content is included and one takes the combined total, ozone also ranks as a member of the first group.) The gases of the first group will be discussed individually in this section, whereas the second group of trace constituents will be dealt with in summary form.

As Table 2 indicates, the residence times of the p.p.m. range trace gases are of the order of 1 year, so that their cycles must be fairly active in comparison with geological time-scales. These gases have been the subject of intensive research in recent years, hence the current status of each of these constituents may be reported here by giving an account of their individual sources and sinks.

METHANE

An overview of measurements of atmospheric methane during the past 25 years is given in Table 3. Methane was first detected in the Earth's atmosphere by its infrared absorption feature in the telluric spectrum of the sun (Migeotte, 1948); and, for a number of years optical spectroscopy remained the only method of study. The more recent measurements have been performed in the main by gas chromatography. In some measurements rather large fluctuations were found. It is not clear whether these fluctuations were real, caused, for example, by the vicinity of sources or sinks, or whether the employed sampling and detection techniques were inadequate. The measurements over the oceans, however, show fluctuations much smaller by comparison and these data, therefore, may serve to indicate the atmospheric background level. The absolute values reported are only as reliable as the instrument calibration techniques permit; nevertheless, it appears that 1.4 p.p.m. is a representative background value. This would also agree with the observations of Cavanagh et al. (1969) in Alaska.

Wilkness et al. (1973) have recently demonstrated a small but statistically significant difference of the mixing ratios in the two hemispheres and this observation is also entered in Table 3. In the cities, methane-mixing ratios as high as 3.5 p.p.m. have been recorded. For example, the average CH_4 mixing ratios in Los Angeles in 1965 was 3.2 p.p.m.. The increase above the background level appears to be due mainly to automobile traffic and industrial activities, i.e. sources of local pollution.

It is generally assumed that methane in the atmosphere derives predominantly from anaerobic fermentation of organic matter by bacteria. Koyoma (1963) investigated the extent of methane production from lake sediments and paddy soils and concluded that only from shallow lakes and paddy fields could methane escape to a significant extent. From his experimental data and the total area of paddy fields in the world, he estimated a production rate of 2.7×10^{14} g/year, including also contributions from such minor sources as upland fields, forests, coal fields, natural-gas wells, and enteric fermentation of animals. Ehhalt (1967) pointed out that this list does not include swamps, which cover more than twice the area of paddy fields in the world, thus he suggested that Koyoma's estimate of the CH_4 source strength be increased accordingly. Robinson and Robbins (1968) speculated that the hot, humid tropical areas of the world might also provide a significant source of methane. The total tropical areas are larger by about a factor of 10 than the areas of paddy fields and swamps combined but the production rate might be smaller by a similar factor thus simply doubling the total source strength. Clearly, the current estimates of the annual production rate of natural methane are so uncertain that any value in the range $(2-14) \times 10^{14}$ g/year appears acceptable.

Methane occurs in the surface layers of the oceans at concentrations exceeding the value expected from the dissolution of atmospheric methane according to its solubility coefficient. The excess methane is probably due to microbiological processes. Maximum concentrations occur at a depth of about 100 m, but close to the surface, the supersaturation of methane is small. The associated flux of methane into the atmosphere can be estimated from boundary layer considerations (see section on general concepts and criteria) to occur at a global rate of 0.1×10^{14} g/year. This rate is at least an order of magnitude less than the production rate from continental sources and may be neglected by comparison.

The anthropogenic contribution to atmospheric hydrocarbons has been estimated by Robinson and Robbins (1968) as 0.88×10^{14} g/year. About 40 per cent of this amount is due to the incomplete combustion of gasoline. Other significant sources are said to be incin-

TABLE 3. Summary of measurements of atmospheric methane

Author	Range of mixing ratios (p.p.m. vol.)
Migeotte (1948)	1.5
Ehhalt (1967), variable	0.6–1.3
Cavanagh et al. (1969), Alaska	1.4–1.5
Shearer (1969)	0.9–1.2
Swinnerton et al. (1969), Gulf of Mexico	1.24±0.03
Lamontagne et al. (1971), Pacific ocean	1.4 average
Wilkness et al. (1972), northern hemisphere	1.44±0.04
Wilkness et al. (1973), southern hemisphere	1.36±0.04

erators and degradation of solvents which evaporate during use. Since it is unlikely that all the hydrocarbons emitted are converted to methane, the amount of 0.88×10^{14} g/year probably represents the upper limit of the anthropogenic CH_4 contribution. From the source list and from the high urban methane concentrations, it appears that automobiles provide the major source of pollution methane. The contribution from this source can be estimated by a comparison of the measured urban concentration levels of methane and carbon monoxide in several cities in the United States with heavy traffic (see Air Quality Criteria for Hydrocarbons, 1970) combined with the anthropogenic source strength for CO, which is known to derive predominantly from car exhaust (see below). The global production rate of anthropogenic methane obtained by this procedure is 0.25×10^{14} g/year. This value agrees approximately with the estimate by Robinson and Robbins (1968) from the same source and should be considered a minimum value. It comprises about one-tenth of the minimum natural production rate.

Additional information about the origin of atmospheric methane might have been obtained from isotope analysis (Ehhalt, 1967) but, unfortunately, all the available measurements were made on air samples obtained from air liquefaction plants. Since these plants are located in heavily industrialized areas, a contamination from anthropogenic sources is unavoidable. So long as the extent of contamination is unknown, it is difficult to interpret the data in a meaningful way.

Methane is transported by turbulent mixing from the troposphere into the stratosphere where it is destroyed by oxidation. The vertical concentration profile of methane in the stratosphere has been measured several times, mostly by balloon-borne instrumentation (Ehhalt and Heidt, 1973; Ackerman and Müller, 1973). Generally, the observations indicated a decline of the methane mixing ratio with increasing altitude. At about 50 km altitude, the mixing ratio decreased to 0.25 p.p.m. (Ehhalt et al., 1972). The consumption in methane in the stratosphere appears to be caused mainly by reaction with OH radicals and excited oxygen atoms, both of which are produced photochemically (Wofsy et al., 1972). The complete mechanism of methane oxidation has not yet been elucidated. The loss of methane from the troposphere by upward transport can be estimated from the observed averaged concentration gradient above the tropopause, the density of the atmosphere at the average tropopause height of 12 km, and the approximately known vertical eddy diffusion coefficient in the lower stratosphere (see section on general concepts and criteria). A value of 0.6×10^{14} g/year is thus obtained for the rate of loss of methane from the troposphere by this mechanism.

A second sink for methane, established by theoretical considerations, is the loss caused by reaction with OH radicals in the troposphere. These radicals are produced by the reaction of water vapour with excited oxygen atoms, O (^{1}D), generated in the photolysis of ozone by ultra-violet solar radiation:

$$O_3 + h\nu \longrightarrow O(^1D) + O_2$$
$$O(^1D) + N_2 \text{ or } O_2 \longrightarrow O(^3P) + N_2 \text{ or } O_2$$
$$O(^1D) + H_2O \longrightarrow OH + OH.$$

Less than 10 per cent of the excited oxygen atoms react with water vapour, the remainder is collisionally deactivated to the ground state, O (^{3}P), which is not capable of reacting with water vapour. The daily and seasonally averaged production rate of OH radicals by this mechanism has recently been computed by Warneck (1975) as a function of latitude. A maximum rate occurs in the region near the equator, because both the ultraviolet solar radiation intensities and water vapour densities are greatest in the equatorial belt. Although OH radicals react with many atmospheric trace gases, the steady-state concentration appears to be determined mainly by the reactions with CO, CH_4 and NO_2. Consequently, the methane loss in a tropospheric volume element is determined by the partitioning of OH radicals between methane and the other reactants and the time-averaged OH production rate.

The world-wide distribution of methane and carbon monoxide is known fairly well now but the concentration of nitrogen dioxide is known only by order of magnitude. If the mixing ratio of NO_2 is assumed to equal 1 p.p.b. uniformly, one calculates the global annual destruction of methane by OH radicals to be about 0.7×10^{14} g/year. This loss rate must be considered a minimum rate, because there exist reaction chains, first pointed out by Levy (1971), involving product radicals derived from the various OH reactions whereby OH radicals are regenerated. A simple example is given by the reaction chain following the oxidation of CO:

$$OH + CO \longrightarrow CO_2 + H$$
$$H + O_2 \longrightarrow HO_2$$
$$HO_2 + NO \longrightarrow NO_2 + OH.$$

The reaction of OH with methane similarly regenerates OH, but the scheme is more complex (Levy, 1971, 1972). The increase of the minimum stationary OH concentrations can be taken into account by an amplification factor, f, whose value depends on the adopted reaction mechanism and is estimated by us as $f \approx 3$. According to Levy (1972) it can reach values of about $f = 20$. The photochemical sink of methane is more effective than indicated above by at least a factor of 3, so that the total rate of loss of methane by the tropospheric and the stratospheric photochemical sinks increases to a minimum value of 2.7×10^{14} g/year.

In addition to these chemical sinks there may exist a biological sink. Its importance cannot be specified at present. Several species of aerobic methane-consuming bacteria are known (for a discussion with respect to ocean waters see Seiler and Schmidt, 1974a). Soil is a

C. E. Junge and P. Warneck

TABLE 4. Methane budget estimates

Production and destruction rates (10^{14} g/year)	Atmospheric CH_4 residence time (years)	Method
$2.0 < Q < 14$	$2.2 < \tau < 12$	Source estimate
$S \geqslant 2.7$	$\tau \leqslant 11.5$	Sink estimate
$Q_N = 2.3;\ Q_S = 1.0$	$\tau \approx 9$	Two-box model
$Q_N + Q_S = 3.3$		

potent biological sink for hydrogen and carbon monoxide, but appears to be ineffective as a sink for methane. The oceans have already been identified as a small source so that few possibilities for a biological sink remain.

From the above discussion it is apparent that atmospheric methane is determined by one major source, anaerobic fermentation, and one major sink, destruction by OH radicals. In the steady state, the strengths of both should match, but the present estimates for the sources and sinks are too uncertain to provide more than an order of magnitude value for the size of the budget. The range of CH_4 residence times calculated from the range of values indicated is 2.2–12 years. The residence time must be at least a few years because of the even distribution of methane mixing ratios in both hemispheres. The fact that the CH_4 mixing ratio is slightly higher in the northern hemisphere suggests that the sources are larger there compared with the southern hemisphere, or the sinks smaller. However, the photochemical sink would be fairly evenly distributed between both hemispheres, so that there should be an excess of the source in the north. This conclusion is reasonable in view of the greater land mass in the northern hemisphere. Due to the even distribution of mixing ratios in each hemisphere, one can treat the exchange of methane between the two hemispheres as a two-box model—using the auxiliary assumptions that the sources are distributed in proportion to the division of the land masses, whereas the sinks are distributed evenly. Including also the anthropogenic source of methane in the north and utilizing the difference of methane mixing ratios in both hemispheres observed by Wilkness *et al.* (1973), one derives the source strengths listed in Table 4 which, for comparison, contains also the individual source and sink strength estimates obtained above in a different manner. The two-box model gives a total source strength of 3.3×10^{14} g/year, which falls inside the range given; the associated CH_4 residence time is 9 years. While a reasonable accord of the data is thereby achieved, it is evident that the quantitative aspects of the methane budget require additional research effort.

CARBON MONOXIDE

Like many other gases in the atmosphere, carbon monoxide was first identified by its terrestrial absorption feature in the infra-red solar spectrum (Migeotte, 1949). A variety of not very systematic measurements made up to 1969 and summarized by Pressman and Warneck (1970) established the atmospheric CO mixing ratio to be approximately 0.1 p.p.m.. However, a considerable variability was also apparent and was attributed to the influence of anthropogenic sources. Robinson and Robbins (1968) were the first to measure CO mixing ratios in background air by a sensitive technique utilizing the reduction of mercury oxide to mercury with the subsequent detection of mercury by atomic absorption spectrometry. Seiler and collaborators (Seiler and Junge, 1970; Seiler and Schmidt, 1974b; Seiler, 1974) perfected this technique and made measurements at all accessible latitudes over the Atlantic Ocean to obtain the latitudinal distribution of CO. Further data on the latitudinal CO distribution were obtained recently by Wilkness *et al.* (1973) over the Pacific Ocean by means of a gas-chromatographic technique. Simultaneously with the shipboard measurements in surface air, Seiler and co-workers also made measurements of CO on board commercial aeroplanes at altitudes of about 10 km. In the northern hemisphere, on the average, the CO mixing ratio is lower in the upper troposphere compared with ground levels, but in the southern hemisphere, in the region adjacent to the equator, the situation is reversed. The distribution of CO mixing ratios with latitude, in the upper troposphere and at the surface is shown in Figure 4, together with the averaged tropospheric distribution derived from such data weighted with the appropriate atmospheric density values. From these measurements, an appropriate two-dimensional distribution of

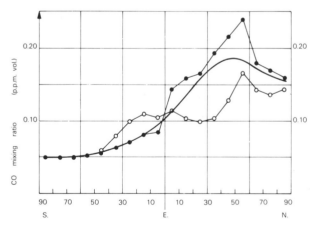

FIG. 4. Averaged latitudinal distribution of carbon monoxide: — ▪ — ▪ — from measurements at sea level; —O—O— from measurements in the upper troposphere; ——— tropospheric average obtained from the individual measurements by appropriate weighting. (From Seiler and Schmidt. 1974b. Reproduced with permission.)

154

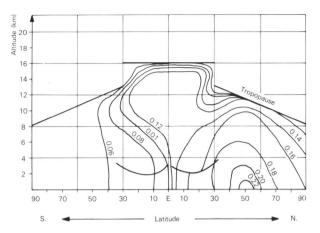

FIG. 5. Distribution of CO mixing ratios with latitude and altitude in the two hemispheres adapted from observations. Tropopause locations are idealized. (From Seiler and Schmidt, 1974b. Reproduced with permission.)

CO mixing ratios was also derived and is shown in Figure 5. The most interesting aspect of these data is the flux of CO directed upwards at 50° N and transport into the southern hemisphere occurring mainly at altitudes around 6 km. In the section on general concepts and criteria it was pointed out that the time for transport of air across a 45° latitudinal belt was approximately 0.3 years. The observed gradient of CO within one hemisphere thus indicates a CO residence time in the atmosphere of the same order of magnitude. Certainly it cannot be appreciably more than a year, which is the time for exchange of air between the two hemispheres. In the northern hemisphere, the variability of CO mixing ratios is fairly high, which in connection with Figure 1 also indicates a CO residence time somewhat less than a year. In the southern hemisphere, however, at high latitudes, the CO mixing ratio is very stable at 0.05 p.p.m..

TABLE 5. Summary of the sources and sinks of atmospheric carbon monoxide (in 10^{14} h/year)

	Northern hemisphere	Southern hemisphere	Total
Source			
Anthropogenic	5.4	1.0	6.4
Oceans	0.4	0.6	1.0
Burning of agricultural waste	0.4	0.2	0.6
Oxidation by CH_4	—	—	4–40
Sinks			
Uptake by soil	3.0	1.5	4.5
Stratosphere	0.9	0.2	1.1
Oxidation by OH	—	—	8–50

Until 1969, the anthropogenic production of CO, mainly from automobile exhaust, was the only definitely established source and no sink mechanism was known. Since then, a number of natural sources as well as sinks have been found. Also, the anthropogenic production estimates have been refined. The main sources and sinks, and the associated CO production and loss rates, respectively, are summarized in Table 5. The individual sources and sinks will be discussed separately.

Since the anthropogenic CO production is dominated by automobile exhaust and since the world's car population has increased over the last decades, the estimates for the anthropogenic CO production rate have been continually revised upwards. In 1952, Bates and Witherspoon obtained 1.2×10^{14} g/year as a value for the global anthropogenic CO source, whereas the more recent values were 4×10^{14} g/year by Jaffe (1968), 2.5×10^{14} g/year by Robinson and Robbins (1963), and 7.2 $(\pm 0.8) \times 10^{14}$ g/year by Seiler (1974). The recent values are based mostly on the world's fuel consumption. An estimate of the latitudinal distribution of the anthropogenic CO source can be obtained on the basis of the population distribution. Accordingly, more than 80 per cent of this source is located at middle latitudes in the northern hemisphere.

A variety of measurements of CO dissolved in sea water by various research groups have established that the surface layers of the oceans are highly supersaturated with CO compared with the equilibrium concentrations expected from the CO mixing ratios in air observed by simultaneous sampling. Excess CO concentrations up to 100 times the equilibrium value calculated from Henry's law were observed in the entire Atlantic Ocean and in the northern Pacific Ocean. A summary of the results has recently been given by Seiler and Schmidt (1974a). Since the excess concentration of CO causes a flux of carbon monoxide from the ocean's surface to the atmosphere, the world's oceans serve as an extended source of atmospheric CO. The associated global source strength was estimated by Seiler and Schmidt (1974a), by the method outlined earlier, and gave a production rate of about 1×10^{14} g/year. Although much less than the total anthropogenic production rate, it must be taken into account in the southern hemisphere where the anthropogenic production rate is comparable.

Two relatively minor sources also entered in Table 5; these are the burning of agricultural waste and the photochemical oxidation of hydrocarbons other than methane—derived from the exhalation of forests, etc. These sources are very uncertain, but do not contribute greatly to the total CO production. The tropospheric photo-oxidation of methane, by contrast, is now recognized as contributing significantly to the CO budget. Compared with our knowledge of the production of CO from the other sources, however, our knowledge concerning the production of CO from methane in the atmosphere is based almost entirely on theoretical considerations. As discussed in the section on methane, the photochemical destruction of methane occurs via its reaction with OH radicals. Probably, formaldehyde is an intermediate product and it appears likely that formaldehyde found in rain derives from the destruction of methane. Undoubtedly, the gaseous formaldehyde is photo-oxidized to form CO (Calvert *et al.*, 1972) before CO is further oxidized to CO_2, but the fraction of formaldehyde subjected to this route compared with that removed from the atmosphere by rain is not known.

Even if all the methane attacked by OH radicals is converted to carbon monoxide, the production rate of CO from this source can be estimated only approximately, because present estimates of the average diurnal OH concentrations are inadequate. Clearly, the total CO-production rate from this source cannot amount to more than the equivalent methane destruction rate by OH radicals. On the basis of the minimum averaged OH concentration computed by Warneck (1975) and a complete conversion of CH_4 to CO, the global CO production rate is 1×10^{14} g/year. Levy (1972) and Wofsy *et al.* (1972) have calculated OH number densities up to 20 times higher, so that the production rate would increase accordingly. Because of the characteristic distribution of radiation intensities centred at the equator and the corresponding distribution of OH number densities, the CO production from the methane oxidation cycle is also highest in the equatorial belt.

The main natural sinks for CO that have been established in recent years are the uptake by soil, destruction in the stratosphere, and tropospheric oxidation by OH radicals. The uptake by soil is due to the action of micro-organisms. Measurements in the laboratory (Seiler and Junge, 1970; Ingersoll *et al.*, 1974; Liebl, 1971) and in the field determined the dependence of the rate on the CO-mixing ratio, the soil temperature, and the type of soil. From such data Seiler (1974) calculated a global sink strength of 4.5×10^{14} g/year, and also gave the latitudinal distribution based on the distribution of the land surface. The difficulty in deriving this estimate was that at temperatures above 40 °C, soil can act as a source of CO if its mixing ratio in air is less than 0.2 p.p.m.. This complication has not been taken into account. Additional uncertainties stem from the differences in soil activities, so that the estimate given here presents only an order of magnitude value for the global removal rate of CO by this sink.

From a limited number of measurements of the vertical profile of CO mixing ratios in the vicinity of the tropopause, all made in the northern hemisphere, it was shown that the CO mixing ratio decreased above the tropopause and reached a new constant value of about 0.05 p.p.m. in the stratosphere (Seiler and Warneck, 1972; Warneck *et al.*, 1973). The resulting flux of CO into the stratosphere constitutes a loss of CO from the troposphere, the extent of which can be estimated on the basis of the approximately known vertical transport coefficients in the lower stratosphere. The estimated net flux on the global scale is 1.1×10^{14} g/year.

The processes responsible for the stratospheric CO sink have not yet been determined with certainty, but the most probable chemical reaction is that with OH radicals generated photochemically. Weinstock (1969) first suggested that the consumption of carbon monoxide by the homogeneous reaction with OH radicals could be an important loss process also in the troposphere.

In fact, OH radicals react with CO about 20 times faster than with methane. The globally integrated sink strength for CO by OH radicals is nevertheless only about 50 per cent higher than the CO production rate from methane, because the average CO mixing ratio is considerably lower than that of methane. In regions where the CO production from the methane oxidation is the dominant source and the oxidation of CO by OH radicals the dominant sink, the steady-state CO mixing ratio is determined only by the ratio of the reaction rate coefficients, which demand $m(CO) = 0.055$ p.p.m.. This value for the CO mixing ratio is observed over the ocean in the southern hemisphere, so that in this region the atmospheric CO budget may be determined entirely by the two reactions:

$$OH + CH_4 \longrightarrow CO + \text{other products, and}$$
$$OH + CO \longrightarrow CO_2 + H .$$

Clearly, this situation can arise only if the CO production rate from atmospheric methane is much higher than either the oceanic source strength or the consumption rate by soil and this conditions demands quite appreciable OH number densities. Nevertheless, model calculations by Levy (1972) and by Wofsy *et al.* (1972) predicted OH concentrations of a sufficient magnitude.

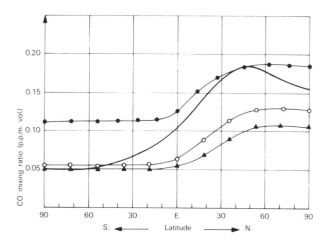

FIG. 6. Latitudinal distributions for CO obtained from model calculations; — — — minimum OH concentration assumed; —●—●— amplification factor for OH concentration assumed to be $f = 10$; —▶—▶— amplification factor for OH concentration assumed to be $f = 20$; ——— tropospheric average obtained from the individual measurements by appropriate weighting. (From Seiler and Schmidt, 1974b. Reproduced with permission.)

It must be realized, however, that if methane oxidation is the preponderant source in the southern hemisphere, an equivalent rate of CO production from the same source would also occur in the northern hemisphere. Since here the CO mixing ratio is much higher than 0.05 p.p.m., the rate of CO production must be matched by a different source. If the source strength estimates in Table 5 are correct, the anthropogenic source is the only additional source of sufficient strength available in the northern hemisphere to provide the observed increase, even though a large portion of it is taken up by destruction of CO in soil. On the basis of the latitudinal distribution estimates for the various sources and sinks and known transport coefficients, Seiler and Schmidt (1974b) calculated the steady-state distribution of CO mixing ratios at various latitudes. The results are compared with the observed averaged latitudinal concentration profile in Figure 6. Three cases corresponding to minimum, maximum and intermediate concentrations of OH radicals were treated. The first case makes anthropogenic CO production dominant, in the second case the CO production from methane oxidation is dominant and in the third case the magnitude of both sources is comparable. None of these cases result in a distribution of CO mixing ratios even approximately compatible with the observations. Accordingly, either the distribution of the sources and the sinks considered is incorrect, or other as yet unknown sources and/or sinks exist. However, Figure 6 makes clear that the two extreme cases assumed are limiting cases in that they provide the upper and lower limits, respectively, to the observational CO distribution curve.

HYDROGEN

The subject of atmospheric hydrogen has recently been reviewed by Schmidt (1974). From a variety of measurements of limited accuracy, performed during the last three decades on samples frequently obtained from air-liquefaction plants, the atmospheric mixing ratio of hydrogen was established to be in the 0.5 p.p.m. range. The considerable variations observed may have been due not only to the inadequacy of the measurement techniques, but also to the influence of local pollution.

TABLE 6. Comparison of atmospheric hydrogen mixing ratios in the two hemispheres

Location	Northern hemisphere	Southern hemisphere
Stratosphere	0.548 ± 0.014	—
Upper troposphere	0.558 ± 0.018	0.545 ± 0.015
Lower troposphere	0.585 ± 0.029	0.552 ± 0.01
Mainz, Federal Republic of Germany	0.800 ± 0.160	—

Schmidt (1974), using a continuous recording device based upon the chemical conversion of mercury oxide to mercury by the reaction with hydrogen, has recently provided a complete set of measurements for hydrogen in background air in the northern and southern hemispheres. A summary of his results is given in Table 6. These data establish that hydrogen is fairly evenly distributed in both hemispheres, although the mixing ratio in the northern hemisphere slightly exceeds that in the southern. In the cities, however, as indicated by the measurements in Mainz (Federal Republic of Germany), the hydrogen mixing ratio increases by as much as 40 per cent over the background value. This behaviour is undoubtedly caused by local human activities. In the northern hemisphere, the average mixing ratio observed in the upper troposphere is also slightly lower (by 4.5 per cent) than at the surface level, indicating a pattern of transport similar to that observed for CO, i.e. from the surface in the northern hemisphere towards the upper troposphere and further into the domain of the southern hemisphere.

The sources and sinks of atmospheric hydrogen have also been discussed by Schmidt (1974). The three main sources appear to be the oceans, photochemical production from methane and anthropogenic activity, whereas the sinks are those of its destruction by soil and its reaction with OH radicals. The individual source strength estimates are listed in Table 7.

The marine source of hydrogen is consistent with the observations by Schmidt (1974), that the surface layers of the entire Atlantic Ocean contain more dissolved hydrogen than the equilibrium with the hydrogen mixing ratio in air would allow. The observed supersaturations are ascribed to microbiological processes and the average supersaturation factor is about 3. Proceeding as for the estimate of the flux of CH_4 and CO from the ocean to the atmosphere (see the section on general concepts and Criteria), one obtains for hydrogen a flux of 1.7×10^{10} molecular/cm^2/sec, which globally leads to a source strength of 1.6×10^{12} g/year for the northern hemisphere, and of 2.4×10^{12} g/year for the southern hemisphere.

The production rate of hydrogen under anaerobic conditions in lake sediments and paddy soils has been studied by Koyama (1963) in conjunction with his work on methane; he estimated the corresponding global H_2 production rate from these sources at 1.3×10^{10} g/year. Enteric fermentation of animals, coal fields, etc. make only a small additional contribution. A subsequent estimate of the amount of volcanic emanation (Koyama, 1964) increased his estimate from the sum of these sources to 10^{11} g/year. This rate is still small compared with the oceanic source strength and hence is not listed in Table 7.

The oxidation of methane by OH radicals—according to the photochemical mechanism discussed in the previous sections—may also produce hydrogen, because when it is subjected to photolysis, formaldehyde largely

TABLE 7. Sources and sinks of atmospheric hydrogen (in 10^{12} g/year)

	Northern hemisphere	Southern hemisphere
Sources		
Oceans	1.6	2.4
Oxidation of methane	2.0–3.5	3.5–5.5
Anthropogenic activities	14.0	4.0
Sinks		
Oxidation by OH radicals	1.6–2.8	3.0–5.5
Soil	8.0	4.0
Global production	27.0–31.0	
Global consumption	10.0–19.0	

decomposes, yielding up H_2 and CO. It has already been discussed in an earlier section on carbon monoxide that not all the formaldehyde expected from the oxidation of methane is photolysed but that a portion is rained out. Moreover, since estimation of the rate of methane oxidation and hence the rate of formaldehyde production is rather uncertain, similar uncertainties apply to the estimates derived for the rate of hydrogen production from this source—hydrogen production is, in fact, estimated to range from 5 to 9×10^{12} g/year.

The largest source, at least in the northern hemisphere according to the data in Table 7, appears to derive from man's activities. The results of measurements at Mainz, given in Table 6, show considerably higher values and larger fluctuations than the data obtained for surface air over the Atlantic Ocean. It is not known with certainty which of the anthropogenic sources contributes most to the H_2 emission, but it appears reasonable to assume that automotive sources dominate. Since CO is also derived mainly from this type of source, Schmidt (1974) has compared data for the mixing ratios of both gases and has used the approximately known anthropogenic CO world production rate to derive an estimate for the anthropogenic H_2 production rate on the basis of proportionality considerations. His results are shown in Table 7.

The consumption of atmospheric hydrogen occurs mainly by bacteria in soil. According to laboratory measurements by Liebl (1971), a variety of soils remove hydrogen very effectively from an isolated volume above the soil. From the measured removal rates, the sink strength expected globally has been calculated at 12×10^{12} g/year. It must be considered a very rough estimate. A second sink of importance is the destruction of hydrogen by the reaction

$$OH + H_2 \longrightarrow H_2O + H,$$

which is equivalent to the oxidation of methane and carbon monoxide, respectively, by OH radicals. The dif-

ficulties in obtaining a good estimate for the OH radical concentrations have been discussed repeatedly in the previous sections so that the resulting sink strength can be only an order of magnitude estimate.

It is evident from Table 7 that the estimated total production and consumption rates of atmospheric hydrogen match only within a factor of 2, the sources exceeding the sinks. The apparent imbalance demonstrates the accuracy (or rather, inaccuracy) currently involved in the procedures for obtaining source and sink strength estimates. The atmospheric residence time derived from the hydrogen budget is approximately 5 years. The dominance of the anthropogenic source in the budget implies that the atmospheric H_2 mixing ratio must have increased over the preceding decades in parallel with the increase in anthropogenic activity. This conclusion cannot be verified from the previous observations, however, because of the uncertainties inherent in these earlier measurements.

For some time it was believed (see the summary by Suess, 1966) that the photo-decomposition of water vapour in the upper stratosphere and the mesosphere provided a significant source of atmospheric hydrogen. The recent data by Schmidt (1974) and Ehhalt and Heidt (1973) showed, however, that the hydrogen mixing ratios in the vicinity of the tropopause exhibited essentially no gradient, so that a significant influx of H_2 from the stratosphere into the troposphere could not take place. Undoubtedly, hydrogen is produced in the upper stratosphere by the photo-dissociation of water vapour and also by the oxidation of methane. It is also destroyed by several radical reactions deriving from the photochemical mechanism. Yet the hydrogen budget in the stratosphere and mesosphere appears to be detached from the main budget of hydrogen in the troposphere. The latter budget, as discussed above, is controlled to large extent by biochemical processes.

NITROUS OXIDE

Nitrous oxide has been known as a minor constituent of the atmosphere since Adel (1939) first discovered its absorption bands in the solar spectrum. Interest in atmospheric N_2O has been slight, presumably because it is neither a pollutant, nor does it exhibit any particular chemical activity. In fact, N_2O is chemically so inert that according to present knowledge the stratosphere is the only atmospheric domain where N_2O undergoes chemical reactions to a significant extent. In this region, however, N_2O has recently been recognized as the main source of the higher nitrogen oxides and a corresponding influence on ozone chemistry. The average N_2O mixing ratio in surface air may be taken as 0.26 ± 0.03 p.p.m., where the variability given corresponds to the standard deviation. Goody (1969) has reported slightly higher mixing ratios from spectroscopic measurements near Boston, Massachusetts. From measurements over a two-year period, Goody inferred a cyclic variation of the mixing ratio with the seasons. This effect has not been substantiated by any of the subsequent measurements. Schütz *et al.* (1970) have also demonstrated from correlations with wind directions at Mainz that there is no detectable anthropogenic influence. There appears to occur a slight decrease in the N_2O mixing ratio with altitude in the troposphere, indicating a vertical flux of N_2O directed upwards. Balloon soundings showed that in the stratosphere the N_2O mixing ratio decreases considerably with height. Cryogenic rocket sampling at altitudes between 40 and 60 km by Martell (1960) failed to indicate N_2O, so that at these levels the stratospheric consumption of N_2O must be complete. Since the work of Arnold (1954), it has been well known that soil is an important source of N_2O. As shown by Verhoeven (1952) and more recently by Matsubara and Mori (1968), N_2O is produced in soil by denitrifying bacteria. Usually the reduction of nitrates and nitrites leads, via the various nitrogen oxides, to nitrogen as the end-product, but under certain conditions, the reduction step from N_2O to N_2 can be inhibited; N_2O is thus released to the atmosphere. From measurements of the production rate of different soils, Albrecht *et al.* (1970) estimated the average annual N_2O flux to be of the order of 10^{-12} grams N_2O/cm^2 sec. This estimate took into account that the production rate of moist soils was about 8 times higher than that of more arid soils. Over the entire land surface of the Earth one can thus expect a rate of N_2O production of about 0.3×10^{14} g $N_2O/year$.

A second source of similar magnitude is apparently provided by the oceans. A series of measurements summarized by Hahn (1972, 1974) made it evident that the surface layers of the Atlantic Ocean contained N_2O dissolved in proportions much higher than would have been expected from the equilibrium of N_2O in air. The excess N_2O in the marine environment is apparently due to the same processes that produce N_2O in soil. The flux of N_2O from the ocean's surface to the atmosphere was estimated by Seiler and Schmidt (1974a) and (independently) by Hahn (1974) using the assumption that the supersaturations of N_2O observed in the Atlantic Ocean applied equally to the other world oceans. The derived source strength was approximately 1.6×10^{14} g/year. From the total source strength of the ocean and land areas combined, and the total atmospheric N_2O content of 2×10^{15} grams, one then calculates the value $\tau \approx 10$ years for the tropospheric residence time of N_2O. This value compared well with the residence time derived from the observed variability of the N_2O mixing ratio in Figure 1.

The annual loss of tropospheric N_2O to the stratosphere can be estimated from the vertical concentration profile of N_2O in the stratosphere as measured by Schütz *et al.* (1970), and the approximately known vertical eddy diffusion coefficients on the basis of the theoretical treatment of the photochemical N_2O destruction provided by Bates and Hayes (1967). The resulting flux

of N_2O to the stratosphere is such that the tropospheric residence time of N_2O is 70 years if no other sink exists. Clearly, the stratospheric sink is not sufficient to take up the entire amount of N_2O produced, so additional sinks must be operative. To date, we have no information about the nature of the processes providing the required additional N_2O consumption. The available data on the production of N_2O in soils indicate that in arid, sandy soils, N_2O production is minimal, but consumption has not been observed. (The homogeneous destruction of N_2O in the troposphere by a photochemical mechanism similar to that in the stratosphere was shown by Bates and Hayes (1967) to be insignificant.)

OZONE

The origin of atmospheric ozone lies in the stratosphere, where it results from the photo-dissociation of oxygen due to its absorption of ultraviolet radiation. The occurrence of ozone in the atmosphere, therefore, is the direct consequence of the presence of oxygen on Earth.

The predominant fate of ozone is destruction near the locus of origin by a variety of gas phase reactions. A much smaller portion of ozone escapes from the principal domain of photochemical activity—located in the equatorial region at altitudes near 30 km—and is transported towards the lower stratosphere from where it enters the troposphere. Transport of air in the stratosphere occurs predominantly in a lateral direction, so that ozone is brought from the equatorial region to higher latitudes. The interplay of photochemical activity and stratospheric air motion results in an ozone layer with a maximum concentration at about 25 km altitude, increasing in thickness with latitude. This behaviour is illustrated in Figure 7. From systematic measurements of ozone at stations located at various latitudes in the northern hemisphere, Hering and Borden (1967) have obtained seasonally and annually averaged concentra-

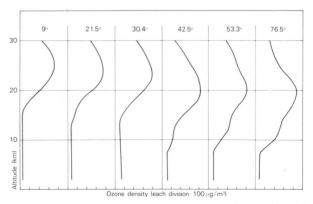

FIG. 7. Annually averaged vertical concentration profiles of ozone at different latitudes over the North American continent, using data from Hering and Borden (1967), obtained by balloon sondes at the stations Balboa (Canal Zone), Grand Turk, Tallahassee (Fla.), Bedford (Mass.), Goose Bay (Lab.), Thule (Greenland).

tion altitude profiles. Figure 7 shows annually averaged concentration altitude profiles of ozone for five latitudes selected so that the effect of accumulation of ozone in the lower stratosphere at higher latitudes is made apparent. From this reservoir, ozone enters the troposphere in several ways. The first is transport across the entire tropopause layer by small-scale turbulent mixing. A second influx path is via the tropopause gaps located at about 30° and 60° latitude; here, stratospheric and tropospheric air are directly exchanged. Finally, there exists a seasonally dependent mode of entry; this takes place in the spring and early summer when the tropopause at high and middle latitudes is relocated to greater heights by action of the heat transfer processes continually reestablishing the tropopause. In this way a portion of stratospheric air is simply added to the tropospheric reservoir.

In the troposphere, ozone is lost mainly by destruction at the Earth's surface. Since the transport of ozone from the tropopause to the Earth's surface is not instantaneous, a small but measurable vertical gradient of the ozone mixing ratio is set up. From the gradient the vertical flux of ozone can be estimated, albeit only roughly since the transport coefficients needed to obtain such an estimate are not precisely known. Similarly, it is difficult to estimate the total annual influx of ozone into the troposphere via the entry modes described above with the desired accuracy. The best estimates for the ozone flux that are currently available probably derive from estimates of the surface destruction rate based on experimentally determined destruction rates on a variety of surfaces, coupled with theoretical considerations of ozone transport across the surface friction layer. The largest destruction rate occurs on land; water and ice being less efficient. From the measured destruction rates, Fabian and Junge (1970) estimated a global annual ozone flux of about 5.5×10^8 tonnes/year, with the destruction in the northern hemisphere contributing about two-thirds of the total ozone destroyed. More recent measurements of the specific ozone destruction rate of the ocean surface by Tiefenau and Fabian (1972) indicated that the figure for the global rate of ozone destruction probably required a slight revision upwards (Regener, 1974). From the total amount of ozone in the troposphere and the ozone flux given above, one calculates a tropospheric residence time of 0.25 year. This is short compared with the exchange time of air between the two hemispheres, so that the ozone budgets in both hemispheres are fairly independent of each other.

Up to this point, photochemical processes in the troposphere have been neglected. For almost three decades it has been known that ozone is produced photochemically in surface air polluted as a result of heavy automobile traffic (so-called photochemical smog). In addition to chemically inert gases, the exhaust fumes of an automobile contain small amounts of nitric oxide formed by the oxidation of nitrogen in the combustion chambers of the engine, as well as unburned or partially oxidized hydrocarbons from the fuel components. In the air, nitric oxide, NO, is rapidly oxidized to nitrogen dioxide, NO_2. The speed of this process has baffled researchers for a long time since the direct oxidation of NO by atmospheric oxygen is a slow reaction at the observed NO concentration levels. It now appears that the oxidation is effected by hydrogen peroxy radicals, HO_2, generated in the photo-oxidation of the hydrocarbons, i.e.

$$NO + HO_2 \longrightarrow NO_2 + OH .$$

The OH radicals thereby formed react again with the hydrocarbons to regenerate HO_2 radicals. The reaction mechanisms are too complex to be discussed here in detail and they are not, in any case, fully understood at the present time. What is well understood is the rapid generation of ozone from NO_2 owing to its photo-dissociation:

$$
\begin{aligned}
NO_2 + h\nu &\longrightarrow NO + O \qquad \lambda < 400 \text{ nm} \\
O + O_2 + M &\longrightarrow O_3 + M \\
NO + O_3 &\longrightarrow NO_2 + O_2 .
\end{aligned}
$$

Since the nitrogen oxides are general constituents of the atmosphere, although at much smaller levels than in polluted areas, these processes occur also in regions of the troposphere unspoiled by pollutants. The last of the three reactions is then important since it reconverts ozone to nitrogen dioxide. This reaction is quite rapid. In the photochemical-smog areas, however, it is replaced by the more rapid reaction of NO with HO_2 radicals generated from the hydrocarbon pollutants.

The major hydrocarbon constituent in the unspoiled atmosphere is methane. As discussed in the section on methane, this component is attacked by OH radicals derived photochemically. Like the oxidation of any hydrocarbon in the smog areas, the oxidation of methane will also lead to the generation of HO_2 radicals which can then react with NO to produce NO_2 and, subsequently, ozone. This excess formation of ozone in the natural troposphere is compensated by other photochemical loss reactions; for example, the reaction of ozone with NO_2 followed by photo-decomposition of the produced NO_3 molecule. As a result of these recent developments, some authors claim that the ozone budget in the troposphere is determined mainly by this type of photochemistry rather than by the influx of stratospheric ozone and destruction at the ground (see, for example, Chameides and Walker, 1973). In this case, however, it is difficult to understand the seasonal and latitudinal variations of the ozone mixing ratio in the troposphere. At the present time, the importance of tropospheric ozone photochemistry is not established with certainty.

While the above discussion has concentrated on the role of ozone in the troposphere, it should be noted briefly that the ozone chemistry in the stratosphere has also received renewed attention, even though it had been an active research subject for forty years. The impetus for the recent activity in the field was supplied by the advent of high-flying passenger aircraft, the so-called supersonic transport (SST), and the possibility of the pollution of the stratosphere by a larger commercial fleet of such aircraft. Any decrease of stratospheric ozone concentrations caused by such a pollution would lead to an undesirable increase of ultraviolet radiation at the Earth's surface. The recent research thus initiated has brought to light the considerable influence on stratospheric ozone chemistry by trace gases such as methane, carbon monoxide, the nitrogen oxides, nitric acid and others. Specifically, ozone consumption by nitric oxide, $O_3 + NO \rightarrow O_2 + NO_2$, has been demonstrated to exert a significant effect.

Nitric oxide, NO, is produced in the stratosphere by the oxidation nitrous oxide, N_2O, which in turn has its origin in the troposphere and is brought to the stratosphere by turbulent transport (see the section on 'The main constituents of Earth's atmosphere'). A second source of NO may be the ionic oxidation of nitrogen on account of cosmic-ray-produced ionization. The produced nitrogen oxides are mainly returned to the troposphere also by turbulent transport. This new insight into stratospheric trace gas chemistry, initially obtained by model calculations, have in the meantime been backed up by measurements of a variety of important stratospheric trace components. A brief summary of the present status has been given recently by Schiff and McConnell (1973). The general literature on the subject is voluminous. The Climatic Impact Assessment Program of the United States Department of Transportation has spent a considerable effort in preparing a very detailed review concerning all the aspects of stratospheric ozone and the SST problem.

Most recently, chlorine has been added to the list of substances capable of affecting the ozone layer. The chlorofluoromethanes CF_2Cl_2 and $CFCl_3$ are released into the atmosphere from spray cans and refrigeration units. The CFM appear to be inert in the troposphere, but are destroyed by photodecomposition at altitudes near 35 km whereby Cl atoms are formed. Chain reactions of Cl with ozone cause a potentially harmful reduction of the ozone layer.

TRACE GASES IN THE P.P.B. CONCENTRATION RANGE

Under this heading fall a variety of atmospheric trace gases. Only a few of them are listed in Table 2. Most important at the present time appears ammonia, NH_3; the nitrogen oxides, NO and NO_2, and their respective acid vapours, HNO_2 and HNO_3; sulphur dioxide, SO_2, and reduced sulphur compounds such as hydrogen sulphide, H_2S, and dimethyl sulphide, $(CH_3)_2S$, respectively, as well as many hydrocarbons other than methane; for example, ethylene or formaldehyde. The practical problems inherent in developing reliable analytical techniques for the atmospheric monitoring of all these trace gases, already well pronounced for the group of gases in the p.p.m. concentration range, become enormous for

the trace gases at p.p.b. concentrations or less. It is not surprising, therefore, that only a restricted number of quantitative and reliable data exist for these gases. Correspondingly meagre is the information about their cycles. For most p.p.b. gases, the low atmospheric abundance is due not to the unimportance of the sources but rather to the high chemical reactivity of the trace gas itself. Many of them are involved in the processes of aerosol formation in which lies their significance. Estimates show that, on a global basis, about one half of the aerosol matter is produced by conversion of these gases within the atmosphere—both polluted and natural. The details of the process of aerosol formation from the gas phase are not at all understood. The removal of aerosols occurs, in addition to dry deposition, predominantly by their incorporation into rain droplets followed by rain-out. Accordingly, precipitation is an important channel in the overall cycle of many trace gases in the p.p.b. concentration range.

Sulphur, in the form of sulphate, appears to make up a substantial fraction of aerosols by mass, both in polluted regions and in the clean atmosphere. Hence, this section briefly summarizes some aspects of the atmospheric sulphur cycle, i.e. mainly sulphur dioxide, SO_2, as a typical representative of the group of p.p.b. range trace gases. A recent summary of the atmospheric sulphur budget has been given by Robinson and Robbins (1970a). The geochemical sulphur cycle has been discussed by Kellog et al. (1972) and by Friend (1973).

Over the continents, measurements of surface air indicate SO_2 volume mixing ratios ranging from a few p.p.b. to about 100 p.p.b., the higher values undoubtedly being due to sources of local pollution. The main sources of pollution are associated with the combustion of coal and fuel oil used for power generation and for the heating of small houses in the winter season. Georgii and Jost (1964) investigated the vertical distribution of atmospheric SO_2 in the lower troposphere over Europe. The average mixing ratios declined from surface values of 10 p.p.b. in the summer and 15 p.p.b. in the

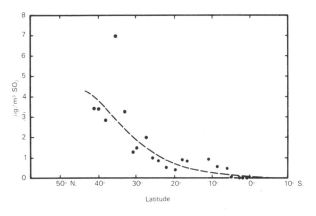

FIG. 8. SO_2 concentrations observed along 30° W. over the northern Atlantic Ocean as a function of latitude (after Georgii, 1970).

winter to approximately 4 p.p.b. at 5 km altitude. At higher altitudes, the SO_2 concentrations are probably still lower but no data exist to support this. Isolated measurements over the American continent show SO_2 mixing ratios of 0.3 to 1 p.p.b. in background air. Only one comprehensive set of SO_2 surface measurements has been obtained for maritime regions. They are shown in Figure 8. The observations, made over the Atlantic Ocean indicated a decrease of SO_2 concentration with decreasing latitudes such that in the tropical region they become unmeasurably low. Since it can be expected that in the upper troposphere the SO_2 distribution is fairly uniform at mixing ratios of about 1 p.p.b., these data indicate that the oceans act as a sink; this is in agreement with our knowledge that SO_2 is fairly rapidly oxidized in aqueous solutions containing oxygen and catalytically active metal ions. The latitudinal gradient over the ocean and the vertical gradient of the European continent suggest that the sources of SO_2 are located to a large extent on the continents in the northern hemisphere. Robinson and Robbins (1970a) estimated that about a half of the atmospheric SO_2 is generated anthropogenically, the other half naturally from the biosphere, most likely over land.

The origin of biogenic SO_2 is not well established. It is generally assumed that, initially, sulphur is released as hydrogen sulphide, H_2S, which is then rapidly oxidized in the atmosphere. It is true that under anaerobic conditions sulphur occurs as H_2S, but it has recently been shown (Rasmussen, 1974) that under aerobic conditions the major sulphur compounds in the gaseous emissions were dimethyl sulphide and related substances, H_2S occurring only when the pH of the medium was lowered to 4.5. It is not known how these substances are converted to SO_2, a process that must be rapid, or whether SO_2 can also be released directly.

In the atmosphere, one prominent sink for SO_2 is the oxidation to sulphuric acid or sulphate which is then incorporated into aerosols and finally precipitated. There are a number of mechanisms by which the oxidation of SO_2 can be accomplished in nature: (a) photochemical oxidation in the gas phase; (b) nonphotochemical gaseous reactions; and (c) heterogeneous oxidation on aerosol and in the liquid phase of cloud and fog droplets. The various possibilities and their relative influence on SO_2 removal from the atmosphere are summarized in Table 8 (after Junge, 1972). During the last twenty years, a remarkable effort has been made towards the laboratory investigation of these mechanisms. The principal result of these studies has been to show the complexity of the involved chemistry. For example, the photochemical oxidation has been elucidated in detail only recently (Sidebottom et al., 1972). The mechanism was shown to involve two excited states of SO_2 which, however, are mainly quenched physically without undergoing a chemical reaction. The SO_3 quantum yields are of the order of 10^{-3} or less (Friend et al., 1973) so that the SO_2 to SO_4^{2-} conversion efficiency is correspondingly low. In the atmosphere, this direct photo-oxidation process cannot be very important (Cox, 1972). If, however, other photochemically active species such as NO_2 are present—as is frequently the case in polluted air—the photochemical mechanism will be modified and the photo-oxidation of SO_2 by the modified mechanism may after all become important (Cox and Penkett, 1971a, b).

According to Table 8, the most effective route of SO_2 oxidation is by reactions in the liquid phase, i.e. in clouds and fog droplets. The oxidation is catalysed by the presence of heavy metal ions (which are a frequent component of the aerosols) acting as the nucleus for cloud droplet formation. The formation of sulphuric acid leads to a lowering of pH which impedes the further oxidation of SO_2 by this route. However, the incorporation of ammonia with the consequent formation of ammonium sulphate would moderate the pH decrease. Scott and Hobbs (1967) have made model calculations for this type of process the results of which have recently been confirmed by Beilke (1974). The resulting SO_2 lifetimes are of the order of a few hours. Unfortunately, as Beilke has shown, the model is not overly realistic, since it does not take into account the initial presence of SO_4^{2-} in the aerosol particle serving as the nucleus for droplet formation. Accordingly, the SO_2 lifetime thus calculated must be considered as lower limit. On the basis of sulphate deposits (after deduction of the sea-spray component), coupled with estimates of trajec-

TABLE 8. Routes of oxidation of SO_2 in the atmosphere

	Reactions	Rate (per cent h^{-1})	SO_2 lifetime
Photo-chemical oxidation under normal sunshine	Cox, 1972. Sidebottom et al., 1972; Friend et al., 1973	0.03–0.2	140–21 days
	In presence of NO:0.3–0.7 p.p.m. Olefins: 0.1–1.0 p.p.m. (Cox and Penkett, 1971a)	2–11	50–9 hours
Non-photo-chemical homo-geneous gas reactions	SO_2:0.1 p.p.m. O_3:0–0.05 p.p.m. Olefins: 0–0.05 p.p.m. (Cox and Penkett (1971b)	0.4–3	10–1 day
	No NH_3 present 50 g/m³ Mn in aerosol (Cheng et al., 1971)	~ 2	~ 2 days
Liquid phase reactions in clouds and fog	With NH_3 present depending on NH_3 and catalyst present (e.g. Scott and Hobbs 1967; Beilke, 1974)	10–100	10–1 hour

tories of air masses originating from distinct anthropogenic SO_2 source areas, Rhode (1972) derived the time for oxidation and rain-out to range from 30 to 70 hours. Bolin et al. (1974) have further computed the rate of dry deposition of both SO_2 and oxidized sulphur to be of the same magnitude, so that the atmospheric residence time of SO_2 will be 1–2 days. It must be emphasized that these results pertain to continental conditions where anthropogenic influences are strong and the SO_2 concentrations are correspondingly high. The SO_2 oxidation and removal rate may well be concentration dependent such that the rate is higher than first order.

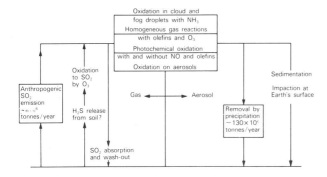

FIG. 9. The atmospheric cycle of sulphur (mainly SO_2) in excess of the sea-salt component. On the left are the gases, on the right the aerosols. The two rates given are the only rates for which reasonably accurate estimates are available.

The SO_2 residence time would then increase with decreasing concentration. This aspect has not yet been sufficiently investigated. It is also clear that the rate of oxidation in the liquid phase, and the subsequent removal by precipitation, depend critically on the degree of cloud formation and the associated precipitation pattern. In a dry climate, the major effective removal path remaining for SO_2 is gaseous deposition and absorption by soil.

Figure 9 summarizes the important routes of the atmospheric cycle of sulphur (in excess of the sea-spray component) emphasizing the better known channels. Only for two routes, namely anthropogenic emission and removal by precipitation, are reasonably reliable flux estimates available. These data show that even on a global basis, the anthropogenic sulphur emissions play an important role and it is for this reason the sulphur cycle is now receiving more attention than previously.

If our knowledge of the cycle of SO_2 and the other sulphur-containing components in the atmosphere is still unsatisfactory, despite the enormous research effort spent on this problem area, it appears that the present status of knowledge concerning the nitrogen compounds, ammonia and the nitrogen oxides, as well as the hydrocarbons, is even less satisfactory. In fact, the available data are so fragmentary that their detailed discussion is not warranted here. As regards the nitrogen compounds, Wlotzka (1972) and Robinson and Robbins (1970b) have given brief accounts and the interested reader is referred to their surveys.

References

ACKERMAN, M.; MÜLLER, C. 1973. Stratospheric Methane and Nitrogen Dioxide from Infrared Spectra. Pure & Appl. Geophys., Vol. 106–8, p. 1325–35.

ADEL, A. 1939. Note on the Atmospheric Oxides of Nitrogen. Astrophys. J., Vol. 90, p. 627.

Air Quality Criteria for Hydrocarbons. 1970. United States Department of Health, Education and Welfare, Public Health Service, Environmental Health Service, National Air Pollution Control Administration. (Publication No. AP–64.)

ALBRECHT, B.; JUNGE, C.; ZAKOSEK, H. 1970. Der N_2O Gehalt der Bodenluft in drei Bodenprofilen, Z. Pflanzenernährung und Bodenkunde, Vol. 125, 205–11.

ARNOLD, P. W. 1954. Losses of Nitrous Oxide from Soil. J. Soil Sci. Vol. 5, p. 116.

BATES, D. R.; HAYES, P. B. 1967. Atmospheric Nitrous Oxide. Planet. Space Sci., Vol. 15, p. 189–97.

BATES, D. R.; WITHERSPOON, A. E. 1952. The Photochemistry of some Constituents of the Earth's Atmosphere (CO_2, CO, CH_4, N_2O). Mon. Notic., Roy. Astron. Soc. Vol. 112, p. 101–24.

BEILKE, S. 1974. Meteorological Institute, University of Frankfurt, Federal Republic of Germany (private communication).

BIRCH, F. 1965. Speculations on the Earth's Thermal History. Geol. Soc. Amer. Bull., Vol. 76, p. 133–54.

BOLIN, B.; ASPLING, P.; PERSSON, C. 1974. Residence Time of Atmospheric Pollutants as Dependent on Source Characteristics. Atmospheric Diffusion Process and Sink Mechanism. Tellus, Vol. 16, p. 185–95.

BOLIN, B.; KEELING, C. D. 1963. Large-scale Atmospheric Mixing as Deduced from Seasonal and Meridional Variations of Carbon Dioxide. J. Geophys. Res., Vol. 68, p. 3899–920.

BROWN, H. 1949. In: G. P. Kuiper (ed.), The Atmospheres of the Earth and the Planets, p. 258. Chicago, Ill., University of Chicago Press.

CALVERT, J. G.; KERR, J. A.; DEMERJAN, K. L.; McQUIGG, R. G. 1972. Photolysis of Formaldehyde as a Hydrogen Atom Source in the Lower Atmosphere. Science, Vol. 175. p. 751–2.

CAVANAGH, L. A.; SCHADT, C. F.; ROBINSON, E. 1969. Atmospheric Hydrocarbon Monoxide Measurements at Point Barrow, Alaska. Environm. Science Techn., Vol. 3, p. 251–7.

CHAMEIDES, M.; WALKER, J. C. G. 1973. A Photochemical Theory of Tropospheric Ozone, J. Geophys. Res., Vol. 78, p. 8751–60.

CHENG, R. T.; CORN, M.; FROHLINGER, J. O. 1971. Contribution to the Reaction Kinetics of Water-soluble Aerosols and SO_2 in Air at ppm Concentrations. Atmosph. Envir., Vol. 5, p. 987–1008.

COX, R. A. 1972. Quantum Yields for the Photo-oxidation of Sulfur Dioxide in the First Allowed Absorption Region. *J. Phys. Chem.*, Vol. 76, p. 814–20.

COX, R. A.; PENKETT, S. A. 1971a. Photo-oxidation of Atmospheric SO_2. *Nature*, Vol. 229, p. 486–8.

——. 1971b. Oxidation of Atmospheric SO_2 by Products of the Ozone-Olefin Reaction. *Nature*, Vol. 230, p. 321–2.

CZEPLAK, G.; JUNGE, D. 1974. Studies of Interhemispheric Exchange in the Troposphere by a Diffusion Model. *Advances in Geophysics*. Vol. 18B, p. 57–72.

EHHALT, D. H. 1967. Methane in the Atmosphere. *J. Air. Poll. Contr. Assoc.*, Vol. 17, p. 518–19.

——. 1974. The Atmospheric Cycle of Methane. *Tellus*, Vol. 26, p. 58–70.

EHHALT, D. H.; HEIDT, L. E. 1973. The Concentrations of Molecular H_2 and CH_4 in the Stratosphere. *Pure & Appl. Geophys.* (PAGEOPH), Vol. 106–8, p. 1352–60.

EHHALT, D. H.; HEIDT, L. E.; MARTELL, E. A. 1972. The Concentration of Atmospheric Methane Between 42 and 62 km Altitude. *J. Geophys. Res.*, Vol. 77, p. 2193–6.

FABIAN, P.; JUNGE, C. E. 1970. Global Rate of Ozone Destruction of the Earth Surface. *Arch. Met. Geophys. Biokl. Ser.*, Vol. A 19, p. 161–72.

FRIEND, J. P. 1973. The Global Sulphur Cycle. In: S. I. Rasool (ed.), *Chemistry of the Lower Atmosphere*. New York, Plenum Press.

FRIEND, J. P.; LEIFER, L.; TRICHON, M. 1973. On the Formation of Stratospheric Aerosols. *J. Atm. Sci.*, Vol. 30, p. 465–79.

GANAPATHY, R.; ANDERS, E. 1974. Bulk Composition of the Moon and Earth, Estimated from Meteorites. *Proceedings Fifth Lunar Science Conference (1974)*.

GEORGII, H. W. 1970. Contribution to the Atmospheric Sulphur Budget. *J. Geophys. Res.*, Vol. 75, p. 2365–71.

GEORGII, H. W.; JOST, D. 1964. Untersuchungen über die Verteilung von Spurengasen in der freien Atmosphäre [Investigations on the Distribution of Trace Gases in the Free Atmosphere], *Pure and Appl. Geophys.*, p. 217–24.

GOODY, R. M. 1969. Time Variations in Atmospheric N_2O in Eastern Massachusetts. *Planet. Space Sci.*, Vol. 17, p. 1319.

HAHN, J. 1972. Nitrous Oxide in Air and Seawater over the Atlantic Ocean. In: D. Dyrssen and D. Jagner (eds.), *The Changing Chemistry of the Oceans*, p. 53–69. Stockholm, Almquist & Wibsell.

——. 1974. The North-Atlantic Ocean as a Source of Atmospheric N_2O. *Tellus*, Vol. 26, p. 160–8.

HERING, W. S.; BORDEN, T. R. 1967. Ozonosonde Observations Over North America. *Environm. Res. Paper 279 AFCRL-64-30 (IV)*. Bedford, Mass., Air Force Cambridge Research Laboratories.

HOLLAND, H. D. 1974. Atmospheric Oxygen and the Isotopic Geochemistry of Carbon. (Unpublished manuscript.)

HUNTEN, D. M. 1973. The Escape of Light Gases from Planetary Atmospheres. *J. Atmosph. Sci.*, Vol. 30, p. 1481–94.

INGERSOLL, R. B.; INMAN, R. E.; FISHER, W. R. 1974. Soil's Potential as a Sink for Atmospheric Carbon Monoxide. *Tellus*, Vol. 26, p. 151–9.

ISRAEL, H. 1962. Die Natürliche und Künstliche Radioaktivität der Atmosphäre, in Israel und Krebs. *Nuclear Radiation in Geophysics*, p. 76–96. Berlin–Göttingen–Heidelberg, Springer-Verlag.

JAFFE, L. S. 1968. Ambient Carbon Monoxide and its Fate in the Atmosphere. *J. Air Poll. Contr. Assoc.*, Vol. 18, p. 534–40.

JUNGE, C. F. 1972. The Cycle of Atmospheric Gases—Natural and Man Made. *Quart. J. Roy. Met. Soc.*, Vol. 98, p. 711–29.

JUNGE, C. E.; CZEPLAK, G. 1968. Some Aspects of the Seasonal Variation of Carbon Dioxide and Ozone. *Tellus*, Vol. 20, p. 422–34.

KEELING, C. D. 1973. The Carbon Dioxide Cycle: Reservoir Models to Depict the Exchange of Atmospheric Carbon Dioxide with the Oceans and Land Plants. In: S.I. Rasool (ed.), *Chemistry of the Lower Atmosphere*. New York–London, Plenum Press.

KELLOG, W. W.; CADLE, R. D.; ALLEN, E. R.; LAZRUS, A. L.; MARTELL, E. A. 1972. The Sulphur Cycle. *Science*, Vol. 175, p. 587–96.

KOYAMA, T. 1963. Gaseous Metabolism in Lake Sediments and Paddy Soils and the Production of Atmospheric Methane and Hydrogen. *J. Geophys. Res.*, Vol. 68, p. 3971–3.

——. 1964. Biochemical Studies on Lake Sediments and Paddy Soils and the Production of Atmospheric Methane and Hydrogen. In: *Recent Researches in the Field of Hydrosphere, Atmosphere and Nuclear Geochemistry*. Tokyo, Maruzen.

LAMONTAGNE, R. A.; SWINNERTON, J. W.; LINNENBOM, V. J. 1971. Non-equilibrium of Carbon Monoxide and Methane of the Air-Sea Surface. *J. Geophys. Res.*, Vol. 76, p. 5117–21.

LEMAIRE, J.; SCHERER, M. 1973. Kinetic Models of the Solar and Polar Winds. *Rev. Geophys. Space Phys.*, Vol. 11, p. 427–68.

LEVY, H. 1971. Normal Atmosphere: Large Radical and Formaldehyde Concentrations Predicted. *Science*, Vol. 173, p. 141–3.

——. 1972. Photochemistry of the Lower Troposphere. *Planet Space Science*, Vol. 20, p. 919–35.

LI, Y. 1972. Geochemical Mass Balance Among Lithosphere, Hydrosphere and Atmosphere. *Amer. J. Sci.*, Vol. 272, p. 119–37.

LIEBL, K. H. 1971. Der Boden als Senke und Quelle für das Atmosphärische Kohlenoxid. Federal Republic of Germany, University at Mainz. (Diploma thesis.)

MACDONALD, G. J. F. 1964. The Escape of Helium from the Earth's Atmosphere. In: P. J. Brancazio and A. G. W. Cameron (eds.), *The Origin and Evolution of Atmospheres and Oceans*. New York, Wiley.

McELROY, J.; DONAHUE, T. M. 1972. Stability of the Martian Atmosphere. *Science*, Vol. 177, p. 986–8.

MAROV, M. Y.; ADUEVSKY, V. S.; BORODIN, N. F.; EKONOMOV, A. P.; KERZHANOVICH, V. V.; LYSOV, V. P.; MOSHKIN, B. Y.; ROZHDESTVENSKY, M. K.; RYABOV, O. L. 1973. Preliminary Results on the Venus Atmosphere from the Venera 8 Descent Module. *Icarus*, Vol. 20, p. 407–21.

MARTELL, E. A. 1970. Transport Patterns and Residence Times for Atmospheric Trace Constituents Versus Altitude. Proc. ACS Symposium on Radionuclides in the Environment. *Advances in Chemistry*, Vol. 93, p. 138–57.

MATSUBARA, J.; MORI, T. 1968. Studies on Denitrification. *J. Biochem.*, Vol. 64, p. 863.

MIGEOTTE, M. V. 1948. Spectroscopic Evidence of Methane in the Earth's Atmosphere. *Phys. Rev.*, Vol. 73, p. 519.

——. 1949. The Fundamental Band of Carbon Monoxide at 4.7 Microns in the Solar Spectrum. *Phys. Rev.*, Vol. 75, p. 1108–9.

PRESSMAN, J.; WARNECK, P. 1970. The Stratosphere as a Chemical Sink for Carbon Monoxide. *J. Atm. Sciences*, Vol. 27, p. 155–63.

RASMUSSEN, R. A. 1974. Emission of Biogenic Hydrogen Sulphide. *Tellus,* Vol. 26, p. 254–60.

RASOOL, S. I.; DEBERGH, C. 1970. The Runaway Greenhouse and the Accumulation of CO_2 in the Venus Atmosphere. *Nature,* Vol. 226, p. 1037–9.

REASONER, D. L. 1973. Auroral Helium Precipitation. *Rev. Geophys. Space Phys.,* Vol. 11, p. 169–80.

REGENER, V. H. 1974. *Arch. Met. Geoph. Bioklim.* Series A, Vol. 23, p. 131–5.

RHODE, H. 1972. A Study of the Sulphur Budget for the Atmosphere over Northern Europe. *Tellus,* Vol. 24, p. 128–8.

ROBINSON, E.; ROBBINS, R. C. 1968. Sources, Abundance and Fate of Gaseous Atmospheric Pollutants. Palo Alto, Stanford Research Institute. (Final Report Project PR.–6755.)

——. 1970a. Gaseous Sulphur Pollutants from Urban and Natural Sources. *J. Air Poll. Contr. Assoc.,* Vol. 20, p. 233–5.

——. 1970b. Gaseous Nitrogen Compound Pollutants from Urban and Natural Sources. *J. Air Poll. Contr. Assoc.,* Vol. 20, p. 303–6.

RONOV, A. B. 1968. Probable Changes in the Composition of Seawater During the Course of Geological Time. *Sedimentology,* Vol. 10, p. 25–43.

RUBEY, W. W. 1951. Geological History of Seawater, An Attempt to State the Problem. *Geol. Soc. American Bull.,* Vol. 62, p. 1111–47.

RUTTEN, M. G. 1971. *The Origin of Life by Natural Causes.* Elsevier, Amsterdam.

SCHIDLOWSKI, M.; EICHMANN, R.; JUNGE, C. E. 1975. Precambrian Sedimentary Carbonates: Carbon and Oxygen Isotope Geochemistry and Implications for the Terrestrial Oxygen Budget. *Precambrian Res.* Vol. 2, p. 1–69.

SCHIFF, H. I.; MCCONNELL, J. C. 1973. Possible Effects of a Fleet of Supersonic Transports on the Stratospheric Ozone Shield. *Rev. Geophys. Space Phys.,* Vol. 11, p. 925–34.

SCHMIDT, U. 1974. Molecular Hydrogen in the Atmosphere. *Tellus,* Vol. 26, p. 78–90.

SCHÜTZ, H.; JUNGE, C. E.; BECK, R.; ALBRECHT, B. 1970. Studies of Atmospheric N_2O. *J. Geophys. Res.,* Vol. 75, p. 2230–46.

SCOTT, W. D.; HOBBS, P. V. 1967. The Formation of Sulphate in Water Droplets. *J. Atm. Sci.,* Vol. 24, p. 54–7.

SEILER, W. 1974. The Cycle of Atmospheric CO. *Tellus,* Vol. 26, p. 116–35.

SEILER, W.; JUNGE, C. E. 1970. Carbon Monoxide in the Atmosphere. *J. Geophys. Rev.,* Vol. 75, p. 2217–26.

SEILER, W.; SCHMIDT, U. 1974a. Dissolved Nonconservative Gases in Seawater. In: E. D. Goldberg (ed.), *The Sea.* Vol. 5. *Marine Chemistry,* p. 219–43. New York, Wiley.

——. 1974b. New Aspects on CO and H_2 Cycles in the Atmosphere. *Proceedings Int. Conf. on Structure Composition and General Circulation of the Upper and Lower Atmospheres and Possible Anthropogenic Perturbations. Melbourne, Australia, January 1974,* p. 192–222.

SEILER, W.; WARNECK, P. 1972. Decrease of the Carbon Monoxide Mixing Ratio at the Tropopause. *J. Geophys. Res.,* Vol. 77, p. 3204–14.

SELLERS, W. D. 1965. *Physical Climatology.* Chicago, University of Chicago Press.

SHEARER, E. C. 1969. Relationship Among Atmospheric Formaldehyde, Methane and Krypton. Fayetteville, Ark., University of Arkansas. (Ph.D. thesis.)

SIDEBOTTOM, H. W.; BADCOCK, C. C.; JACKSON, G. E.; CALVERT, J. G.; REINHARDT, G. W.; DAMON, E. K. 1972. Photooxidation of Sulphur Dioxide. *Environment. Sci. Techn.,* Vol. 6, p. 72–7.

SIGNER, P.; SUESS, H. E. 1963. Rare Gases in the Sun, in the Atmosphere and in Meteorites. In: J. Geiss and E. D. Goldberg (comps.), *Earth Science and Meteorites,* p. 241–72. Amsterdam, North Holland Publishing Co.

SILLEN, L. G. 1963. How has seawater got its Present Composition. *Svensk Kemisk Tidskrift,* Vol. 75, p. 161–77.

SLADE, D. H. (ed.). 1968. *Meteorology and Atomic Energy,* p. 204. Washington, D.C., United States Atomic Energy Commission.

Smic Report on Inadvertent Climate Modification. 1971. Cambridge, Mass., The MIT Press.

SUESS, H. E. 1949. Die Häufigkeit der Edelgase auf der Erde und im Kosmos. *J. Geolog. Res.,* Vol. 57, p. 600.

——. 1966. Some Chemical Aspects of the Evolution of the Terrestrial Atmosphere. *Tellus,* Vol. 18, p. 207–11.

SWINNERTON, J. W.; LINNENBOOM, V. J.; CHEEK, C. H. 1969. Distribution of Methane and Carbon Monoxide between the Atmosphere and Natural Waters. *Environment. Science Techn.,* Vol. 3, p. 836-8.

TIEFENAU, H.; FABIAN, P. 1972. The Specific Ozone Destruction of the Ocean. Surface and its Dependence on Horizontal Wind Velocity from Profile Measurements. *Arch. Met. Geophys. Biokl. Ser.,* Vol. 21, p. 309-412.

TSUNG-HUNG PENG. 1973. Determination of Gas Exchange Rates Across Sea-Air Interface by the Radon Method. New York, Faculty of Pure Science, Columbia University. (Ph.D. thesis.)

VERHOEVEN, W. 1952. Aerobic Spore Forming Nitrate Reducing Bacteria. Delft (Netherlands). (Dissertation.)

WARNECK, P. 1972. Cosmic Radiation as a Source of Odd Nitrogen in the Stratosphere. *J. Geophys. Res.,* Vol. 77, p. 6589–91.

——. 1975. OH Production Rates in the Troposphere. *Planet. Space Sci.* Vol. 23, p. 1507–18.

WARNECK, P.; JUNGE, C.; SEILER, W. 1973. OH-Radical Concentrations in the Stratosphere. *J. Appl. Geophys.,* Vol. 106-8, p. 1417–30.

WASSERBURG, G. J.; MAZOT, E.; ZARTMAN, R. E. 1963. Isotopic and Chemical Composition of Some Terrestrial Natural Gases. In: J. Geiss and E. D. Goldberg (eds.), *Earth Science and Meteorites.*

WEINSTOCK, B. 1969. Carbon Monoxide: Residence Time in the Atmosphere. *Science,* Vol. 166, p. 224–5.

WILKNESS, P. E.; LAMONTAGNE, R. A.; LARSON, L. E.; SWINNERTON, J. W.; THOMPSON, T. 1973. *Nature,* Vol. 245, p. 45–7.

WLOTZKA, F. 1972. Nitrogen. In: R. H. Wedepohl (ed.), *Handbook of Geochemistry,* Vol. II, No. 3. Berlin, Heidelberg, New York, Springer-Verlag.

WOFSY, S. C.; MCCONNELL, J. C.; ELROY, M. B. 1972. Atmospheric CH_4, CO and CO_2. *J. Geophys. Res.,* Vol. 77, p. 4477–93.

ZARTMAN, R. E.; WASSERBURG, G. J.; REYNOLDS, J. H. 1961. Helium, Argon and Carbon in Some Natural Gases. *J. Geophys. Res.,* Vol. 66, p. 277–306.

ZIMEN, K. E.; ALTENHEIM, F. K. 1973. The Future Burden of Industrial CO_2 on the Atmosphere and the Oceans. *Z. f. Naturf.,* Vol. 28a, p. 1747–52.

The unpolluted atmosphere

L. B. Ronca and R. B. Furlong

Department of Geology,
Wayne State University,
Detroit, Michigan (United States of America)

Introduction

Pollution is commonly defined as the process of 'impairing the purity' of a system. It is evident that, in the case of air pollution, it is necessary to state clearly what is meant by 'pure atmosphere'. From a layman's point of view, a 'pure atmosphere' is rich in oxygen, low in irritating or poisonous constituents and with a low absorption of visible light. Some simple considerations, however, necessitate modification of this definition. The atmosphere may have constituents which, though contrary to the layman's definition of pure air, are still either essential to the ecology of a region or an obvious corollary to a natural situation. For example, large amounts of water vapour may produce fog and mist while the air in the vicinity of certain vegetable communities may have high concentrations of pollen and spores; or dust may be abundant because of nearby deserts or volcanoes, and so on. The resulting atmosphere may not be the pleasant 'pure air' of the layman but it would be difficult to deny that any technology that would try to 'purify' such an atmosphere by eliminating the above-mentioned constituents would be in grave danger of changing ecological equilibria. The concept of 'pure air' is not a simple one, and it is necessary to understand what this idea of 'pure air' implies to the organisms that must inhabit it before contemplating any radical alteration of its components.

We can readily conclude that we may not use the comfort of *Homo sapiens* as the operational definition for a 'pure', or unpolluted, atmosphere because the ideal human atmosphere is rich in oxygen, diluted with one pure gas and with nothing else. In this chapter we will refer to air pollution as simply the effects, direct or indirect, of human, and especially industrial, activities, upon the atmosphere of the planet. The unpolluted atmosphere may be considered, then, as displaying any conditions that can naturally occur without human interference.

What are the possible compositions of our unpolluted atmosphere? Conceivably this could be determined by analysing a large number of samples collected from points all over Earth and then correcting the analysis with appropriate considerations about distances from industrial or population centres.

There is a basic objection to this approach; namely, are the results thus calculated representative of a steady state of the atmosphere, or are they simply points in trends, or oscillations, involving hundreds, or thousands of years?

It is well known that the atmosphere is not a closed system, and that some constituents are continuously added and subtracted by a wide variety of processes. A process that adds components to the atmosphere is said to constitute a 'source', while a process that removes a component from the atmosphere is called a 'sink'. In order to understand the behaviour of an atmospheric constituent as a function of time it is necessary to know whether the constituent has reached an equilibrium between its source input and its sink output and whether the rates of either the input or the output or both are changing.

The length of time since careful records have been kept is too short to permit any accurate deductions. Whether or not natural (i.e. not resulting from human pollution), long-time changes occur in the atmosphere is, however, of fundamental importance in any discussion of the effects of pollution. The purpose of this chapter is to discuss some of the theories proposed concerning the Earth's atmospheric composition as it has evolved through geologic time and some of the conclusions that might be drawn from these theories.

Vicissitudes of oxygen, water and carbon dioxide

From man's point of view, the most important constituent of the Earth's atmosphere is undoubtedly oxygen, and the degree of stability of oxygen in its various pools is a question of great importance. The evolution and development of the oxygen content in the atmosphere, hydrosphere and crust has been the subject of study by many authors, for example, Urey (1959), Rubey (1961), Holland (1963), Cloud (1968), Berkner and Marshall (1971), Van Valen (1971), and others (see references).

Oxygen is the most abundant element in our crust, where it constitutes approximately 63 per cent of all the atoms of crustal rocks (Mason, 1966). The greatest mobility displayed by oxygen, however, is in the atmosphere, hydrosphere and biosphere, where this element is present mainly as a component of water, molecular oxygen, carbon dioxide and organic compounds. Strahler and Strahler (1973) gave the orders of magnitude of the amounts of oxygen present, calculated from data supplied mainly by Rubey (1961).

Oceanic waters consist of about 7×10^{22} moles of H_2O which is the largest pool of oxygen at the surface of the Earth. Continental waters contain an additional 4.6×10^{20} moles of H_2O. In addition, there are about 7×10^{17} moles of H_2O in the atmosphere. With regard to molecular oxygen, there are approximately 1.5×10^{20} moles in the atmosphere and about 2×10^{17} moles dissolved in the oceans.

The greatest amount of CO_2 is found in the oceans in which about 6×10^{18} moles are dissolved. In addition, the atmosphere contains about 10^{17} moles of CO_2. A considerable amount of oxygen is also bound in organic compounds and it has been estimated that about 10^{17} moles of organic compounds are present in the biosphere.

Oxygen is highly reactive and is continuously being transferred from one pool to another. Although it is unlikely that all transfers are known, the biological cycle of H_2O, CO_2 and O_2 by photosynthesis and respiration is probably the process that produces the greatest rate of oxygen transfer from pool to pool.

The major sink for oxygen is burial of weathered rocks and organic compounds, and the major source is vulcanism.

Of the three major inorganic compounds of oxygen, i.e. water, carbon dioxide and molecular oxygen, the last is the most difficult to explain. Its relationship with life processes and its absence in the atmospheres of the other terrestrial planets in our solar system suggests strongly that molecular oxygen owes its presence in our atmosphere to biological activities.

A question that immediately comes to mind is whether the oxygen content of the atmosphere can be changed by changes in biological activities. If this is possible then changes in the habitat caused by human endeavours may have disastrous consequences.

The answer to the question is a definite affirmative for the extreme case, i.e. with no life whatsoever. On the pre-biotic Earth the following considerations indicate that the amount of molecular oxygen was very low (Perry *et al.*, 1970). The amino-acids and other materials that presumably were the building blocks of life would have been oxidized and burned up in the present oxygen concentration. This, in fact, presents one of the most interesting paradoxes of our existence. Life could not have originated with our present atmosphere but, at present, life could not survive with a different atmosphere.

The only non-biological source of oxygen seems to be photochemical breakdown of water by ultraviolet radiation as no free oxygen is released by volcanic gases. The amount of oxygen that can be produced by this process is open to debate, and ranges from 0.1 per cent of the present level (Berkner and Marshall, 1971) to 25 per cent (Brinkman, 1969). Could this process have accumulated the oxygen until the present level was reached? The so-called 'Urey's effect' would seem to forbid it, for the oxygen produced by the photo-dissociation of water automatically shields water molecules from further dissociation.

Berkner and Marshall (1971), from geochemical and palaeontological evidence, developed a very interesting history of the development of the oxygen content of the atmosphere through geological time. Early in the pre-biotic history of the Earth, the photo-dissociation of water was the only source of molecular oxygen. When life evolved and photosynthesis developed, the Urey's limit was passed and the amount of oxygen began to increase. When the oxygen reached approximately 1 per cent of the present content, the 'Pasteur' point was reached, and fermentation changed to respiration. Respiration is energetically more effective than fermentation and it is likely that life took advantage of this fact. Berkner and Marshall set the time for this basic change at the beginning of the Palaeozoic era. As the oxygen content continued to increase, the growing ozone layer became more of a shield to ultraviolet rays, permitting the development of life near the surface of water bodies and, finally, on land. During the late Silurian era, the oxygen content reached about 10 per cent of the present level when, it is believed, the ozone layer would have developed sufficiently to allow for life on land.

By the Carboniferous era, the oxygen content may have exceeded the present value, depleting the carbon dioxide content and cooling the Earth because of the decrease of the greenhouse effect.

The Permian period ice ages would then be the ultimate product of the extreme photosynthesis. The inclement climate would have reversed the trend to decreasing oxygen. The authors speculated that the many extinctions at the end of the Mesozoic era might have been due to a loss in oxygen content.

The above theory, if true, indicates that large variations in oxygen content may occur under natural conditions with almost catastrophic consequences. The time constant for these changes appears to be extremely long by human standards and, if such a change is building up for the next million years, plenty of warning signs should be available.

The above interpretation has been criticized by Van Valen (1971). He presented a line of reasoning that indicated that the amount of oxygen should actually have been pretty constant, at least since the Palaeozoic era. Van Valen pointed out that photosynthesis, like any unified set of chemical reactions, does not produce a net change in oxidation. Oxygen is indeed produced by photosynthesis but a stoichiometrically equal quantity of reduced carbon is also produced. Almost all the oxygen is eventually used to oxidize the reduced carbon, the exception being the amount of reduced carbon buried as peat and in euxinic sediments. A reduction in photosynthesis would result in a new steady state between production of oxygen and its consumption in the oxidation of carbon. Van Valen calculated that this new steady state would be reached with a reduction of no more than 1 per cent of the present oxygen content, even if the decrease in photosynthesis were sudden and drastic. In other words, a reduction in photosynthesis decreases the source of oxygen, but also decreases one of the main sinks, i.e. the oxidation of carbon. Inorganic sinks would obviously remain unchanged by such a decrease in photosynthesis. Holland (1964) calculated that, over a minimum period of 10^9 years, about 2×10^{22} grams of oxygen have been taken by inorganic oxygen sinks (oxidation of volcanic products and possibly of hydrogen from the solar wind). This is a very slow loss indeed.

If Van Valen's interpretation is correct, then one can assume that the amount of oxygen in the atmosphere has been a stable quantity, at least since the beginning of the Cambrian period. A similar statement may also be true of carbon dioxide. Holland (1968b) believed that the magnesium carbonate buffer system of the oceans has kept the atmospheric concentration of carbon dioxide very close to the present level since the development of land plants in the Silurian period.

Nitrogen, the great diluter

Nitrogen totally dominates the gases of the atmosphere, comprising 78.03 per cent of its volume; while oxygen, the second most abundant constituent, makes up only 20.99 per cent. To the layman, nitrogen may seem to be little more than a dilutant to the life-giving oxygen. The immediate, continual need for oxygen in respiration tends to promote such a view. On the contrary, nitrogen is just as essential to life on this planet as oxygen. After carbon, hydrogen and oxygen there is no other element so intimately associated with the reactions of living organisms. Just how essential nitrogen is to life on this planet can be seen from the fact that 65 per cent of the dry weight of multicellular animals and many micro-organisms is composed of nitrogen compounds and nearly all the biochemical processes of

these organisms are catalysed by nitrogen-containing enzymes (Stevenson, 1972).

The primary source for nitrogen, and, in fact, for all atmospheric gases on earth with the exception of oxygen, is volcanism. Berkner and Marshall (1971) pointed out that the gases and water vapour that have been released through volcanism over 3,000 million to 4,000 million years are more than sufficient to account for the oceans, plus the nitrogen, and all other gases in the atmosphere excepting only oxygen. As pointed out above, no free oxygen is released by volcanic gases and the development of oxygen in the atmosphere has been due, principally, to photosynthesis.

Most of the terrestrial nitrogen is found in the lithosphere and particularly in the mantle of the Earth, that zone of the Earth extending from about 30 km to about 2,900 km deep and where approximately 97 per cent of all terrestrial nitrogen is believed stored. This leaves only about 1 per cent of the total terrestrial nitrogen near the surface and in the active cycle.

While the primary source of atmospheric nitrogen is volcanic gas, most nitrogen now being added to the atmosphere and to the active nitrogen cycle comes from the action of micro-organisms on mineral nitrogen found in the soil as nitrate. This implies that even if there were a considerable increase in volcanism in the future, there would not be much of a change in the nitrogen content of the atmosphere. In the denitrifying action of micro-organisms on mineral nitrogen, nitrate is converted through two intermediate reduction steps to nitrous oxide and N_2 gas. The bacteria producing denitrification are ubiquitous and highly adaptive (facultative anaerobes). They are found in sediments, soils and water and are both autotrophic (able to synthesize food from organic material) and heterotrophic. The heterotrophic denitrifiers are many, though most belong to the genera *Pseudomonas, Micrococcus, Achromobacter,* and *Bacillus.* These organisms are also active in plant and animal decay.

The autotrophic denitrifers include *Micrococcus denitrificans* and *Thiobacillus denitrificans.* The latter is an anaerobe which oxidizes sulphur while reducing nitrate (Stevenson, 1972). Stevenson points out that some scientists believe that the reason nitrogen is the principal constituent of the Earth's atmosphere is because of the increasing action of denitrifying bacteria throughout geological time.

The nitrogen exchange between biosphere and atmosphere ranges from 0.017 to 0.034 mg cm^{-2} year^{-1} (Hutchinson, 1944) and nitrogen would complete the active cycle from atmosphere to biosphere and back again every 44–220 million years (Stevenson, 1972). This means that, with the exception of the small quantity of juvenile nitrogen added by volcanic activity, most of the nitrogen in the atmosphere will have passed at least once through the denitrification cycle. It is likely that this is also true for much of the nitrogen and nitrogen compounds released as volcanic gas; for, as a result of recycling and remelting of mineralized nitrogen, much of the nitrogen now released results from recycling and only a small quantity will be juvenile nitrogen.

The principal sink for nitrogen is through nitrogen fixation by bacteria and the blue-green algae. Atmospheric nitrogen occurs as N_2, except for minute amounts of N_2O, NH_3, NO_2^-, NO_3^-, and organically bound nitrogen. The N_2 molecule contains a triple bond and is very stable chemically, requiring a temperature of over 4,000 °C to destroy the covalent bonds and decompose the N_2 molecule into its constituent atoms. The only natural means of converting diatomic gaseous nitrogen into a form that is usable in the biosphere is through nitrogen fixation. The process of biochemical nitrogen fixation is relatively complex but consists mainly of converting N_2 into ammonia (NH_3), which in turn is converted into amino-acids, and finally into protein. Once in organic compounds, the nitrogen remains in combined form until acted upon by micro-organisms, where it is incorporated into microbial protoplasm, primarily in the form of proteins and nucleic acids (Stevenson, 1972).

Subsequently, upon the death of the micro-organisms this nitrogen is converted to ammonia. This process of converting organically bound nitrogen to ammonia is called 'ammonification'. The ammonia so formed is then either converted by oxidizing bacteria into nitrate or assimilated by organisms for re-use. The conversion from ammonia to nitrate is a step-wise process, proceeding from NH_3 (ammonia) to NO_2^- (nitrite), principally through the action of the bacteria *Nitrosomonas*. This nitrite is then oxidized to nitrate (NO_3^-), principally through the action of the bacteria *Nitrobacter*. This process of oxidation from ammonia to nitrate is known as 'nitrification'.

Once in the nitrate form, nitrogen may then follow one of two paths: (a) denitrification (discussed above), in which bacteria convert the nitrate to N_2; or (b) assimilation, in which plants and micro-organisms extract the nitrate from the soil and convert it to ammonia, whereby it can be utilized in the formation of the two amino-acids, aspartic and glutamic acids. These acids are, in turn, converted into other nitrogen compounds in living matter. The denitrification of nitrate to N_2 completes the nitrogen cycle.

It is important to emphasize that until the development of nitrogen-fixing bacteria and algae, there existed no process for removing nitrogen from the atmosphere. During the early stages of atmospheric development, it is quite likely that all nitrogen escaping from the Earth through volcanic activity remained in the atmosphere; initially in its most reduced form (NH_3) and then, as small amounts of oxygen became available through photo-dissociation of water, the ammonia was converted to N_2. As time progressed and free oxygen and biochemical fixation developed, the more oxidized forms began to appear. It is noteworthy, however, that even at the present time nitrogen can be incorporated

by organisms in the reduced state as ammonia or, if an organism absorbs mineral nitrogen in the form of nitrate, the conversion to the reduced state takes place in the cells of the organism. In any case, the importance of nitrogen in the development of primeval life in a reducing ammonia-containing atmosphere is supported by the necessity for organisms to convert nitrogen in a more oxidized form to its lowest oxidation state before it can be used.

So nitrogen, which may seem at first glance to be an abundant but unimportant component of the atmosphere adding only to its volume, may have been the most significant component at the first awakenings of life.

Trace gases in the unpolluted atmosphere

In addition to the principal gases in the atmosphere previously discussed, many other gases occur. Their low amounts justifies referring to these components as 'trace gases'. The principal trace gases and their percentage abundance in dry air are: argon, 0.94 per cent; hydrogen, 0.011; neon, 0.0015; helium, 0.0005; krypton, 0.0001; xenon, 0.00009; and radon, less than 0.00001 per cent.

In addition to these trace gases, some others are present as combined molecular gases in very small quantities. These are most significant in that they represent some of the major pollution gases and therefore a knowledge of the atmospheric background concentration is important. These gases and their natural concentration in the unpolluted atmosphere (expressed in p.p.m. or p.p.b.) are (after Hidy, 1973): sulphur dioxide (SO_2), 0.2 p.p.b.; hydrogen sulphide (H_2S), 0.2 p.p.b.; sulphate (SO_4^{2-}), 1.5 $\mu g\ m^{-3}$ (surface air (Friend, 1973)); carbon monoxide (CO), 0.1 p.p.m.; nitric oxide (NO), 0.2–2 p.p.b.; nitrogen dioxide (NO_2), 0.5–4 p.p.b.; ammonia (NH_3), 6–20 p.p.b.; nitrous oxide (N_2O), 0.25 p.p.m.; hydrocarbons (CH_4), 1.5 p.p.m.; and hydrocarbons (non-CH_4), 1 p.p.b.

Still others such as HCl, HF, I_2, Br, F and NaCl are so variable in time and space as to make accurate estimates difficult. Sodium chloride (NaCl) is an important temporary constituent of the atmosphere and 20–200 kg/hectare are precipitated annually on the Earth's surface (Mason, 1966). Most of this NaCl derives from sea spray. The HCl and HF are so variable that estimates of their concentration are impossible, except that they would be present as less than 1 p.p.b. Fluorine is likely to be an industrial contaminant and not found as a natural component of the atmosphere. It is possible that some small quantity of fluorine may be contributed by the decomposition of hydrogen fluoride from volcanic exhalations. Bromine and iodine are universally present in the atlosphere in very small quantities, but again it is difficult to separate the natural contributions of these elements from industrial contamination sources.

There can be little question but that volcanic exhalations are the principal natural source of most trace gases. Some, however, have a significant component from other sources. A summary of the sources and sinks of the trace gases is given below.

ARGON

Argon is by far the most abundant rare gas in the Earth's atmosphere. Most of the argon in the atmosphere is the ^{40}Ar isotope, whereas the ^{36}Ar isotope is believed to be the most abundant argon isotope in the universe. The $^{40}Ar:^{36}Ar$ ratio in the Earth's atmosphere is a little less than 300:1. It has been postulated that the predominance of ^{40}Ar is a result of the decay of its radioactive isobar ^{40}K. The ^{40}K decay in rocks produces ^{40}Ar which is subsequently released to the atmosphere through volcanic activity, the weathering and erosion of rocks and metamosphism. Because Argon is chemically inert once released to the atmosphere it remains there, and estimates of the amount of argon in the atmosphere accurately reflect the amount calculated to have been released to the atmosphere by various processes throughout geologic time (Wasserberg et al., 1963).

HYDROGEN

The second most abundant trace gas in the atmosphere is hydrogen, comprising approximately 0.11 per cent of the lower atmospheric gases. While hydrogen totally dominates the substance of the solar system (92 per cent of atoms and 74 per cent by mass), it constitutes only about 15 per cent of the atoms of the Earth. It is only the ninth most abundant element in the Earth and only comprises about 14 per cent of the crust of the Earth by weight (Mason, 1966). In the atmosphere, hydrogen is found only in small quantities ranging from 1 part in 15,000 to 1 p.p.m. (Andrews and Trusan, 1972). The low abundance of hydrogen in the atmosphere is the result of two facts: (a) hydrogen escapes from the upper atmosphere into space; and (b) molecular hydrogen (H_2) is readily oxidized by oxygen to form H_2O.

Although degassing of the Earth, and other geological processes (weathering, erosion and metamorphism) constitute an important source of hydrogen throughout geologic time, the principal source for hydrogen is thought to be the result of photo-dissociation of water in the upper atmosphere (Rubey, 1962). The principal sink for hydrogen is in the formation of H_2O, for although hydrogen is one of the principal elements found in animal and vegetable matter, most organic hydrogen is extracted from the water in organisms rather than from the atmosphere.

Hydrogen offers the possibility of being a strong, extraplanetary agent for terrestrial atmospheric events. The solar wind is primarily constituted by protons, i.e. hydrogen nuclei, and the Earth is at present 'bathed' in the solar wind. It appears, however, that the Earth is not gaining any hydrogen from the solar wind, but is in fact losing it.

Fluctuations of the solar wind in the geological past may have resulted in the Earth gaining rather than losing hydrogen. Such an event will result in a sink for oxygen and an increase in water vapour content in the atmosphere. Research is being conducted by the authors to determine whether such a possible event may be a cause for the glacial ages on Earth and the 'water ages' on Mars.

INEET (NOBLE, RARE) GASES

Because of its abundance in the atmosphere, argon was discussed above. The other inert gases, helium, neon, krypton and xenon, are also present in the atmosphere. The approximate percentage composition of these gases in the atmosphere has been presented above.

The relative abundances of the inert gases in the atmosphere provides us with important information about the early history of the Earth. For instance, krypton and xenon are a million times less abundant than immediate neighbours in the periodic table (such as bromine and iodine). These gases should be present in large quantity unless some early event stripped away the early atmosphere (Mason, 1966). It has been proposed (Krauskopf, 1967; Sylvester-Bradley, 1971; Lepp, 1973; and others) that the primeval or original atmosphere was either removed by some early event or else never developed. Urey (1952) proposed that gases such as the heavier inert gases were lost during the development of the Earth from the small asteroid-sized planetismals. His theory proposes that gases could escape from the gravity of these smaller bodies and in this way there was little, if any, of these gases present on the early Earth. As gases with similar molecular weights to the inert gases (H_2, HCl, NH_3, CH_4, CO_2, H_2S and SO_2) were not lost in the same proportion as the inert gases, these others must have been bound up in compounds or preserved as frozen particles and only those gases which would not form compounds and had a very low freezing point were retained. Another theory (Sylvester-Bradley, 1971; Lepp, 1973) supposes that an early, primeval atmosphere did develop for the Earth but this atmosphere was later lost through some catastrophic event which stripped it away. Lepp (among others) proposed that this event may have been a hotter phase of the Earth; a time when the Earth was hot enough to cause the gases to be driven into space. The development of the inert gases which *are* found in the atmosphere may have come about as a result of radioactive decay, which produced the gases as stable end-products of such decay. Subsequent volcanic activity then released the gases into the atmosphere. An example of

TABLE 1. Sources and sinks of molecular trace gases in the atmosphere

Gas	Natural sources	Estimated annual emission (tonnes)	Atmospheric concentration from natural sources	Calculated residence time	Principal sink
SO_2	Volcanoes	2×10^6	0.2 p.p.b.	4 days	1. Oxidation to sulphate by ozone. 2. Absorbed on solid and liquid aerosols then oxidized to sulphate.
H_2S	1. Volcanoes 2. Biological action in swamp areas	100×10^6	0.2 p.p.b.	2 days	Oxidation to SO_2.
SO_4^{2-}	Sea spray	44×10^6	$1.5 \mu g\ m^{-3}$	?	1. Washed out by precipitation. 2. Dry fall-out.
CO	1. Forest fires 2. Terpene reactions	75×10^6	0.1 p.p.m.	< 3 years	Ocean and soil (?)
NO/NO_2	Bacterial action in soil	NO: 430×10^6 NO_2: 658×10^6	NO: 0.2–2 p.p.b. NO_2: 0.6–4 p.p.b.	5 days	1. Oxidation to nitrate after sorption by solid and liquid aerosols. 2. Hydrocarbon photochemical reactions.
NH_3	Biological decay	$1,160 \times 10^6$	6–20 p.p.b.	7 days	1. Reaction with SO_2 to form $(NH_4)_2 SO_4$ 2. Oxidation to nitrate.
N_2	Biological action in soil	590×10^6	0.25 p.p.m.	4 years	1. Photo-dissociation in stratosphere. 2. Biological action in soil.
Hydrocarbons	Biological processes	480×10^6	CH_4: 1.5 p.p.m. non-CH_4: <1 p.p.b.	16 years (CH_4)	Photochemical reaction with NO/NO_2, O_3.

Source: Hidy, 1973.
© 1973 Plenum Publishing Corp. Reprinted with permission.

such radioactive decay was described above for ^{40}Ar, which is produced from the decay of ^{40}K.

There are no sinks for the inert gases. As mentioned above, their non-reactivity means that, once in the atmosphere, they will remain there and cannot be withdrawn by combination with other elements—as is the case for most of the other atmospheric gases.

MOLECULAR TRACE GASES

The other gases found in trace quantities in the atmosphere are found as combined elements. These gases together with their principal source(s) and sink(s) are listed in Table 1.

In general, the rare or trace gases of the atmosphere are relatively stable; that is, gases are added and removed at a fairly constant rate such that an atmospheric equilibrium has been reached and the amount of these gases in the unpolluted atmosphere is constant. The one principal exception to this, explained above, are the inert gases.

Non-gaseous components of the atmosphere

A common classification of particulate components of the atmosphere is to separate them into a viable (derived from life processes) and a non-viable group.

The major natural source of non-viable particles are forest fires, volcanoes, and wind-scoured dust. The major sinks are removal by gravitational sedimentation for particles larger than approximately 1 mm, and removal by precipitation for smaller particles (Hobbs *et al.*, 1974). Non-viable particles may include a variety of materials such as carbon, fine volcanic ash (mostly silica), clay, etc. Viable particles are spores, pollen, small organisms and fragments of organisms.

The importance of particulate components in the atmosphere far outweighs their amounts. It is well known that the development of clouds is directly affected by the quantity of particulate matter in the atmosphere. This atmospheric particulate matter may change the whole thermal balance of the planet producing changes in albedo and thermal absorption and thus produce a profound change in global climate (Wilson and Mathews, 1971).

The effects of particles on cloud formation and development appear to be related to temperature (Hobbs *et al.*, 1974). Cold clouds, i.e. clouds at temperatures below 0 °C, are effected by small particles that act as ice nuclei. The concentration of natural ice nuclei is quite low (only one particle out of 10^8 may act as an ice nucleus; this corresponds to about one per litre of air). Unfortunately, there appear to be many industrial processes emitting potential ice nuclei such as particulates from automobile emissions. This is one case in which human activities can truly overwhelm natural activities. In general, it appears that concentrations of up to one ice nucleus per litre (approximately the maximum natural concentration) enhances precipitation; higher amounts, however, tend to reduce precipitation.

If cloud temperatures are higher than 0 °C, the hygroscopic properties of the particles become important. The term cloud-condensation nuclei is used for these particles.

In addition to changes in albedo caused by varying cloud covers, particulate matter in the atmosphere may have other climatic effects. Heating rates of a body or air are dependent on the imaginary part of an aerosol's index of refraction and on the particle-size distribution. Hobbs *et al.* (1974) reported that in the troposphere—a location where human effects are insignificant—an aerosol content of about 10^{-8} kg/m^3 produced a heating rate of about 0.5° per 12-hour day. Increased amounts of aerosol, either natural or man-made, would, therefore, increase the heating rate considerably.

In pre-industrial-revolution days, the amount of natural particulate matter in the atmosphere may have played a role in the development of climates. There are no reliable estimates on the constancy of volcanic eruption, forest fires and dust storms rates in the geological past. There is little doubt, however, that human activities may be an important contribution in the future. The following chapters will discuss in detail these potentially dangerous conditions.

Summary and conclusions

It is evident that a considerable amount of research must be done before a complete model of unpolluted atmosphere is available.

Drastic natural changes may be possible in the cases of oxygen and particulate matter. Natural changes in oxygen content, if occurring at all, are likely to be operating with long time constants, involving hundreds of thousands of years or more. The possibility that the oxygen content of the atmosphere may be altered by human activities is still open to debate. A considerable decrease in photosynthesis (by polluting the oceans, for example) has strong effects if the model proposed by Berkner and Marshall (1971) is correct. On the other hand, if Van Valen's model (1971) is correct, then no drastic changes should follow.

A long period of increased volcanic activity will have its most noticeable effect in the amount of particulate matter in the atmosphere. The relationship between weather and particulate matter is very stringent and ice ages have been explained by increases in volcanic activity, It is probably in this context that human activities have the highest potential for provoking planetary-wide events.

In conclusion, it may be said (even if a complete model is not available) that the composition of the Earth's atmosphere is due to a large extent to complicated physico-chemical equilibria. This makes its constancy much more vulnerable to interferences from human activities. Even if, in absolute scale, the amounts of energy employed by man are considerably less than the amounts of energy operating in the atmosphere (an average hurricane involves energy more than two orders of magnitude larger than the largest atomic bomb), a well-placed interference may move an equilibrium to new values which may be disastrous from a human point of view. Other chapters in this book are dedicated to possible human interferences in the bio-atmo-hydro-spheric equilibria.

References

ANDREWS, A. C.; TRURAN, J. W. 1972. Hydrogen: Element and Geochemistry. In: R. W. Fairbridge (ed.), *The Encyclopedia of Geochemistry and Environmental Sciences,* p. 503–8. New York, Van Nostrand Reinhold.

BARANOV, V. I.; KNORRE, K. G. 1969. Age of the Crust and Entry of Radiogenic He^4 and Ar^{40} into the Atmosphere, *Geochem. Int.,* Vol. 4, No. 6, p. 1121–9.

BASHARINA, L. A. 1973. The Effect of Volcanism on the Atmosphere and Surface Waters of Kamchatka Peninsula. In: *Symposium on Hydrogeochemistry and Biogeochemistry,*

Vol. 1, Hydrogeochemistry, p. 574–80. Washington D.C., Clark Co.

BERKNER, L. V.; MARSHALL, L. C. 1971. Oxygen and Evolution. In: *Understanding the Earth,* p. 143–9. Cambridge, Mass. MIT Press.

BEWERS, J. M. 1972. The Global Circulation of Halogens in Nature. In: *Geochemistry—Géochimie, Section 10, Int. Geol. Congr., Proc. Congr. Geol. Int., Programme,* No 24, p. 273–81.

BLOCH, M. R.; LUECKE, W. 1970. The Origin of Fixed Nitrogen in the Atmosphere. *Isr. J. Earth-Sci.*, Vol. 19, No. 2, p. 41–9.

BOYCHENKO, Y. A.; SAYENKO, G. N.; UDEL'NOVA, T. M. 1972. Changes in Metal Ratios in Evolution of the Plant Biosphere. In: *Ocherki Sovremennoy Geokhimii i Analiticheskoy Khimii.* p. 454–8. Moscow, Nauka.

BRINKMANN, R. T. 1969. Dissociation of Water Vapour and Evolution of Oxygen in the Terrestrial Atmosphere. *Journal of Geophysical Research,* Vol. 74, p. 5355–68.

CHENEY, S. 1971. Coelomates, Subduction, and the History of Atmospheric Oxygen. *Geol. Soc. Am., Bull.,* Vol. 82, No. 11, p. 3227–9.

CLOUD, P. E. 1965. Significance of the Gunflint (Precambrian) Microflora. *Science,* Vol. 148, p. 27–35.

CLOUD, P. E. 1968. Atmospheric and Hydrospheric Evolution of the Primitive Earth. *Science,* Vol. 160, p. 729–36.

——. 1971. The Primitive Earth. In: Gass, Smith and Wilson (eds.). *Understanding the Earth,* p. 151–5. Cambridge, Mass. MIT Press.

DAMON, P. E. 1970. Climatic Versus Magnetic Perturbation of the Atmospheric C^{14} Reservoir. In: Radiocarbon Variations and Absolute Chronology, *Tex. Mem. Mus., Misc. Pap.,* p. 571–93.

DAVITAYA, F. F. 1971. Pollution of the Atmosphere and Changes in its Composition. *Akad. Nauk SSSR, Izv., Ser. Geogr.,* No. 4, p. 5–17.

DUCE, , R. A.; QUINN, J. G.; OLNEY, C. E.; PIOTROWICZ, S. R.; RAY, B. J.; WADE, T. L. 1972. Enrichment of Heavy Metals and Organic Compounds in the Surface Microlayer of Narragansett Bay, Rhode Island. *Science (AAAS),* Vol. 176, No. 4031, p. 161–3.

FANALE, F. P.; CANNON, W. A. 1971. Physical Absorption of Rare Gas on Terrigenous Sediments. *Earth Planet. Sci. Lett.,* Vol. 11, No. 5, p. 362–8.

FLINT, R. F. 1973. *The Earth and Its History,* p. 86–106. New York, W. W. Norton.

FREYER, H. D.; WAGENER, K. 1970. Investigation of Ancient Air Samples From Zechstein Salt of Germany. *Z. Naturforsch.,* Vol. 25A, No. 10, p. 1427–30.

FRIEND, J. P. 1973. The Global Sulfur Cycle. In: S. I. Rasol (ed.). *Chemistry of the Lower Atmosphere,* p. 177–201. New York, Plenum Press.

GALIMOV, E. M.; KUZNETSOVA, N. G.; PROKHOROV, V. S. 1968. Ancient Atmospheric Composition of the Earth Based on Results from Isotope Analysis of Carbon from Precambrian Carbonates. *Geokhim.,* No. 11, p. 1376–81.

GREGOR, B.; VAN VALEN, L. 1971. Carbon and Atmospheric Oxygen. *Science (AAAS),* Vol. 174, No. 4006, p. 316–17.

GREY, D. C. 1969. Geophysical Mechanisms for ^{14}C Variations. *J. Geophys. Res.,* Vol. 74, No. 26, p. 6333–40.

HAYS, P. B.; ROBLE, R. G.; SHAH, A. N. 1972. Terrestrial Atmospheric Composition from Stellar Occultations. *Science (AAAS),* Vol. 176, No. 4036, p. 793–4.

HIDY, G. M. 1973. Removal Processes of Gaseous and Particulate Pollutants. In: S. I. Rasool (ed.). *Chemistry of the Lower Atmosphere,* p. 121–73. New York, Plenum Press.

HOBBS, P. V.; HARRISON, H.; ROBINSON, E. 1974. Atmospheric Effects of Pollutants. *Science,* Vol. 183, p. 909–15.

HOLLAND, H. D. 1962. Model for the Evolution of the Earth's Atmosphere. *Geol. Soc. Am.,* p. 447–77.

——. 1963. On the Chemical Evolution of the Terrestrial and Cytherean Atmospheres. In: *The Origin and Evolution of Atmospheres and Oceans,* p. 86–101. New York, Wiley.

——. 1964. In: P. J. Brancazio, A. G. W. Cameron (eds.), *The Origin and Evolution of the Elements,* p. 86–91. New York, Wiley.

——. 1968a. In: C. N. Ahren (ed.) *Origin, and Distribution of the Elements,* p. 949–54. Oxford, Pergamon Press.

——. 1968b. The Abundance of CO_2 in the Earth's Atmosphere through Geologic Time. In: *Origin and Distribution of the Elements. Int. Ser. Mon. Earth Sci.,* Vol. 30, p. 949–54.

——. 1973. Ocean Water, Nutrients and Atmospheric Oxygen. In: *Symposium on Hydrogeochemistry and Biogeochemistry.* Vol. 1: *Hydrogeochemistry,* p. 68–81. Washington, D.C., Clark Co.

HUTCHISON, G. E. 1944. Nitrogen in the Biochemistry of the Atmosphere. *American Scientist,* Vol. 32, p. 178–95.

JUNGE, C. E. 1963. *Air Chemistry and Radioactivity.* New York, Academic Press.

KAZANSKIY, Y. P.; KATAYEVA, V. N.; SHUGUROVA, N. A. 1970. Gas and Liquid Phases in Inclusions as Relicts of Former Atmospheres and Hydrospheres. *Int. Geol. Rev.,* Vol. 12, No. 9, p. 1150–3.

KELLOGG, W. W.; GADLE, R. D.; ALLEN, E. R.; LAZRUS, A. L.; MARTELL, E. A. 1972. The Sulphur Cycle. *Science (AAAS),* Vol. 175, No. 4022, p. 587–96.

KRAUSKOPF, K. B. 1967. *Introduction to Geochemistry.* New York, McGraw-Hill.

KROOPNICK, P.; CRAIG, H. 1972. Atmospheric Oxygen; Isotopic Composition and Solubility Fractionation. *Science (AAAS),* Vol. 175, No. 4017, No. 4017, p. 54–5

LAL, D.; VENKATAVARADAN, V. S. 1970. Analysis of the Causes of C^{14} Variations in the Atmosphere. In: *Radiocarbon Variations and Absolute Chronology, Tex. Mem. Mus., Misc. Pap.,* p. 549–69.

LASAGA, A. C.; HOLLAND, H. D.; DWYER, M. J. 1971. Primordial Oil Slick. *Science (AAAS),* Vol. 174, No. 4004, p. 53–5.

LEPP, H. 1973. *Dynamic Earth,* p. 442–63. New York, McGraw-Hill.

MARTELL, E. A. 1973. Hydrogen Compounds in the Stratosphere and Mesophere. In: *Symposium on Hydrogeochemistry and Biochemistry.* Vol. 1: *Hydrogeochemistry,* p. 27–39. Washington, D.C., Clark Co.

MASON, B. 1966. *Principles of Geochemistry.* 3rd ed. Table 3.4., p. 48. New York, Wiley.

MEADOWS, A. J. 1970. Surface Temperature of the Early Earth and the Nature of the Terrestrial Atmosphere. *Nature,* Vol. 226, No. 5249, p. 927–8.

NAKAI, N.; JENSEN, M. L. 1967. Sources of Atmospheric Sulphur Compounds. *Geochem. J. (Geochem. Soc. Jap.),* Vol. 1, No. 4, p. 199–210.

NISHIMURA, M.; NAKAYA, S.; TANAKA, K. 1973. Boron in the Atmosphere and Precipitation; Is the Sea the Source of Atmospheric Boron? In: *Symposium on Hydrogeochemistry and Biogeochemistry.* Vol. 1: *Hydrogeochemistry,* p. 547–57. Washington D.C., Clark Co.

PERMAN, J.; PERMAN, E. 1969. Spectographic Analysis of Trace Metals in the Atmosphere. In: *Coloquio Espectroscopico Internacional, XV; Comunicaciones sobre Geoquimica y Cosmoquimica. Inst. Geol. Minero, Bol. Geol. Miner* (Spain), Vol. 80, No. 5, p. 476–84.

PERRY, E. C.; MONSTER, J.; REIMER, T. 1970. Sulphur Isotopes in Swaziland System Barites and the Evolution of the Earth's Atmosphere. *Science,* Vol. 171, p. 1015–16.

RAFF, R. A.; MEABURN, G. M. 1969. Photochemical Reaction Mechanisms for Production of Organic Compounds in a Primitive Earth Atmosphere. *Nature,* Vol. 221, No. 5179, p. 459–60.

RAMANATHAN, K. R. 1969. Evolution of the Oxygen and Nitrogen of the Earth's Atmosphere. In: *Physics of the Solid State,* p. 523–8. London; New York, Academic Press.

RASOOL, S. I.; McGOVERN, W. E. 1966. Primitive Atmosphere of the Earth. *Nature,* Vol. 212, No. 5067, p. 1225–6.

RONOV, A. B. 1971. General Trends in the Compositional Evolution of the Outer Shell of the Earth. *Dtsch. Ges. Geol. Wiss., Ber., Reihe A. Geol. Palaontol.,* Vol. 16, Nos. 3–5, p. 331–50.

RONOV, A. B.; MIDGISOV, A. A.; YAROSHEVSKY, A. A. 1973. The Main Stages of the Geochemical History of the Outer Shells of the Earth. In: *Symposium on Hydrogeochemistry and Biogeochemistry.* Vol. 1: *Hydrogeochemistry,* p. 40–53. Washington, D.C., Clark Co.

RUBEY, W. W. 1961. Geologic History of Sea Water. *Geol. Soc. of Amer. Bull.,* Vol. 62, p. 1111–48.

——. 1962. Development of the Hydrosphere and Atmosphere. In: J. F. White (ed.), *Study of the Earth,* p. 363–86. Englewood Cliffs, N.J., Prentice-Hall.

RYDELL, H. S.; PROSPERO, J. M. 1972. Uranium and Thorium Concentrations in Wind-Borne Saharan Dust over the Western Equatorial North Atlantic Ocean. *Earth Planet. Sci. Lett.,* Vol. 14, No. 3, p. 397–402.

SCHIDLOWSKI, M. 1971. Problems Relating to the Evolution of the Precambrian Atmosphere. *Geol. Rundsch.,* Vol. 60, No. 4, p. 1351–84.

SHARMA, R. N. 1965. Geochemical Probe into the CO_2 Economy in the Earth. *Vasundhara (Univ. Saugar, Geol. Soc., J.),* Vol. 1, p. 64–72.

SINITSYN, V. M. 1970. Fossilization of Gases in Sediments and its Influence on the Evolution of the Earth's Atmosphere. Leningrad, Univ., Vestn. *Geol. Geogr.,* Nos. 1, 6, p. 46–55.

——. 1971. Fossilization of Gases in Sedimentary Shell and its Significance in Evolution of Earth's Atmosphere. *Int. Geol. Rev.,* Vol. 13, No. 3, p. 365–71.

SOKOLOV, V. A. 1971. *Geochemistry of Natural Gases.* Moscow, Izd. Nedra.

STEVENS, C. M.; KROUT, L.; WALLING, D.; VENTERS, A.; ENGELKEMEIR, A.; ROSS, L. E. 1972. The Isotopic Composition of Atmospheric Carbon Monoxide. *Earth Planet. Sci. Lett.,* Vol. 16, No. 6, p. 147–65.

STEVENSON, F. J. 1972. Nitrogen: Element and Geochemistry, and The Nitrogen Cycle. In: R. W. Fairbridge, (ed.), *Encyclopedia of Geochemistry and Environmental Sciences,* p. 503–8. New York, Van Nostrand Reinhold.

STRAHLER, A. N.; STRAHLER, A. H. 1973. *Environmental Geoscience.* Hamilton.

SUESS, H. E. 1970. The Three Causes of the Secular C^{14} Fluctuations, their Amplitudes and Time Constants in Radiocarbon Variations and Absolute Chronology. *Tex. Mem. Mus., Misc. Pap.,* p. 595–605.

SYLVESTER-BRADLEY, P. C. 1971. An Evolutionary Model for the Origin of Life. In: Gass, Smith and Wilson (eds.), *Understanding the Earth,* p. 123–42. Cambridge, Mass, MIT Press.

TRENDALL, A. F. 1966. Carbon Dioxide in the Precambrian Atmosphere. *Geochim. Cosmochim. Acta,* Vol. 30, No. 4, p. 435–7.

TUREKIAN, K. K. 1969. The Oceans, Streams and the Atmosphere. In: K. H. Wedepohl (executive ed.) *Handbook of Geochemistry,* Vol. 1, p. 297-323. New York, Springer-Verlag.

——. 1972. *Chemistry of the Earth.* New York, Holt, Rinehart & Winston.

UREY, H. C. 1952. *The Planets: Their Origin and Development.* New Haven, Conn., Yale University Press.

——. 1959. The Atmospheres of the Planets. *Handbuch der Physik,* Vol. 52, p. 363–418.

VALYASHKO, M. G. 1972. Geochemistry of Continental Waters. In: *Geochemistry—Géochimie, Section, Int. Geol. Congr., Proc. Congr. Geol. Int., Programme,* No. 24, p. 331–5.

——. 1973. Natural Waters, their Origin and the Main Formation Regularities of their Composition. In: *Symposium on Hydrogeochemistry and Biochemistry.* Vol. 1: *Hydrogeochemistry* p. 54-67. Washington, D.C., Clark Co.

VAN VALEN, L. 1971. The History and Stability of Atmospheric Oxygen. *Science,* Vol. 171, p. 439–43.

VERHOOGEN, J. 1970. *The Earth—An Introduction to Physical Geology.* New York, Holt, Rinehart & Winston.

VINOGRADOV, A. P. 1967. Gas Regime of the Earth. *Chem. Earth's Crust,* Vol. 2, p. 1–19.

——. 1972. Changes of the Atmosphere Under the Influence of Human Activity. *Geokhim. (Akad. Nauk SSSR),* No. 1, p. 3–10.

VINOGRADOV, V. I.; IVANOV, I. B.; LITSAREV, M. A.; PERTSEV, N. N.; SHANIN, L. L. 1969. Age of the Oxygen Atmosphere of the Earth. *Akad. Nauk SSSR, Dokl.,* Vol. 188, No. 5, p. 1144–47.

WASSERBERG, G. J.; MAZOR, E.; ZARTMAN, R. E. 1963:. Isotopic and Chemical Composition of some Terrestrial Natural Gases. In: J. Geiss, E. O. Goldberg (eds.), *Earth Science and Meteorites,* p. 219–43. Amsterdam, North Holland Publishing Co.

WILKNISS, P. E.; BRESSAN, D. J. 1971. Geochemical Aspects of Inorganic Aerosols Near the Ocean–Atmosphere Interface. *Am. Chem. Soc., Div. Water, Air, Waste Chem.,* Vol. 11, No. 2, p. 156–8.

WILSON, C. L.; MATHEWS, W. H. (eds.) 1971. *Inadvertent Climate Modifications.* Cambridge, Mass., MIT Press.

ZOLLER, W. H.; GORDON, G. E.; GLADNEY, E. S.; JONES, A. G. 1971. The Sources and Distribution of Vanadium in the Atmosphere. *Am. Chem. Soc., Div. Water, Air, Waste Chem.,* Vol. 11, No. 2, p. 159–65.

Atmospheric particulate 'pollutants'

Paul R. Harrison

Meteorology Research Inc.,
Altadena, California (United States of America)

K. A. Rahn

Graduate School of Oceanography,
University of Rhode Island,
Kingston, Rhode Island (United States of America)

Introduction

When addressing the 'pollution problem' one usually thinks immediately of the aesthetic degradation caused by loss of visibility, soiling of clothes and house dusts. Chronic effects on vegetation and structures, and effects of odours are secondary. Thus, we are usually aware of the presence of an over-abundance of particles in the air before noticing other pollutants. As any home-maker knows, no place on Earth is free of 'dust', that temporary resting-place of those suspended solids floating around us.

If one ponders further, it soon becomes evident that we are surrounded by a suspension of particles of various sizes, composition, shape and origins. These suspended particles are found everywhere. Even in the remoteness of Antarctica, levels of $0.5\ \mu g/m^3$ (200 times less than those values usually found in urban areas) are present (Rahn, 1971). The sources and effects are manyfold, making the study of particulate pollutants complex and challenging.

The purpose of this chapter is to cover the basic concepts of: classification with respect to source; methods of movement (transport); methods of capture and parameterization; methods of determining basic consistency and shape (analysis); methods of prevention (control); and, finally, some discussion concerning recent important theories and discoveries. Some of the latter concepts are yet to be proved but enough evidence is available to demand their inclusion.

Atmospheric aerosols are defined as solid or liquid matter dispersed in air. Particulates are the solid fraction of aerosols. The term most used to describe particles in the air (other than water), is 'total suspended particulates' (TSP). TSP is operationally defined as the mass per unit volume of those particles that are captured by pulling a large volume of air through a filter (usually glass fibre), excluding water. A typical high-volume sampler (HiVol) is shown in Figures 1

and 2. By virtue of the design of the samplers, the sizes of particles sampled are $30–40\ \mu m$ in diameter and smaller (see *Guidelines for Development of a Quality Assurance Programme*, 1973). Figure 3 is a typical sample of material taken from a HiVol glass-fibre filter.

As one can see, both the physical and operational definitions of particulates are very broad and inclusive, especially when contrasted to gaseous pollutants such as

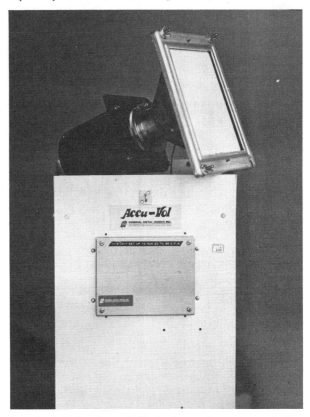

FIG. 1. A high-volume (HiVol) air sampler motor and filter holder resting on its shelter. A clock timer is usually used to program the time and date of sampling.

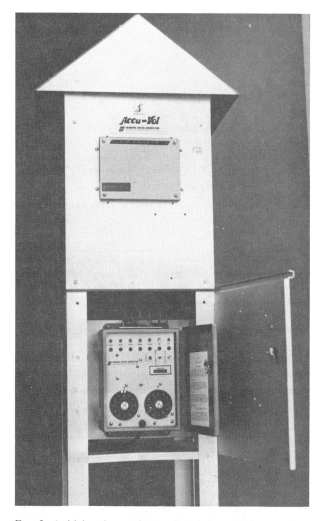

FIG. 2. A high-volume air sampler shelter with a constant flow-controlling device.

SO_2 which are chemically unique and specific. Particulates, like hydrocarbons, have a large number of forms, chemical compositions and sources. Further classifications according to origin, morphology, chemical species and content, size, shape, etc., are available to the point of being an unlimited number of subsets and permutations.

Sources

There are many sources of particulates. Some particles are man-made, others are from entirely natural sources, and still others are formed by secondary reactions emanating from both natural and man-made origins.

Natural sources include volcanoes, pollens, organic debris from insects, plants and micro-organisms. Wind erosion and re-suspension account for large masses of particles of various sizes, shape and consistency. Forest fires also emit large amounts of both gases and particulates into the atmosphere. Table 1 lists some estimates of the amounts of particulate emissions from various sources.

Some man-made sources are obvious. Factory and home chimneys are prime examples. Lesser-known sources include emissions caused by agriculture, construction and demolition. Much of the fuel used by man is used in such a way as to release hydrocarbons and oxides of nitrogen. By various proposed formulae these gases are driven by sunlight to form small particulates. Much of the sulphur contained in these fuels is oxidized and emitted to the atmosphere. The end result is reaction chains forming particulate sulphates. Similar reactions form nitrates and oxyhydrocarbons. These three compounds constitute most of the secondary particulates. It becomes clear, then, that there are particles

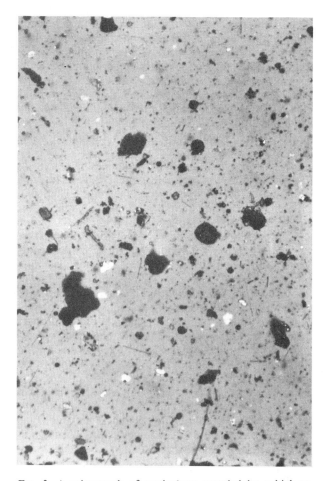

FIG. 3. A micrograph of particulates sampled by a high-volume air sampler using a glass-fibre filter media. The sample was taken at a federal CAMP station in Chicago (United States). TSP = 170 μg/m^3, 24-hour sample, 3 December 1972. Magnification 160 $\times$.

TABLE 1. Significant sources of particulate pollution in the United States

Source	Quantity of emissions (thousand tons/year)	Per-centage
Natural dusts	63,000	51
Forest fires		
Wildfire	37,000	
Controlled fire		
Slash burning	6,000	
Accumulated litter	11,000	
Agricultural burning	2,400	
	56,400	46
Transportation		
Motor vehicles		
Petrol	420	
Diesel	260	
Aircraft	30	
Railway	220	
Water transport	150	
Non-highway use		
Agriculture	79	
Commercial	12	
Construction	3	
Other	26	1,200	1
Incineration		
Municipal incineration	98	
On-site incineration	185	
Wigwam burners	35	
Open dump	613	931	1
Other minor sources		
Rubber from tyres	300	
Cigarette smoke	230	
Cosmic dust	24	
Aerosols from spray cans	390	
Ocean salt spray	340	1,284	1
TOTAL	122,815	100

Sources: Fennelly, 1975; Environmental Protection Agency, 1971; Langren and Paulus, 1975.

TABLE 2. Estimate of global emission of all particulates

Emission	Quantity (million tons/year)	Per-centage
Man-made		
Particles	92	31
Gas-particle conversion:		
SO_2	147	50
NO_x	30	10
Photochemical from hydrocarbons	27	9
TOTAL	296	100
Natural		
Soil dust	200	8.6
Gas-particle conversion:		
H_2S	204	8.8
NO_x	432	18.8
NH_3	269	11.6
Photochemical from terpenes, etc.	200	8.6
Volcanic	4	0.2
Forest fires	3	0.2
Sea salt	1,000	43.2
TOTAL	2,312	100.0

Sources: Fennelly, 1975; Robinson and Robbins, 1977.

TABLE 3. Size distribution of a typical atmospheric dust sample

Particle range (μm)	Typical number per m^3 [1]	Percentage by count	Percentage by volume (weight)
30–10	1,000	0.0005	28
10–5	35,000	0.175	52
5–3	50,000	0.25	11
3–1	214,000	1.07	6
1–0.5	1,352,000	6.78	2
0.5–0	18,280,000	91.70	1

1. Normalized to 1,000 particles in the 30–40 μm range.
Source: University of Minnesota, 1968.

of primary origin emitted directly into the air and there are particles of secondary origin formed from gases. The result is frequently the biomodal distribution of particles shown in Figure 4. Table 2 shows another estimation of the gross amounts and sources of the global emissions for most particulates.

When inspecting Tables 1 and 2, one should exert some caution in the interpretation of the data. For example, natural dusts account for about half of the tonnage of particles emitted (Table 1). These dust particles generally have sizes greater than 10 μm and are dense. The result is a particle that settles out quickly and has little effect on health but a large effect on TSP values.

Table 3 represents data taken by the University of Minnesota showing a 'typical' atmospheric dust sample normalized to 1,000 particles in the 20 μm range. Figure 5 shows examples of the size ranges by source. One can easily see that a few large particles can account for a large portion of the total weight in a sample, even though there are thousands of smaller particles present. Of course, there is a large variation from place to place according to local industrial, agricultural and wind conditions. Table 4 is a clear example of the relative abundance of selected constituents with respect to location or proximity to local sources.

TABLE 4. Selected particulate constituents as percentages of gross suspended particulates (1966–67)

| Constituent | Urban (217 stations) | | Non-urban | | | | | |
| | | | Proximate (5) | | Intermediate (15) | | Remote (10) | |
	$\mu g/m^3$	Percentage	$\mu m/m^3$	Percentage	$\mu m/m^3$	Percentage	$\mu g/m^3$	Percentage
Suspended particulates	102.0		45.0		40.0		21.0	
Benzene soluble organisms	6.7	6.6	2.5	5.6	2.2	5.4	1.1	5.1
Ammonium ion	0.9	0.9	1.22	2.7	0.28	0.7	0.15	0.7
Nitrate ion	2.4	2.4	1.4	3.1	0.85	2.1	0.46	2.2
Sulphate ion	10.1	9.9	10.0	22.2	5.29	13.1	2.51	11.8
Copper	0.16	0.15	0.16	0.36	0.078	0.19	0.06	0.28
Iron	1.43	1.38	0.56	1.24	0.27	0.67	0.15	0.71
Manganese	0.073	0.07	0.26	0.06	0.012	0.03	0.005	0.02
Nickel	0.017	0.02	0.008	0.02	0.004	0.01	0.002	0.01
Lead	1.11	1.07	0.21	0.47	0.096	0.24	0.022	0.10

Source: After McMullen *et al.* (1969).
Reprinted with permission of the *Journal of the Air Pollution Control Association.*

Most mechanical operations such as grinding, milling, agriculture and construction are incapable of producing large amounts of aerosols less than 10 μm; most of the particulates formed in this manner are large and fall rapidly to the ground. The resultant airborne distribution is somewhere between 1–10 μm diameter.

Figure 6 is a classical size distribution presented in Junge (1963) from non-polluted marine and continental air. Some natural mechanisms seem to perpetuate such a distribution. Agglomeration is a possible mechanism but if left to Brownian motion alone it is too slow. More probable mechanisms are: (a) the exercising of the hygroscopic nature of most particles or (b) deposition–saltation–reflotation. A device that humidifies a parcel of sampled air by a sudden drop in pressure has shown that recently combustion-formed aerosols are high in their hygroscopic ability and readily form many cloud particles. Conversely, old aerosols of continental origin are low in condensation nuclei (CN). If one humidifies the same parcel several times, the CN count decreases (*Humidigraph Institution Manual IM-189, 1975*).

Another source of aerosols is the ocean. The wave action tends to entrap air bubbles which, in turn, rise to the surface and burst. This mechanisms, graphically represented in Figure 7, is responsible for the injection of a tremendous amount of water droplets containing the dissolved salts of the parent ocean or sea (Junge, 1963). These droplets are called jet drops. They are sources of an atmospheric charge (negative) and water vapour (most evaporate). They also contain enrichments of organic compounds and other ocean surface active material (e.g. ammonia).

Conversely, transportation emissions account for about 1 per cent of the total aerosol mass, but particulates have mass mean diameters (spherical approximation) of less than 2 μm. The particulates can be inhaled easily by the lungs, contain known harmful compounds and in addition, settle out very slowly—at rates less than 1 m/min.

The conclusion is that 'availability' and 'toxicity' is far more important than 'mass' of emission. The current tendency to incinerate industrial effluents is an example of energy used to decrease mass emissions but incineration in turn creates small, concentrated particles of a respirable nature that can be transported over large distances.

Gaseous reactions, driven by electromagnetic radiation in the form of sunlight, are a primary source of the very small, secondary particulates. The basic mechanism is a three-component reaction represented by:

Reactant gas + oxygen + inert gas →
oxidized gas or particle + inert gas.

(Reaction rates for SO_2 are somewhere between 0.5 and 3 percent per hour in the free atmosphere). There are three over-simplified but classical sets of formulae for these reactions. They are as follows:

For SO_2:

$$SO_2 + h\upsilon \longrightarrow SO_2^*;$$

$$SO_2^* + O_2 + \text{inert gas (I)} \longrightarrow SO_4^{2-} + (I);$$

$$SO_4 + SO_2 \longrightarrow 2\,SO_3;$$

$$SO_4 + O_2 \longrightarrow SO_3 + O_3;$$

(also a catalytic reaction)

$$SO_2 + H_2O \xrightarrow{\text{catalyst}} H_2SO_3;$$

$$2\,H_2SO_3 + O_2 \longrightarrow 2\,H_2SO_4.$$

Then:

$$H_2SO_4 + \text{anion} \longrightarrow \text{sulphate} \downarrow + H_2 \uparrow.$$

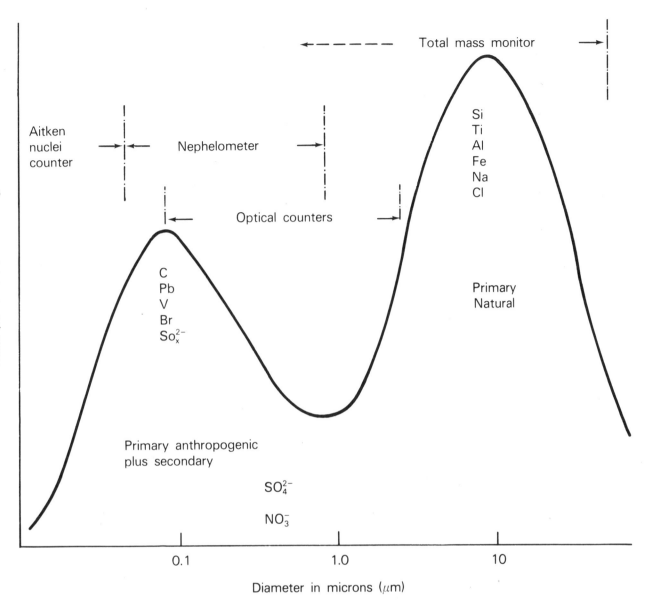

FIG. 4. Idealized mass to size distribution for urban aerosols. Approximate size ranges for various instruments and probable chemical constituents are also shown. (Hidy *et al.*, 1975. Reproduced with permission of the *Journal of the Air Pollution Control Association*.)

It is not clear whether this is primarily a gas-gas reaction to form a particle or a gas–particle reaction (surface reaction).

For a hydrocarbon or OH radical:

$$NO_2 + h\upsilon \longrightarrow NO + O;$$

$$O + O_2 + (I) \longrightarrow O_3 + (I);$$

$$O_3 + olefins \longrightarrow RO^{\bullet} + O_2;$$

$$RO^{\bullet} + olefins \longrightarrow R-R'O \text{ or } R\overset{\overset{\displaystyle O}{\diagup \diagdown}}{-}R' \text{ particulate}.$$

About 5 per cent of atmospheric organic compounds form particulates.

For nitrates:

$$NO_2 + O_3 \longrightarrow NO_3 + O_2;$$

$$NO_2 + NO_3 \longrightarrow N_2O_5;$$

$$N_2O_5 + H_2O \longrightarrow HNO_3;$$

or

$$NO_2 + OH + (I) \longrightarrow HNO_3 + (I).$$

181

Paul R. Harrison and K. A. Rahn

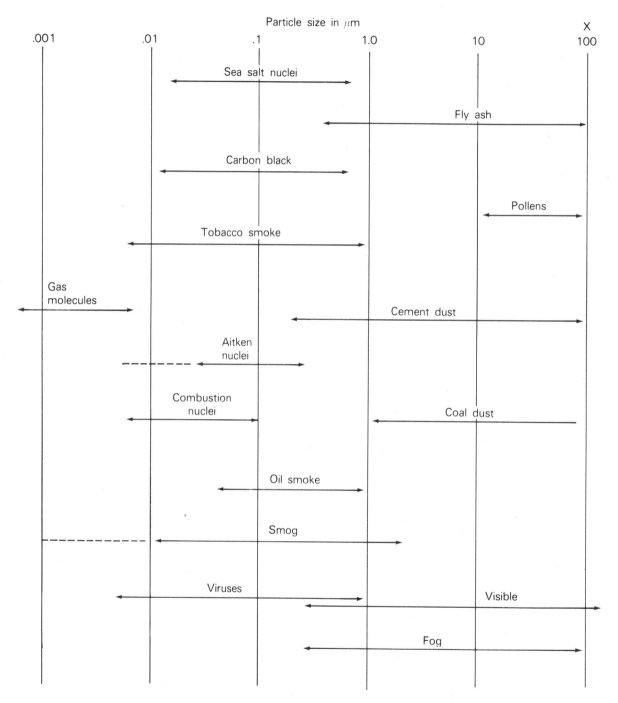

Fig. 5. Approximate size ranges of particles from selected sources. (Reproduced with permission of the *Journal of the Air Pollution Control Association*.)

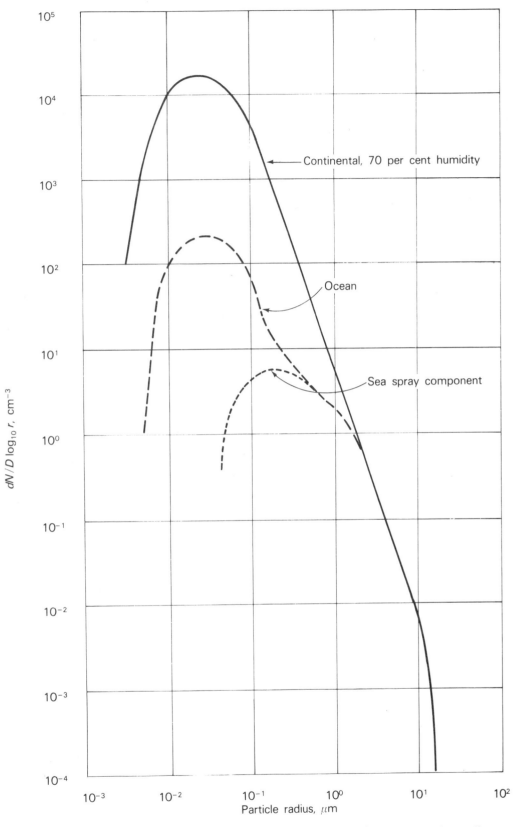

FIG. 6. Classical size distribution for atmospheric aerosols. N is the total number of particles less than radius r. (From Junge, 1963. © Academic Press, New York. Reproduced with permission.)

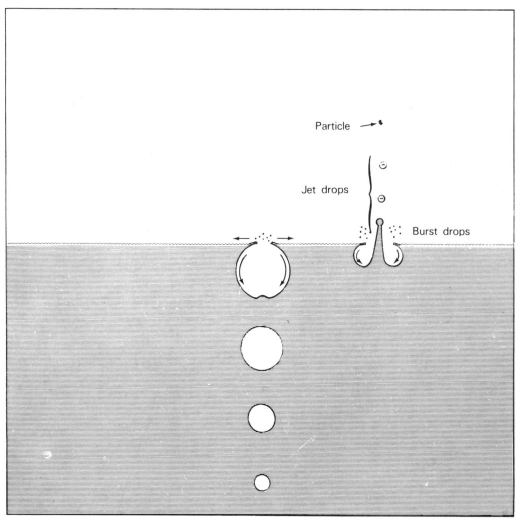

Particle

Jet drops

Burst drops

FIG. 7. Bursting bubbles cause water droplets to be injected into the air. These subsequently form salt particles which are transported some distance by the wind.

Then to form particulates, we can have:

$$NH_3 + HNO_3 \longrightarrow NH_4NO_3;$$

$$Anion + HNO_3 \longrightarrow \text{nitrate, particulate};$$

or

$$NaCl + HNO_3 \longrightarrow NaNO_3 + HCl.$$

The above reactions are presented as a plausible example of primary reactions of gaseous effluents both man-made and natural (e.g. turpines). Many other permutations can be presented. In summary, the reactions, rates and final product form are dependent upon too many factors to be completely predictable, both in quantity and method.

A final but most important point should be made. When examining the global sources of particulates, one is easily convinced that natural sources account for the overwhelming majority *by weight*. One must be fully aware, however, that the major concern is for health effects and aesthetics (visability). This immediately implies that those particles affecting these concerns are of prime importance. Most particles greater than $5 \mu m$ either settle out rapidly or are impacted on surfaces such as the nasal passages or lung tree. Smaller particles between 0.1 and $5 \mu m$ are easily inhaled and can be transported to the lung alevioli—where they have a high probability of retention. Smaller particles still behave like air molecules and are expired with the air from the lungs. Some particles retained in the lungs will dissolve in the fluids of the tissue and can account for elevated levels of certain elements (e.g. lead). Other components are left for chronic irritation and cellular destruction (e.g. silica and asbestos). Ambient combined with personal (cigarettes) pollution exacerbates the effect on health by the inhalation of small particles. In addition, the respirable range is the same range that results in maximum light scattering (b_{scat}) or loss of visibility (Carlson *et al.*, 1974). Most of the smaller particulates result from man's activities, either directly or by reactions similar to those described above.

In conclusion, we must realize that urbanization without control of anthropogenically produced fine particulates will cause health and visibility problems far beyond the effect that their consideration on a weight-for-weight basis with naturally occurring particulates would suggest.

Transport

Describing the concentration, the form and the effects of particulates at the receptors, given an estimate of the emission strengths and locations, is not a trivial task. Assessment of the source strength or receptor concentration is not difficult, but the task of accurately estimating one when given the other requires a thorough knowledge of the transport phenomena. Figure 8 depicts the classical 'emission–transport–receptor' division of disciplines. The particle is emitted into the atmosphere and, after a relatively short period of time, equilibrates its carrier-gas temperature with its surroundings. It is then transported as the wind and its eddies move, all along the way entraining new or 'cleaner' air. At some point downwind, it is deposited, impacted, or washed out by rain.

During this time it may be modified. When it is deposited, a receptor is 'exposed' to it. Obviously the story does not necessarily end at that point; the environmental pollutant can go through the cycle several times (Harrison, 1975; 1976a). For example, if a particle is deposited in a positive sink (i.e. grass), the particle becomes absorbed into the local biota and completely loses its individual characteristics. In other cases, such as deposition on a paved surface, the particles may coagulate with others and/or rest these until wash-off transports it to another location and form (Newman *et al.*, 1976; Rahn and Harrison, 1976). Or a high wind can re-suspend the particle in its original or modified form, transporting it to another location and possibly another modification. It is extremely important to understand that unless the 'pollutant' goes through a transformation or is deposited in a positive sink—where it is held and absorbed into the local ecosystems (such as absorption into plants, dissolved into a liquid form, etc.)—it is available for re-emission from a 'natural' source.

Since most aerosols are carried by air currents, we should understand the turbulent nature of the atmosphere. Warm, less dense air rises. Thus, a hot parcel of air will rise until it is cooled by adiabatic expansion to the temperature and density of the surrounding atmosphere. Some overshoot can occur. Effluents (particles or gas) are carried with that parcel and are transported in the direction of the ambient air currents.

Under low wind and stable conditions, there are few vertical currents. Effluents are then transported in a roughly horizontal meandering manner, with the larg-

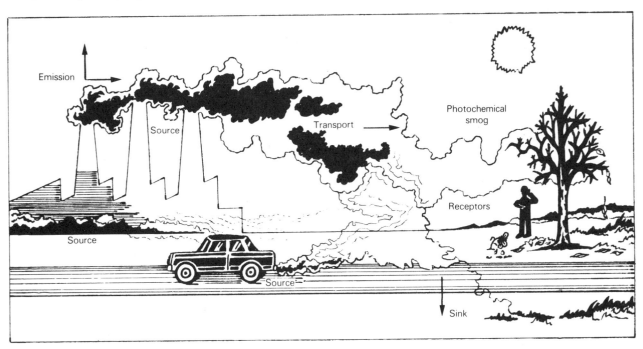

FIG. 8. Example of subsets of particulate movements.

Paul R. Harrison and K. A. Rahn

FIG. 9. Early-morning photograph showing low-level emissions into a radiation temperature inversion at Riverside, Calif. Note the lack of diffusion indicated by the discrete line between the particulate cloud and clean air.

er particles slowly falling as they are carried over longer distances (Fig. 9). In most estimates of dispersion, except for large particles close to the source, this falling motion is not considered. For the sake of completeness, the following equations are shown (*The Frank Chart,* 1937), governing various settling rates of large to small particles (Spheres of specific gravity: 1.0 gm/cc):

$d > 200 \ \mu m$

$$c = \sqrt{\frac{2 g d s_1}{3 K s_2}} = 0.15 \sqrt{d S_1}.$$

$1 < d < 200 \ \mu m$
(Stokes' law)

$$c = \left(\frac{2 r^2 g}{9}\right)\left(\frac{s_1 - s_2}{\eta}\right).$$

$0.1 < d < 1 \ \mu m$
(Stokes–Cunningham law)

$$c = c'\left(1 + K_1 \frac{\lambda}{r}\right)$$

where c' = Stokes' settling velocity

$\left(1 + K_1 \dfrac{\lambda}{r}\right)$ is Cunningham's correction factor, and

$$K_1 = 0.8 \text{ to } 0.86.$$

186

$d < 0.1\ \mu\mathrm{m}$

(Brownian motion)

$$A = \sqrt{\frac{RT}{N}\frac{t}{3\pi\eta r}}.$$

where c = setting velocity (cm/sec.),
$\quad\quad d$ = particle diameter (cm),
$\quad\quad r$ = particle radius (cm),
$\quad\quad g$ = gravitational constant (981 cm/sec^2),
$\quad\quad s_1$ = particle density,
$\quad\quad s_2$ = air density ($\ll s_1$),
$\quad\quad \eta$ = viscosity of air ($1{,}814 \times 10^{-7}p$ at $70\,^\circ$F),
$\quad\quad \lambda$ = mean free path of gas molecules (10^{-5} cm),
$\quad\quad A$ = distance travelled,
$\quad\quad R$ = gas constant (8.316×10^7),
$\quad\quad N$ = number of molecules in 1 mole of gas (6.06×10^{23}),
$\quad\quad K$ = constant,
$\quad\quad t$ = time interval.

As a rough guide, a $20\ \mu\mathrm{m}$ particle of density 1.0 will settle in still air at a rate of approximately 1 m/min. If, however, we encounter turbulent air motion, the particles are rapidly carried both aloft and downward. The plume is continually being torn apart by the circular motion (eddies) to the point where the effluents 'diffuse' rapidly, spreading over a wide horizontal distance. The effluents are also brought to ground level more rapidly, not by natural sedimentation or settling but by turbulent motion of the transport fluid itself.

Figure 10 presents graphical representations of some of the various diffusion characteristics of the air currents classically used by air pollution meteorologists. Since the vertical temperature profile determines the largest part of this diffusion, it is important to understand these profiles when examining the type of parcel motion expected. It is obvious that under turbulent conditions the effluents are brought to the surface in greater concentrations and at points closer to the emission point. However, due to the lack of rapid diffusion under stable conditions, the effluent concentrations maintain their integrity as a continuous plume much longer.

No discussion of diffusion is complete unless one explains that mechanical diffusion (turbulence) is the overwhelmingly dominant factor in the transport and modification to the characterization of the effluent. Brownian diffusion, as studied by most chemists and physicists, is an insignificant factor and should not be considered in such studies.[1]

There are two mathematical approaches to the calculation of the dispersion of gaseous and particulate pollutants, both of which treat the particles as gases. A

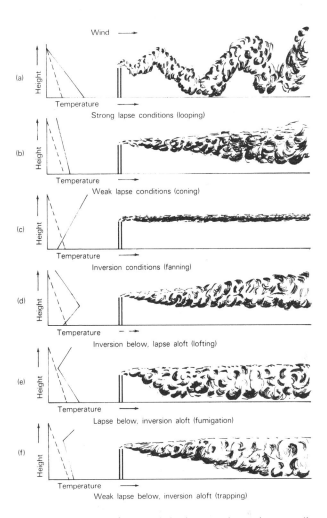

FIG. 10. Six types of plume behaviour, under various conditions of stability and instability. At left: broken lines, dry adiabatic lapse rate; full lines, existing lapse rates (Hewson, 1964).

fall-out term can easily be added, however, for close-to-the-source calculations.

The Gaussian diffusion equation is the primary basis for calculating concentrations at various distances and heights downwind. The most used equation is:

$$X = \frac{Q}{\bar{u}}\left[\frac{1}{2\pi\sigma_y\sigma_z}\left\{\exp\left[-\frac{1}{2}\left(\frac{Z-H}{\sigma_z}\right)^2\right] + \exp\left[-\frac{1}{2}\left(\frac{Z+H}{\sigma_z}\right)^2\right]\right\}\right],$$

1. I recommend such references as *Workbook of Atmospheric Dispersion Estimates* (Turner, 1969). In this workbook one can find a clear and concise discussion of the various phenomena to be considered in the estimation of the transport and diffusion of gaseous and particulate emissions in the predicted downwind concentrations. Formulae and graphs are presented which are extremely easy to use and which will facilitate estimations of concentrations downwind from emission points.

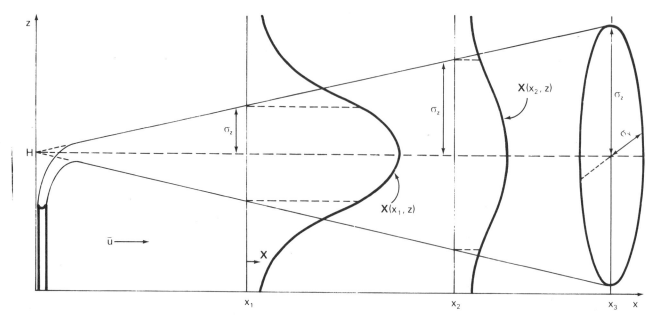

FIG. 11. Idealized Gaussian plume. The y plume plane is also shown. σ_y is usually not the same as σ_z (Turner, 1969).

where X = concentration (g/cm³),

Q = emission rate (gm/sec),

$\bar{u}$ = average horizontal wind speed (m/sec),

$\left.\begin{array}{l}\sigma_y \\ \sigma_z\end{array}\right\}$ = standard deviations of plume concentration distributions in the horizontal and vertical planes (m),

and H = effective stack height or height of the centre-line of the plume.

The units given in parentheses are those most widely used. Figure 11 is a graphical depiction of a 'classical' Gaussian distributed plume.

It should be cautioned that estimates of σ_y and σ_z are based on statistical analyses of several plumes under various conditions. Recent research concerning high-capacity power-plant plumes has shown that, under stable conditions, these equations can underestimate the initial plume-rise dilution and overestimate the downwind diffusion. It is also known that the Gaussian equations can underestimate the maximum concentrations under stable conditions and overestimate the concentrations under turbulent conditions.

In conclusion, we should point out that the use of such equations and formulae is sufficient for most work but not for rough terrain and conditions of marginal compliance with standards. The solution to this dilemma is the use of hydrodynamic or fluid-flow models. These physical models are more exact, but they require large computers and are more expensive to run. They also require multiple passes to arrive at concentrations at each point.

As the particulate/gaseous effluent is transported, certain modifications or evolutions can take place. As pointed out in the previous section, if the plume con-

tains hydrocarbons, oxides, nitrogen and (possibly) water vapour in certain proportions, photochemical reactions may occur. These are triggered by the energy imparted to the gas by sunlight and go through various complex reactions that are as yet not completely understood. The final result is the formation of very fine particulates. Since these reactions are responsible for the formation of a large number of very small particles, they also modify the size distribution of atmospheric aerosols (Lillis and Young, 1975). If one assumes that naturally occurring aerosols take on the form proposed

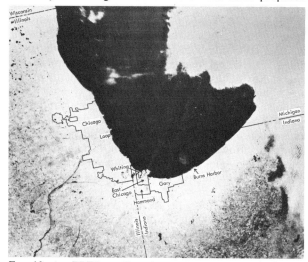

FIG. 12. An ERTS satellite photograph of particulate plumes emanating from southern Lake Michigan industry into a cloud bank 80 km over the lake. Note the obvious modification to the cloud bank. Wind is from the south-west (Lyons, 1973. Reproduced with permission of the American Meteorological Society.)

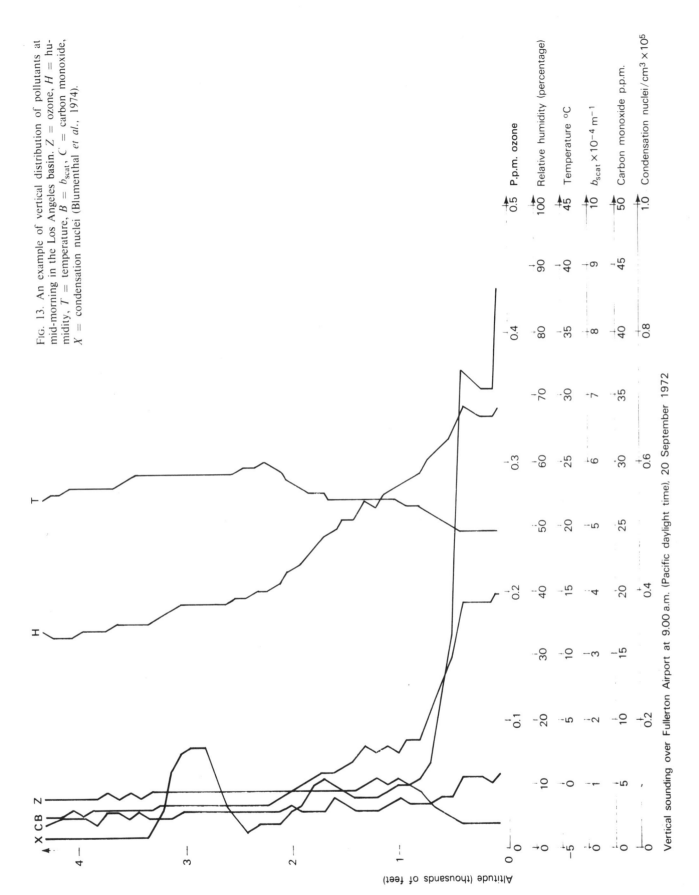

FIG. 13. An example of vertical distribution of pollutants at mid-morning in the Los Angeles basin. Z = ozone, H = humidity, T = temperature, $B = b_{scat}$, C = carbon monoxide, X = condensation nuclei (Blumenthal *et al.*, 1974).

P.p.m. ozone

Relative humidity (percentage)

Temperature °C

$b_{scat} \times 10^{-4}\ m^{-1}$

Carbon monoxide p.p.m.

Condensation nuclei/cm³ $\times 10^5$

Vertical sounding over Fullerton Airport at 9.00 a.m. (Pacific daylight time), 20 September 1972

Altitude (thousands of feet)

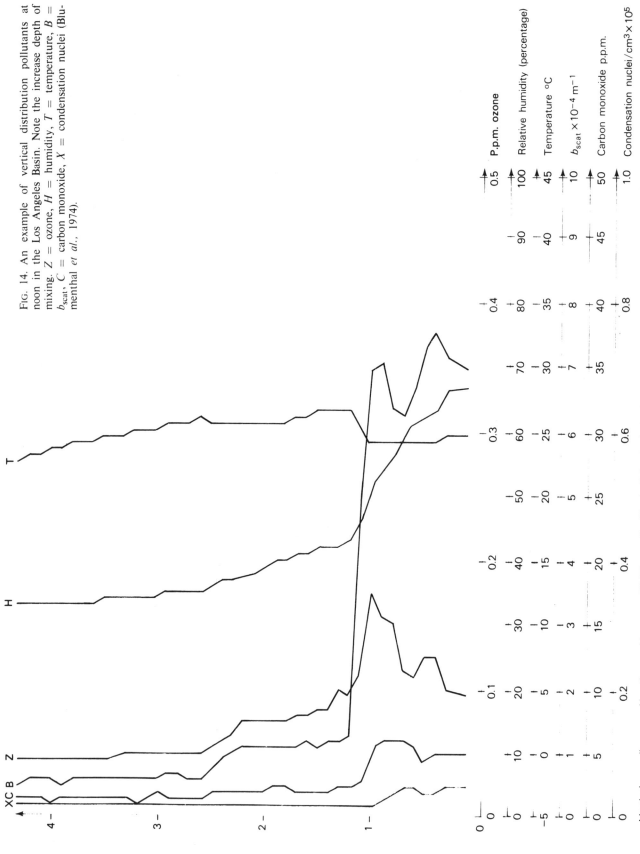

FIG. 14. An example of vertical distribution pollutants at noon in the Los Angeles Basin. Note the increase depth of mixing. Z = ozone, H = humidity, T = temperature, B = b_{scat}, C = carbon monoxide, X = condensation nuclei (Blumenthal *et al.*, 1974).

Vertical sounding over Hawthorne Airport at 12.56 p.m. (Pacific daylight time), 20 September 1972

Altitude (thousands of feet)

P.p.m. ozone

Relative humidity (percentage)

Temperature °C

$b_{scat} \times 10^{-4}\ m^{-1}$

Carbon monoxide p.p.m.

Condensation nuclei/$cm^3 \times 10^5$

190

by Junge (Fig. 6) in 1963 and that emissions of gaseous materials form a large abundance of smaller particles only in the submicron range, we would assume that a bimodal distribution would develop with maxima in the submicron and micron range, as in Figure 4. This is indeed found to be the case. Due to the small size of these particles, they tend to have a long residence time, affect visibility and diminish suspended mass except in those areas where large particles are not present. The average lifetime and ultimate fate of these particles is still not completely understood. It is known that particles tend to agglomerate, forming larger particles which are more susceptible to natural removing processes such as impaction, sedimentation and wash-out by thunderstorms; the latter is known to be an extremely efficient dynamic electrostatic precipitator (Gunn, 1963). In the winter, snow seems to be an equally efficient particle remover due to the condensation of the ice material onto the particle and subsequent snow-out. Figure 12 is a satellite photograph of industrial particulates being transported with only minor dispersion over many kilometres before being modified by a cloud bank.

The concept of air mass trajectory is an important one (Harrison, 1970, 1971a, b, 1973). On a large scale (a few hundred kilometres) on the surface of the Earth, parcels of air separated by frontal zones are somewhat conservative phenomena. Water vapour and potential energies are slowly varying with respect to the more local, short-term perturbations. Thus, effluents injected either by direct or by photochemical means tend to stay with the air mass until the air mass is completely broken up. This, particulates are carried by these air masses for long distances and can build up and slowly change their characteristics as they pass over other source areas. If the air mass is very slow-moving and particle concentrations increase from the regional emissions, slow but serious deterioration in visibility and increase in the submicron or respirable fraction of particulates can occur. The Los Angeles basin is a prime example of a non-mixing air mass that is moved back and forth and around the basin but it is not transported significantly into areas where wash-out would clean it or where cleaner air would significantly dilute it. Thus particles build up both by photochemical reactions and direct emission to a state of equilibrium with fall-out and impingement, with very little, if any, wash-out occurring in the summer months. It is important to understand that this long-range transport of particles can occur even at the lower levels of the atmosphere.

When a particle or gas is injected above a mixing layer of approximately 3 km, it finds itself in a different transport regime where the air mass concept begins to break down. Thus, if the effluents are emitted or transferred to high levels, they become caught in a faster moving and more direct transport system which can carry the particles over very large distances in relatively short lengths of time. Effluents from volcanic eruptions and atomic bomb tests are emitted into these high levels and are rapidly transported. Fall-out and turbulence carry the particles slowly into the lower atmosphere where other mechanisms can accelerate deposition. It should be remembered that the smaller the particle, the less likely are its chances of deposition; and, the greater the altitude of the original emission, the longer the period before deposition occurs. Figures 13 and 14 show a vertical profile of various pollutants at two different times of day in Los Angeles, California. The scattering coefficient, b_{scat}, is the inverse of visibility and is a good measure of smaller, respirable particulates. Concentration of condensation nuclei is a measure of newness of sources from combustion. Note the increase in the mixing depth between the two measurements and the loss of visibility (increased back-scatter).[1]

Finally, the concept of short-range reflotation of particles should be clearly understood. Particles becoming available for re-suspension such as those from construction sites, demolitions, or impervious surfaces (concrete, roof-tops, roads, and such) tend to allow both mechanical and wind turbulent re-suspension (Newman et al., 1976; Harrison, 1976a; Rahn and Harrison, 1976; Harrison 1976b; Harrison et al., 1974). Mechanical re-suspension is accomplished by lifting the bulk material to some elevation; whereas the large particles drop back directly to the surface, the middle and smaller sized particles are carried farther away by the wind.

The larger particles can impart their momentum to other surface particles causing a surface dust cloud which again allows middle and smaller particles (i.e. $<10 \mu m$) to be carried aloft by turbulent motion, where such motion exists. This latter concept of larger particles being rolled, tumbled, or lifted a short distance to be dropped again and thereby imparting their momentum to smaller particles which are more permanently

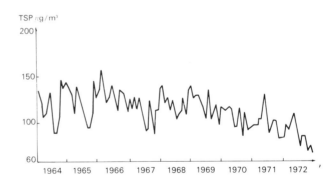

FIG. 15. Average monthly variation of total suspended particulates (TSP) taken from 20 HiVol stations sampled three times a week in Chicago.

1. I recommend the text *Introduction to Meteorology* (Petterssen 1958) as an introduction to the basic concepts of meteorology and transport phenomena. Many other texts are available pertaining specifically to transport of particulate effluents in the short range.

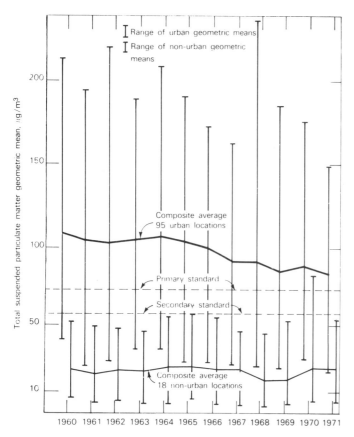

FIG. 16. Composite annual means of TSP at urban and Non-urban National Air Surveillance Network Stations, United States (McCormick and Holzworth, 1976).

refloated is called 'saltation' and is an important emission/transportation mechanism. Agricultural erosion and a large amount of urban-suspended particulate is derived from this mechanism often in conjunction with the mechanical turbulence caused by the tilling of the soil or by the passage of motor vehicles.

Figure 15 is an example of the monthly average variations of TSP in a continental, cold-winter climate where much coal was originally used. As the use of coal decreased, the yearly maxima shifted towards the late spring months when agricultural activities were responsible for particulate formation. Figure 16 presents a composite of data from the United States showing the range and means of urban and non-urban TSP levels.

Sampling and analysis techniques

SAMPLING

Almost all of the techniques for particulate capture use the physical properties of the aerosol. For example, fil-

tration relies on the size and electrostatic properties for retention. Figures 17, 18 and 19 are photo-micrographs of, respectively, a glass-fibre filter, a membrane filter, and a nuclepore filter. One should note that for a fibre filter, large particles are physically prevented from passing through the filter while smaller particles become attached by diffusion and electrostatic forces to the fibres themselves. They do so even though it is physically possible for them to pass through the pad of fibres by a circuitous route. Figures 1 and 2 show an actual sampling device commonly used. In the case of the membrane and nuclepore filters, the particles are too

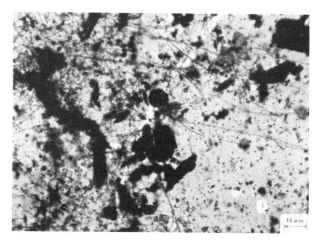

FIG. 17. Photo-micrograph of a sample extracted from a glass-fibre filter. Note that small particles cling to the fibres themselves, suggesting an electrostatic action. The spherical particles are probably oil soot, the larger particles are rubber-tyre dust, and the very smallest are auto-exhaust related. The whitest particles are calcite. Magnification is $163 \times$.

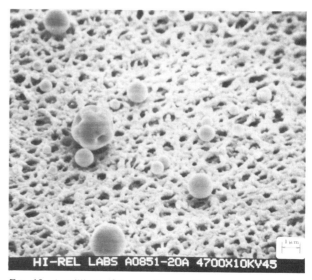

FIG. 18. A millipore filter media sampling an ammonium fluorescein test aerosol.

FIG. 19. An example of nuclepore filter for the smaller of the particles. Note that the smaller pore sizes will limit the flow rate proportionately. Test pores are 0.5 μm.

large to pass due to the extremely small holes in the thin filter pads (Figs. 18 and 19).

When sampling for an aerosol, one must ensure that the particle of interest must end up on the sampling substrate and not in the prior ducting or apparatus. For example, most atmospheric sampling devices are made with shelters so that the large particles and debris are not sampled. To accomplish this, various covers are used so that the larger particles do not enter the sampling ports. Figure 2 shows a simple rain cover in place that allows particles of around 40 μm or less to be sucked into the side openings. The non-aerodynamically shaped and lighter aerosols can enter as larger particles (*Quality Control Practices in Processing Air Pollution Samples,* 1973; Stern, 1976, 1968).

Some devices use charged plates, wires, rods, etc. of various shapes and sizes to 'plate out' the particles by electrostatic induction or direct attraction. The current sensed across the plates is measured and assumptions are made relative to the average charge carried by the 'average' particles. This measurement technique is used mostly to estimate the number of smaller sized particles. Figure 20 is a prototype collection device using a charged screen followed by a rod collector.

FIG. 20. A disassembled electrostatic sampler. The particles are white and are charged in the chamber shown in the foreground. They then 'plate out' on the centre rod. The long, thin, white lines are a lighting aberration. The back filter shows the flow uniformity for uncharged particles. This device is for the collection of very small particles.

FIG. 21. Inertial impactor device for ambient and in-stack source samples.

Figure 21 represents a cascade inertial impactor used to capture mean size cuts of a particle distribution. Figure 22 depicts the inertial impaction principle involved. The inertial impactor devices normally do not provide discrete size cut-offs. They tend to have an overlapping distribution of sizes from stage to stage, as shown in Figure 23. In some cases, the larger particles bounce, shatter, or shed smaller particles, to end up on subsequent stages. To minimize this possibility, 'sticky' substances are applied to the collection surfaces, or a technique of virtual impaction is used, where the particle is 'impacted' into a cavity of still air which slows the particle before it actually strikes a surface. Particles are collected from the cavity by a slow air flow through a filter. This technique also eliminates 'blow off'.

No discussion of capture would be complete if the 'dust fall jar' were not mentioned (Stern, 1976). Briefly, the principle is to simply allow dust to fall into a wide-mouthed jar for subsequent weighing. The problem is that the small particles simply flow over the jar and, if a liquid is not used, a high wind can blow part of the sample out of the jar. The data are simply indicative of large particle loading and not of the respirable, small sizes.

Finally, simply scrubbing the air stream with a liquid in a turbulent motion can capture the aerosol for subsequent analysis. Cryogenics, the nearly complete freezing of the gases and particulates, has also been used.

All of the above techniques modify, lose or change the samples in some way and are applicable only for specific size distributions and properties. Certain passive systems using nuclear radiation, light transmission and light scattering, are promising techniques of recent use to obtain size distributions as well as opacity detection (Fig. 24).

In all cases except light scattering, the principle is to concentrate the particles on a substrate for later quantification or analysis.

Two inferior methods of particulate detection commonly found utilize the light transmittance and reflect-

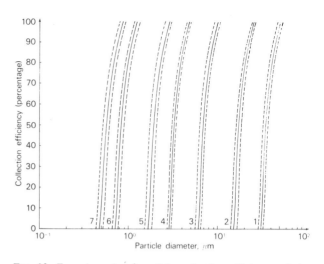

FIG. 23. Error bounds of particle collection efficiency of the MRI inertial impactor system calculated from the Ranz and Wong equations. Flow rate is 0.28 acfm (0.008 m³/min) with 7 per cent error. The jet hole tolerances are 0 (see Fegley *et al.*, 1975).

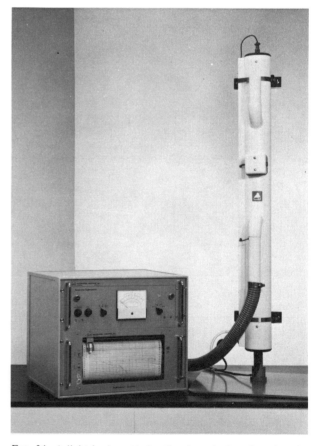

FIG. 24. A light back-scattering (b_{scat}) particulate detection device. This instrument is very sensitive to visibility and gives a good description of the distribution of respirable and near-respirable sized particles.

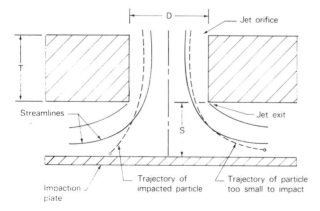

FIG. 22. Principles of a jet impactor sampling system.

ance characteristics of the aerosol sample. The sampler collects the aerosol particles on a white filter tape. A light source and detector are used to measure the diminished transmittance through the tape or the loss of reflectance from the sample surface. The data are represented by the 'coefficient of haze' (COHS) and 'reflective units of dispersion' (RUDS), respectively.

ANALYSIS

The methods of analysis of particulates are somewhat dictated by their method of capture. For example, a filter can be weighed to obtain mass per unit volume. This classical method consists of weighing the sample after storing it in a constant humidity chamber (or an oven), removing it to sample the air, and then reweighing the exposed filter paper after equilibration under similar humidity conditions. The difference is the sample

weight. The weight is divided by the volume of air passed through the filter to obtain mass per unit volume. Classically, a good estimate is given by:

$$\text{volume sampled} = \left(\frac{\text{initial flow rate} + 2 \text{ (final flow rate)}}{3}\right)\text{time}$$

Newer samplers have constant flow devices to eliminate this inaccuracy.

If the filter media is properly selected, optical and electron microscopic analyses can be performed. By use of compendia of micrographs and by experience, much information can be found in this manner. Ordinary light microscopes can be useful, but the art has been markedly improved by, for example, polarized light and emersions. The scanning electron microscope (SEM) is particularly useful for examining submicron-sized particles.

TABLE 5. Example of a properly documented data set.

Statistics about the elemental concentrations in street dust

Element	Minimum concentration p.p.m.	Maximum concentration p.p.m.	Maximum/Minimum	Mean, p.p.m.	σ p.p.m. (percentage)	Number of values
Na	2,600	6,900	2.7	4,410	880 (20)	49
Mg	30,000	72,000	2.4	42,200	9,700 (23)	49
Al	14,300	26,300	1.8	18,500	2,400 (13)	49
Cl	370	2,400	6.6	1,070	580 (54)	49
Sc	2.9	5.9	2.0	4.16	0.75 (18)	49
Ti	740	2,100	2.8	1,430	320 (23)	49
V	25	88	3.6	40.9	10.5 (26)	49
Cr	76	260	3.5	136	39 (29)	49
Mn	390	860	2.2	561	106 (19)	49
Fe	24,000	68,000	2.8	48,300	9,500 (20)	49
Co	5.6	12.6	2.2	8.06	1.72 (21)	49
Ni	< 50	120	> 2.4	84	33 (39)	8
Cu	<150	490	> 3.2	214	101 (47)	29
Zn	420	2,040	4.9	1,090	360 (33)	49
As	< 5	33	> 6.6	12.0	8.4 (70)	37
Se	< 2	5.5	> 2.7	2.9	1.9 (65)	4
Br	15	660	44	263	163 (62)	48
Rb	< 20	190	> 10	45.2	26.5 (59)	41
Ag	< 2	19	> 9.5	8.0	6.3 (79)	7
Sb	7.1	61	8.6	24.3	12.2 (50)	49
Cs	< 0.5	1.9	> 3.8	1.08	0.42 (39)	37
Ba	280	1,310	4.7	574	186 (32)	42
La	10	24	2.4	15.6	3.0 (19)	48
Ce	16	42	2.6	29.4	5.8 (20)	49
Sm	1.5	3.3	2.2	2.34	0.44 (19)	49
Eu	< 0.5	1.05	> 2.1	0.603	0.170 (28)	44
Lu	0.13	0.5	3.8	0.237	0.075 (32)	47
Hf	3.8	56	15	10.4	7.8 (75)	49
W	< 8	43	> 5.4	11.6	9.2 (79)	21
Au	< 0.06	47	>780	3.9	10.5 (268)	36
Hg	< 8	2,080	>260	105	299 (284)	47
Th	1.4	7.8	5.4	3.08	1.18 (38)	49

Sources: Rahn 1971; Rahn and Harrison, 1976; Rahn *et al.,* 1971.

Individual elemental analysis can be performed by many techniques, some of which are capable of nearly total elemental determination. Some of these techniques are the following: atomic absorption (AA), wet chemistry, neutron activation analysis (NAA), X-ray fluorescence, optical microscope, scanning electron microscope (SEM), auger spectroscopy (for cross-sectional analysis), mass spectrometry, ion-excited X-ray emission (IEXE), spark-emission spectroscopy, and polography. The limits of detectability of each method determine the superiority or inferiority of many of the techniques. In turn, each method is determined by the interferences of other elements present or the contamination or background of the elements in question. Also, sensitivities vary from method to method and from element to element.

It is imperative that one follow the outline below to maximize the probability of obtaining a statistically significant sample.

Estimate the levels of each element present and/or determine the minimum sample amount needed.

Secure a sample collection medium that has a blank or contamination level at least a factor of 2 below the minimum estimated in the paragraph above.

Determine the analytical methods available to detect those minimum levels with confidence.

Determine if the capture medium is compatible in the total system to the sensitivity desired and the technique.

Carry a system blank through the total procedure including placement on the collection device (do not sample).

Run a system check, at least in the laboratory, to be certain that the procedure works and that the collection medium is free of contamination and does not have an unacceptable interference.

Always run a duplicate analysis on selected samples to determine statistical variances. (See examples of a data display presented in Table 5.)

Consider that the samples can change in time, especially the organic or carbonaceous fraction (as much as 10 per cent loss for urban samples).

Equilibrate all containers of liquid with similar material, especially for low concentration standards.

Do not report data with excess accuracy, i.e. greater than its variance.

The methods of determining compounds are too many to enumerate here. It should be mentioned, however, that superior physical methods to determine the elemental arrangements in compounds are coming into their own. Many of these techniques, however, still require wet chemistry.

The proper selection of substrates for the first condensation or collection of the airborne particulate cannot be overemphasized. For example, if one wants to look at iron and nickel, one does not use a stainless steel impactor plate and scrape the sample off into an analysis medium.

Control techniques

The techniques of control of particulate matter are similar to those of collection, except applied on larger flow rates and/or volumes (see American Petroleum Institute, 1956, 1961*a, b*; Environmental Protection Agency, 1972*a, b*; *Joy Manufacturing Company Bulletins*). Usual techniques are: (a) impaction (cyclones), (b) scrubbing; (c) electrostatic Precipitation (ESP); (d) filtration; (e) sedimentation; (f) thermal diffusion; (g) sonic; (h) incineration; (i) combinations of the above (e.g. ESP, and scrubbing with water drops). Figures 25 to 29 illustrate some of the devices listed above. Each of the methods have technical advantages and disadvantages. The selection is determined by economic consideration, availability of power and water, exit temperatures, gas volume and secondary environmental effects (disposal). All must be balanced to provide the proper control method.

Each technique is briefly described below. The reader may obtain more detailed descriptions from manufacturers' specifications and the referenced literature.

IMPACTION

The inertial impaction techniques work well with larger particles and are frequently seen on the sides of grain silos and the roofs of many factories.

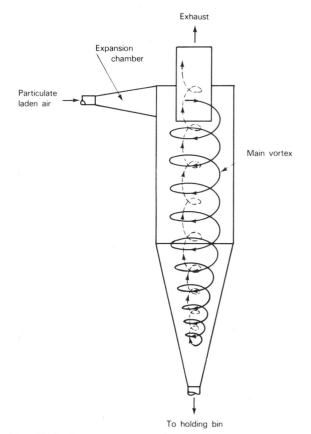

Fig. 25. Basic cyclone air cleaning device.

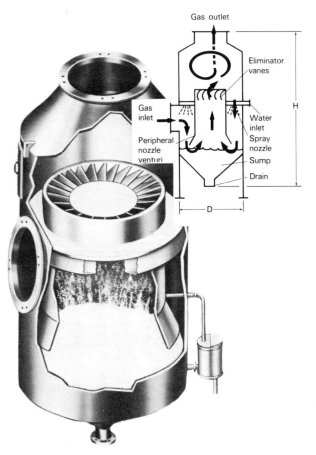

FIG. 26. An example of a low-pressure drop scrubber system. (Courtesy of Western Precipitator Company, Los Angeles, Calif.; *Joy Manufacturing Company Bulletins.*)

Referring to Figure 22, we see that large particles are easily thrown out of a quickly changing air stream. Most dry intertial control devices use a cyclone to whirl the air stream in a circular motion, at ever increasing speeds, by decreasing the radius of curvature. Since there is little fluid motion once the particles reach the side, they fall down into the slower air stream. The proper placement of guides and vanes encourages the proper motion. A hopper below receives the particles by gravity settling (see Fig. 25).

Again, since smaller particles can make the twisting route, this principle is used mostly for large particles from mechanical processes such as milling, grain handling, woodworking, etc., and should only be used as a first-stage eliminator for combustion-type processes. Primary advantages are the ability to handle large quantities of gas and use no liquid. The process is inexpensive and results in a higher plume rise since the effluent temperature is not significantly reduced. Cyclones are excellent pre-cleaners to precipitators and scrubbers to reduce mass loading.

SCRUBBERS

If a plentiful supply of liquid, usually water, is available, and the temperature of the inlet gases is not too high, the principle of impaction can be enhanced by use of a wet scrubber.

Wet scrubbers simply wash the gas stream by breaking it into small bubbles of large surface-area-to-volume ratio. The particles are captured in the liquid directly by impaction due to the increased effective surface exposed and by accretion of water on the particle itself. Figures 26 and 27 show examples of wet scrubbers. Other modifications such as beds of marbles and venturi orifices provide methods of increasing contact and subsequent wash-out.

The advantages are that many fine particles are eliminated and the gas stream cooled. Additionally, the scrubber can handle large quantities of gases. The disadvantages include cost, maintenance, secondary environmental effects (disposal of the slurry) and loss of plume buoyancy due to cooling. Increased power consumption occurs if the effluent needs a forced draught fan to move the gas up the stack.

ELECTROSTATIC PRECIPITATOR

The electrostatic precipitator is a superior device for fine dusts. The device basically consists of high-voltage

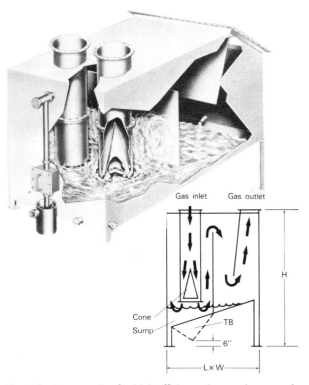

FIG. 27. An example of a high-efficiency, large-volume scrubber. (Courtesy of Western Precipitator Company, Los Angeles, Calif.; *Joy Manufacturing Company Bulletin.*).

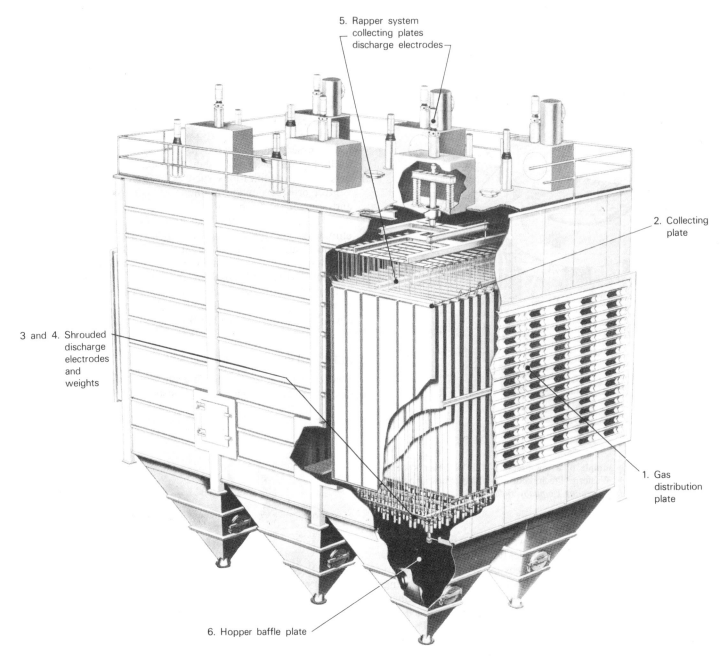

5. Rapper system
collecting plates
discharge electrodes

2. Collecting
plate

3 and 4. Shrouded
discharge
electrodes
and
weights

1. Gas
distribution
plate

6. Hopper baffle plate

FIG. 28. An example of an electrostatic precipitator. (Courtesy of Western Precipitator Company, Los Angeles, Calif.; *Joy Manufacturing Company Bulletins.*)

wires suspended between plates. The component material, spacing and shape of the plates are dependent on the conductivity, temperature and corrosion characteristics of the effluent stream. Figure 28 shows a cutaway of a typical device. These devices can operate at high temperatures and gas volumes but are expensive to maintain if corrosion and arcing are present. Should upsets occur, the wires can arc to the plates and explo-

sions can result if incomplete combustion products are present.

To clean, the plates are 'rapped' with mechanical hammers, which thus remove the dry particles. If the particles are sticky, the plates will retain them and will have to be washed off. Sections are cleaned sequentially with the voltage removed. Increased emissions can occur if there is not a method of recirculation from the temporarily uncontrolled region. In some cases, ESPs can be used with wet gas streams. If care is not taken, increased arcing and current can be encountered. If large particles are present, they should be removed by inertial impaction before entering the precipitator.

Since the collecting material is very fine (about the consistency of face powder) there is a disposal problem. This difficulty is usually solved by a closed, screw-fed removal system to either a recycling facility, a liquid dust suppression system or to a settling pond via a water slurry system.

FILTRATION

In recent years, new temperature-resistant fibres have allowed construction of filter bags for cleaning higher temperature gases. This principle is quite simple and works just like a vacuum cleaner. By altering the tightness of the weave, one can achieve control of most size distributions, releasing only the smallest particles. The major drawback, other than limits on gas temperatures, is the pressure drops encountered. If too few bags are used and the dust is not cleaned off the surface with sufficient frequency, the pressure builds up to unacceptable levels. Usually, a 'bag-house' is constructed in modules and shaken for cleaning in sequence, thereby preserving the flow. For more moist and compressible material, rollers are used for cleaning the bags. Again, the gas from the disabled section is recycled during cleaning. Bag-houses require a large space for deployment but can filter a very large size range of particles with ease. Bag integrity and maintenance is a chronic problem (see Fig. 29) for an example of a small facility).

SEDIMENTATION

For removal of large particles, sedimentation is a simple, straightforward technique. While this technique is obviously inexpensive, it is also ineffective. Low velocities are achieved by an expansion chamber with gas removal near the top of the chamber.

THERMAL PRECIPITATION

For removal of small particles from a low-velocity warm air stream, one can use a cold plate. These devices are also used for removal of low volatile gases such as greases and hydrocarbons from cooking and solvent use. The principle is that hot (warm) gases have a higher kinetic energy (motion) towards a cold object than away from it. The molecular momentum is partially used to impart 'heat' to the cooler object. This energy transport causes the gases and particles to migrate towards the cold object (plate). The particle then impacts on the plate with low energy, but is usually retained. For this reason, in buildings located in cold climates the walls which are adjacent to the outside get dirty faster than the internally located walls—one can illustrate this by looking behind a wall picture.

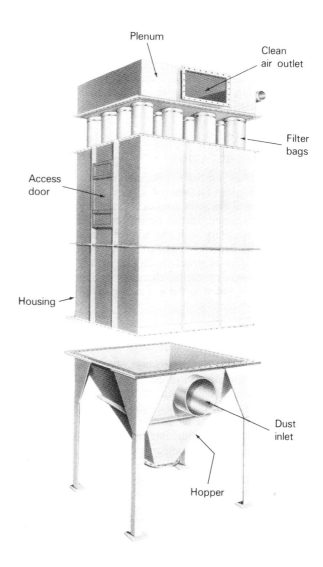

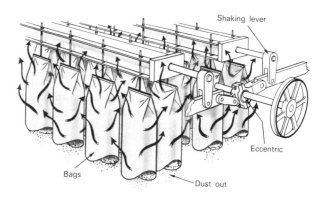

Fig. 29. An example of a small bag-house unit and a cleaning technique commonly used for dry gas streams. (Courtesy of Western Precipitator Company, Los Angles, Calif.; *Joy Manufacturing Company Bulletins*.)

When this situation is accentuated by high temperature differences under low velocities, the removal is accelerated. This diffusion mechanism is effective for the very small particles, but requires a longer time period and a large surface-to-gas-volume ratio. It is used almost exclusively in low-volume, fine-particle and low-volatility gas situations.

SONIC PRECIPITATION

This method is similar to the thermal precipitator except an additional focusing of kinetic energy on a size classification of particles can occur if the sonic frequency is properly tuned. Increasing the molecular motion of the gas by coupling with the sonic transducer imparts energy to the particles, also increasing their kinetic energy. If a surface is struck, collection occurs. The applications are similar to the thermal precipitator except that energy is added to the gas rather than removed by the collection surface.

INCINERATION

Prior to the sudden increased cost of fuels, many effluent streams containing hydrocarbons or organic particulates were simply incinerated. This was accomplished by passing the stream through a properly configured flame curtain maintained at temperatures high enough to assure incineration of burnable gases and particulates. Control of H_2S and visible particulate plumes are examples. This system is costly since large volumes of gas need to be heated, causing waste of fuels and combustibles. In addition, most carbonaceous particulates contain low concentrations of trace elements. During combustion, they are converted to very small particles containing enriched quantities of metals. Even some of these small particles are partially volatilized and are difficult to remove or control. Thus, one could trade a visible plume, easily controlled by other means, into one which looks acceptable, but is actually less acceptable, both from a health and a control point of view. In some cases, however, especially for odour and pure hydrocarbon releases, incineration is the best method.

COMBINATION SYSTEMS

Depending on the problem, combining the control methods in tandem simultaneously can result in distinct cost advantages and efficiency. A pre-cleaning cyclone is often used to relieve ESP from excessive loading of large particles, including liquid droplets. If a final ultra-cleaning of particles is necessary, filtration, scrubbing, thermal precipitation or sonic impaction may be used.

A very recent improvement in ESP is the use of highly charged water droplets. It has been known empirically for some time that thunderstorms are excellent air cleaners. In 1963, experiments were conducted by Gunn in a test chamber designed for still-air experiments. He found that simple sedimentation, electrification, humidification and a combination of humidity (droplets) and electrification, was progressively more efficient time-wise in cleaning the air in the chamber. The range in time to accomplish this goal was from many days to 2–3 minutes. Of course, the problem of increased current and arcing with water droplets has yet to be completely eliminated, but promise is ever increasing.

Recent technology

Three areas of recent aerosol research are of particular interest. They are: (a) elemental enrichment of particulate aerosols (Rahn, 1971; Rahn and Harrison, 1976); (b) aerosol surface adsorption and reaction phenomena; and (c) respirable fractions of total suspended particulates (Harrison, 1973, 1975, 1976a; Newman et al., 1976; Rahn and Harrison, 1976).

ELEMENTAL ENRICHMENT OF PARTICULATE AEROSOLS

In answer to the question, 'Does the elemental content of aerosols change with location and size?' Harrison (1970) showed a definite geographical variance in a heavily polluted area (Fig. 30).

Rahn collected elemental analysis data of more remote aerosols. As a normalization technique, he proposed an overall quasi-uniformity of the chemical composition of 'background' atmospheric aerosols. This technique may be conveniently expressed by comparison of the relative abundance of each element in the aerosol to its abundance on the Earth. In order to obtain such a comparison, each element is normalized to an element universally found in all areas of the world; Al, Si, Fe, and Ti can be used. Rahn chose aluminium due to its sparcity of local sources from industrial activities. Thus:

$$\frac{\text{Concentration of element}}{\text{Concentration of Al}} \equiv \frac{C_x}{C_{Al}},$$

is averaged from all available data on crustal rock and on soil. Then the enrichment factor EF, is represented by:

$$\text{EF}_x = \left(\frac{C_x}{C_{Al}}\right)_{\text{aerosol}} \Bigg/ \left(\frac{C_x}{C_{Al}}\right)_{\text{crust}}.$$

Figure 31 shows the typical enrichment factor for certain elements related from a compilation of aerosol data reported by Rahn (personal communications, 1975–76). As one can see, approximately 29 of the elements are significantly enriched (i.e. EF > 10 per cent). Note also that no elements are significantly depleted. The presence of all elements in the Earth's crust confirms

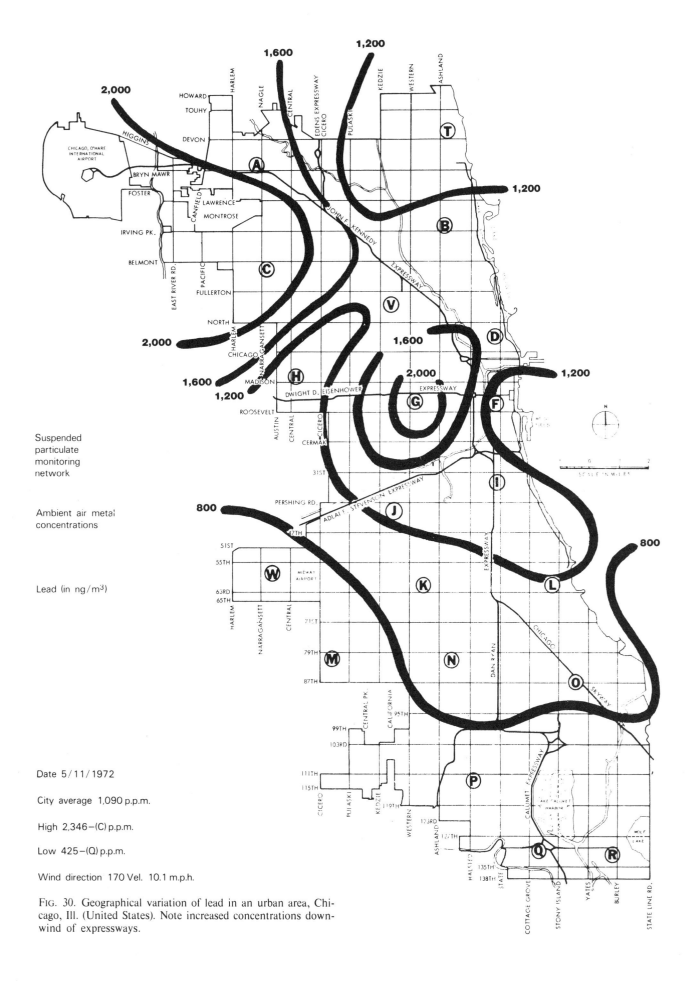

Suspended
particulate
monitoring
network

Ambient air metal
concentrations

Lead (in ng/m³)

Date 5/11/1972

City average 1,090 p.p.m.

High 2,346–(C) p.p.m.

Low 425–(Q) p.p.m.

Wind direction 170 Vel. 10.1 m.p.h.

FIG. 30. Geographical variation of lead in an urban area, Chi-cago, Ill. (United States). Note increased concentrations down-wind of expressways.

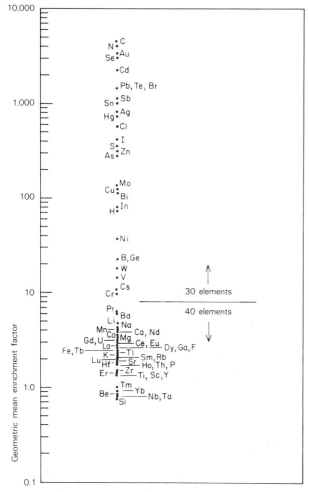

FIG. 31. Geometric mean EF for 69 elements. Note that, in general, the more volatile elements are highly enriched.

ment and lithophilic character. It is extremely interesting to note that the enriched elements lie on the lower right-hand side of the periodic table presented in Figure 32. This discovery immediately raised the question as to the source of this extreme enrichment of certain elements.

Looking at the size distribution, we find that elements have characteristic particle sizes represented as mass-median diameters (MMDs). Figure 33 represents the ranges and medians of the particle diameters for various elements, and Figure 34 summarizes the data. Note that the large particles are less enriched and the smaller ones are highly enriched. The phenomenon is better seen in Figure 35 where the enrichment factors from Figure 31 are plotted as a function of elemental particle size from Figure 34. Note also that the intermediate enrichments are on intermediate sizes. Thus the enrichment is related to the size distribution.

It appears that the large particles are of crustal origin, either directly or indirectly. Soil saltation can account for a large part of this range. The small and intermediate size particles are either of other origins or extremely surface-enriched. It is difficult to explain how this enrichment could occur in the free atmosphere on small crustal particles. Their origin from the gas phase seems the most likely explanation at present, but it should be noted that present theory cannot explain the slope of −4.6 seen in Figure 35. However, the explanation must be in the right direction since several of the elements have known gas phases in the atmosphere (e.g. Si, Se, Ni, Hg, Cl, Br, I). In addition, most of their reactions occur in the volatile form. Finally, if the mechanism is volatilization followed by condensation on an existing particle, there is simultaneous enrichment of an element relative to the parent material and an enrichment of the smaller particles by the volatile element. Thus the exercise is to enrich towards the smaller particles. This is a serious conclusion with respect to health effects since the smaller particles penetrate deeper into respiratory areas.

An exciting area of current research is the investigation into the sources and mechanisms creating these trace element particles. Some possible suggestions are high-temperature combustion sources, either pollution or natural. Examples of pollution sources are combus-

that there is a probable crustal source for these elements. Many elements, however, are somehow enriched. It is important to note that these data come from all areas and both hemispheres of the world.

The enrichment elements are of chalcophilic nature (see Table 6). In fact, there is almost perfect correlation between enrichment and chalcophilic character. Likewise, there is a definite correlation between non-enrich-

TABLE 6. Geochemical classification of the elements and their geometric-mean enrichment factors in the atmospheric aerosol

Chalcophilic	Chalcophilic and lithophilic	Lithophilic
Ag (20) Zn (310) Cd (2,300) Hg (730) In (80) Tl (2.4) Pb (1,400) As (280) Sb (1,100) Bi (110) S (360) Se (3,000) Te (1,500) Mo (140)	Fe (2.4) Co (3.5) Ni (38) Ge (23) Sn (1,000) Cu (130) Ga (2.6)	Li (4.8) Na (4.2) K (2.2) Rb (2.2) Cs (11) Be (0.8) Mg (3.4) Ca (3.8) Sr (1.9) Ba (6.0) B (22) Al (—) Sc (1.3) Y (1.3) La–Lu (0.9–6.1) Si (0.8) Ti (1.3) Zr (1.5) Hf (1.9) Th (1.8) P (1.8) V (14) Nb (0.8) Ta (0.8) Cr (8.0) U (3.1) Mn (4.0)
		H (73) F (2.6) Cl (570) Br (1,500) I (420)

1 H 73*																				2 He –
3 Li 4.8	4 Be 0.83											5 B 22*	6 C 4500*	7 N 4000*	8 O –	9 F 2.6*	10 Ne –
11 Na 4.2	12 Mg 3.4											13 Al –	14 Si 0.8	15 P 1.8	16 S 360	17 Cl 570	18 Ar –
19 K 2.2	20 Ca 3.8	21 Sc 1.3	22 Ti 1.3	23 V 14	24 Cr 9.7	25 Mn 4.0	26 Fe 2.4	27 Co 3.5	28 Ni 38	29 Cu 130	30 Zn 310	31 Ga 2.6	32 Ge 23*	33 As 280	34 Se 3000	35 Br 1500	36 Kr –
37 Rb 2.2	38 Sr 1.9	39 Y 1.3*	40 Zr 1.5*	41 Nb 0.8*	42 Mo 140	43 Tc –	44 Ru –	45 Rh –	46 Pd –	47 Ag 820	48 Cd 2300	49 In 80	50 Sn 1000	51 Sb 1100	52 Te 1500*	53 I 420	54 Xe –
55 Cs 11	56 Ba 6.0	57 La 2.8 **	72 Hf 1.9	73 Ta 0.8*	74 W 18	75 Re –	76 Os –	77 Ir –	78 Pt –	79 Au 3300	80 Hg 730	81 Tl 2.4*	82 Pb 1400	83 Bi 110	84 Po –	85 At –	86 Rn –
87 Fr –	88 Ra –	89 Ac *** –															

**	58 Ce 2.9	59 Pr 6.1*	60 Nd 3.9*	61 Pm –	62 Sm 2.15	63 Eu 3.0	64 Gd 3.0*	65 Tb 2.4*	66 Dy 2.5*	67 Ho 1.7*	68 Er 1.4*	69 Tm 1.0*	70 Yb 0.91*	71 Lu 2.1
***	90 Th 1.8	91 Pa –	92 U 3.1*	93 Np –	94 Pu –	95 Am –	96 Cm –	97 Bk –	98 Cf –	99 Es –	100 Fm –	101 Md –	102 No –	103 Lw –

* Data from 1–6 American cities only (Henry and Blosser, 1971)

FIG. 32. Periodic table with enrichment factors. Note that most of the enrichment factors greater than 7.0 lie on the right side of the table.

tion of fossil fuels, smelting and incineration. Natural sources include volcanoes, forest fires and effects of solar insulation, the latter especially for mercury. Another possible explanation of the world-wide presence of these particles involves biological activities either of animal organisms or plants.

A final fact concerning these phenomena is that there is a progressive, continuous consistency in the enrichment process. There are few outlying enrichment factors. Simply stated, a highly enriched element has virtually no data points lying close to or below unity. Even though there is a high degree of variability, the data are definitely bracketed.

AEROSOL SURFACE ADSORPTION AND REACTION PHENOMENA

The previous discussion suggests the possibility of a phenomenon causing surface enrichment in compounds as well as elements.

It is well known that sulphur and nitrogen compounds emitted from power plants and other combustion sources are converted in the atmosphere to particulate sulphate and nitrates. Recent efforts using auger spectroscopy actually examine a particle as a function of depth from the surface. Thus, for some elements it is almost possible to describe their relative concentration with depth in the particle. So far, there is data highly suggestive of surface enrichment of sulphur and nitrogen components, as well as of some other elements, possibly ammonium sulphate.

Other research is taking place to determine the surface catalytic effects of certain airborne particulates on the gas-to-particle conversions. The theory and information is new and insufficient as yet to make final detailed conclusions on the mechanisms and rates, but there is great evidence that the phenomenon occurs.

RESPIRABLE FRACTIONS OF TOTAL SUSPENDED PARTICULATES[1]

It has been known for some time that there are three

1. From Harrison, 1973, 1975, 1976a; Newman et al., 1976; Rahn and Harrison, 1976.

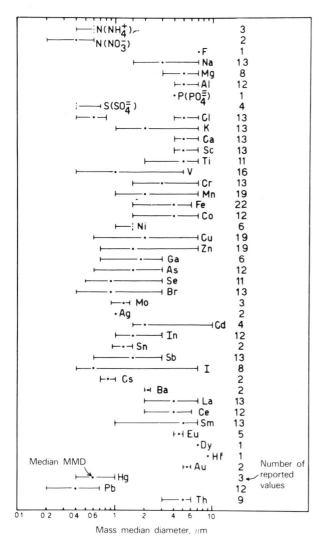

FIG. 33. Range of MMD found in background aerosols as related to elemental composition.

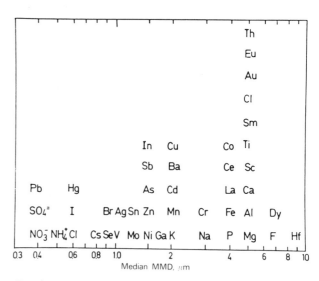

FIG. 34. Position of selected elements relative to size (MMD). Note that the EF is generally inversely related to size.

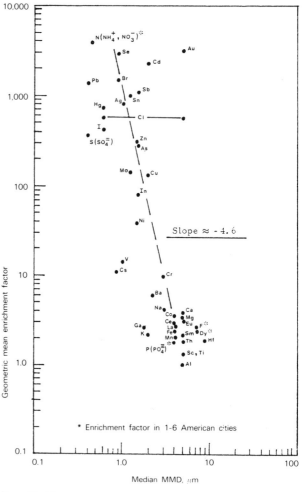

FIG. 35. Mean size as compared with enrichment factor.

physiological size classifications of particles with respect to health effects. They are respirable (0.1–5 μm diameter), sub-respirable (<0.1 μm), and non-respirable (>5 μm).

The large, non-respirable particles are filtered out by nasal hairs or impact in the nasal passage or bronchial tree. These particles are either expectorated or swallowed. Swallowing of this mucus is a possible source of contamination to the host organism but it is less hazardous than inhalation where the particles can be directly absorbed into the bloodstream. The respirable (0.1–5 μm) fraction penetrates beyond the bronchial tree and air passageways—which contain cilia—to the alveoli, where they have a high probability of impaction and residence. Some of the material can be dissolved by natural defence cells. Others such as silicon, asbestos and beryllium destroy cells by causing the alveoli sacs to fill with dead cells and/or fluid. The cell thus ceases to exchange oxygen and lung capacity is diminished.

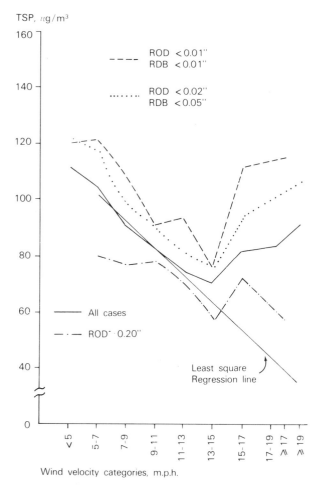

FIG. 36. Total suspended particulate (TSP) loadings averaged over 20 stations for 1964–74 in Chicago, Ill. Note that rain on day (ROD) and rain on day before (RDB) less than 0.01 data show a steep increase after 15 m.p.h. (Newman, 1974).

Cigarette smoke is in this category. Finally, the sub-respirable particles less than 0.01 μm are so small that they act as a gas and are returned to the atmosphere upon exhalation.

There is a definite need to continue efforts to measure the respirable and non-respirable fractions of total suspended particulates, especially in urban areas. These measurements are necessary for the determination of damage to health and for obtaining an understanding of the sources for their ultimate control.

Recent studies by Harrison (1975, 1976a) have shown the following conclusions concerning urban particulates:

Studies and analysis over the last three years have shown, in the city of Chicago that:
1. The highest and lowest elevation of sampling stations are different by about a factor of 2.
2. Siting a sampler over a large area of grass can halve the particulate loading by comparison with adjacent high-activity areas.
3. Wind speeds in excess of 12 m.p.h. are efficient re-suspension velocities.
4. Much of the material in urban areas is related to automobile activities, both direct emissions (tyre dust and exhausts) as well as providing a wind field for re-suspension.
5. The general trend follows rainfall patterns and agricultural activities.

Many of the above observations lead to the conclusion that street washing and provisions for foliation of vacant land are desired and effective control strategies.

A most important result in the correlation of concentrations with wind speed and soil moisture is shown in Figure 36. Although the particles are larger, a high wind can generate TSP values equal to those found under stagnation inversion conditions.

Conclusions

The purpose of this chapter is to try to span the total field of atmospheric aerosols without pausing for the more intricate details. Obviously, even a partial attempt to explain all the details of phenomena affecting the science of aerosols would span a large volume in itself. We hope that we have accomplished the goal of imparting a basic understanding of the highly complex but exciting field of particulate aerosols. For further information the reader is referred to the references at the end of the chapter.

Acknowledgements

George J. Woffinden assited in the technical editing of this chapter while Diana Johnston was instrumental in the editing and typing of the manuscript. The authors express their sincere thanks and appreciation to them and to any others who assisted.

References

AMERICAN PETROLEUM INSTITUTE. 1956. *Cyclone Dust Collectors*. New York, Division of Refining.

——. 1961a. *Electrostatic Precipitators*, New York, Division of Refining.

——. 1961b. *Gravity, Inertial, Sonic, and Thermal Collectors*, New York, Division of Refining.

BLUMENTHAL, D. L.; SMITH, T. B; WHITE, W. H.; MARSH, S. L.; ENSOR, D. S.; HUSAR, R. B.; MCMURRY, P. S.; HEIS-

LER, S. L.; OWENS, P. 1974. Three-dimensional Pollutant Gradient Study—1972–1973 Program. *Meteorology Research, Inc. Technical Report No. MRI 74 FR-1262,* for California Air Resources Board, Sacramento, California.

CHARLSON, J.; VANDERPOL, A. H.; COVERT, D. S.; WAGGONER, A. P.; AHLQUIST, N. L. 1974. $H_2SO_4/(NH_4)_2SO_4$ Background Aerosol: Optical Detection in St. Louis Region. *Atmospheric Environment,* Vol. 8, p. 1257–67.

ENVIRONMENTAL PROTECTION AGENCY. 1971. *Particulate Pollutant System Study.* Vol. I: *Mass Emissions.* Midwest Research Institute. (Project No. 3326–C; a report prepared for Air Pollution Control Office. Contract No. CPA 22-69-104.).

——. 1972a. *Field Operations and Enforcement Manual for Air Pollution Control.* Vol. II: *Control Technology.* Research Triangle Park, N.C. (Technical Information Services APTD-1101).

——. 1972b. *Field Operations and Enforcement Manual for Air Pollution Control.* Vol. III: *Inspection Procedures.* Research Triangle Park, N.C. (Technical Information Services APTD-1102).

FEGLEY, M. J.; ENSOR, D. S.; SPARKS, L. E. 1975. *The Propagation of Errors in Particle Size Distribution Measurements Performed Using Cascade Impactors.* For presentation at the 68th Annual Meeting of the Air Pollution Control Association, Environmental Protection Agency.

FENNELLY, P. F. 1975. Primary and Secondary Particulates as Pollutants. *JAPCA,* Vol. 25, p. 697–704.

The Frank Chart. 1937. Louisville, Ky., American Air Filter Co.

Guidelines for Development of a Quality Assurance Program. spended Particulates 1973. (EPA No. R4-73-0286. Washington, D.C., USEPA.

GUNN, R. 1963. *The Electrical Conductivity and Electric Field Intensity Over the North Atlantic and Its Bearing on Changes in the World Wide Pollution of the Free Atmosphere.* Washington, D.C., American University. (Library of Congress No. 63-20346; NSF Contract C-183).

HARRISON, P. R. 1970. Area-Wide Distribution of Lead, Copper, Cadmium and Bismuth in Atmospheric Particulates in Chicago and Northwest Indiana: A Multi-Sample Application and Anodic Stripping Voltammetry. Ann Arbor, Mich., University of Michigan. (Ph.D. Thesis.)

——. 1973. Air Pollution by Lead and Other Trace Metals. In: S. K. Dhar (ed.), *Metal Ions in Biological Systems.* Plenum Press.

——. 1975. Particle Resuspension in Urban Atmospheres. *Energy and the Environment, Proceedings of the Third Nat'l Conference.* Dayton, Ohio, AIChE.

——. 1976a. Identification and Impact of Chicago's Ambient Suspended Dusts. *ERDA Conference 740921,* Springfield, Va., Nat'l Tech. Info. Serv. (NTIS).

——. 1976b. *Particle Resuspension in Urban Atmospheres.* Submitted, Meteorology Research, Inc.

HARRISON, P. R.; GEORGII, H. W.; TER HAAR, G. L.; *et al.* 1974. *Analysis of Industrial Air Pollutants.* Vol. III. New York, MSS Information Corp.

HARRISON, P. R.; MATSON, W. R.; WINCHESTER, J. W. 1971b. Time Variations of Lead, Copper and Cadmium Concentrations in Aerosols in Ann Arbor, Michigan. *Atmospheric Environment,* Vol. 5, p. 613.

HARRISON, P. R.; RAHN, K. A.; DAMS, R.; ROBBINS, J. A.; WINCHESTER, J. W.; BRAR, S. S.; NELSON, D. M. 1971a. Areawide Trace Metal Concentrations Measured by Mutielement Neutron Activation Analysis, A One-Day Study in Northwest Indiana. *APCA,* Vol. 21.

HENRY, W. M.; BLOSSAR, R. R. 1971. *Compounds in Particulate Matter Collected from Ambient Air.* Battelle Columbus Laboratories, Columbus, Ohio. (T.R. CPA-70-159.)

HEWSON, E. W. 1964. *Industrial Air Pollution Meteorology.* Ann Arbor, Mich., Department of Atmospheric and Oceanographic Sciences, University of Michigan.

HIDY, G. M.; APPEL, B. R.; CHARLSON, R. J.; *et al.* 1975. Summary of the California Aerosol Characterization Experiment. *JAPCA,* Vol. 25, p. 1106–14.

Humidograph Instruction Manual IM-189. 1975. Vol. I. Atmospheric Research Group, Meteorology Research, Inc.

Joy Manufacturing Co. Bulletins, Nos. FK-375, P-150, DF-1024, and WPS-100. Los Angeles, Calif.

JUNGE, C. E. 1963 *Air Chemistry and Radioactivity.* p. 177. New York, Academic Press.

LANGREN, D. A.; PAULUS, H. J. 1975. The Mass Distribution of Large Atmospheric Particles. *JAPCA,* Vol. 25, p. 1227–31.

LILLIS, E. J.; YOUNG, D. 1975. EPA Looks at Fugitive Emissions. *JAPCA,* Vol. 25, p. 1015.

LYONS, W. J. 1973. *Bulletin of the Meteorological Society,* Vol. 54. (Cover story.)

McCORMICK, R. A.; HOLZWORTH, G. C. 1976. Air Pollution Climatology. In: A. C. Stern (ed.), *Air Pollution,* 3rd. ed., Vol. I, Chap. 12. New York, Academic Press.

McMULLEN, T. B.; FADRO, R. B.; MORGAN, G. B. 1969. *Profile of Pollutant Fractions in Non-Urban Suspended Particulate Matter.* Pittsburgh, Pa. (APCA Paper No. 69–165.)

NEWMAN, J. E.; ABEL, M. D.; HARRISON, P. R.; YOST, K. J. 1976. Wind as Related to Critical Flashing Speed versus Refloatation Speed by High-Volume Sampler Particulate Loading, *ERDA Conference on Atmospheric-Surface Exchange of Particulate and Gaseous Pollutants.*

PETTERSSEN, S. 1958. *Introduction to Meteorology.* New York, McGraw-Hill.

Quality Control Practices in Processing Air Pollution Samples. 1973. Research Triangle Park, N.C. EPA Technical Information Service. (APTD-1132).

RAHN, K. A. 1971. Sources of Trace Elements in Aerosols—An Approach to Clean Air. Ann Arbor, Michigan University of Michigan (Ph.D. Thesis, COO-1705-a.)

——. HARRISON, P. R. 1976. The Chemical Composition of Chicago Street Dust. *ERDA Conference 740921,* Springfield, Va. Nat'l Tech. Info. Serv. (NTIS).

ROBINSON, E.; ROBBINS, R. C. 1971. Emission Concentration and Fate of Particulate Atmospheric Pollutants. *Final Report.* Menlo Park, Calif. Stanford Research Institute, (SRI Project SCC-8507.)

STERN, A. C. 1968. *Air Pollution,* Vols. I–III, 2nd ed. New York, Academic Press.

——. 1976. *Air Pollution,* Vol. I, 3rd ed. New York, Academic Press.

TURNER, D. B. 1969. *Workbook of Atmospheric Dispersion Estimates,* United States Department of Health, Education and Welfare. (PHS Publication No. 999-AP-26.)

The problems of energy and mineral resources[1]

M. H. Govett

Govett & Govett Pty Ltd
Blakehurst (Australia)

G. J. S. Govett

School of Applied Geology,
University of New Soth Wales,
Kensington (Australia)

Introduction

Changing patterns of mineral production and consumption, growing concern for environmental damage, and impending shortages of mineral reserves, as was a ses-contribute to the increasing awareness of the economic vulnerability of both the developed and developing countries; the energy crisis of 1973–74 was dramatic evidence of the problems caused by world industrialization and the geographic concentration of mineral resources. Geologists—and geochemists—are becoming increasingly aware of the economics of world mineral consumption and production. The theme of the Jackling lecture at the 1973 Annual Meeting of the American Institute of Mining, Metallurgical and Petroleum Engineers (Hart, 1973) was the role of geologists in meeting the impending shortages of mineral reserves, as was a session of the Annual Meeting of the Canadian Institute of Mining and Metallurgy in Montreal, in 1974. At the Fifth International Symposium on Exploration Geochemistry in April, 1974, the opening speech was devoted to the subject of mineral scarcity (McLaren, 1974); at the same meeting, another lecture also dealt with the limits to growth imposed by world-wide resource shortages (Nichol, 1974). The address to the Association of Exploration Geochemists at the Sixth International Symposium on Exploration Geochemistry, in Sydney (Australia), was entitled 'World Mineral Supplies—The Role of Exploration Geochemistry (Govett, 1977b).

The view of the future ranges from the conservative optimism of the United States Bureau of Mines (1970) to the doomsday conclusions of the Club of Rome (Meadows et al., 1972), with the middle ground held by works such as those of Connelly and Perlman (1975) and Govett and Govett (1976a). The optimists and the pessimists differ in their answers to three main questions:

1. At what rate will the demand for minerals grow in the developed and developing countries?

2. How adequate are mineral supplies to meet projected demand?
3. What effects will environmental restrictions have on mineral exploration and exploitation?

The geologist can do little to answer the first question. The rise in demand for minerals in the past few decades and the implications for the future of a continuation of the current rate of growth, given increasing population and the legitimate aspirations of the developing countries, cannot be denied. Even those who are reluctant to accept some of the models of exponential growth and the crises predicted by the models, admit that a continuation of the current rate of increase of consumption is a cause for concern. In the past three decades, the United States alone consumed more minerals than the world used in its entire recorded history. If the American and Western European rates of growth of consumption were reduced, or at least held constant, the rest of the world, where population is growing at an annual rate of nearly 3 per cent, would still strive to industrialize, thereby increasing the demand for raw materials. It is refreshing to find some economists advocating a change from the idea that consumption of material things is the *sine qua*

---

1. This chapter was completed in May 1974; in September 1977 revisions were made, with current data added where necessary. The text was not substantially changed. Further minor changes in the text were made and the following note added in December 1978:

The extreme difficulty in presenting up-to-date resource date (and hence making policy recommendations and decisions) is illustrated by the dramatic changes that have occurred since this chapter was first written: the OPEC-inspired petroleum 'crisis' occurred during that time, whereas it is only subsequent to the 1977 revisions that the magnitude of the petroleum resources of Mexico have become evident—they are likely to exceed those of Saudi Arabia.

The disruptions to the world economy in the last few years have caused some revisions in economic growth projections discussed in this chapter; nevertheless, we believe that the broad trends that we have identified are still valid.

Projections of resource demand and supply are quoted from Mineral Facts and Problems (United States Bureau of Mines, 1970); these should be read in conjunction with the more recent edition of volume (United States Bureau of Mines, 1976).

non of economic development (e.g. Boulding, 1966), but it is difficult to apply a philosophy of limited growth or 'zero-growth' to the developing two-thirds of the world.

In answer to the second question, the world's supply of mineral resources is enormous—considering the crustal abundance of the elements, the potential of the ocean floor and the seas, and the possibilities of harnessing new energy sources; however, known reserves of a number of minerals are critically small. The present aim of the geologist must be to find new ore-bodies at greater depths and in unexplored areas. Whether new exploration techniques can be developed quickly enough to meet the projected increases in demand for a number of minerals is uncertain; extraction and processing technology for low-grade ores, which are not currently considered as reserves, may be of singular importance in the future. The extreme remedy of mining whole rock and sea water for their elements (Brown *et al.*, 1957; Nolan, 1955) is so distant a possibility that it is beyond practical consideration.

Most observers—including geologists—agree that something drastic must be done about the pollution of the Earth and the seas and the deterioration of the quality of life. At the same time, exploration and mining geologists are becoming justifiably concerned about environmental protection measures to the extent that they result in increased exploration and mining costs. Research into the effect of pollution abatement measures on the mining industry is only beginning, but there are already indications that the economic effects will be substantial, both directly—as they increase mining and processing costs—and indirectly—through restriction of exploration and mining in certain geographical areas and through the curtailment of technological advances.

The optimism among some economists about the future—based on the belief that science and technology can produce the necessary substitutes for scarce raw materials, will find new mineral deposits, will develop new metallurgical and extraction techniques, and will find a way to reduce environmental damage without restricting the growth of the mining industry—is very flattering to the geologist and the engineer; however, the conclusion that the era of modern alchemy is at hand, given adequate supplies of energy (Spengler, 1961), needs to be tempered by geologists and geochemists. The view that present, known reserves of most metals will be exhausted within 50 years (Ecologist, 1972) must also be questioned, although it seems unavoidable that, at present and projected rates of world mineral consumption, a large number of important minerals may become increasingly costly.

This chapter surveys the implications of current and future mineral consumption in both the industrialized and, as yet, non-industrialized countries in terms of technical, environmental and economic limits on resources. The first section discusses mineral consumption and world mineral supplies; the second section surveys the factors affecting future reserves. The final sections consider the present and future problems in the developed and developing countries.

The adequacy of world mineral resources

WORLD MINERAL DEMAND

In the past two centuries, technical progress seemed to prove Malthus wrong in his predictions that population growth would bring about disaster unless checked either by 'moral restraint' or 'natural phenomena', but the sudden rise in population in the past few decades has again raised the Malthusian spectre. Harrison Brown (1954) issued one of the first warnings of the consequences of unchecked increases in world population at a time when the revolution of rising expectations in the developing countries of the world was putting increasing strain on the world's supply of raw materials. The Club of Rome (Meadows *et al.*, 1972) and the *Blueprint for Survival* (Ecologist, 1972), with their dire predictions for the future, are only the best known in a growing literature on the limits to growth set by population increase and mineral resource shortages.

The population of the world is estimated at approximately 4,000 million; the rate of annual increase is 1.9 per cent (United Nations, 1976*a*). The implications of this rate of growth form the bases of many of the recently published exponential growth models; as population grows, so too does the demand for economic development and the raw materials necessary for agriculture and industry. The long-term effects of this pattern are unquestionable: if it is assumed that world population will continue to grow (especially in the developing areas where the average annual rate of population growth is in excess of the world average) and material well-being will remain the general goal, the demand for minerals must necessarily continue to increase.

Should world population reach or exceed 6,000 million by the year 2000, the 1970 level of world mineral production would have to double, even if per capita mineral consumption did not increase at all. A very modest increase in per capita mineral consumption (say, 2 per cent annually, which is considerably less than the present annual rate of growth of gross national product (GNP) in many countries) would result in a further doubling of consumption in 35 years. A more realistic rate of increase of 4 per cent annually would cause a doubling in only 18 years and a fourfold increase in 36 years. Such an exponential increase is extremely likely, at least for the next 50 to 75 years (Lovering, 1969).

The most extreme models show that if mineral consumption in the United States remained at 1970 levels (these have already been surpassed) and the developing countries increased their per capita consumption by the end of the century to the same level as the United States now enjoys, annual mineral production by the year 2000 would have to increase 30 times (Govett and

Govett, 1972). The implications of such growth models—particularly those that project rates of growth far into the future—are rejected by many on the grounds that steps will be taken before the process gets out of hand (e.g. Maddox, 1972). Also, there is evidence that the rate of population growth has been slowing down, not only in the developed, but also in some of the developing countries (see Perlman, 1977).

Nevertheless, exponential growth is a treacherous and misleading process. A variable can continue to double many times without seeming to be significant, but within one or two more doubling periods it may become overwhelming. An annual rate of growth of 4 per cent causes a doubling in 18 years; a rate of growth of 7 per cent annually (the demand for energy in Japan has grown in excess of this rate for the past few years) will cause a doubling every 10 years unless something is done to halt the process; within the first 30 years of a man's lifetime an annual rate of growth of 7 per cent would result in only an eightfold increase in demand, but if he lived to be 70 years old he would see an increase of 128 times. The recent energy crisis—although it was largely politically motivated—may be the first example of the consequences of ignoring the implications of exponential growth.

Meadows et al. (1972) calculated a 'static index' to forecast the number of years' supply remaining of a given mineral, based on present known reserves and the current level of demand and compared it with an 'exponential index', which takes into account the rate of increase in consumption of the mineral. Although, as Zwartendyk (1974) had pointed out, the life index of a mineral reserve is largely a 'statistical mirage' and should be used, if at all, with great caution, the results of the comparison by the Club of Rome are dramatic. The number of years' supply left after 1970, based on the static index, for nickel, aluminium, copper, lead and zinc are 150, 100, 36, 26 and 23 years, respectively; the exponential index for these minerals is 53, 31, 21, 21 and 18 years, respectively. While these calculations and the conclusions based on them can be criticized on the grounds that projections of demand beyond two or three decades are relatively meaningless (since prices will increase and steps will be taken to augment reserves before serious depletion occurs), the data on which the model is based—the projections of the United States Bureau of Mines—cannot be dismissed.

The United States Bureau of Mines (1970) forecasts (which covered 88 minerals) were based on past trends, projections of GNP, and the changing patterns of consumption both in the United States and the rest of the world. Total United States demand for the minerals considered is expected to increase between 3 to 5 times the 1968 level by the year 2000 (an annual rate of growth of between 3.4 and 5.5 per cent); total world demand is expected to grow at about the same average rate, taking into consideration the higher rates of growth in Japan, Europe, and the centrally planned

economies, and the lower rates of growth in some industrialized countries and many of the other regions. The high range of the forecasts do not differ significantly from the more recent work contained in a United Nations-sponsored study by a group headed by Leontief (United Nations, 1976b).

The United States demand for energy is expected to grow within the range 2.7–5.2 per cent annually, while that for the rest of the world is forecast to increase at an annual rate of between 3.6 and 5.3 per cent. Projections of a slower rate of growth in the developed countries is based, in part, on the assumption that once a country reaches a high level of economic development, its per capita use of energy does not grow as fast as it did during the earlier periods of industrialization. Unfortunately, this assumption ignores the fact that as lower-grade ore and less conventional sources of minerals have to be exploited, a greater amount of energy is required for the same output; energy requirements to process low-grade and non-conventional ores will be higher; pollution abatement measures are energy-intensive (see Cohen, 1976). Therefore, the high range of the Bureau of Mines forecast for the United States is probably realistic, and the low range is perhaps too optimistic (this would apply as well to most of Western Europe), especially in view of the rising costs of imported fuels and resulting conservation measures.

In the 1950s, the Paley Commission (United States President's Materials Policy Commission, 1952) drastically underestimated United States energy demand for the decades 1950–70 based on an annual rate of growth of GNP of 3 per cent, a rate which was exceeded. Notwithstanding the current depressed state of the world economy, the Bureau of Mines estimates, based on an assumed 4 per cent annual rate of growth of GNP, are probably reasonable for the United States; this growth rate will certainly be exceeded in the centrally planned economies and in some Latin American countries.

CHANGING PATTERNS OF WORLD CONSUMPTION

Between the turn of this century and 1970, energy consumption in the United States increased thirteenfold; in the early years of the 1970s it was increasing at an annual rate of about 3 per cent. In the rest of the world the demand for energy has been expanding at a rate nearly double that experienced in the United States, with the highest rates of growth in the industrialized countries of Europe and in Japan, the centrally planned countries and a limited number of development areas. Notwithstanding the increase in the price of oil and general economic depression in the 1970s, it does not seem reasonable to expect the long-term demand for energy to grow at a significantly reduced rate. What can be expected is a change in the pattern of energy consumption away from petroleum. The speed with which this change takes place will depend on a number of factors, especially the acceptance of the inevitability of the growth of nuclear power and the technological advances which will allow the exploitation of today's non-conventional sources of energy.

In the past two decades, Japan has reproduced the phenomenal rate of growth of energy consumption experienced in the United States during its period of rapid industrialization. Japanese energy consumption nearly quadrupled in the 14 years between 1950 and 1964; between 1960 and 1973 it again quadrupled. Total energy consumption in Japan now surpasses that in the United Kingdom; even on a per capita basis it exceeds the level

of consumption in Italy, Ireland, Portugal, Greece and Spain and is almost as high as that of Switzerland (UNCTAD, 1976). If the Japanese rate of growth is a model for the developing countries, the approximately 10 per cent of world energy consumed in the less-developed parts of the world must increase significantly. There is every reason to believe that some Latin American and Asian countries will experience a rate of growth which, if it does not meet the Japanese levels, may well outstrip some of the countries of Western Europe.

Table 1 presents data for the growth of per capita energy consumption in a number of countries between 1961 and 1971. Rates of growth in per capita consumption in countries like Japan, Greece, Iran, Thailand and the oil-producing countries of the Middle East are perhaps exceptional, but even in the less industrialized and heavily populated countries of Asia (India, Pakistan and China), the rate of increase exceeded that of the United Kingdom—Pakistan and China nearly matched the United States rate of growth. Calculations of world energy consumption (United Nations, 1973) reinforce the conclusions of a dramatic change in the pattern of energy use. For the world as a whole, the demand for energy rose by nearly 70 per cent in the period 1961–71; for the developing countries, it rose by 95 per cent. Per capita energy consumption in these countries also increased relative to that in the developed countries. In spite of the increasing cost of petroleum, this pattern of growth continues (United Nations Conference on Trade and Development, 1976).

Table 2, based on the United States Bureau of Mines projections (1970), illustrates how the pattern of

TABLE 1. Per capita energy consumption, 1961 and 1971 (million tonnes coal equivalent)

Country	1961	1971	Percentage change
Developed			
Australia	3,857	5,448	41.2
Canada	5,614	9,327	66.1
France	2,512	3,928	56.4
Germany, Federal Republic of	3,639	5,226	43,6
Greece	462	1,470	218.2
Japan	1,312	3,267	149.0
Sweden	3,479	6,090	75.0
United Kingdom	4,903	5,507	12.3
United States of America	8,068	11,241	39.3
Yugoslavia	902	1,610	78.5
Developing			
Algeria	273	499	82.8
Bolivia	144	207	43.8
Brazil	340	515	51.5
Chile	928	1,487	60.2
Egypt	278	285	2.5
Ghana	92	186	102.2
India	154	186	20.8
Iran	323	1,026	217.6
Libya	268	567	115.7
Mexico	963	1,270	31.9
Nigeria	38	59	55.3
Pakistan	60	81	35.0
Peru	508	615	21.1
Philippines	158	292	84,8
Saudi Arabia	248	963	288.3
Sudan	51	119	133.3
Thailand	66	312	372.7
Venezuela	2,667	2,494	− 6.5
Zambia	458	470	2.6
Centrally planned			
Bulgaria	1,498	4.133	175.9
China	408	560	37.3
Czechoslovakia	5,078	6,612	30.2
Germany, Democratic Republic of	4,937	6,312	27.9
Hungary	2,250	3,289	46.2
Poland	3,182	4,374	37.5
U.S.S.R.	2,891	4.537	56.9

Source: United Nations, 1972, 1973.

TABLE 2. Ratios of mineral demand in 1968 to forecast mineral demand in 2000 for selected minerals in the United States and the rest of the world

| Mineral | United States | | Rest of the world | |
	High projection	Low projection	High projections	Low projection
Petroleum	3.3	1.5	5.6	3.0
Uranium	29.1	25.0	24.8	22.6
Chromium	2.8	1.9	2.8	1.8
Iron ore	2.1	1.5	2.2	1.6
Nickel	3.5	2.4	n.a.	2.3
Aluminium	9.5	4.8	10.6	4.9
Copper	1.3	− 0.8	6.1	2.9
Lead	3.1	1.5	2.1	1.7
Mercury	2.5	1.6	2.7	2.0
Silver	3.1	1.6	3.5	1.6
Tin	1.7	1.2	2.1	1.0
Zinc	2.8	1.5	2.8	2.2
Phosphorus	2.8	1.6	7.8	4.0
Tungsten	5.7	3.7	2.4	1.9
Vanadium	6.5	4.4	3.6	2.6

Source: United States Bureau of Mines, 1970.

demand for energy and also for metals is expected to change in the future. The minerals essential for agriculture (represented by phosphorus) and energy and electrical transmission (petroleum and copper) show the highest rates of increase in countries outside the United States. In contrast, the demand for minerals such as tungsten and vanadium is expected to increase relatively more in the United States. The demand for uranium is only marginally higher in the United States, reflecting the belief that a number of European and developing countries will develop nuclear power facilities in the next three decades. Although there have been changes in the forecasts for individual commodities since 1970, the relation between rates of change in the United States and the rest of the world have not changed significantly (see Perlman, 1977).

The overall conclusion that the demand in the rest of the world will grow more rapidly for a number of minerals than in the United States is supported by the study by Malenbaum et al. (1973). Based on an appraisal of the differences in the various rates of growth in consumption of minerals in the developed and developing countries, they found that at an early stage of development the demand for all minerals is very high; it reaches a peak as per capita income rises (e.g. it peaks for energy at an income of about U.S.\$ 2,000 per capita) and thereafter declines relatively. Intensity-of-use indices (derived from the ratio of the quantity of the mineral consumed to GNP) for iron ore, copper, aluminium, zinc, fluorspar, sulphur and energy fuels were calculated. The index reached a peak in 1966–69 in Western Europe, Japan, Australia, Canada, the U.S.S.R. and Eastern Europe; it is expected to continue to increase

only for fluorspar, sulphur and the energy fuels in Japan, Canada, Australia and the U.S.S.R.; otherwise, it will decline over the next 30 years in the above countries. In Africa (excluding South Africa), Asia, Latin America and China the index is expected to continue to increase until the year 2000 for all minerals considered.

The non-United States share of total world mineral consumption is forecast to expand by about 10 per cent: demand for iron ore in the rest of the world is expected to increase from 79 per cent of the total to 87 per cent by the year 2000; for copper and aluminium the demand in the rest of the world is forecast to grow from 71 per cent and 60 per cent respectively, to 78 per cent and 67 per cent. Natural gas demand outside the United States is likely to show one of the largest increases—from the present 37 per cent of the total, it should rise to 55 per cent in 2000. Only the demand for nuclear energy is expected to grow more rapidly in the United States than in other countries.

The projections from the Malenbaum study are compared with those of the United States Bureau of Mines in Table 3. While the two forecasts differ significantly in detail, the Malenbaum forecasts generally fall within the range of the high and low forecasts of the United States Bureau of Mines, except in the case of iron ore where Malenbaum's estimates are considerably higher. This may be due to the fact that the intensity-of-use index for iron ore in the developing countries is expected to increase very rapidly in the next three decades.

Both studies came to approximately the same conclusions with respect to the mineral situation in the United States by the end of the century. The United

TABLE 3. Comparison between forecasts for mineral demand in the year 2000

	Iron ore (million tonnes)	Copper	Aluminium	Zinc	Sulphur
		(thousand tonnes)			
United States of America					
Malenbaum	141	4,389	15,657	2,822	26,648
United States Bureau of Mines					
High	117	7,131	33,385	3,629	33,566
Low	87	4,445	16,783	1,896	20,866
Rest of the world					
Malenbaum	945	15,304	31,086	10,626	73,776
United States Bureau of Mines					
High	758	31,661	61,690	10,161	113,400
Low	537	15,241	28,577	7,983	77,112
World total					
Malenbaum	1,086	19,693	46,743	13,448	100,424
United States Bureau of Mines					
High	875	38,792	95,075	13,790	146,966
Low	624	19,686	45,360	9,879	97,978

Source: Malenbaum *et al.,* 1973; United States Bureau of Mines, 1970.

States Bureau of Mines (1970) predicted that: 'The trend for a growing number of primary minerals is toward high costs, a leveling off of domestic production, losses of traditional markets, recourse to substitutes, and increasing dependence on foreign sources of supply.' Malenbaum *et al.* (1973) similarly concluded that, for the minerals considered: 'U.S. requirements are expected to exceed domestic production in 2000. Expanded requirements abroad and such import needs by the United States can only be met through growth in production in the rest of the world.'

These conclusions for the United States are probably valid for much of the industrialized world. At the same time as the developed countries will have to increase mineral imports to satisfy growing demand, it is reasonable to predict that the rate of growth in consumption in the developing countries will continue to surpass that of the Western countries and the centrally planned economies, consequently putting increased pressure on the world's mineral reserves. Indeed, the New York-based Commodities Research Unit (Perlman, 1977) concluded that Malenbaum's projections of the rate of growth in the development areas was probably too low—but with the important assumptions that there would be moderate success in reducing the rate of population growth, and that the real price of primary products would increase. By the year 2000 they predict that Latin American consumption patterns would be similar to those in Western Europe in the mid-1960s, and the intensity of Africa's mineral consumption could be similar to that in Latin America in the early 1950s.

CURRENT ESTIMATES OF WORLD MINERAL SUPPLIES

Attempts to determine the adequacy of mineral supplies in the face of growing world demand can be either alarming or reassuring. Adequacy—and shortage—are relative concepts, depending not only on the absolute figures for reserves and resources but also on forecasts of future economic and technical changes (see Govett and Govett, 1974, 1976b).

Based on almost any estimate of *current* reserves, the situation is critical for a number of important minerals; world reserves of manganese, molybdenum, tungsten, aluminium, copper, lead, zinc, tin, titanium and sulphur are all inadequate to meet projected demand in the next 50 years (Govett and Govett, 1972; Fishman and Landsberg, 1972; Meadows *et al.*, 1972; Fischman, 1977). Even if current reserves of these minerals are arbitrarily assumed to increase by some factor (such as 5) to reflect decreases in economically mineable grade and future discoveries, the calculations of the number of years' supply remaining change very little.

In a study in 1972 (Govett and Govett, 1972), we found that comparing the United States Bureau of Mines (1970) cumulative demand forecasts for 1968 to 2000 with 1968–70 reserve estimates for 11 metals (tin,

zinc, copper, lead, tungsten, molybdenum, titanium, nickel, cobalt, vanadium and chromium), there were serious problems for tin, copper, zinc, lead, tungsten and molybdenum; if reserves were arbitrarily increased 5 times (to reflect the expansion in supply by the year 2000), there would still be serious problems of adequacy of tin, copper and zinc. In a more recent study (Govett and Govett, 1977a), we compared the demand forecasts with estimates of identified resources (i.e. specific bodies of mineral-bearing rocks whose existence and location are known) compiled by the United States Geological Survey (Brobst and Pratt, 1973) and found that there were serious problems of adequacy by the year 2000 only for tin, copper and uranium (unfortunately uranium was omitted from the earlier study due to lack of reliable data).

Fischman (1977) has also compared the United States Bureau of Mines demand forecasts with 1974 reserves and with the 'recoverable resource potential' of 19 minerals (including petroleum, natural gas and uranium); on the basis of 1974 reserve estimates, he foresaw serious problems with the same metals as we did in the 1972 study; in the case of the energy minerals, the materials balance in the year 2000 was negative for uranium and barely positive for petroleum. In contrast, he saw no serious problems of adequacy of supply for any of the 19 minerals when demand projections were compared with the so-called resource potentials (defined as 'producible at today's prices, but yet to be discovered').

Clearly, conclusions about the adequacy or scarcity of mineral supplies in the future depend very much on the 'resource' estimate used (see below for a fuller discussion of reserves and resources). Whatever data are used, there are a number of geologists who are optimistic about future mineral supply adequacy; this optimism is based on a belief that lower-grade ores can be mined and that new processing techniques will be developed. This view necessarily assumes that current patterns of world energy use—with petroleum and natural gas accounting for some 60 per cent of consumption—will change, and that the long-term supply of energy is virtually limitless (given technological solutions to the problems of harnessing solar and nuclear energy).

If all the technological restrictions are ignored, if oil shales and tar sands are fully developed, and if nuclear energy is fully exploited, it is argued that a 1,000-year world energy supply is assured (Darmstadter, 1972). More immediately, the United States Bureau of Mines (1970) concluded that a *current* technological levels, supplies of energy fuels for the world will probably be adequate at least until the year 2000, assuming that nuclear power supplies between 37 to 54 per cent of the electricity generation in the year 2000 in the United States, the world's chief energy-consuming nation.

As recently as 1976 Bowie concluded that known world uranium resources would only be adequate to meet world demand by the 1990s if there was an enormous increase in world uranium exploration. This has, of course, occurred and it is becoming clear that uranium is not a particularly rare element: in Canada alone reserves have increased by two-thirds since 1975; a similar dramatic increase in reserves can be expected from renewed exploration in Australia if that country favourably resolves its restrictions on uranium mining (limited development of some deposits was authorized in 1978); furthermore, the development of conversion technology for advanced reactors, especially the development of the breeder reactor, which enhances the raw material use by a factor of 50 or more, would significantly reduce the problem of the adequacy of uranium supplies. On the other hand, concern for the safety of nuclear installations, particularly the use of reprocessed material for nuclear weapons, may result in the unwillingness of uranium-producing countries to allow foreign sales, especially to developing nations.

Whether one is optimistic or pessimistic about the long-term energy supply, the immediate problem remains for countries like Japan, much of Europe and even the United States to procure and guarantee supplies of imported petroleum. The technical and political problems of nuclear energy, coal gasification, environmental barriers to the development of the oil shales and tar sands, and the political disputes over Arctic deposits and off-shore mineral rights (not to mention mining the ocean-floors), will all have to be solved quickly if the present rate of use of conventional fossil fuels continues.

If there is an energy crisis now it is actually a crisis in the pattern and the distribution of petroleum supplies, the lead-times necessary to develop new technology and new energy sources, and the difficulties likely to emerge in the development of nuclear power. The drive towards self-sufficiency—increasingly the reaction of countries which were affected by the oil shortages of 1973–74—may be harmful to the cause of world trade, but hopefully it may also result in stepped-up exploration and the development of new methods for exploiting unconventional energy sources.

Factors affecting future mineral supplies

THE CONCEPT OF MINERAL RESOURCES

While the world may face a reserve crisis for some minerals, it does not necessarily face a 'resource' crisis. Controversy about the adequacy of world mineral supplies is intensified by the problem of defining exactly what is being measured. Zimmermann (1933), Blondel and Lasky (1956), Flawn (1965), Pruitt (1966); Lovering (1969), McKelvey (1972, 1973), Zwartendyk (1972), Brooks (1973) and Govett and Govett (1974, 1976b) are among those who have tackled this problem.

The reserve of a mineral is that quantity of ore that is exploitable under present economic and technical conditions. Resources, in contrast, include the total amount of an element—both known and unknown—down to some defined grade that is higher than its crustal abundance but lower than the present economically mineable grade. Total resources—as a concept—must include not only reserves plus all known deposits that are not now economically recoverable, but also all undiscovered deposits in the Earth's crust. The share of reserves in total resources changes constantly as new discoveries are made, the mineable grade of ore changes, new extraction and processing techniques are developed and prices change. As Boyd (1973) pointed out: 'total resources are large enough to stagger the imagination ... reserves will always appear to be limited.

McKelvey (1973) suggested the division of all resources except reserves into two categories:
1. Known but currently uneconomic deposits—divided into 'paramarginal resources' (which include deposits which could be profitably worked at recovery costs up to 1.5 times the present costs) and 'submarginal resources' (which could possibly be recovered at higher prices or after technological advances).
2. Undiscovered resources—subdivided into 'hypothetical resources' (those which may be reasonably expected to exist but which are not yet discovered in known mining districts) and 'speculative resources' (those which may be found in areas which are geologically favourable for a particular type of mineral deposit, but where no deposits have yet been found).

Figure 1 presents a conceptual model of the relation between reserves and resources—similar models have been proposed by McKelvey (1972) and Zwartendyk (1972). The area of triangle ABC represents the toal resources of any particular mineral within a nation, a region, or the world. The absolute amount represented by ABC will depend on some defined lower limit for the economic grade of the mineral (i.e. point B) and a defined limit to the thickness of the Earth's crust which, realistically, may be explored and mined.

The area ABD represents unknown resources and includes all mineral deposits not yet discovered (the 'hypothetical' and 'speculative' resources of McKelvey). The area of the smaller triangle, ADC, delineates the portion of known resources. Line EF shows the division between reserves and non-exploitable resources and will shift as improved extraction techniques are developed, prices change and the economic grade is lowered.

Whether dealing with known or unknown resources, all estimates necessarily depend upon what grade ore is considered economically recoverable. Since economic grade is itself a function of price and technology (both of which will change to an unimaginable extent with time), it is clearly unrealistic to expect a finite estimate for total world resources. Attempts to make such estimates based on crustal abundance of a particu-

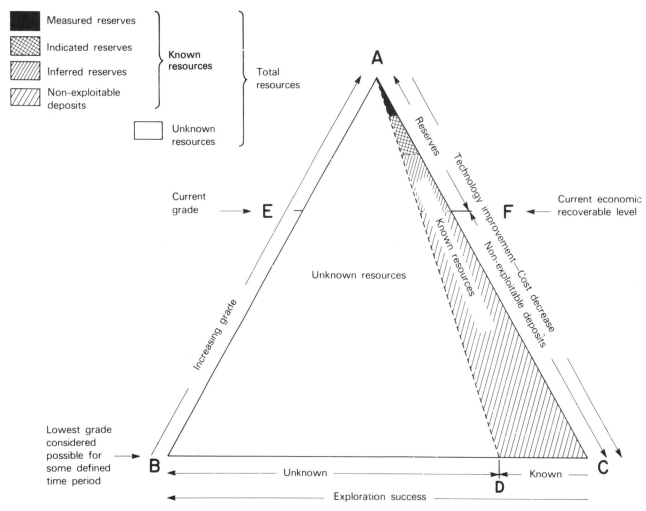

Fig. 1. Schematic conceptual model of the relation between reserves and resources. (From Govett and Govett, 1974; reproduced with the permission of *Resources Policy*.)

lar element are relatively meaningless; there is enough copper in a one-mile thickness of the Earth's crust in the United Kingdom alone to supply the world, at present production rates, for a period of 10,000 years (Govett and Govett, 1974).

A more realistic approach is based on geological projections which assume present day cut-off grades. These geological techniques fall into two broad categories. One extrapolates, on an area or volume basis, known deposits in a well-explored region to similarly favourable but unexplored geological regions (Nolan, 1950; De Geoffroy and Wu, 1970). Using this approach Lowell (1970) has made some order-of-magnitude estimates of the number of copper porphyry deposits which may be awaiting discovery in geologically favourable areas. In Chile and Peru, where 50 per cent of the favourable belt has been prospected and 12 deposits found, it is assumed that 12 more deposits may be found in the remaining unprospected area; 13 copper porphyry

deposits have been found in the 10 to 15 per cent of the geologically favourable region of British Columbia in Canada that has been prospected, leaving about 100 additional deposits to be found.

The second approach to assessing total resources uses the apparently linear relation between known reserves and crustal abundance of the elements. This relation was first remarked and quantified by McKelvey (1960), who showed that for long-sought-after elements in a well-explored region—such as the United States—reserves in tonnes, (R), equalled crustal abundance in per cent, (A), times 10^9 to 10^{10}, i.e.

$$R = A \times 10^9 \sim 10^{10}.$$

If it is assumed that the United States is a representative sample of the Earth's crust, then world reserves should be 17.3 times as great, so,

$$R = 1.7\,A \times 10^{10} \sim 10^{11}.$$

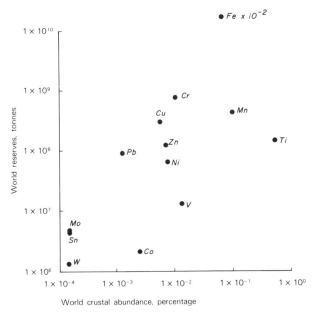

FIG. 2. Relation of reserves to average crustal abundance of some industrially important elements.

In Figure 2, world reserves for 13 industrially important elements are plotted against their crustal abundance; 9 of these elements (based on 1970 reserve estimates) show a linear relation expressed by:

$$R = 2.3 \, A \times 10^{10} \sim 10^{11},$$

which is extraordinary close to an extrapolation of world estimates from McKelvey's figure for the United States and is similar to the figures of

$$R = A \times 10^{10} \sim 10^{11} \times (1.7 \sim 37)$$

deduced by Sekine (1963) and the average figure given by Ovchinnikov (1971), i.e.

$$R = 3.2 \, A \times 10^{10}.$$

This relation (discussed by McKelevey, 1960, 1972) shows that, where data are adequate, there is a gross relationship between elements concentrated as mineable deposits and the absolute concentration in the Earth's crust, and that there is also a general correlation between present workable grade and crustal abundance; it does not indicate, however, a reasonable upper limit for the proportionality factor since this is a function of technology and prices. The relationship has value in assisting in the prediction of expected resources in relatively unexplored parts of the world, given present mining conditions, and may also assist in predicting resources of metals which have only been sought for a short period of time (e.g. uranium) and which do not conform to the linear relation to abundance (McKelvey, 1972).

Increasingly sophisticated geo-statistical models are being developed to aid in resource assessments (see Vogley, 1975; Ridge, 1974; Fabbri, 1975; Agterberg, 1977). Nevertheless, we are still far from having even regional resource estimates for the metals, although the work of the United States and Canadian Geological Surveys is providing an increasing amount of information on regional metal occurrences.

A less sophisticated approach to estimating total resources is simply to consider mining at greater depths; Grossling (1970) has calculated that present reserves could be increased by as much as 50 to 100 times, depending on the depth reached. The technology is available for mining at great depths now; this does not, unfortunately, solve the problem of *finding* deeply buried deposits—or even deposits covered by only tens of feet of surficial overburden and rock (Govett and Govett, 1972). Moreover, as pointed out by Skinner and Barton (1973), there is growing evidence that most hydrothermal-type deposits are restricted to the upper one to two miles of the Earth's crust.

The undoubtedly vast resources of the ocean and its floor could provide almost limitless supplies of the essential elements—the equally vast magnitude of finding, recovering, and processing these resources is frequently overlooked (Wang and McKelvey, 1976). Furthermore, there are potentially disastrous economic consequences for traditional producers of minerals such as nickel and manganese if a vast new supply of these minerals can be recovered from the oceans; if the virtually unlimited resources of nickel in the sea nodules could be economically mined, the Canadian position in the world nickel market—Canada currently produces approximately 32 per cent of the world's nickel—would be seriously jeopardized (Govett and Govett, 1973). Other countries would find themselves in similar situations.

CURRENT PROBLEMS FOR THE MINING INDUSTRY

The geologist is interested in the projected demand for minerals and the implications such demand may have on the finance available for research, exploration and mining. How fast will the demand for such minerals as copper, lead and zinc grow? What are the present supplies of these minerals under different economic and technological circumstances? What effect will an increase in demand have on price and therefore on the mineable economic grade of an ore deposit? What kind of investment climate will prevail in the countries of Asia, Africa and Latin America? What kind of advances in extraction and processing techniques are likely? What effect will such progress have on the kind of ore deposits that can be exploited? Will environmental restraints make costs so high that many areas will not be considered economically attractive to the mining industry? These, and similar questions, are currently being asked in mining circles throughout the world.

In terms of total resources there will probably be no serious problems for the mining industry in the immediate future, provided that new exploration is undertaken in both known and unprospected areas of the globe, that new exploration techniques for deeply buried deposits are developed, and that research is under way for new extraction and processing methods. There may be problems if environmental restraints on the mining industry prove to be severe, particularly if they impede exploration. Pollution control may prove to be one of the most important limiting factors in meeting the world's demand for minerals.

In all countries, as growth takes place the pollution of the Earth, the atmosphere and the seas increases. Ecology, the environmental sciences (including environmental geology) and pollution control have all gained respectability in the past two decades. It is perhaps significant that one of the acknowledged founders of modern exploration geochemistry, J. S. Webb, chose the subject of environmental geochemistry for his keynote address to the Fifth International Exploration Geochemistry Symposium (Webb, 1974), and that a considerable part of this book was devoted to environmental geochemistry. Current interest in the environment can be gauged by the number of publications that have appeared in the past few years: a bibliography published by OECD (Block, 1973) listed 430 separate books, articles and reports on the environmental aspects of economic growth.

While concern for the environment and the resulting measures for pollution control will have a broad economic effect, the most immediate effect will be on the mining industry; any decrease in net earnings as a result of the costs of pollution-abatement measures will directly reduce exploration budgets. Measures such as those currently in force in the United States to close off large areas of public lands to mineral exploration through creation of wilderness 'park' areas, delays in public land-lease sales and the issuance of drilling permits, cancellation or refusal to grant mineral patents and increased prices for exploration rights will curtail exploration (Flawn, 1973). Even when exploration programmes are not seriously affected by such measures, the uncertainties about future possibilities of exploiting mineral deposits in many areas may act as a deterrent to active exploration. Deposits most commonly exploited economically by open-pit or strip mining (e.g. coal, copper porphyries) have already become less attractive exploration targets as measures to restrict or eliminate open-pit mining spread; the same will be the case with deposits which would require large-scale underground explosions.

While the actual costs of environmental protection in terms of exploration forgone cannot be estimated, there is no doubt that uncertainty about future government policy will deter exploration expenditures. Research into the long-term effects of pollution control on mineral exploration is urgently needed.

At the same time as environmental concern is reducing exploration in some areas, the political environment in many countries in which the exploration geologist and the mining engineer must work is becoming increasingly unfavourable. In an article in *World Mining,* Derry et al. (1974) concluded that: 'Life for the explorationist is not as simple as it was in the late 1960s; recent events have shown us that exploration cannot be viewed in isolation and merely as a response to anticipated demand.' In both Canada and Australia—which, together with South Africa and the United States, accounted for 80 per cent of the world's exploration activity in the last decade (United Nations, 1975)—expenditures on exploration dropped sharply after 1970 as a result of changes in investment and taxation policies and new legislation.

Until this decade Canada was at the forefront of world exploration, in part because of the policies at both the federal and the provincial level which encouraged the expansion of the mining industry. In 1972 new federal taxes were introduced, and a number of the provinces followed suit (for example, British Columbia increased taxes to a level considered by the mining industry as 'confiscatory' (see Govett and Govett, 1976b). The Canadian mining industry has also been threatened by increasing government participation; in the case of potash in Saskatchewan there has been an outright takeover of part of the industry. The fall in exploration expenditures in general in Canada, and particularly in the provinces where taxes were increased, is dramatic evidence of the direct effect government policy can have on exploration investment; in British Columbia exploration expenditures fell from $46 million in 1970 to $22 million in 1975 (British Columbia, 1976). Most mining companies in Canada share the view of the President of Falconbridge Nickel Mines Ltd in his address to the Plenary Session of the CIM Annual General Meeting in 1977: 'With a relaxation of some of the more oppressive measures under which it now operates, the industry could soon begin to regenerate momentum. Even then, dramatic expansion would not be evident overnight. It would, in my view, take a strong ten-year effort to get the industry's productive machine running again at full capacity.' (Cooper, 1977, p. 72).

The unfavourable investment climate in Australia and Canada came at a time when exploration costs were rising. Roscoe (1971) calculated a ten-fold increase in Canadian exploration costs between 1951 and 1969; it is now estimated to cost on the average $25–30 million to discover an ore deposit (Derry et al., 1974; Derry and Booth, 1977). Furthermore, the ratio of discoveries to exploration tries is increasing as it becomes necessary to seek deeper deposits and to explore relatively unprospected areas.

The rising costs and unfavourable investment climate in the developed countries could give added impetus to increased exploration in the developing nations, and it may well be asked whether exploration should not, in fact, be increasingly concentrated on the less well prospected continents of Latin America, Asia, and Africa, especially those with large land areas and geological potential (e.g. Brazil, Argentina and Indonesia). Although there is a definite correlation between land area and the number and value of minerals produced in the well-prospected parts of the world (see Govett and Govett, 1977b), size and geology alone are not the only factors in exploration success (this is indicated by the fact that the correlation between land area and mineral production which we found for the developed and centrally planned countries did not hold for most developing countries). Political and institutional factors are frequently as important as geological potential in determining whether an area will be explored.

Although there is little question that world mineral supplies could be increased significantly if exploration were to be undertaken in a number of these areas (see Govett and Govett, 1978), unfortunately, many of these countries are hostile to foreign investment in the mining industry and have neither the capital nor the expertise to develop their own exploration programmes. Technical assistance such as that provided by the United Nations (see Harkin, 1976) is inadequately funded at present; service contracts, through which international mining companies provide the expertise, may be used increasingly (see Ridge, 1973; Tovar, 1973; Waelde, 1977).

Even if new types of partnership arrangements are developed and technical assistance increases, the present policies of governments—in both the developing and the industrialized countries—is a hinderance to the development of new mineral deposits. In any particular case, the avowed reason for a particular policy may be laudable (e.g. to provide funds to remedy social injustice, to assure that profits from mineral production remain in the home country, or to protect the environment); the long-term effects on the world's supply of minerals of such policies are often not fully appreciated.

The direct contribution of the mining industry to a country's national income seems to be small; too few people in policy-making positions realize that the entire basis of a modern industrial society is ultimately minerals—iron ore (steel for bridges, automobiles and bicycles), limestone (cement for buildings and roads), phosphate rock and potash (fertilizer), petroleum (energy and raw materials for plastics and synthetic fibres). The total dependence on minerals demands a vigorous mining industry; both direct and indirect government policy should be designed to encourage its growth.

Problems for the developed countries

The developed countries *consume* the major share of the world's minerals. They also *produce* the major share of

the world's minerals (M. H. Govett, 1975, 1976b); the greater part of the mineral imports by the developed countries come from other developed countries (see Govett and Govett, 1977b). The United States, with only 6 per cent of the world's population, consumes one-third of the world's total energy output; 20 tonnes per person of new mineral supplies are needed each year by the American economy (United States Secretary of the Interior, 1972). North America and Western Europe together consume nearly two-thirds of the world's petroleum and more than 70 per cent of the natural gas (Darmstadter, 1971).

The industrialized countries are becoming increasingly dependent on imported minerals while the exporting countries are demanding higher prices. In most of the developed countries of the world, mineral imports now average about a quarter of total merchandise imports—Canada, South Africa, Australia and the U.S.S.R. are the only major countries which do not have to depend on imports for the more important minerals.

The United States is now a net importer of most minerals. Of the 5 metals—aluminium, zinc, lead, iron ore and copper—which account for 88 per cent of the total value of all metals consumed in the United States, nearly all primary aluminium, one-half of the zinc, more than one-third of the lead and iron ore and approximately one-fifth of the copper is imported (United States National Commission on Materials Policy, 1973). All primary supplies of rutile, chromite, tin, tantalum, mica and columbium are imported; also imported are 90 per cent of all primary antimony, cobalt, manganese and platinum and more than one-half of the asbestos, cadmium, beryl and fluorspar.

Japan is the most extreme case of mineral import-dependence; 40 per cent of Japan's total foreign trade is in minerals. Japan imports all of its bauxite, phosphate, nickel and cobalt; more than 95 per cent of its iron ore and petroleum; more than three-quarters of its tin; and one-half of its copper, manganese and coal, as well as significant quantities of tungsten, molybdenum, chromium and potash. In Japan, energy imports as a percentage of total energy consumption rose from zero in 1925 to 80 per cent in the 1970s. Between 1925 and the early 1970s, energy imports in Western Europe increased from 2 per cent to more than 60 per cent; Eastern Europe, which exported 15 per cent of its energy production in 1925 now imports energy fuels (Ridker, 1972).

It may be argued that the energy crisis and its effect on importing countries is a special situation; however, to the extent that producing countries are able to band together to create cartel-type organizations, price increases for other minerals could result. The call for a 'new economic order', which aims at achieving 'equality of opportunity' for the world's primary producers (see Tinbergen, 1976) is a new and potentially serious factor in world mineral trade. A number of producer associations have been formed since 1970, and the developing nations have increasingly threatened to use 'resource diplomacy' to redress the imbalance in wealth between them and industrialized world (see Drolet, 1977). Although it can be argued that the potential for cartels in most metals is not promising (e.g. Fried, 1976; Govett and Govett, 1977*b*), the development of a policy to achieve mineral self-sufficiency by a number of industralized countries (especially the United States) could have serious repercussions on world trade. Furthermore, while it may be possible for countries with vast resources such as the United States and Canada to develop new sources of mineral supplies within their borders, it is clearly impractical for many of the European countries and Japan.

The consequences of pressing mineral demand in the developed countries has given rise to a school of thought which advocates 'slow-growth' or 'zero-growth' (e.g. 'The Alternatives to Growth Programme' sponsored by the Club of Rome). While it is difficult to imagine the North American and European countries intentionally slowing down their rates of growth as a matter of policy to conserve mineral resources, limits to that growth are, in fact, being imposed by pollution (see United Nations, 1976).

The United States Council on Environmental Quality (1973) in its *Fourth Annual Report* concluded that: 'Environmental controls are among the most significant factors responsible for the current energy shortages and those predicted for the near future'. Flawn (1973) predicted that concern for the environment in the United States will have a serious negative effect on the mining industry. A study by 'Resources for the Future' concluded that environmental concern could upset the generally optimistic appraisal that they made for American mineral supplies in the next three decades (Ridker, 1972). A report prepared for the United States Commission on Population Growth and the American Future (Ayers and Gutmanis, 1972) estimated that the total cost of air pollution abatement measures alone between 1967 and 1979 for the primary metal industry would reach U.S.$ 677 million; this was the highest cost for air pollution control estimated for any sector of the economy.

Conservation of minerals by increased recycling and the use of substitutes for scarce metals may, in some cases, add to the pollution problem (see Pearce and Walter, 1977). Nevertheless, recycling of scrap is an important factor in reducing the demand for primary metals. Scrap reclamation is already a major source of American metals; 35 per cent of the lead consumed, 28 per cent of the iron and between 20 and 30 per cent of the copper, nickel, antimony, mercury, silver, gold and platinum are recovered as scrap (United States Secretary of the Interior, 1972). A test of the potential effects of increased recycling of 5 metals in the next three decades—iron, copper, lead, zinc and aluminium—showed that an active recycling policy would re-

duce United States demand for primary aluminium and zinc significantly, the demand for primary copper modestly, and the demand for primary iron and lead hardly at all; the limited effect on iron and lead demand is largely a function of the already high portion of these metals reclaimed in the United States. The savings in terms of cumulative mineral requirements in the United States in the next 50 years was estimated to be of the order of 30 per cent for aluminium, 20 per cent for zinc and copper, and 6 and 3 per cent for lead and iron respectively (Ridker, 1972). Unfortunately, the additional energy requirements needed to achieve this level of recycling were not calculated. In spite of such indirect costs, an extension of recycling throughout the industrialized world would clearly contribute significantly to a reduction in the demand for primary metals, especially in those countries where scrap-metal reclamation and recycling are not widely practised.

Although recycling seems an obvious remedy to the pressures on mineral supplies, promotion of the use of substitutes—the other major remedy proposed for mineral shortages—can have negative as well as positive results. The use of non-returnable containers and synthetics pose serious pollution threats; the number of beer bottles which have to be disposed of in the United States increased sixfold in the past two decades; detergent phosphorous effluent has increased by 1,845 per cent since 1946 and inorganic fertilizer nitrogen by 648 per cent (Commoner, 1972).

The environmental costs of technological development in the industrialized countries have been so high and the effects so damaging that it can be argued that massive new technologies should be developed to reverse the present ecologically disasterous trends (Commoner, 1972). While many of the solutions offered to the world's environmental problems are Utopian in terms of what mankind would be willing to do and, most importantly, do without, some of the less radical proposals deserve serious consideration. Measures such as a raw materials tax, amortization taxes proportionate to the life of a product, and power taxes to penalize power-intensive manufacturing processes may help to conserve resources. When it is realized that the energy required to build a motorway is three to four times greater than that required for a railway serving the same population centres and that the amount of land needed for a motorway is of an order of four times as much as for a railway, proposals to limit motorway construction, reduce the use of private cars by higher taxation and provide alternative public transport seem reasonable. Calculations of thermal waste energy, emissions of pollutants into the atmosphere and waters of the Earth, the effect of pesticides and heavy metal pollution from sewage and industrial discharge, and oil pollution of the world's seas and oceans, all point to the need for immediate action.

To the extent that environmental protection measures can be devised without directly or indirectly reducing exploration and production in the mining industry, the pressures within the developed countries will be eased—although the costs will be substantial. This does not mean that the industrialized countries can continue to increase their consumption of raw materials at the present rate without severe repercussions on the world's supply of minerals. While it is not necessary to agree with the conclusion that unless the developed countries reduce their consumption considerably there is no hope that the developing nations can markedly improve their living standards (Ecologist, 1972), there is a serious challenge to the policy of unbridled economic growth in the industrialized world. As Hubbert (1973) concluded in a survey of world energy resources.

The real crisis confronting us is... not an energy crisis but a cultural crisis. During the last two centuries, we have evolved what amounts to an exponential-growth culture, with institutions based on the premise of an indefinite continuation of exponential growth. On of the principal consequences of the cessation of exponential growth will be an inevitable revision of some of the tenets of that culture.

How the challenge in the developed countries will be met is uncertain. The economist and the politician are necessarily most concerned with broad problems of the environment, international trade and economic growth within their own countries. Any impending environmental or resource crisis is viewed within a national framework, notwithstanding the efforts for international co-operation such as the Stockholm conference. The points of view understandably differ between Washington, London, Tokyo and New Delhi.

The present and future for the developing countries

It is plausible to argue that the developed countries should learn to reduce their consumption of relatively scarce mineral resources; it is much more difficult to apply the same philosophy to those which are developing. While tackling the problem of population growth would probably be the single most important measure to ease the pressures on world resources, in the next few decades the demand for minerals in these countries will rise at an increasing rate, and their growing share of world mineral consumption will probably be one of the most significant changes in the world resources picture.

Although a number of these developing States are major mineral producers, the majority of the countries of Asia, Africa and Latin America depend on imports of most industrially important minerals and, especially, import of energy fuels—the increase in petroleum prices in 1973–74 was felt most severely in countries like India and the Philippines. It is not surprising that the most vocal complaints against high petroleum prices come from these countries (see Fried and Schultze, 1975).

Individual countries may benefit (as did the Middle Eastern countries in the 1973–74 energy crisis), but to the degree that the developing countries are net mineral importers, mineral price increases will worsen their balance-of-payments position. A two-price system—one for the developed and one for the developing—is impractical; similarly, proposals for some form of subsidy to poorer countries by the richer mineral-producing countries of the less-developed world is probably wishful thinking.

Higher mineral prices will make development either more costly or slow down the rate of growth; the only viable alternative is an expansion of mineral supplies to keep prices from increasing too rapidly. Exploration and increased mineral production in the less-developed areas of the world—many of which are geologically favourable and still relatively unexplored—seems an obvious course of action.

The need for new arrangements whereby foreign expertise and capital can be invested in these areas without necessitating foreign ownership of mineral deposits has been discussed earlier in this chapter. While the objections to foreign domination of mineral resources is one of the main barriers to exploration and development that needs to be overcome, governments are also adverse to development of the mining industry on the grounds that the internal economic effects of using scarce capital to develop mining are less advantageous than investment in manufacturing industries. Primary production is regarded as 'degrading', producing 'hewers of wood' rather than an educated technical labour force. It is alleged that infrastructure is developed solely to serve the mining industry and that capital should be spent instead on roads, railways and ports that would serve the whole economy rather than just one sector—ignoring the fact that the mining industry provides indirect services and generates wealth far beyond the direct contribution that it makes to national income or the balance of payments. The point was made most forcibly at the Canadian Provincial Mines Minister's Conference in 1973:

Cities like Toronto and Vancouver would be shadows of their present entities without the contributions from mining and exploration and... our north country would largely be a vacuum had large areas not been opened up by mineral exploration (Nowlan, 1973).

The further argument that the mineral-rich countries should 'conserve' their resources now in order to get a better price at some future date, ignores the fact that reduction of exports (or a slower rate of growth of exports) would have a directly adverse effect both on national income and on a country's balance of payments. More importantly, there is no certainty that if resources are 'saved' they will find a ready market some decades hence. Vast new sources of mineral supplies may be developed as technological advances make it possible to exploit the ocean floors, extract oil from the tar sands and shales, and extract aluminium from clays. Manufacturing 30 years from now may be based on different metals or fuels than those used today; fairly radical changes in the energy mix can be expected in the near future. Also, while the prices of some minerals will certainly rise, the prices of most other commodities—particularly agricultural—may also rise. In an inflationary world it is dangerous to 'save' resources, hoping that the returns from sales in the future will be greater.

The argument that the developing countries are 'handicapped' by specialization in mining because the long-term terms of trade are against primary products (Spengler, 1961) continues to be heard. The issue was first raised by Prebisch (United Nations, 1950) and has since been discussed by economists at length. Morgan (1973), in reviewing a number of statistical price series covering a period of 150 years, found no decisive evidence of adverse terms of trade for mineral products; he concluded: 'The data as a whole underline the prudence of *not* trying to predict from the experience of one country or region, what is likely to be the experience of others.' Both Tims (1975) and Connelly and Perlman (1975) have pointed out that terms of trade indices for the developing areas as a whole are relatively meaningless. Between 1970 and 1975 the terms of trade index (based on 1970 = 100) for the developed countries fell to 89, while in the developing States as a whole it rose to 166; however, among these countries, the index for the oil-exporting countries increased to 318, while for all others it fell to 88 (and for those with a GNP of less than U.S.$ 200 it was 76—UNCTAD, 1976). Predictions of future trends are, at best, risky; it can be persuasively argued that mineral prices must rise because of continuing high rates of growth of demand; it can equally be argued that technology will provide the means to allow the long-term trend of little increase in real prices to continue.

Any historical tendency for some mineral prices to be stable or to decline relative to those for other commodities is probably due in large part to the rate of discovery of new mineral deposits, changes in the economically mineable grade of the deposits and technological advances in extracting and processing. Furthermore, in the first decades of this century the demand for most minerals did not increase at anything like the current world-wide rate. In a study of the movement of copper prices in the United States between 1870 and 1957, which showed that the long-run price of copper (in constant dollars) was remarkably stable, it was found that major technical changes in mining methods and in the processing of copper were important factors in price stability, as was the fact that the demand for copper after the First World War grew slowly due to the increasing use of aluminium and plastics and the increase in the reclamation and recycling of scrap copper (Herfindahl, 1959).

Much of the concern for the terms of trade of minerals—reflected in the policy of discouraging mineral production for export in the developing countries—implies concentration on one or a few export commodities, e.g. copper in Zambia, tin in Bolivia, petroleum in Venezuela, and the criticism that such concentration makes a country especially vulnerable to world price changes. There is no doubt that the mineral export industry is more vulnerable to price swings during the business cycle than the manufacturing industry, and that if a country is dependent on one or two export goods it can be seriously affected by a price decrease for those goods. However, such price volatility not only results in losses on the downswing, but also in gains on the upswing; good fiscal and monetary management can alleviate much of the cyclical effect.

It is necessary to separate the argument against concentration on one or two export commodities from the general stand against primary production in the developing countries. The neo-classical economic approach to international trade still has a good deal to recommend it—sell those goods in which you have a comparative advantage and invest the proceeds in diversifying the economy and developing new industries. As pointed out by Abramovitz (1961): '... natural resources have contributed to differences among the rates of growth of those countries that have been prepared to make good use of their resources'. Surely one should not be surprised, as are some economists when they view the development record of Latin America (e.g. Grunwald, 1964) that a number of the Latin American countries have advanced primarily through the exploitation of their natural resouces; the Venezuelan policy of 'sowing the petroleum' has much to recommend it despite the arguments of Page, 1976, and others.

Canada, Australia and South Africa—all primary producers—have large and growing incomes. While there is not necessarily a high correlation between the size of a country's mining sector and its per capita income, there is clearly a high correlation between a country's income and the impact that technology has on its resources, and an even higher correlation between income and the manner in which revenues from the mining industry are reinvested in the economy.

The decline in the relative importance of the mining industry in countries like the United States and the United Kingdom is sometimes taken as evidence that as economic development proceeds the relative role of primary production must decline. This does not mean, as some deduce, that the mining sector produces less, but rather that the manufacturing and services sectors expand at an even more rapid rate than the mining industry.

Historical analogies are of very limited use in making future projections or in deriving a design for development for the developing countries. In the past, industrialization depended to a large extent on a country having either a local supply of raw materials or an assured source from abroad. The extent of world trade and technological developments now make it possible for a country to industrialize with only a very limited supply of domestic resources. Japan is a recent example of how an economy can become a major industrial power using large quantities of imported raw materials. For those countries with relatively limited mineral resources it may now pay to be a 'late-comer' to economic development.

The pattern of North American and Western European growth does not have to be repeated in the developing regions. Most of these countries are unlikely ever to catch up by attempting to emulate the traditional development pattern of the Western world; a radical approach is required, whereby they can profit rather than be handicapped by their late start in development—much as Germany, with its shattered industrial base after the Second World War was not bound to outdated plant and processes and was therefore able to adopt the most modern and efficient equipment and techniques. Today it is those countries which 'go nuclear' that will have the advantage.

In the industrialized countries, energy requirements are still filled almost entirely by fossil fuels which are essentially limited in supply, unless coal can become an environmentally acceptable fuel. Many of the developing countries which do not yet have a major investment in conventional energy generating plants can avoid the thermal stage of generation to a large extent and establish nuclear plants in the next few decades; as early as the 1950s, Harrison Brown (1954) was advocating this approach for the less-developed world. Even those countries which have abundant sources of fossil fuels would do well to sell petroleum to the industrialized countries while they are still largely committed to thermal power generation, and invest the proceeds from these sales in nuclear-power development. Arrangements whereby the oil-producing countries 'trade' petroleum for technical expertise in nuclear-power plant construction and operation would be mutually advantageous to both the developed and developing countries; other minerals could similarly be traded for technical assistance for economic development. Such an approach could not be implemented quickly; more research into the problems involved is urgently needed.

This approach—which admittedly requires some daring and imaginative development planning in both the developed and developing countries—can be extended. The developing nations can upgrade their mineral raw materials, insisting on more processing of materials before sale for export, when it is economically reasonable for them to do so. They can and should learn from the mistakes of the Western world and avoid the worst consequences of industrialization in the future; the environmental damage now apparent in the developed countries can be avoided, in part, if steps are taken now to design manufacturing plants to minimize pollution. The research under way in the industrialized

countries should be applied to the developing regions too; the sprawling urban blight already reaching crisis proportions in many of the large cities of Asia and Latin America must be controlled before deterioration reaches the levels found in cities like Tokyo.

The social and political means for achieving environmental control in the developing countries depends on a determination to tackle the problem now. The resolve must come from within; the expertise can be imported from the industrialized countries. In a world of increasing pollution and dwindling mineral reserves, the greatest well-being for the most people will probably be fastest achieved by an exchange of raw materials from the developing countries for the enhanced technical and management aid from the developed countries. This would allow the developing countries to 'leap-frog' to material well-being, without the disastrous consequences to the environment experienced in the industrialized countries.

While population growth and the resulting pressure on the world's supply of minerals is ultimately the most pressing and difficult problem facing the world today, the future relations between the industrialized and the non-industrialized countries—and between the 'have' and 'have not' countries of Asia, Latin America and Africa—may become an increasingly severe world problem. To solve this problem there must be a redressing of the current imbalances without condemning some of the non-industrialized countries to the role of raw-material suppliers to a more affluent world, nor depriving the more developed countries or those developing which must import the raw materials of mineral resources.

Conclusions

World demand for minerals will continue to increase at an accelerating rate unless there is a drastic change in world population growth. This demand can be met by:

Vastly increased mineral supplies—which is largely the responsibility of the exploration geologist and the mining industry.

The use of much lower grades of ores—which implies dramatic changes in technology.

More comprehensive recycling of reclaimed materials—which may ultimately require government legislation.

Greater flexibility in trade and aid relations between the industrialized and the less-industrialized countries—which will require new political and economic initiatives.

A reduction in the apparently inexorable acceleration of consumption of minerals—which implies a change in present consumption patterns in the developed countries and enlightened planning in the developing countries to avoid the worst excesses of the industrialized world.

It is essential that planning policy clearly distinguish between reserves and resources. There is an impending shortage of reserves which could reach crisis proportions within the present exploration and exploitation pattern. In terms of resources there should be no long-term deficiencies; there is no absolute shortage of any particular element, provided that new exploration and extraction processes are developed.

The short-term problem is one of providing greatly increased reserves of minerals of grades not dissimilar to those being mined today; the long-term problem is one of developing extractive techniques to handle much lower grade materials in the future and the technology to exploit new sources of minerals (e.g. the oceans and seas) and nuclear energy. The geochemist has much to contribute to the short-term problem, not only by developing new exploration techniques, but also by formulating genetic hypotheses of the origins of ore deposits and the factors important in their localization and distribution (Govett, 1976). Important contributions can also be sought from other Earth scientists—geophysicists, economic geologists, theoretical petrologists and mineralogists. However, the pace of development is not likely to be rapid enough to meet the immediate problems of the next 25 years unless massive national and international finance is available to sponsor research and unless the mining industry is encouraged to finance exploration.

The future contribution of the geologist will be proportionately less as ore grades decrease and ocean mining develops and the roles of the chemical engineer and extractive metallurgist will become progressively more important. Although very-low-grade-ore processing techniques are not necessary immediately, major research efforts to develop them should already be under way. In the normal commercial order of life there is little incentive to develop techniques to process 0.25 per cent copper 'ore' while there is an abundant supply of 0.5 per cent ore; unfortunately, in a world of exponential growth there is too little warning of impending depletion of currently economically mineable grade ore to allow enough time for demand pressures to force the pace of technological development.

Even with the most generous projections of undiscovered mineral deposits and increasing ability to use lower grade ores, there may not be enough reserves of some minerals for the developing countries to achieve anything approaching parity with Western Europe and Japan in the next few decades if consumption also continues to grow in the industrialized world. The argument that these countries should avoid primary production (or that they should leave their ores in the ground to increase in value as world shortages develop) ignores the fact that as mineral-consuming nations develop alternative sources of supply and new technology they will not necessarily require the conventional high-grade reserves so carefully stored up for the future. The developing countries will serve their long-term interests best—and at the same time assist the industrialized countries in their short-term problems—by selling their minerals now for the best price possible and investing the proceeds of these sales in the next phase of industrial technology.

The developing areas can profit by their relatively late start in industrialization by basing their energy requirements on nuclear power rather than on the fossil fuels, thus avoiding the inefficiency and costs of changing to nuclear power later and saving the foreign exchange demanded by the oil-exporting countries for increasingly expensive petroleum. The philosophy of 'go nuclear' can be applied to all phases of development, particularly to the problems of pollution and urban development. Traditional marketing arrangements and foreign investment patterns will have to change; the international mining companies should be encouraged to invest 'expertise' rather than capital in these non-industrialized nations.

While the immediate, and even the long-term problems of adequate mineral supplies can probably be solved by geology, technology and new political and financial approaches to marketing and production, the problems of world population pressure and its impact on the environment will require more radical solutions. Geochemists are now aware of the need for their expertise in the development of environmental sciences. All Earth scientists—indeed, all scientists—will have to cooperate with the politician and the economist in the search for a new approach to man's relation to his world.

References

ABRAMOVITZ, M. 1961. Comment. In: J. J. Spengler (ed.), *Natural Resources and Economic Growth*, p. 9–16. Washington, D.C., Resources for the Future.

AGTERBERG, F. P. 1977. Statistical Methods for Regional Resources Appraisal. *C.I.M. Bull.*, Vol. 70, No. 778, p. 96–8.

AYERS, R. U.; GUTMANIS, I. 1972. Technological Change, Pollution and Treatment Cost Coefficients in Input-Output Analysis. *In:* R. G. Ridker, (ed.), *Population, Resources, and the Environment*, Vol. III, p. 313–37. Washington, D.C., United States Government Printing Office.

BLOCK, D. 1973. *Environmental Aspects of Economic Growth in Less Developed Countries; An Annotated Bibliography.* Paris, OECD. 111 p.

BLONDEL, F.; LASKY, S. G. 1956. Mineral Reserves and Mineral Resources. *Econ. Geol.*, Vol. 51, p. 686–97.

BOULDING, K. 1966. The Economics of the Coming Spaceship Earth. In: H. Jarrett (ed.), *Envrionmental Quality in a Growing Environment*, p. 3–14. Baltimore, Md., Johns Hopkins.

BOWIE, S. H. U. 1976. Whither Uranium? *Trans. I.M.M.*, Vol. 85, p. B163–9.

BOYD, J. 1973. Minerals and How We Use Them. In: E. N. Cameron (ed.), *The Mineral Position of the United States 1975–2000*, p. 1–8. Madison, Wis., University of Wisconsin Press.

BRITISH COLUMBIA, MINISTRY of MINES and PETROLEUM RESOURCES. 1976. *Exploration in British Columbia, 1975.* 225 p.

BROBST, D. A.; PRATT, W. P. (eds.). 1973. *United States Mines Resources.* Washington, D.C., United States Government Printing Office. 722 p.

BROOKS, D. B. 1973. *Minerals: An Expanding or a Dwindling Resource?* Ottawa, Department of Energy, Mines and Resources. 17 p.

BROWN, H. 1954. *The Challenge of Man's Future.* New York, Viking. 290 p.

BROWN, H.; BONNER, J.; WEIR, J. 1957. *The Next Hundred Years.* New York, Viking. 193 p.

COHEN, H. E. 1976. Mineral Supplies for the Future—The Role of Extraction and Processing Technology. In: G. J. S. Govett and M. H. Govett (eds.), *World Mineral Supplies—Assessment and Perspective*, p. 419–37. Amsterdam, Elsevier.

COMMONER, B. 1972. The Environmental Cost of Economic Growth. In: R. G. Ridker (ed.), *Population, Resources and the Environment*, Vol. III, p. 343–63. Washington, D.C., United States Government Printing Office.

CONNELLY, P.; PERLMAN, R. 1975. *The Politics of Scarcity.* London, Oxford University Press. 162 p.

COOPER, M. A. 1977. The Canadian Mineral Industry in a World Context. *C.I.M. Bull.*, Vol. 70, No. 784, p. 72–5.

DARMSTADTER, J. 1971. *Energy in the World Economy.* Baltimore, Md., Johns Hopkins. 876 p.

——. 1972. Energy. In: R. G. Ridker (ed.), *Population, Resources, and the Environment*, Vol. III, p. 103–49. Washington, D.C., United States Government Printing Office.

DE GEOFFROY, J.; WU, S. M. 1970. A Statistical Study of Ore Occurrences in the Greenstone Belts of the Canadian Shield. *Econ. Geol.*, Vol. 65, p. 496–504.

DERRY, D. R.; BOOTH, J. K. B. 1977. *Mineral Discoveries and Exploration Expenditure—A Revised Review 1966–1976.* (Paper presented at the 79th Annual General Meeting of the C.I.M., Ottawa, April 1977).

DERRY, D. R.; MICHENER, C. E.; BOOTH, J. K. B. 1974. World Mining Exploration Trends. *World Mining*, Vol. 27, p. 46–50.

DROLET, J. P. 1977. *Resource Diplomacy Revisited.* (Paper presented at McGill University, Montreal, 14 Feb. 1977).

ECOLOGIST, 1972. *Blueprint for Survival.* Harmondsworth (United Kingdom), Penguin. 139 p.

FABBRI, A. G. 1975. Design and Structure of Geological Data Banks for Regional Mineral Potential Evaluation. *C.I.M. Bull.*, Vol. 68, No. 760, p. 91–8.

FISCHMAN, L. L. 1977. Adequacy of World Mineral Supplies. *C.I.M. Bull.*, Vol. 70, No. 785, p. 75–9.

FISCHMAN, L. L.; LANDSBERG, H. H. 1972. Adequacy of Non-fuel Minerals and Forest Resources. In: R. G. RIDKER, (ed.), *Population, Resources, and the Environment*, Vol. III, p. 79–101. Washington, D.C., United States Government Printing Office.

FLAWN, P. T. 1965. Minerals: A final Harvest or an Endless Crop? *Engng Min. J.*, Vol. 166, p. 106–7.

——. 1973. Impact of Environmental Concerns on the Mineral Industry 1975–2000. In: E. N. Cameron (ed.), *The Mineral Position of the United States 1975–2000*, p. 95–108. Madison, Wis., University of Wisconsin Press.

FRIED, E. R. 1976. International Trade in Raw Materials: Myths and Realities. *Science*, Vol. 191, No. 20, p. 641–6.

FRIED, E. R.; SCHULTZE, C. L. (eds.). 1975. *Higher Oil Prices and the World Economy.* Washingon, D.C., Brookings. 284 p.

GOVETT, G. J. S. 1976. Development of Geochemical Exploration Techniques. In: G. J. S. Govett and M. H. Govett (eds.), *World Mineral Supplies—Assessment and Perspective*, p. 343–76. Amsterdam, Elsevier.

——. 1977. World Mineral Supplies—The Role of Exploration Geochemistry. *J. Geochem. Explor.*, Vol. 8.

GOVETT, G. J. S.; GOVETT, M. H. 1972. Mineral Resource Supplies and the Limits of Economic Growth. *Earth Sci. Rev.*, Vol. 8, p. 275–90.

——. 1973. Mineral Resources and Canadian-American Trade—Doubled-edged Vulnerability. *C.I.M. Bull.*, Vol. 66, p. 66–71.

——. 1974. The Concept and Measurement of Mineral Reserves and Resources. *Resources Policy*, Vol. 1, No. 1, p. 46–56.

——. (eds.). 1976a. *World Mineral Supplies—Assessment and Perspective.* Amsterdam, Elsevier. 472 p.

——. 1976c. The Canadian Minerals Industry: A Perspective. *Resources Policy*, Vol. 2, No. 1, p. 4–10.

——. 1977b. The Inequality of the Distribution of World Mineral Supplies. *C.I.M. Bull.*, Vol. 70, No. 784, p. 59–71.

GOVETT, M. H. 1975. The Geographic Concentration of World Mineral Supplies. *Resources Policy*, Vol. 1, No. 6, p. 357–70.

——. 1976. Geographic Concentration of World Mineral Supplies, Production, and Consumption. In: G. J. S. Govett and M. H. Govett (eds.), *World Mineral Supplies—Assessment and Perspective*, p. 99–145. Amsterdam, Elsevier.

GOVETT, M. H.; GOVETT, G. J. S. 1976b. Defining and Measuring World Mineral Supplies. In: G. J. S. Govett and M. H. Govett (eds.), *World Mineral Supplies—Assessment and Perspective*, p. 13–36. Amsterdam, Elsevier.

——. 1977a. Scarcity of Basic Materials and Fuels—Assessment and Implications. In: D. W. Pearce and I. Walter (eds.), *Resource Conservation: Social and Economic Dimensions of Recycling.* New York, New York University Press.

——. 1978. Geological Supply and Economic Demand—The Unresolved Equation. In: *Proc. Resources Policy Conference '78*, Oxford, March 1978.

GROSSLING, B. F. 1970. Future Mineral Supply. *Econ. Geol.*, Vol. 65, p. 348–54.

GRUNWALD, J. 1964. Resource Aspects of Latin American Economic Development. In: M. Clawson (ed.), *Natural Resources and International Development*, p. 307–26. Baltimore, Md., Johns Hopkins.

HARKIN, D. A. 1976. Mineral Exploration and Technical Co-operation in the Developing Countries. In: G. J. S. Govett and M. H. Govett (eds.), *World Mineral Supplies—Assessment and Perspective*, p. 317–41. Amsterdam, Elsevier.

HART, L. H. 1973. Mineral Science and the Future of Metals. *A.I.M.E. Trans.* Vol. 254, p. 105–10.

HERFINDAHL, O. C. 1959. *Copper Costs and Prices 1870–1957.* Baltimore, Md., Johns Hopkins. 260 p.

HUBBERT, M. K. 1973. Survey of World Energy Resources. *C.I.M. Bull.*, Vol. 66, p. 37–53.

LOVERING, T. S. 1969. Mineral Resources from the Land. In: *Committee on Resources and Man (National Academy of Sciences and National Research Council), Resources and Man*, p. 109–33. San Francisco, Calif., W. H. Freeman.

LOWELL, J. D. 1970. Copper Resources in 1970. *Min. Eng.*, Vol. 22, p. 67–73.

MADDOX, J. 1972. *The Doomsday Syndrome.* London, Macmillan. 248 p.

223

McKELVEY, V. E. 1960. Relation of Reserves of the Elements to their Crustal Abundance. *Am. J. Sci.*, Vol. 258A, p. 234–41.

——. 1972. Mineral Resources Estimates and Public Policy. *Am. Scient.*, Vol. 60, p. 32–40.

——. 1973. Mineral Potential of the United States. In F. N. Cameron (ed.), *The Mineral Position of the United States, 1975–2000*, p. 67–82. Madison, Wis., University of Wisconsin Press.

McLAREN, D. J. 1974. *Address of Welcome. Fifth International Geochemical Exploration Symposium.* Amsterdam, Elsevier.

MALENBAUM, W.; CICHOWSKI, C.; MIRZABAGHERI, F. 1973. *Material Requirements in the United States and Abroad in the Year 2000.* Washington, D.C., National Commission on Materials Policy. 30 p.

MEADOWS, D. H.; MEADOWS, D. L.; RANDERS, J.; BEHRENS, W.W. 1972. *The Limits to Growth.* New York, Universe Books. 205 p.

MORGAN, T. 1963. Trends in Terms of Trade, and their Repercussions on Primary Producers. In R. Harrod; D. Hague (eds.), *International Trade Theory in a Developing World,* p. 52–95. London, Macmillan.

NICHOL, I. 1974. *Presidential Address to Annual Meeting of the Association of Exploration Geochemists. Fifth International Geochemical Exploration Symposium.* Amsterdam, Elsevier.

NOLAN, T. B. 1950. The Search for New Mining Districts. *Econ. Geol.,* Vol. 45, p. 601–8.

——. 1955. The Outlook for the Future—Non-renewable Resources. *Econ. Geol.,* Vol. 50, p. 1–8.

NOWLAN, J. P. 1973. Canadian Mineral Policy. *C.I.M. Bull.,* Vol. 66, p. 15–17.

OVCHINNIKOV, L. N. 1971. Estimate of World Reserves of Metals in Terrestrial Deposits. *Dokl. Akad. Nauk SSSR,* Vol. 196. (American Geological Institute Translation 200–3.)

PAGE, W. 1976. Mining and Development: Are They Compatible in South America? *Resources Policy,* Vol. 2, No. 4, p. 235–46.

PEARCE, D. W.; WALTER, I. (eds.). 1977. *Resource Conservation: Social and Economic Dimensions of Recycling.* New York, New York University Press.

PERLMAN, L. M. 1977. Population Growth and Mineral Demand. *C.I.M. Bull.,* Vol. 70, No. 785, p. 80–4.

PRUITT, R. G. 1966. Mineral Terms—Some Problems in their Use and Definition. *11th Annual Rocky Mountain Mineral Law Institute.* p. 1–34. New York, Matthew Bender.

RIDGE, J. D. 1973. Minerals From Abroad: The Changing Scene. In: E. N. Cameron (ed.), *The Mineral Position of the United States 1975–2000,* p. 127–52. Madison, Wis., University of Wisconsin Press.

——. 1974. Mineral-resource Appraisal and Analysis. *U.S. Geol. Surv. Prof., Paper (921).* p. 12–16.

RIDKER, R. G. 1972. Resource and Environmental Consequences of Population Growth in the United States. The Economy, Resource Requirements, and Pollution Levels. In: R. G. Ridker (ed.), *Population, Resources, and The Environment.* Vol. III, p. 19–57. Washington, D.C., United States Government Printing Office.

ROSCOE, W. E. 1971. Probability of an Exploration Discovery in Canada. *C.I.M. Bull.,* Vol. 64, p. 134–7.

SEKINE, Y. 1963. On the Concept of Concentration of Ore-forming Elements and the Relationship of their Frequency in the Earth's Crust. *Int. Geol. Rev.,* Vol. 5, p. 505–15.

SKINNER, B. J.; BARTON, P. B. 1973. Genesis of Mineral Deposits. *A. Rev. Earth and Planet. Sci.,* Vol. 1, p. 183–211.

SPENGLER, J. J. (ed.). 1961. *Natural Resources and Economic Growth.* Washington, D.C., Resources for the Future. 306 p.

TIMS, W. 1975. The Developing Countries. In: E. R. Fried; C. L. Schultze (eds.), *Higher Oil Prices and the World Economy,* p. 169–96. Washington, D.C., Brookings.

TINBERGEN, J. 1976. *Reshaping the World Order.* New York, E. P. Cutton. 325 p.

TOVAR, O. N. 1973. The New Service Contract for Petroleum Development in Venezuela. *Colo. Sch. Mines Q,* Vol. 168, No. 4, p. 55–71.

UNITED NATIONS. 1950. *The Economic Development of Latin America and its Principal Problems,* by R. Prebisch, New York, United Nations. 59 p.

——. 1972. *World Energy Supplies 1961–1970.* New York, United Nations. 373 p.

——. 1973. *World Energy Supplies 1968–1971.* New York, United Nations. 187 p.

——. 1975. Problems of Availability and Supply of Material Resources. Report by the Secretary-General to the Committee on Natural Resources, Tokyo, 24 March–4 April 1975. (Mimeo.)

——. 1976a. *Statistical Yearbook, 1975.* New York, United Nations, 914 p.

——. 1976b. *The Future of the World Economy.* New York, United Nations.

UNITED NATIONS CONFERENCE ON TRADE AND DEVELOPMENT. 1976. *Handbook of International Trade and Development Statistics.* New York, United Nations. 657 p.

UNITED STATES BUREAU OF MINES. 1970. *Mineral Facts and Problems.* Washington, D.C., United States Government Printing Office. 1291 p.

UNITED STATES COUNCIL ON ENVIRONMENTAL QUALITY. 1973. *Environmental Quality: 4th Annual Report.* Washington, D.C., United States Government Printing Office. 499 p.

UNITED STATES NATIONAL COMMISSION ON MATERIALS POLICY. 1973. *Material Needs and the Environment Today and Tomorrow.* Washington, D.C., United States Government Printing Office.

UNITED STATES PRESIDENT'S MATERIALS POLICY COMMISSION. 1952. *Resources for Freedom.* Washington, D.C., United States Government Printing Office, 5 vol.

UNITED STATES SECRETARY OF THE INTERIOR. 1972. *First Annual Report of the Secretary of the Interior under the Mining and Minerals Policy Act of 1970.* Washington, D.C., United States Government Printing Office, 142 p.

VOGELY, W. A. (ed.). 1975. *Mineral Materials Modelling.* Washington, D.C., Resources for the Future. 404 p.

WAELDE, T. M. 1977. Lifting the Veil from Transnational Contracts. *Natural Resources Forum,* Vol. 1, p. 277–84.

WANG, F. F. H.; McKELVEY, V. E. 1976. Marine Mineral Resources. In: G. J. S. Govett and M. H. Govett (eds.), *World Mineral Supplies—Assessment and Perspective,* p. 221-86. Amsterdam, Elsevier.

WEBB, J. S. 1975. Environmental Problems and the Exploration Geochemist. In: I. L. Elliott; W. K. Fletcher (eds.), *Geochemical Exploration, 1974,* p. 5–17. Amsterdam, Elsevier.

ZIMMERMANN, E. W. 1933. *World Resources and Industries.* New York, Harper. 832 p.

ZWARTENDYK, J. 1972. *What is 'Mineral Endowment' and How Should We Measure It?* Ottawa, Department of Energy, Mines and Resources. 17 p.

——. 1974. The Life Index of Mineral Reserves—A Statistical Mirage. *C.I.M. Bull.,* Vol. 67, p. 67–71.

Geochemical aspects of the origins of ore deposits

James R. Craig

Department of Geological Sciences,
Virginia Polytechnic Institute
and State University, Blacksburg,
Virginia (United States of America)

Introduction

The Earth consists of a large, virtually closed, chemical system in which approximately ninety elements serve as active participants. However, within the geologic realm of the Earth's crust only a few chemical elements, notably O, Si, Al, Fe, Ca, Mg, Na and K, in order of decreasing abundance, constitute more than 99 per cent of all rocks. The remainder of the elements, including most of economic consequence, are widely

TABLE 1. Concentrations of common ore metals in igneous and sedimentary rocks (in p.p.m.)[1]

	Earth's crust A	Ultramafic TW	Ultramafic V	Mafic TW	Mafic V	Inter-mediate V	Grano-diorite V	Granite High Ca TW	Granite Low Ca TW	Syenite TW	Clay + shale V	Shale TW	Sand-stone TW	Car-bonate TW
Ag	0.07	0.06	0.05	0.11	0.1	0.07	0.05	0.05	0.04	0.0X	0.1	0.07	0.0X	0.0X
Al	82,300	20,000	4,500	78,000	87,600	88,500	7,700	82,000	72,000	88,000	104,500	80,000	25,000	4,200
As	1.8	1	0.5	2	2	2.4	1.5	1.9	1.5	1.4	6.6	13	1	1
Au	0.004	0.006	0.005	0.004	0.004	—	0.005	0.004	0.004	0.00X	0.001	0.00X	0.00X	0.00X
Bi	0.17	—	0.001	0.007	0.007	0.01	0.01	—	0.01	—	0.01	—	—	—
Cd	0.2	0.X	0.05	0.22	0.19	—	0.1	0.13	0.13	0.13	0.3	0.3	0.0X	0.04
Co	25	150	200	48	45	10	5	7	1	1	20	19	0.3	0.1
Cr	100	1,600	2,000	170	200	50	25	22	4.1	2	100	90	35	11
Cu	55	10	20	87	100	35	20	30	10	5	57	45	X	4
Fe	56,000	94,300	98,500	86,500	85,600	58,500	27,000	29,600	14,200	36,700	33,300	47,200	9,800	3,800
Hg	0.08	0.0X	0.01	0.09	0.09	—	0.08	0.08	0.08	0.0X	0.4	0.4	0.03	0.04
Mn	950	1,620	1,500	1,500	2,000	1,200	600	540	390	850	670	850	X0	1,100
Mo	1.5	0.3	0.2	1.5	1.4	0.9	1	1	1.3	0.6	2	2.6	0.2	0.4
Ni	75	2,000	2,000	130	160	55	8	15	4.5	4	95	68	2	20
Pb	12.5	1	0.1	6	8	15	20	15	19	12	20	20	7	9
Pd	0.01	0.12	0.12	0.02	0.02	—	0.01	0.00X	—	—	—	—	—	—
Pt	0.005	—	0.2	—	0.1	—	—	—	—	—	—	—	—	—
S	260	300	100	300	300	200	400	300	300	300	3,000	2,400	240	1,200
Sb	0.2	0.1	0.1	0.2	1	0.2	0.26	0.2	0.2	0.X	2	1.5	0.0X	0.2
Sn	2	0.5	0.5	1.5	1.5	—	3	1.5	3	X	10	6	0.0X	0.0X
Ti	5,700	300	300	13,800	9,000	8,000	2,300	3,400	1,200	3,500	4,500	4,600	1,500	400
U	2.7	0.001	0.003	1	0.5	1.8	3.5	3	3	3	3.2	3.7	0.45	2.2
V	135	40	40	250	200	100	40	88	44	30	130	130	20	20
W	1.5	0.77	0.1	0.7	1	1	1.5	1.3	2.2	1.3	2	1.8	1.5	0.6
Zn	70	50	30	105	130	72	60	60	39	130	80	95	16	20

1. X = 1–9 p.p.m., 0.X = 0.1–0.9 p.p.m., 0.0X = 0.01–0.09 p.p.m.
Source: From Ahrens, 1965 (A); Turekian and Wedepohl, 1961 (TW); Vinogradov, 1962 (V).

dispersed in trace amounts, generally with abundances of only a few tens of parts per million or less (Table 1). The development of economically exploitable concentrations of various elements into 'ore bodies' thus requires enrichment of the elements by one or more geological and/or geochemical processes. These processes are as diverse as the geologic realm within which they work; the present chapter represents an attempt to present a concise and up-to-date view of these processes, where and how they act, and their limitations.

It is important to recognize that the concept of an 'ore body' is determined by more than geological factors. In spite of the obvious importance of mineralogy, tenor or concentration and structural setting, these factors are often overshadowed by such non-geological parameters as market value, taxation, transportation costs, mining and milling technology, ownership and politics. Although the economic aspects of ores or potential ores may vary abruptly on a daily basis, this chapter will be devoted to the more slowly changing parameters of the geologic realm. The conciseness required here necessitates that many topics will be incompletely discussed; accordingly the reader is referred to the following references which contain more extensive discussions: Barnes (1967), Barton (1970), Skinner and Barton (1973), Park and MacDiarmid (1970), Stanton (1972), Wilson (1968), Ridge (1968) and in various chapters of Wedepohl (1969). Additional references to works on specific aspects of ore genesis will be given throughout this chapter.

There are innumerable approaches which can be used to consider the diverse origins of the world's ore deposits. The geochemical processes involved in ore genesis will be treated here as they operate to form ores of magmatic, sedimentary and metamorphic affiliations. Each of these is subsequently divided according to commonly observed ore types. Such a classification is to some extent inadequate because one is attempting to place arbitrary boundaries on processes which actually represent a broad continuum. The entire topic of ore genesis, in spite of countless man-years of research, remains in many ways enigmatic. In spite of the recognition of the major influences and processes, the deciphering of the fine details of ore formation remains to be accomplished.

General geochemical affinities

The various occurrences of the metals which constitute ore deposits depend on a number of geochemical parameters—terrestrial abundance, atomic and ionic size, electronegativities, ionization potentials, bond type, relative stabilities of their compounds, etc.

Since virtually all ores exist concentrated as sulphide and/or oxide masses in host rocks composed chiefly of silicates and/or carbonates, the processes we

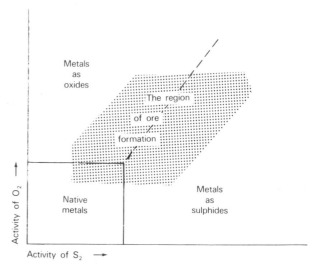

FIG. 1. Schematic diagram of some primary ore types in terms of the activities of oxygen and sulphur. Under conditions of high oxygen activity the metals will exist as oxides; high sulphur activity results in formation of sulphides. Because the boundaries between oxide, sulphide and native metals are different for each element, many ores consist of mixtures of various metal oxides and sulphides. Secondary reactions such as oxidation of sulphides or reaction with carbonate ions can lead to formation of sulphates, carbonates, etc.

must consider are those that determine whether an element will be broadly dispersed and inaccessible in silicate or carbonate mineral solid solutions, or in some manner concentrated as native elements, oxides and/or sulphides (Fig. 1). This approximates to an examination of the affinities of the elements for oxygen (in the form of silicates or oxides) and for sulphur (in the form of sulphides). Goldschmidt's (1937) simplified classification of terrestrial elements as siderophile (tending to be associated with metallic iron), chalcophile (tending to be bound with sulphur), or lithophile (tending to be bound to oxygen) is still useful but must, as pointed out by Burns and Fyfe (1966), be used with caution. Basically, however, exploration for the metal ores is a search for areas in which useful and economically valuable elements have been concentrated either by their introduction or their preferential enrichment as other elements are removed.

In an attempt to maximize the usefulness of this chapter to those who do not already have a moderate background knowledge of ore deposits, the metallic elements to be discussed are itemized and briefly annotated in Table 2.

Ores of magmatic affiliation

Igneous activity has long been recognized as an integral part of some ore-forming processes and has directly or indirectly been accounted as responsible for the exis-

TABLE 2. Ore metals, their mineralogy and modes of occurrence

Ore metal	Mineralogy	Mode of occurrence
Aluminium	Bauxite (mixture of): gibbsite, $Al_2O_3 \cdot 3\,H_2O$; boehmite, $Al_2O_3 \cdot H_2O$; diaspore, $Al_2O_3 \cdot H_2O$	Deeply weathered tropical (or previously tropical) soils
Antimony	Stibnite Sb_2S_3; tetrahedrite, $Cu_{12}Sb_4S_{13}$	Low-temperature hydrothermal deposits
Arsenic	Realgar, AsS; orpiment, As_2S_3; arsenopyrite, FeAsS; enargite, Cu_3AsS_4	Wide variety of hydrothermal veins and replacements
Bismuth	Bismuth, Bi; Bismuthinite, Bi_2S_3	Hydrothermal ores commonly with tin when associated with granitic bodies or with Co, Ni, As, Ag when associated with mafic bodies
Cadmium	Cadmium sphalerite, (Zn, Cd)S	Low-temperature hydrothermal stratiform ores
Chromium	Chromite $(Mg, Fe)O \cdot (Cr, Al, Fe)_2O_3$	Cumulate layers in layered ultramafics
Cobalt	Cobaltite, CoAsS; skutterudite, $(Co, Ni, Fe)As_{3-x}$; linnaeite, Co_3S_4	Hydrothermal veins, reduzate sediments
	Pentlandite, $(Fe, Ni, CO)_9S_8$	Immisible Fe–Ni rich sulphide melts separated from ultrabasic magmas
Copper	Copper, Cu; chalcocite, Cu_2S; chalcopyrite, $CuFeS_2$; digenite, Cu_9S_5; covellite, CuS; bornite, Cu_5FeS_4	Broadly distributed in many associations—minor constituent of immiscible Fe–Ni sulphide melts in ultrabasic rocks—many hydrothermal ores especially associated with intermediate to acid porphyry bodies—in black shales, red beds, volcanogenic sequences—supergene zones
Gold	Gold, Au; petzite, Ag_3AuTe_2; calaverite, $AuTe_2$; sylvanite, $AuAgTe_4$; krennerite, $(Au, Ag)Te_2$	Hydrothermal veins, pegmatites, placer deposits
Iron	Magnetite, Fe_3O_4	Massive segregations in anorthosites—magmatic injections—contact metasomatic deposits—high-grade metamorphic zones (?)
	Hematite, Fe_2O_3, 'limonite' Fe-hydroxides', siderite, $FeCo_3$	Closed marine (?) basin sediments (especially Pre-Cambrian)
Lead	Galena, PbS	Hydrothermal veins and replacements
Manganese	Pyrolusite, MnO_2; cryptomelane, $K(Mn)_8O_{16}$; psilomelane, (Mn oxides)	Secondary in highly weathered zones frequently with iron—marshes and lakes—ocean floor nodules
Mercury	Mercury, Hg; cinnabar, HgS; metacinnabar, HgS	Low temperature hydrothermal, hot springs and associated volcanics
	Sphalerite, (Zn, Hg)S	Trace quantities in hydrothermal ores
Molybdenum	MoS_2	Disseminated in intermediate porphyritic rocks
Nickel	Pentlandite, $(Fe, Ni)_9S_8$	Immiscible Fe–Ni sulphide melts separated from mafic magmas
	Violarite, $FeNi_2S_4$	Supergene zones of pyrrhotite–pentlandite ores
	Garnierite, $(Mg, Fe, Ni)_3Si_2O_5(OH)_4$	Lateritic deposits formed on serpentines
Platinum group	Native metals, (Pt, Pd, Ir, Os, Ru, Rh) and alloys; cooperite, PtAsS; laurite, $(Ru, Os)S_2$; sperrylite, $(Pt, Ir)As_2$	Cumulates in ultramafic rocks; placer-deposits
Silver	Silver, Ag; argentite, Ag_2S; tennantite, $(Cu, Fe, Ag)_{12}(As, Sb)_4S_{13}$; galena, $Pb(AgBi)S_2$	Hydrothermal veins; supergene zones
Tin	Cassiterite, SnO_2; stannite, Cu_2FeSnS_4	Hydrothermal veins; placer deposits
Titanium	Rutile, TiO_2; ilmenite, $FeTiO_2$	Subsilicic rocks, especially anorthosites; placers
Tungsten	Wolframite, $(Fe, Mn)WO_4$; scheelite, $(CaWO_4)$	Late magmatic hydrothermal veins generally with Sn- and Mo-contact metasomatic ores
Uranium	Uraninite (pitchblende), UO_2; coffinite, $(USiO_4)_{1-x}(OH)_{4x}$	Hydrothermal veins and sandstone 'roll-type' deposits

TABLE 2 (*continued*)

Ore metal	Mineralogy	Mode of occurrence
Vanadium	Roscoelite, $K(V, Al)_3Si_3O_{10}(OH)_2$; montroseite, $(V, Fe)O \cdot OH$	Sandstone 'roll-type' deposits
	Vanidiferous spinel, $(V, Fe)_3O_4$	Layered intrusives
Zinc	Sphalerite, ZnS; wurtzite, ZnS; zincite, ZnO	Hydrothermal veins stratabound ores
	Willemite, Zn_2SiO_4; franklinite, $(Fe, Zn, Mn)(Fe, Mn)_2O_4$	High-grade metamorphic deposits

tence of most economically viable deposits now known. Our present state of knowledge of magmatism and related activities now permits an arbitrary but convenient segregation of ore-forming igneous processes into three groups based on the general nature of the parent magma: (a) mafic and ultramafic rocks; (b) felsic rocks, and (c) acid rocks and hydrothermal fluids.

ORES ASSOCIATED WITH MAFIC AND ULTRAMAFIC ROCKS

The ore deposits associated with mafic and ultramafic igneous rocks are recognized to be of two general types: (a) cumulates forming during normal crystallization and (b) oxide and/or sulphide masses which have segregated from silicate magmas as separate immiscible melts.

Cumulate ores

Cumulate deposits, generally comprising stratified accumulations of chromite, ilmenite and/or magnetite and (rarely) platinoids are known from layered and alpine mafic and ultramafic intrusions around the world: Muskox intrusion, Bushveldt complex, Stillwater complex, and the Great Dike (Wager and Brown, 1967; Wyllie, 1967; Wilson, 1968).

Iron, primarily in the Fe^{2+} state, constitutes one of the essential elements in the formation of the silicate and oxide minerals in mafic and ultramafic rocks. The classic work of Osborne (1959) demonstrated that the activity of oxygen at the time of crystallization of mafic magmas dictated whether the iron present would crystallize in the Fe^{2+} state incorporated in olivenes and pyroxenes or whether it would be raised to the Fe^{3+} state and crystallize as the iron spinel magnetite (Fe_3O_4). Magnetic cumulates of this type, though widespread, do not generally constitute economic deposits because ore grade is not sufficiently high; however, the incorporation of other elements such as chromium may change this picture. Chromium, occurring in nature principally as the Cr^{3+} ion (0.62 Å),[1] readily substitutes in silicate structures, especially clinopyroxenes for Al^{3+} (0.53 Å), or in spinel structures as chromite $(Mg, Fe)O \cdot (Cr, Al, Fe)_2O_3$ for trivalent iron, Fe^{3+} (0.65 Å). The separation of chromium into the early

crystallized mafic rocks in layered intrusives is well demonstrated in the Muskox intrusion where in the following concentrations are observed (Smith, 1972): chilled marginal gabbro, 110–1,000 p.p.m. Cr; lower ultramafic rocks, 3,000–5,000 p.p.m. Cr; upper gabbros, 225–1,000 p.p.m. Cr; and granophyric rocks, 25 p.p.m. Cr.

Typical chromite occurrences consists of alternating discontinuous thin layers of olivine, olivine and chromite, olivine and orthopyroxene, and orthopyroxene and clinopyroxene. Chromite layers rarely exceed 10 or 20 cm but the individual silicate bands may reach tens of metres in thickness.

Laboratory studies on crystallization conditions of chromite from mafic magmas (Ulmer, 1968) and field observations (i.e. Cameron and Desborough, 1968, and Jackson, 1968) have mutually concluded that the principal factor in causing the development of chromite bands is variation in oxygen activity in the melt. The cyclic intimate banding of chromite layers with olivine and ortho-pyroxene layers on the centimetre scale suggests a pulsing of oxygen into the magma, possibly from assimilation of foreign rock, by multiple intrusion, by incorporation of water, or by loss of hydrogen. With each pulse of increased oxygen activity, chromite begins to crystallize and settle; as the oxygen activity reverts to its initial level, precipitation of chromium as the spinel ceases and crystallization reverts to olivine and orthopyroxene ($\pm$ plagioclase). If chromite crystallization has occurred near the top of a chamber, there may be partial resorption as it settles through less oxygen-rich lower zones. Both experimental and field observation showed that an increase in oxygen activity during chromite crystallization resulted in increased Mg^{2+}:Fe^{2+} ratio in the chromite and the co-precipitated olivines (Ulmer, 1968; Jackson, 1968). The regularity of this variation has led Jackson (1968) to suggest that an olivene–chromite geothermometer might be established when sufficient thermodynamic data are available. The actual accumulation and distribution of chromite layers in ultramafic sequences appears also to be a function of

---

1. The ionic radii used throughout this chapter are the 'effective ionic radii' of Shannon and Prewitt (1969) which are based on r ($^{VI}O^{-2}$) = 1.40 Å.

Stage   1   2   3   4   5   6

Minerals forming: 98 per cent olivine + 2 per cent chromite; 100 per cent orthopyroxene; 25 per cent orthopyroxene + 75 per cent clinopyroxene

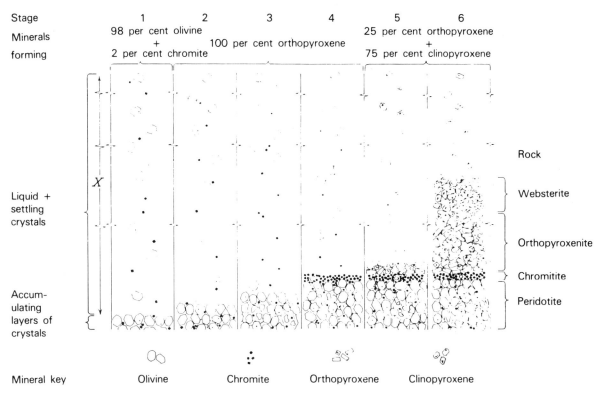

FIG. 2. A model for the origin of chromite-rich layers based on crystallization relations and apparent settling velocities of cumulus minerals in layered ultramafic intrusives (from Irvine and Smith, 1968).

thermal currents in the magma chamber (as attested to by scour channels and cross-bedding) and the differences in settling rates of chromite, olivine and pyroxene (Irvine and Smith, 1968). A schematic model incorporating these considerations is presented in Figure 2.

A most unique occurrence of an economic chromite cumulate is the platinum-rich zone of the Bushveld intrusion, known as the Merensky Reef Unit. The reef, which extends 250 km, constitutes a macrorhythmically banded sequence of pyroxenitic to anorthositic rocks some 175 m thick. Of five chromite-bearing horizons, two (occurring at the base of the Merensky and Pseudo Reefs) are platinum bearing. The actual ore-bearing horizon or 'reef' varies from about 1 to 5 m in thickness. The ore minerals present include ferroplatinum (FePt), nickeliferous braggite ((Ni, Pt, Pd)S$_2$), cooperite (PtAsS), laurite ((Ru, Os)S$_2$), stibiopalladinite (Pd$_3$Sb) and sperrylite (Pt, Ir)As$_2$; other minerals include gold, pentlandite, chalcopyrite, pyrrhotite and vallerite. The overall percentage proportions of the platinum group elements and gold are: Pt, 60; Pd, 27; Ru, 5; Rh, 2.7; Os, 0.6, and Au, 4. Hess (1960) and Cousins (1968) have suggested that the platinum-rich zone represents the first cumulates (possibly from an immiscible sulphide liquid from a fresh surge of basic magma). Wager and Brown (1967), while willing to accept some sort of immiscible sulphide liquid as the source of the platinum minerals, felt that the Merensky Reef Unit rocks, because of their gradational nature with rocks above and below, were merely part of the overall crystallizing intrusive.

Immiscible sulphide and oxide ores

The immiscibility of silicate and sulphide or sulphide-oxide melts has long been recognized in the smelting of numerous kinds of ores. The possibility of this immiscibility serving as a generative mechanism for some ores in nature was first seriously considered by Vogt (1894) and subsequently by many others. It is now accepted that many of the massive Ni-Cu sulphide ores that lie at the base of differentiated mafic and ultramafic masses (e.g. Sudbury, Ontario; Lynn Lake, Manitoba; Kambalda, Australia; Insizwa, South Africa; Norilsk, U.S.S.R.) crystallized from sulphide liquids which separated from their host silicate melt before, during or after emplacement.

The first unequivocal confirmation of natural immiscible sulphide and silicate melts came from a detailed study of the crystallization of the Alae lava lake in Hawaii. Skinner and Peck (1968) found that at an advanced state of crystallization, small globules of a sulphide-rich liquid segregated from the basaltic melt. The approximate composition of the globules is shown in Table 3.

229

James R. Craig

TABLE 3. Approximate composition of globules

Element	Weight percentage	Minerals	Volume percentage
Fe	61	Pyrrhotite	75
Cu	4	Magnetite	15
Ni	<1	Chalcopyrite	10
S	31	Bornite	Trace
O	4	Pentlandite	Trace

Subsequently, similar drop-like structures of sulphides have been observed in lunar basalts (Roedder and Weiblen, 1970) and in eclogite nodules from Kimberlite pipes (Desborough and Czamanske, 1973).

The development of our understanding of the geochemistry of sulphur in silicate melts was derived simultaneously from geological investigations (Ol'Shansky, 1951; Yazawa and Kameda, 1953) and metallurgical studies (Hilty and Crafts, 1952; Fincham and Richardson, 1954; Richardson and Fincham, 1954; Taylor and Stabo, 1954; Abraham et al., 1960; Dewing and Richardson, 1960; Nagamori and Kameda, 1965). Fincham and Richardson (1954) demonstrated that the only ox-

idation states of sulphur which must be considered in silicate melts were the sulphide (at low oxygen activities) and sulphate (at high oxygen activities). Basaltic magmas of the type associated with massive sulphide ores have extremely low oxygen activities (Sato and Wright, 1966); hence the sulphur is present in the sulphide form. The actual amount of sulphur which can dissolve in a silicate melt (and analogous magmas) is primarily a function of the FeO content and, to a much lesser degree, of the CaO, MgO and Na_2O contents. Fincham and Richardson (1954) and subsequent studies have shown that, just as in chromite precipitation as previously discussed, an increase in oxygen activity oxidized iron, thus lowered the FeO content of the melt, reduced the sulphide solubility, and heralded the possible development of an immiscible sulphide liquid. The large size of the sulphide ion (1.84 Å) relative to oxygen (1.40 Å) prohibits its incorporation into silicate structures in the subsolidus region. Haughton et al. (1974) noted that at a constant ratio of $f_{o_2}:f_{s_2}$, an increase in temperature of 100 °C increased the capacity of mafic magmas to dissolve sulphur by 5 to 7 times.

In recent years our knowledge of the limits of sulphur solubility, the primary crystallization fields, and the crystallization paths, in immiscible liquids and in the sulphide-rich liquid once it segregates have been ex-

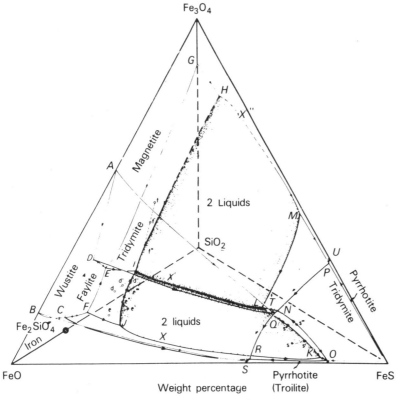

FIG. 3. Liquidus relations in the FeS–FeO–Fe_3O_4–SiO_2 system. The stippled portion is a region of liquid immiscibility between a silicate-rich melt and a sulphide-rich melt (after MacLean, 1969).

tended by the studies of Naldrett (1969), MacLean (1969), and Shamazaki and Clark (1973).

MacLean (1969) found that there existed a large volume of liquid immiscibility in the interior of the FeO–Fe$_3$O$_4$–SiO$_2$–FeS system (Fig. 3). He demonstrated that a homogeneous silicate melt containing small amounts of sulphur might, during crystallization, become saturated in sulphide, thus segregating a separate sulphide liquid. The crystallization path followed depended on the initial composition and whether there had been a decreasing, constant or increasing oxygen activity during crystallization. In the FeO–Fe$_3$O$_4$–SiO$_2$–FeS system, the minimum temperature at which two liquids coexist is 1,140 °C. The amount of sulphur that can be dissolved in this simplified basic magma is about 4 per cent and the amount of silica in the sulphide oxide liquid is about 1 per cent. Crystallization of sulphide minerals from this liquid does not begin until crystallization of the silicates is complete. The presence of Na$_2$O, however, depresses the melting-point of the silicate liquid such that the sulphides may crystallize before the silicates (Shimazaki and Clark, 1973).

Natural basaltic liquids contain only on the order of 0.03 per cent by weight sulphur (MacLean, 1969; Skin-

ner and Peck, 1968), or 100 times less than the maximum value found in the FeO–Fe$_3$O$_4$–SiO$_2$–FeS system. Apparently, the much lower contant of FeO in natural magmas as compared with the synthetic melt, the large amounts of CaO, MgO and Al$_2$O$_3$, and the incorporation of the FeO in silicates, all greatly reduce the activity of FeO. The reduced FeO activity would reduce the solubility of sulphur in the melt, as observed by Fincham and Richardson (1954).

Once a sulphide melt has segregated from a silicate melt, the globules will begin to settle in a silicate melt because of their much greater density. Obviously, segregation of sulphide-rich melts into massive deposits such as those at Sudbury required that the silicate mass was still largely fluid at the time of segregation. The cause of immiscible sulphide liquid segregation in a basic magma could be decreased temperature, normal crystallization, or increased oxygen activity. Regardless of the mechanism by which the sulphide-oxide liquid is generated, once it has formed its crystallization path it is independent of the silicate melt. The initial composition of typical sulphide-oxide melts are plotted in Figure 4 which also illustrates the phase relations in a portion of the Fe–S–O system. These melts lie in the primary crys-

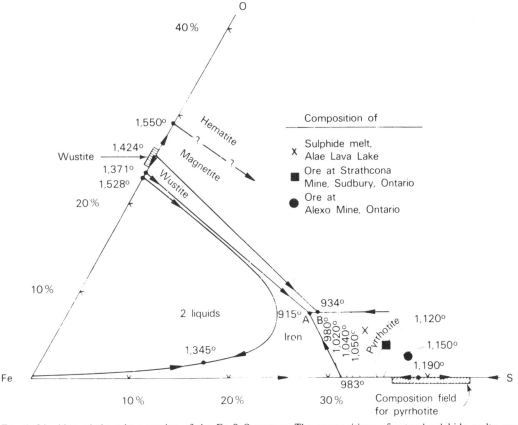

FIG. 4. Liquidus relations in a portion of the Fe–S–O system. The compositions of natural sulphide melts, separated immiscibly from their host silicate melts, are indicated. Initial crystallization of pyrrhotite results in change of the remaining liquid towards the pyrrhotite and magnetite cotectic whereupon both minerals crystallize. (After Naldrett, 1969, and Skinner and Barton, 1973.)

231

tallization field of pyrrhotite; this is compatible with Skinner's and Peck's (1968) observation of simultaneous separation of pyrrhotite and an Fe–S–O liquid. The Fe–S–O liquid ultimately crystallizes to a mixture of magnetite and pyrrhotite as is found in many of the world's massive sulphide ores.

Segregation of purely magnetite–pyrrhotite masses would result in formation of few ore deposits. The principal value of ores such as those at Sudbury lies in the presence of a few per cent nickel and copper. Observations of the Sudbury-type ores and their host indicates that nickel, copper, cobalt, and the platinoid metals are strongly partitioned towards the sulphide melt (other metals such as Pb, Zn and Sr are retained in the silicates). Nickel and—to a lesser degree—copper are readily taken into silicate structures (orthopyroxene and

olivine) as substitutes for iron ($Fe^{2+} = 0.77$ Å, $Ni^{2+} = 0.70$ Å, $Cu^{2+} = 0.73$ Å) in the early stages of crystallization; in fact, the Ni–O bond is stronger than the Fe–O bond. Skinner and Barton (1973) pointed out that the nickel content of an immiscible melt thus reflected the extent to which the parent magma crystallized before segregation of the sulphide melt. The greater the extent of silicate crystallization prior to separation of the sulphide melt and the higher oxygen activity during that time, the greater the tendency for Ni to be incorporated into silicate minerals and the lesser the tendency for Ni to be incorporated into the sulphide melt. This is diagramatically illustrated in Figure 5. Cobalt exhibits behaviour very similar to that of nickel, whereas copper and the platinoid metals are much more chalcophilic than nickel and are more readily partitioned into the sulphide melt. Major problems regarding the genesis of immiscible sulphide melts from mafic magmas include the nature and distribution of sulphur and the ore metals in the mantle.

ORES ASSOCIATED WITH FELSIC ROCKS

The felsic igneous rocks, varying over the broad compositional range from carbonatites and anorthosite to granodiorite, serve as host and apparent source for three major types of ores: (a) carbonatites—rare-earth concentrations; (b) anorthosites—iron-titanium bodies, and (c) quartz monzonite—granodiorite—porphyry Cu–Mo deposits.

Carbonatites

Few types of rocks have been the source of more conjecture than the carbonatites (Heinrich, 1966; Tuttle and Gittins, 1966). In terms of ores, the carbonatites are unique with locally intense concentration of: (a) niobium—primarily in the form of pyrochlore (Ca, Na, Cl)(Nb, Ti, Ta)$_2$(O, OH, F)$_7$ wherein the Nb^{5+} (0.64 Å) substitutes for Ti^{4+} (0.61 Å); (b) rare earths (RE)—primarily in the form of bastnaesite ((Ce, La)CO$_3$F) which at Mountain Pass, California, contains La$_2$O$_3$—29.6 per cent; CeO$_2$—50.3 per cent; Nd$_2$O$_3$—14.3 per cent; Pr$_6$O$_{11}$—4.4 per cent; Sm$_2$O$_3$—1.3 per cent, and occasionally as monazite (Ce, La) PO$_4$ and cerian goyazite [Sr(Al, Ce)$_3$(PO$_4$)$_2$(OH)$_5$H$_2$O]; (c) uranium and thorium—present in uneconomic amounts of thorium bearing pyrochlore and uranothorianite ((Th, U)O$_2$ which have proved useful in locating carbonatites; and (d) sulphides—large, low-grade concentrations of Cu, Fe, Pb, and Zn sulphides usually in uneconomic quantities (<1 per cent).

The greatest economic interest has centred on the niobium and the rare earths. Niobium is concentrated in late stages of magmatic crystallization appearing in granite pegmatites and nepheline syenites in concentrations of 100–300 p.p.m. and in carbonatites up to nearly 1 per cent (the great Oka deposit in Quebec contains

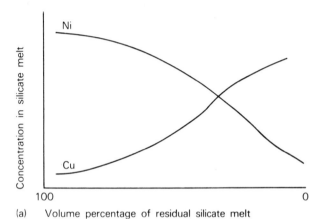

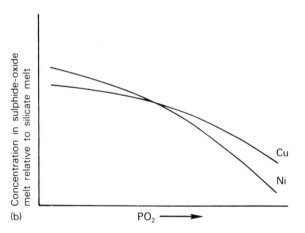

Fig. 5. (a) Schematic representation of the fractionation of Ni and Cu between crystallizing silicate minerals and residual silicate melt. Nickel tends to be incorporated in silicate minerals whereas copper is strongly concentrated in the melt; thus the Ni:Cu ratio of a separating immiscible oxide–sulphide melt depends upon the degree of crystallization of the magma up to that time; (b) Schematic diagram of the effects of oxygen pressure on the Ni:Cu ratio of an immiscible sulphide melt separating from a basis magma. (After Skinner and Barton, 1973. © 1973 Annual Reviews Inc. Reproduced with permission.)

225 million tons of 0.25–0.6 per cent Nb in pyrochlore). This behaviour has led Gornitz and Warde (1972) to conclude that niobium is not present in magmas as free positive ions but rather as complex oxy-anions. Otherwise, such a rare ion in a high oxidation state would be expected to enter a crystal structure during early stages of crystallization in place of a common low-valency ion.

The rare earths occur in most rocks only in trace quantities but show a distinct enrichment during magmatic crystallization and are currently used as an index of differentiation (Schilling and Winchester, 1967). The large size of the tetravalent rare-earth ions (1.14–0.85 Å) limits their substitutions for major cations to Ca^{2+} (1.00 Å). Their concentrations in late magmatic liquids apparently result from the more covalent nature of RE–O bonds than Ca–O bonds and the difficulty of charge balance for tetravalent ions in eightfold co-ordination. As noted above, generally the rare earths are accompanied by high concentrations of other large, highly charged cations (e.g. Th^{4+}, U^{4+}, Zr^{4+}, Hf^{4+}, Nb^{5+}).

Anorthosites

Anorthositic masses occur in two types of geological settings: extensive thin sheets in basic layered complexes (as noted in the proximity of the Merensky Reef Unit of the Bushveldt) and plutonic masses, located in Pre-Cambrian gneissic terrains. Concentrations of iron-titanium oxides are associated with both types of anorthosites but reach significant economic size only in the plutonic masses. Typical of this latter type of occurrence are the Lac Tio deposit at Allard Lake, Quebec, and the Tahawus deposit in the Adirondack mountains. The ores commonly include: magnetite (Fe_3O_4), haematite (Fe_2O_3), maghemite ($\gamma - Fe_2O_3$), ilmenite ($FeTiO_3$), ulvöspinel (Fe_2TiO_4), rutile (TiO_2), spinel ($MgAl_2O_4$), minor sulphides and apatite ($Ca_5(PO_4)_3(F, Cl, OH)$).

Titanium—with ionic sizes of 0.90 Å (Ti^{2+}) and 0.68 Å (Ti^{4+})—substitutes only to a minor extent for Al^{3+} (0.53 Å), Si^{4+} (0.41 Å), Ca^{2+} (1.00 Å), Mg^{2+} (0.72 Å), and Fe^{3+} (0.65 Å) in silicates such as amphiboles, pyroxenes and biotites. In contrast, extensive substitution of titanium for iron occurs in oxides (e.g. complete solid solution from Fe_3O_4 to Fe_2TiO_4, from Fe_2O_3 to $FeTiO_3$ and from Fe_2TiO_5 to $FeTi_2O_5$, Figure 6(a)). Buddington and Lindsley (1964) found that the compositions of coexisting magnetite ($mt_{90}usp_{10}$ to $mt_{20}usp_{80}$) and ilmenite ($hm_{15}ilm_{85}$ to hm_3ilm_{97}) provided a unique solution to the temperature and oxygen fugacity at the time of crystallization (Fig. 6(b)). They applied this 'geothermometer' to various natural magnetite-ilmenite pairs and estimated temperatures of equilibration at 600–800 °C and oxygen fugacities at $10^{-13.6}$–$10^{-18.5}$ atm.

The most broadly accepted origin for these ores has been that of residual liquid segregation as proposed by Bateman (1951) and shown schematically in Figure 7. He summarized the process in the following manner.

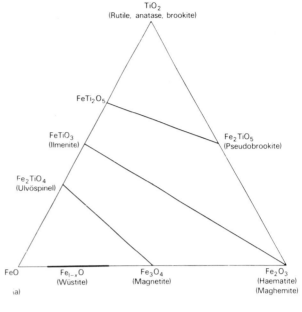

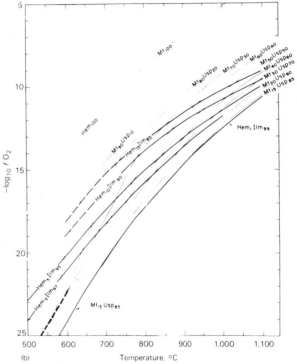

FIG. 6. (a) Phases in the FeO–TiO–Fe_2O_3 system; (b) fO_2–T projection of conjugate fO_2–T–X surfaces of coexisting ulvöspine–magnetite and haematite–ilmenite solid solutions (from Buddington and Lindsley, 1964).

With progressive crystallization and enrichment of the residual liquid, its density would come to exceed that of the silicate crystals and its composition might reach the point of being chiefly oxides of iron and titanium .... At this stage three possibilities may occur: final freezing may occur to yield a basic

James R. Craig

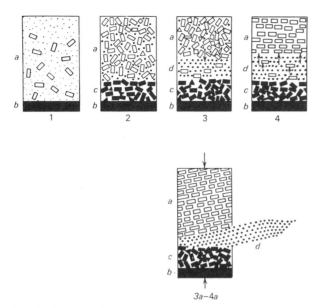

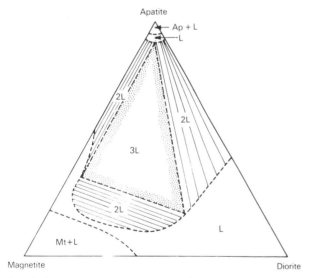

FIG. 7. Idealized diagrammatic representation of late gravitative liquid accumulation of magnetite; 1. Initial crystallization of basic magma *a* after formation of a chill zone *b*; 2. development of a layer *c* of ferromagnesian minerals; 3. mobile, oxide-rich liquid drains down to layer *d*; 4. crystallization of silicate crystals over concordant oxide-rich liquid; *3a–4a*. Mobile oxide-rich melt squeezed out to form late magmatic injections (from Bateman, 1951).

FIG. 8. Schematic isothermal section of magnetite–apatite–diorite system at 1,420 °C showing possible existence of 2- and 3-liquid (L) fields responsible for formation of oxide ores (after Philpotts, 1967).

igneous rock with interstitial oxides, as mentioned above; the residual liquid may be filter-pressed out of the crystal mesh and be injected elsewhere; or the enriched residual liquid may drain downwards through the crystal interstices and collect below to form a gravitative liquid accumulation.

Philpotts (1967), in his examination of Quebec anorthosites, has pointed out the significant quantities of apatite $(Ca_5(PO_4)_3F)$ associated with the Fe–Ti oxides and suggested that the liquid in which the ores were emplaced may have been a eutectic between Fe–Ti oxides and apatite. He supported his argument with experimental data which showed that the presence of apatite permitted the generation of immiscible silicate- and magnetite-rich liquids (Fig. 8) and that the magnetite:apatite ratio in that eutectic liquid was about 2:1 by weight—the same as that observed in the ores. The melting-point of the magnetite is depressed from 1,597 °C for pure Fe_3O_4, to less than 1,200 °C, approaching more geologically reasonable temperatures. Thus it appears that the presence of phosphorus—which is known to be important in separating iron melts from slags (Muan and Osborne, 1965)—and soda (shown by Shimazaki and Clark, 1973, to reduce drastically melting temperatures in the $FeO–FeS–SiO_2$ system) play an important role in the development of iron- and titanium-rich melts in intermediate rocks. Skinner and Barton (1973) further noted that additional components, quite possibly in a gas phase now escaped, might well have contributed to melting-point depression.

The great deposits of iron ore at Kiruna in northern Sweden (Geijer, 1931, 1967) and a nearly pure magnetite lava-like flow at El Laco of northern Chile (Park, 1960)—and several other similar bodies around the Pacific Basin (Park, 1972)—appear to have the same origin as the ores discussed above. Although apparently physically separated from their parent rock, these deposits exhibit well-defined flow structures and are rich in apatite. An unanswered question is why some magnetite ores are titanium-rich and others titanium-poor.

Minor element correlations have been investigated extensively in the coexisting iron- and titanium-bearing oxides of anorthositic ores. Lister (1966) demonstrated that V, Al, and Cr preferentially enter magnetite (substituting for Fe^{3+}) rather than ilmenite. He attributed the difference in behaviour to the lower electronegativity of Mg^{2+} and Mn^{2+} than Fe^{2+} and the greater tendency toward ionic bonding in ilmenite than magnetite.

Porphyry-type deposits

The bulk of the world's copper production in recent years has been derived from low-grade, disseminated copper minerals which occur in and are associated with igneous stocks of generally porphyritic type—the 'porphyry coppers'. Lowell and Guilbert (1970), in an excellent paper, define these prophyry deposits as 'a copper and/or molybdenum sulphide deposit, consisting of disseminated and stockwork veinlet sulphide mineralization emplaced in various host rocks that have been altered by hydrothermal solutions into roughly concentric zonal patterns'. These ores are not magmatic in the sense of previously described cumulates and immiscible liquids, but have rather been deposited from hydrother-

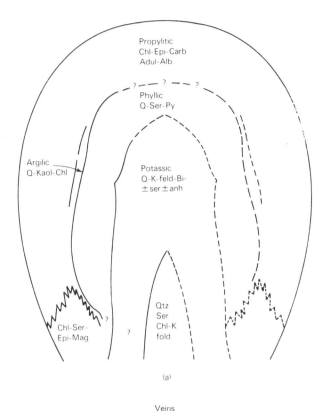

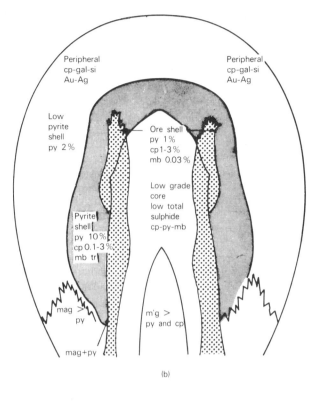

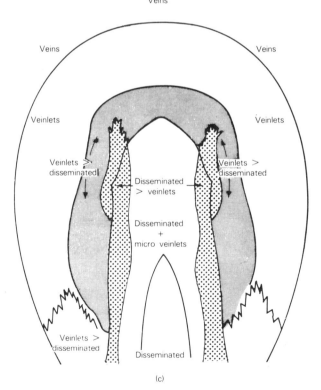

FIG. 9. Concentric alteration–mineralization of typical porphyry-type deposits: (a) alteration zones (b) mineralization zones (c) occurrence of sulphides (after Lowell and Guilbert, 1970).

mal fluids in multiple sets of fractures. However, in contrast to the hydrothermal ores to be discussed later, these ores have remained in or immediately adjacent to their comagmatic host rocks.

The porphyry coppers have been the subject of numerous studies; two collections of excellent papers recommended for further reading are those of Titley and Hicks (1966) and Ridge (1967). The various studies have revealed certain distinctive features common to most (or all) known deposits found in the south-western United States:

Distribution. The deposits appear to show a marked affinity for eugeosynclinal zones of orogenic belts.

Rock type. Although the igneous host rocks are stock-like and range in composition from granite to diorite, the most common porphyritic quartz monzonites and granodiorites.

Ore mineralogy. The ore mineralogy is simple, the primary sulphides being (in order of decreasing abundance) pyrite (FeS_2), chalcopyrite ($CuFeS_2$), bornite (Cu_5FeS_4), molybdenite (MoS_2); secondary sulphides may include covellite (CuS) and chalcocite (Cu_2S). The variation in ore mineralogy appears to represent a complete spectrum from very copper-rich and molybdenum-poor deposits such as Ajo, Arizona, and El Salvador, Chile, to molybdenum-rich, copper-absent bodies such as Climax, Colorado and Questa, New Mexico. Commonly veinlets of Cu, Pb and Zn sulphides and trace quantities of Au and

Ag occur in a broad peripheral zone. The grade of mineralization averages about 0.8 per cent for Cu-rich deposits and ~ 0.1 per cent for Mo-rich deposits. The mineralogy and, to a lesser extent, the mode of occurrence of the minerals exhibits a distinct concentric zonation as shown in Figure 9.

Alteration zones. All known porphyry-type deposits exhibit concentric zones of hydrothermal alteration (Lowell and Guilbert, 1970; Rose, 1970) which are designated from the centre outwards (see Fig. 9) as: (a) potassic—characterized by quartz, K-feldspar, biotite ± sericite ± anhydrite; (b) phyllic—characterized by quartz, sericite, pyrite; (c) argillic—characterized by quartz, kaolinite, chlorite; (d) propylitic—characterized by chlorite, epidote, carbonates, adularia, albite and (e) two deeper-level zones: (quartz, sericite, chlorite, feldspar) or (chlorite, sericite, epidote, magnetite).

Age. Radiometric age-dating has demonstrated that (with the exception of Bisbee) all known porphyry deposits in the south-western United States are between 54 and 72 million years old (Livingston et al., 1968).

Sulphur isotope data (exhibiting near 0-per-mille values) and strontium isotopes (with ratios of $^{87}Sr/^{86}Sr$ of 0.705) suggest that the ultimate source of the prophyritic rocks and the ores has been the upper mantle (Jensen, 1971). The oxygen and deuterium isotope data from rocks of the altered zones strongly suggest heavy meteoric water involvement in the generation of the ores (Hall et al., 1973; Taylor, 1973). The source the metals has generally been assumed to be magmatic, with the intermediate to acid magmas constituting the stocks because copper and molybdenum tend to be concentrated in these types of magmas by normal differentiation processes. Jensen (1971) has suggested, however, that the source of the metals may well be crustal shale beds cut by or affected by the proximity of rising intrusives.

Whatever the ultimate source of the metals, studies of the remnant ore solutions trapped in fluid inclusions (Roedder, 1971) clearly indicated that the solutions were very rich in halides. In fact, some inclusions contain more than 60 per cent salt by weight and are 'virtually a hydrous saline melt' (see Table 4). Whether these separated as the result of late-stage immiscibility between granitic silicate melts and hydrosaline melts as suggested by the findings of Roedder and Coombs (1967) on Ascension Island or are merely the result of concentration of salts by boiling, as suggested by White et al. (1970), is not known. There seems little doubt, however, that saline solutions played an important role in transporting the metals, probably as complexes (discussed in greater detail below). Interestingly, Roedder (1971) has also observed fluid inclusions containing liquid H_2S in addition to an aqueous solution and even some inclusions with three immiscible liquids—aqueous, H_2S, and an unknown (possibly organic volatile). The temperature of ore deposition in porphyry-type deposits

TABLE 4. Approximate compositions and densities of fluid in inclusions at Bingham, Utah (in weight percentage)

	High-density multiphase with NaCl only	Low-density steam inclusions	High-density multiphase with NaCl and KCl	
H_2O	83.3	63.7	42.6	31.9
CO_2	11.4	0.2	0.2	0.1
NaCl	3.3	35.4	38.7	45.8
KCl	1.1	—	12.7	9.0
Fe_2O_3	0.4	0.4	1.2	0.8
$CaSO_4$	—	0.3	1.5	1.4
Unknowns	—	—	3.1	11.0
	100	100	100	100
Density g·cm^{-3}	0.29	1.13	1.21	1.30
Weight percentage solids	5.3	36.1	57.2	68.0

Source: Roedder, 1971.

is probably quite variable but fluid inclusion data suggest a range from 400 to 725 °C at Bingham, Utah (Roedder, 1971) and from 350 to 600 °C at El Salvador, Chile (Gustafson and Hunt, 1971). In both cases peripheral deposits formed at least 100 degrees lower in temperature. The development of hydrothermal alteration zones, though by no means restricted to porphyry-type deposits, is nowhere better exemplified than in these deposits.

Although not all zones of the idealized porphyry deposit illustrated in Figure 9 are present in all ore bodies, all deposits exhibit extensive alteration and all zones are in the indicated relative positions. The complex chemistry of the alteration processes have been extensively examined by Meyer and Hemley (1967). Some of the many reactions which appear to contribute to the development of each of these zones are given below:

Potassic and phyllic zones

$$NaCaAl_3Si_5O_{16} + 2H^+ + K^+ = KAl_3Si_3O_{10}(OH)_2 +$$
andesine muscovite

$$+ Na^+ + Ca^{2+} + 2SiO_2;$$
quartz

$$NaCaAl_3Si_5O_{16} + 3K^+ + 4SiO_2 = 3KAlSi_3O_8 +$$
andesine quartz K-feldspar

$$+ Na^+ + Ca^{2+};$$

Argillic zone

$$KAl_2AlSi_3O_{10}(OH)_2 + H^+ +$$
muscovite

$$+ 3/2 H_2O = 3/2 Al_2Si_2O_5(OH)_4 + K^+;$$
kaolinite

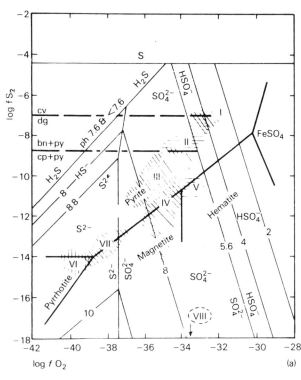

$$1.17\,NaAlSi_3O_8 + H^+ = 0.5\,Na_{0.33}Al_{2.33}Si_{3.67}O_{10}(OH)_2 +$$

albite   montmorillorite

$$+ 1.67\,SiO_2 + Na^+;$$

quartz

$$2K(Mg,Fe)_3AlSi_3O_{10}(OH)_2 + 4H^+ =$$

biotite

$$= Al(Mg,Fe)_5AlSi_3O_{10}(OH)_8 + (Mg,Fe)^{2+} +$$

montmorillonite

$$+ 2K^+ + 3SiO_2;$$

quartz

Propylitic zone

$$2NaCaAl_3Si_5O_{16} + 2SiO_2 + Na^+ + H_2O =$$

andesine   quartz

$$= Ca_2Al_3Si_3O_{12}(OH) + 3\,NaAlSi_3O_8 + H^+;$$

clinozoisite   albite

$$4CaAl_2Si_2O_8 + 5MgSiO_3 + 5H_2O =$$

anorthite   pyroxene

$$= 2Ca_2Al_3Si_{13}O_{12}(OH) + Mg_5Al_2Si_3O_{10}(OH)_8 + 4SiO_2\,.$$

epidote   chlorite   quartz

It is obvious that many factors can control the extent to which these reactions occur, with special attention having been paid by most workers to the pH and Na^+ and K^+ concentrations. The variation in pH is related to the activities of oxygen and sulphur through reactions of the type:

$$2H^+ + SO_4^{2-} = 1/2\,S_2 + 3/2\,O_2 + H_2O;$$
$$S^{2-} + 0.5\,O_2 + 2H^+ = 0.5\,S_2 + H_2O;$$
$$2HS^- + O_2 + 2H^+ = S_2 + 2H_2O;\ \text{and}$$
$$2H_2S + O_2 = 2H_2O + S_2\,.$$

Ore assemblage	Alteration	Examples
I. Covellite, digenite, chalcocite, pyrite	Advanced argillic + alunite, or serictic	Butte, Montana, central zone
II. Chalcocite, bornite, pyrite, chalcopyrite, bornite, pyrite	Zoned serictic and intermediate argillic (fringe propylitic)	Butte, Montana, intermediate zone; some porphyry coppers
III. Chalcopyrite, pyrite	Zoned serictic (K-feldspar): intermediate argillic (propylitic)	Butte, Montana, deep chalcopyrite zone; porphyry copper deposit
IV. Chalcopyrite magnetite, molybdenite (hematite) (pyrite); bornite, magnetite, (molybdenite)	K-silicate (anhydrite) (ankdrite) (serictic) (propylitic)	Butte, Montana, premain stage; some porphyry coppers
V. Chalcopyrite (pyrite); hematite	K-silicate (anhydrite) (ankerite); red feldspars (propylitic)	Chilean coast range copper deposits; Cobalt, Ontario
VI. Pyrrhotite, pyrite, chalcopyrite	Chloritic (sericite) (carbonate); albitization	Gold quartz veins of eastern Canada
VII. Pyrrhotite, pyrite, chalcopyrite, magnatite	Chloritic (sericite) (carbonate)	Noranda, Chibougamou, Quebec
VIII. Native copper, chalcocite (haematite) bornite–chalcocite	Zeolites, chlorite, alkali feldspars carbonates	Michigan copper

FIG. 10. (a) FS$_2$–FO diagram showing the distribution of some typical ore mineral assemblages (shown by shading) in the Cu–Fe–S–O system and associated wall rock alteration. Temperature is 250 °C, total aqueous sulphur = 0.1. (From Meyer and Hemley, 1967. © 1967 Holt, Rinehart & Winston Inc. Reproduced with permission.)

Meyer and Hemley (1967) have presented a series of diagrams (shown here as Figures 10a and 10b) which are particularly useful in explaining the variations in hydrothermal alteration observed in porphyry-type and other ores.

ORES FORMED
BY HYDROTHERMAL PROCESSES

Introduction and origin

A multitude of ores occur as simple fissure fillings, or as veins or more extensive replacements of pre-existing rocks. Although the origin of these ores has been the subject of considerable conjecture in times past, most workers now accept that their formation was from heated aqueous solutions and/or vapours/hydrothermal fluids.

The hydrothermal ores represent the ultimate in diversity with regard to occurrence and mineralogical content. The fluids from which they have been precipitated have ranged from supercritical fluids boiled off from basic igneous melts at temperatures of 1,000 °C, to gently warmed deep circulating groundwaters. The present discussion is brief and is not intended to provide an adequate coverage of this topic—about which a voluminous literature exists. The reader is referred to the following works for more detailed discussion of the numerous aspects of hydrothermal ores and solutions: Barton (1959), Garrels and Christ (1965), Barnes (1967), Ridge (1968), Helgeson (1969, 1970), Skinner and Barton (1973), and to nearly every issue of journals such as *Economic Geology, Geochimica et Cosmochimica Acta,* and *Mineralium Deposita.*

Our knowledge of hydrothermal fluids comes from: (a) the ore minerals left behind; (b) their alteration effect on the host rocks; (c) fluid inclusions in some ore and gangue minerals, and (d) some samples of modern ore fluids themselves. The complexity and variability of the ores is indicative of the complexity and variability of the fluids themselves. Clearly there is no one 'ore-forming fluid' but rather a broad spectrum of fluids, each with the capability of developing ore bodies under certain conditions. White (1974) notes that there are five primary classes of waters—meteoric, ocean, evolved connate, metamorphic and magmatic—which account in varying proportions for hydrothermal fluids.

The mineral assemblages and paragenetic sequences of hydrothermal ores are extremely diverse and the information on each could well fill a book by itself; accordingly, the present attempt is merely to give a broad overview of modern concepts of hydrothermal ores.

The problems are several: what have been the sources of the heat, the water, and the metal (and usually sulphur; sometimes arsenic, selenium, tellurium) which now constitutes the ores? The close proximity of some ores to igneous rocks—even to specific types of rocks, has been noted by many workers in the past and has thus led to a general conclusion that igneous rocks are the principal sources of hydrothermal fluids. For example, tin deposits occur almost exclusively in association with granites (Lindgren, 1933); tantalum is virtually restricted to granites and granite pegmatites (Ginzburg, 1956); rare earths have a strong affiliation for alkalic rocks (Heinrich and Levinson, 1961; Krauskopf, 1967). Some other correlations are less perfect; e.g. copper and molybdenum commonly occur associated with intermediate porphyritic rocks; Au:Ag ratios are low in ores associated with high-potash granites but high in ores associated with high soda granites (Gallagher, 1940). Many types of ores show only regional affiliations at best; e.g. silver associated with granitic rocks in most areas, but with diabase at Cobalt, Ontario.

Krauskopf (1967) has pointed out that the search for correlations between igneous rock type and ore type has led to two diametrically opposed views: (a) an ore deposit may form from a magma because the magma has an unusually high metal concentration to begin with, in which case the rock adjacent to the deposit should retain some vestige of the original high concentration; or (b) the process of ore formation may be envisaged as special form of differentiation by which certain metals are concentrated from an otherwise normal magma, in which case the igneous rock near the ore deposit should be depleted in its ore-metal content. He further points out the diversity of field observations which support and refute both views. For example, tin deposits are usually associated with tin-rich granites (Jedwab, 1957); the great silver deposits at Cobalt are associated with the most silver-rich diabase in eastern Canada (Fairbarn *et al.,* 1953); whereas, Huff (1952) found no enrichment of Cu, Pb, Zn in Front Range granites associated with ores of these metals; and Shrivastava and Proctor (1962) suggested that a Nevada granite was depleted in Cu, Pb, V because they were extracted to form the adjacent deposits.

In recent years isotopic studies have aided in deciphering the origin of many ores, both by providing accurate ages which may be related to igneous bodies and, to some extent, by indicating genetic relationships. Of primary use have been the ratios of radiogenic to nonradiogenic lead; of oxygen, $^{18}O{:}^{16}O$; of hydrogen, D:H; of carbon, $^{13}C{:}^{12}C$; and of sulphur, $^{34}S{:}^{32}S$. The importance and application of the stable isotopes in deciphering the origins of ore deposits has been collectively considered in a symposium, 'Stable Isotopes as Applied to Problems of Ore Deposits', published in *Economic Geology* (Vol. 69, No. 6; 1974).

Many hydrothermal ores occur with no apparent proximal relationship to their source rocks and thus do not have a clear genetic tie with any igneous rock. It has become obvious that many sediments and metamorphic rocks as well as igneous rocks (Table 1) possess sufficient disseminated metals to account for ore deposits and thus become candidates for parenthood of hydrothermal solutions. The problem is not one of sufficiency of metals but one of extraction from source rocks, transport and redeposition. For example, Wedepohl (1969) pointed out that only a ten millionth part of the crust's zinc is present in the form of ore bodies. To compound this, Skinner and Barton (1973) indicated that:

a continuum exists between the four end-member components of hydrothermal fluids, namely, (1) surface water, (2) connate or deeply penetrating groundwaters, (3) metamorphic waters, and (4) magmatic waters ... [and] the similarities between end product solutions give exact specifications of origin little meaning.

Identification of the origin(s) have been possible through the examination of H, O and S isotopes and chemical compositions. These parameters clearly indicate that many, if not most, hydrothermal fluids are mixtures of various waters (Taylor, 1971, 1973; Ohmoto and Rye, 1970). White (1968) noted that the Salton Sea brines were largely meteoric, with perhaps 25 per cent admixed magmatic water. Ohmoto (1972) has pointed out that the isotopic composition of sulphur and carbon, though once considered straightforward indications of the sources of the elements, are very complex parameters. The isotopic composition of a mineral is controlled not only by the isotopic composition of the source fluid, but also by the temperature, the oxygen activity, the pH and the coexisting minerals.

The compositions of hydrothermal fluids have been determined indirectly from the minerals they left behind (Barton and Skinner, 1967) and from associated alteration effects (Meyer and Hemley, 1967) and directly from fluid inclusions (Roedder, 1972), from samples of modern subsurface hydrothermal fluids (White et al., 1963; Skinner et al., 1967; Lebedev and Nikitina, 1968), and from samples of Red Sea brines (Degens and Ross, 1969). Helgeson (1964) pointed out that the inadequacy of pure water as a hydrothermal transport medium had been demonstrated by numerous workers. Two types of fluids have received considerable attention; (a) chloride-rich, on the basis of inclusions and natural brines; and (b) bisulphide-rich, because so many ore minerals are sulphides and because H_2S is known from some hot springs, especially those associated with Hg-mineralization (White, 1967). The analyses of ore solutions and fluid inclusions (Table 5) supports Helgeson's (1964) contention that: '... convincing evidence continues to accumulate suggesting that most hydrothermal solutions are weakly dissociated alkali, chloride-rich, electrolyte solutions'. Fluid inclusions are generally less than 10^{-9} cm^3 in size, although a few larger than tens or even hundreds of cubic centimetres are known (Roedder, 1967a, b). The fluids vary in density from 0.3 to 1.3 gm/cm^3, commonly contain 5–20 per cent by weight) dissolved solids and not infrequently contain daughter products which have separated on cooling—halite, sylvite, haematite, liquid CO_2, and even

TABLE 5. Compositions of natural ore-forming fluids (concentrations in p.p.m.)[1]

	Salton Sea geothermal brine	Cheleken geothermal brine	Cave in Rock District, Illinois fluorite	Creed, Colorado sphalerite
Cl	155,000	157,000	84,000	46,500
Na	50,400	76,140	45,200	19,700
Ca	28,000	19,708	6,000	7,500
K	17,500	490	2,400	3,700
Sr	400	636	—	—
Ba	235	—	—	—
Li	215	7.9	—	—
Rb	135	1.0	—	—
Cs	14	0	—	—
Mg	54	3,080	2,100	570
B	390	—	—	—
Br	120	526.5	—	—
I	18	31.7	—	—
F	15	—	—	—
NH_4^+	409	—	—	—
HCO_3^-	>150	31.9	—	—
H_2S	16	0	—	—
SO_4^{2+}	5	309	3,400	1,600
CO_2	—	—	—	—
Fe	2,290	14	—	—
Mn	1,400	46.5	—	690
Zn	540	3.0	—	1,330
Pb	102	9.2	—	—
Cu	8	1.4	—	140
As	12	0.03	—	—
Ag	1	—	—	—

Sources: Column 1. Muffler and White (1969); Column 2, Lebedev and Nikitina (1968); Column 3, Hall and Friedman (1963); Column 4, Roedder (1965).

© 1966 McGraw-Hill Book Co. Reprinted with permission.

liquid H_2S (Roedder, 1971). Petroleum (Roedder, 1967a) has been noted in some deposits where trapping of a two-phase emulsion has occurred.

The principal ores of most hydrothermal deposits are sulphide minerals; thus, the state of the sulphur in the ore solutions is of considerable importance. Garrels and Naeser (1958) were the first of many workers to examine the variation of sulphur species in solutions (Fig. 11) and found that under various geologically reasonable conditions bisulphate, sulphate, hydrogen sulphide or bisulphide could be the dominating species. Barnes and Czamanske (1967) maintained that the sulphide and bisulphide ligands were major transport species for metals in hydrothermal solutions. If the few natural ore fluids samples (e.g. Salton Sea brine, Cheleken brine, etc., Table 5) are typical, there is a problem concerning the source of the sulphur now combined as the sulphide ore minerals. The fluids cited do not contain sufficient sulphur to precipitate the metals they possess. Sulphur isotope data (Fig. 12) from a wide range of deposits (Jensen, 1967) suggest that many ores

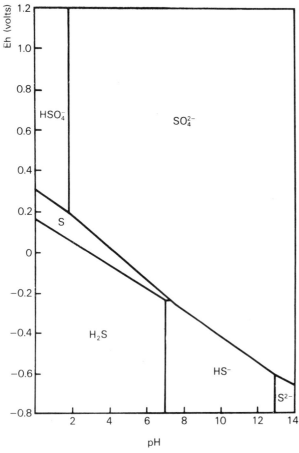

FIG. 11. Regions of dominance of sulphur-bearing species in H–S–O system in terms of Eh and pH at 25°C (after Garrels and Naeser, 1958).

contain sulphur which has been through the sedimentary cycle and hence that it may have been extracted from the sediments by the fluids. For example, Brown (1971) suggested that sulphur from pyritic sediments had been responsible for precipitating copper in the White Pine district of Michigan in the United States; a suggestion supported by sulphur isotope data (Burnie et al., 1972). It is important, however, to bear in mind that numerous complexities are now known to cloud the simple sulphur isotope picture found in Figure 12.

The ultimate source of the metals in hydrothermal solutions is at least as variable as that of the sulphur. The immediate association of Sn veins with granite bodies clearly indicates a comagmatic origin; but the Pb in Utah vein deposits was derived from both intrusives and country rocks (Stacey et al., 1968). On the other hand, the sources of Pb in the Salton Sea brines are the host rocks of the area (Doe et al., 1966) and Pb in the Old Lead Belt of south-east Missouri was extracted from La Motte Sandstone (Doe and Delevaux, 1972).

The problem of just how metals may be transported and deposited has been considered by Skinner and Barton (1973); their example of zinc is considered in Figure 13. Assuming a minimum effective concentration for reasonable transport of zinc as 0.7 p.p.m. at 100°C, the $ZnCl_2$ complex only has sufficient concentration if the activity of HS^- is less than 10^{-10}. However, this

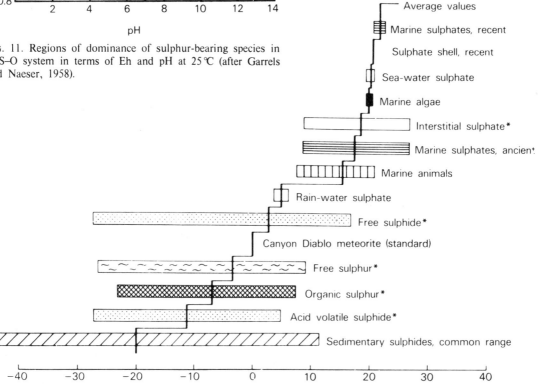

FIG. 12. Ranges of the $\delta^{34}S$ in various sulphur-containing materials. Data with asterisk are from the same environment (after Degens, 1965 from many sources).

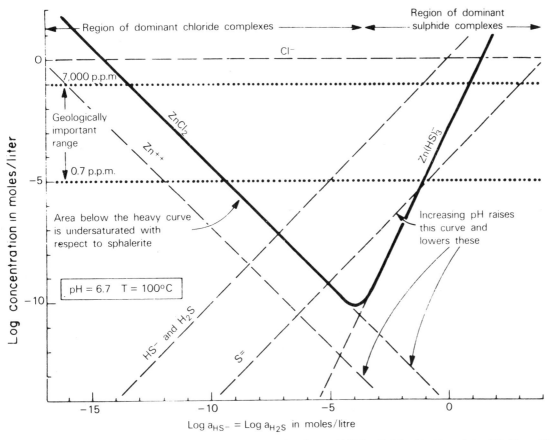

FIG. 13. Comparison of chloride and bisulphide complexes for zinc at 100 °C, pH 6.7, chloride activity 19°. (From Skinner and Barton, 1973. © 1973 Annual Reviews Inc. Reproduced with permission.)

does not provide sufficient sulphur to precipitate the metal unless the sulphur is present as dissolved sulphate, which is reduced to sulphide at the site of deposition. On the other hand, to transport the Zn as $Zn(HS)_3^-$, the HS^- activity must be above 10^{-1}, a condition in which the equilibrium pressure of H_2S would be high enough to convert all iron oxides and silicates to sulphides. Since this does not occur, such sulphide complexes cannot be important in ore transport. Skinner and Barton (1973) indicated, however, that: '... as temperature increases, the separation between the chloride and sulphide complex fields decreases so that at perhaps 300 °C or above, the designation of the proper complex becomes indefinite'. The effectiveness of other complexing ligands—sulphate, hydroxide, carbonate, bicarbonate, fluoride, etc. is probably only locally of importance.

The mechanisms that cause deposition of the ores apparently vary from one deposit to another but include:

Temperature changes. Most sulphides, whatever the means of transport, have decreased solubility at lower temperatures.

Changes in pressure. Holland (1967) showed that decrease in pressure reduced the solubilities of quartz, fluorite and barite—however the effects were not large.

Reaction with wall rocks. This interaction may produce extensive alteration effects as well as changing the chemistry of the fluids. Meyer and Hemley (1967) noted correlation between ore assemblages and the types of alteration in copper deposits (see Fig. 10(a) and Table 2). Some of the pertinent reactions involved in the formation of these altered zones were given earlier in this chapter under the discussion of porphyry-type deposits.

Mixing with other solutions. This was suggested by the isotopic data of Taylor (1973). Skinner and Barton (1973) noted that dilution could reduce the concentration of transporting liquids and could thus destabilize the complexes and cause precipitation.

Many hydrothermal ore types exhibit a distinct zonation with regard to mineralogy and chemistry. Although such zonation was once ascribed to differences in volatility of metals, it is now generally accepted to be the result of differential transport of soluble metal complexes (Barnes, 1962). Metal precipitation may result from cooling, dilution of waters, reaction with dissolved anions, and/or reactions with wall rocks. The different magnitudes of these effects on various metal complexes may result in simultaneous precipitation (complex ores containing several minerals) or sequential deposition (zoned ores) of metal sulphides.

Hydrothermal ore types

Within the realm of hydrothermal ores there exist some fairly well-defined ore types—associations of elements and minerals—which are recurrent the world over. Detailed discussion of each of these ore types is impossible, but a brief review of the mineralogy of the most well-defined associations and references to more detailed works on them are given below.

The associations noted are not always mutually exclusive but are generally individually recognizable. The reader is referred to the excellent, comprehensive listing of ore mineral assemblages by Ramdohr (1969) and to the beautiful atlas of ore associations given by Oelsner (1966).

Lead–zinc–silver (–copper–iron)

The association of sphalerite, galena, pyrite and chalcopyrite probably represents the most widespread hydrothermal ore type. It occurs as vein fillings and replacements in all types of host rocks. Typical ores of this type such as those at Creede, Colorado (Bethke et al., 1973) have formed from deeply circulating, meteoric waters which interacted with wall rocks. The temperature ranged from 190 to 265 °C with deposition of the sulphides from saline (4–12 per cent by weight) near neutral solutions which carried sulphate in excess of sulphide. The narrow sulphur isotope values clustering near O $\delta^{34}S$ ‰ CD^1 have often been considered as evidence of a mantle source for sulphur and metal but are interpreted differently by Bethke et al. (1973).

Although many examples of this ore type are valued for their Cu, Pb and Zn contents, frequently there are also present significant amounts of silver, either dissolved in the galena where it participates in a coupled substitution such as Ag^+ and Bi^{3+} for Pb^{2+}, or included as small grains of argentite or silver-bearing tetrahedrite. In some regions such as the Cœur d'Alene district, Idaho, tetrahedrite becomes the dominant sulphide in sideritic veins. Zonal distribution of metals in ores of this type are common.

Bismuth–cobalt–nickel (–silver–arsenic–uranium)

This association is typified by the rich silver ores of Cobalt, Ontario (Canada), Great Bear Lake, North-west Territories (Canada), Kongsberg (Norway) and Joachimsthal (Czechoslovakia). At Cobalt the ores contain complex masses, botryoidal bodies, rosettes, layers, veinlets and

disseminated grains of Ni, Co and Fe arsenides with associated intergrowths of silver and allargentum and veinlets of native bismuth and galena. Petruk (1971a, b) has subdivided the arsenide assemblages into five categories, i.e. Ni–As, Ni–Co–As, Co–As, Co–Fe–As, and Fe–As.

The origin of the Cobalt ores remains somewhat enigmatic, with two schools of thought prevalent. First, their distillation from the underlying Archean rocks. Boyle and Dass (1971a) stated that: ... support for [this] thesis comes from the fact that a metal source is readily available stratigraphically below the deposits of Cobalt, and that a number of other similar deposits throughout the world show a close genetic relationship to pre-existing sulfides'. Secondly, their derivation from the Nipissing diabase. Jambor (1971) remarked that 'the main argument favouring a direct Nipissing diabase source may be simply stated: where there was no diabase, Ni–Co arsenide–native silver veins have not been found'.

Scott and O'Conner (1971) found that fluid inclusions in the quartz associated with the ores suggested ore deposition from concentrated saline brines (up to 30 per cent NaCl) in the temperature range 195–360 °C. Petruk (1971b) summarized the ore depositional history (Fig. 14).

The Kongsberg and Joachimsthal deposits are similar to those at Cobalt and Great Bear Lake except that Bi is lacking at Kongsberg and native arsenic and proustite are common at Joachimstahl. Both of these deposits contain multiple sets of veins. The origin of these deposits is suggested by Boyle (1968) to be a leaching process from surrounding rocks. Specific references include Bastin (1939), Boyle (1968), and Petruk (1968).

Gold–silver (–tellurium–selenium) or gold quartz

Gold quartz to Au–Ag–Te–Se-type ores have probably intrigued more geologists than any other kind of deposit. Typical deposits, such as those of the 'Gold Belt' of the Superior Province of Canada and the Mother Lode of California (United States), Bendigo (Australia), have a relatively simple mineralogy consisting of pyrite, quartz and disseminated gold. In other deposits, however, such as Kalgoorlie (Australia), Cripple Creek, Colorado (United States), South Porcupine–Timmins (Canada) and Acupan (Philippines), the gold is combined with silver and tellurium in the form of calaverite ($AuTe_2$), krennerite ((Au, Ag)Te_2), sylvanite ($AuAgTe_4$) and petzite (Ag_3AuTe_2). Suggested readings concerning the gold tellurides include Markham (1960), Cabri (1965), Shcherbina and Zer'yan (1964) and Kelly and Goddard (1969).

The processes involved in gold transport and deposition have been points of conjecture. Helgeson and Garrels (1968) suggested that gold was transported in chloride complexes in acid solutions and deposited by

---

1. Refers to the Canon Diablo meteorite (standard).

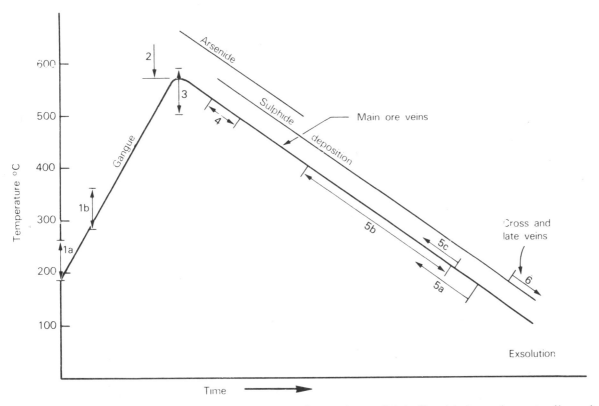

FIG. 14. Schematic deposition sequence and temperatures of ore veins at Cobalt (Ontario). 1a = early quartz, 1b = end of early quartz, 2 = upper stability of chlorite, 3 = rammelsbergite–para rammelsbergite inversion, 4 = stability field of Cu–Pb–S phase A, 5a = lower limit of argentite stability, 5b = stability of field for silver–pyrargyrite–acanthite assemblages, 5c lower limit for galena–matildite solid solution, 6 = upper stability of stephanite, 7 = fluid inclusions in late quartz, 8 = upper stability of smythite, 9 = upper stability of: a = mckinstryite, b = stromeyerite, c = stromeyerite-chalcocite exsolutions (after Petruk, 1971).

posits, such as that at Carlin, Nevada (United States); this concept is supported by Radtke and Scheiner (1970) who found activated carbon components capable of absorbing gold chlorides in the host rocks.

Tin–tungsten (–bismuth–molybdenum)

Tin–tungsten ores are typified by the extensive deposits of the Bolivian Andes—the arc of deposits in the Malay peninsula and the famous Cornwall deposit in England. Although they have been divided into two distinct ore types, i.e. tin–tungsten or tin–silver (Kelly and Turneaure, 1970; Turneaure, 1971), the ores are herein treated together. These ores, closely associated with granitic batholiths and porphyritic stocks and dikes, ex-cooling or by wall-rock reaction. On the other hand, Weissberg (1970) and Seward (1973) have argued that gold was more likely transported as sulphide or bisulphide complexes in alkaline solutions and that deposition occurred because of pH, temperature and Eh changes of the solutions. The role of organic liquids and carbonaceous matter in the transport of gold is not known but Kelly (1972) suggested that it might play an important role in the formation of 'invisible gold' de-

hibit four stages of vein growth (Fig. 15): (a) quartz–cassiterite–wolframite, (b) base-metal sulphides, (c) hypogene alteration of pyrrhotite to pyrite, marcasite and siderite, (d) late veinlets and crusts of siderite, fluorite and hydrous phosphates.

In the typical Bolivian ores, the initial ores contain quartz and cassiterite add minor amounts of arsenopyrite, bismuthinite and wolframite. The base-metal sulphides include pyrrhotite, sphalerite, stannite (Cu_2FeSnS_4), chalcopyrite, franckeite ($Pb_5Sn_3Sb_2S_{14}$) and teallite ($PbSnS_2$). The ore-forming fluids were complex, NaCl-rich brines of low CO_2 content. Depositional temperatures decreased from the 350° to 530 °C range during quartz–cassiterite deposition to less than 70 °C during the last stages of ore modification.

Molybdenite occurs in minor quantities in some tin ores and some authors consider the great porphyry-type molybdenum deposits (e.g. Climax, Colorado) to be closely related to tin–tungsten deposits.

Copper–iron–arsenic

Cu–Fe–As sulphide mineralization is typified by the fabulously rich ore deposits in the United States at Butte,

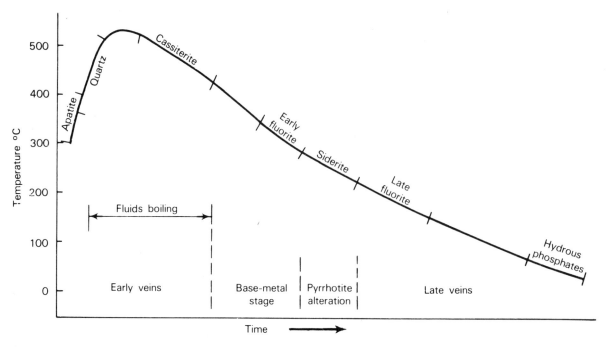

FIG. 15. Schematic sequence and temperatures of Bolivian tin and tungsten deposits (after Kelley and Turneaure, 1970).

Montana, and the rich veins at Magma, Arizona. They are characterized by pyrite–chalcocite–enargite and chalcopyrite–bornite–chalcocite–tennantite assemblages. At Butte the ores are present in complex series of 'horsetail fissures': a crudely concentric zonation is present from the interior outward: copper–zinc, zinc and manganese. The hydrothermal fluids have resulted in development of extensive alteration zones (Sales and Meyer, 1948): propylitic–argillic–kaolinitic–serietic. Recommended readings regarding the copper–iron–arsenic ores include the Meyer *et al.* (1968) discussion of the Butte ores and Gustafson's (1963) consideration of the Cu–Fe–As–S system on the basis of natural mineral assemblages.

Antimony–mercury–arsenic

Mercury–antimony mineralization occurs in low-temperature, hydrothermal deposits and is frequently associated with hot-spring activity. Mercury is generally present in the forms of cinnabar and metacinnabar, HgS, but occasionally is present as the native metal; antimony generally exists as stibnite (Sb_2S_3); and arsenic occurs primarily as the sulphide realgar (AsS) and orpiment (As_2S_3). Commonly the only other sulphide present is pyrite. In the classic mercury deposits such as at Almaden (Spain) and Idrija (Yugoslavia) the cinnabar occurs as a replacement of quartzitic beds. Related deposits include the well-known ores at New Idira and New Almaden (California) and the unique occurrences of the only mercury–antimony–sulphide, livingstonite,

$HgSb_4S_8$, in Mexico. This mineral apparently formed as a result of hydrothermal fluids encountering free sulphur in a gypsum–anhydrite sequence.

The ores included in this type form a spectrum from nearly pure mercury deposits (Almaden), to mercury–antimony deposits (Huitzuco, Mexico), to mercury–antimony–arsenic deposits (Red Devil, Central Kuskokwim region, Alaska), to arsenic–antimony–mercury deposits (Getchell, Nevada). In the last type, arsenic, in the form of realgar and orpiment, greatly exceeds in volume the minor amounts of anitmony and mercury (both of these are present in rare mineral forms—getchellite, $AsSbS_3$, and galkhaite (Hg, Cu, Tl, Zn) (As, Sb)S_2.

On the bases of mercury-depositing, sulphide-containing, hot springs, and experimental investigations of the solubilities of cinnabar and stibnite in alkaline sulphide solutions, Dickson and Tunell (1958, 1968) and Tunell (1964) have concluded that the ore-forming fluids transported the mercury and antimony as sulphide and bisulphide complexes at high pH. Such solutions have the double capability of importing mercury and antimony and aiding in the solution of the replaced quartz. Learned *et al.* (1967) in fact demonstrated that the presence of quartz reduced the solubility of mercury and antimony in solution. White (1967), on the other hand, suggested that saline-acid solutions constituted a more likely type of mercury-transporting fluid. Recommended reading on these ores include: Tunell (1964), Dickson and Tunell (1968), Arnold *et al.* (1973), White (1967) and Krauskopf (1951).

Stratiform lead–zinc (–barite, –fluoride)

These ores which occur as bodies that extend parallel or nearly so with bedding—that is, those which are 'strataform'—constitute the world's primary sources of lead and zinc. Because of their occurrence in Europe and the United States they have commonly become known as 'Alpine' or 'Mississippi Valley' type ores, respectively. Although world-wide in extent, they are best developed in the upper Mississippi Valley, in south-east Missouri, in Tri-State area and in east Tennessee of the United States, Pine Point, North-west Territories (Canada), Bleiberg and Silesia-Cracovian regions of Germany and Poland and the Trento Valley (Italy). The ore mineralogy is usually simple, consisting of galena, low-Fe sphalerite, barite, fluorite, pyrite and marcasite with only traces of other constituents. These minerals occur as replacements and as fissure and breccia fillings; pseudo-colloform bands in the sphalerites are quite common.

Brown (1970) and Stanton (1972) summarized many of the pertinent features of these deposits; some of their important points include:

In any one area, the ores favoured specific facies; there might also be structural control in faulting, gentle folding, karst formation.

Fluid inclusions indicated ore deposition from Na–Ca–Cl solutions in the temperature range 100–200 °C.

The metallic elements were Pb–Zn–Fe (Cu and Ag, if present, were so in small quantities).

The host rocks contained organic matter—petroleum was present in some fluid inclusions.

S-isotopic composition was 'heavy' but lay in a narrow range—not typically igneous or sedimentary.

Pb-isotopes were specific to regions: (a) J-type (future model age)—central United States, Sweden; (b) B-types (model age of ore was older than enclosing sediment)—Alpine ores; (c) normal-type (model age of ore = age of bed)—the English Pennines.

In spite of the remarkable similarities of numerous intensively studied ores there has been, and remains, a great deal of debate regarding their origin. Brown (1970) summarized the views thus: 'North American opinion [is] converging toward acceptance of a dominantly connate marine but epigenetic ore fluid ... [but] ... European opinion is divided almost equally between proponents of syngenesis-diagenesis and magmatic-epigenetic origin'.

Roedder (1967b) and Jackson and Beales (1967) suggested that the metals could have been extracted from sediments by brines as compaction occurred. Boyle and Lynch (1968) offered the interesting speculation that the initial concentration of metals was in the soft parts of marine organisms (oysters, clams, sponges, etc.), and that further concentration was the result of bacterial humification and putrefaction, only final concentration and movement being by migrating brines.

Stanton and Rafter (1966) and Stanton (1972) noted

that the sulphur isotopes are too heavy to have been induced by simple sea-floor bacterial sulphate reduction. Barton (1967) proposed the interesting alternative of an abiological methane reaction of the type:

$$CH_4 + ZnCl_2 + S_4^{2-} + Mg^{2+} + 3\,CaCO_3 = ZnS + CaMg(CO_3)_2 \\ + 2\,Ca^{2+} + 2\,Cl^- + 2\,HCO_3^- + 2\,H_2;$$

such reactions could account for the heavy S isotopes, the precipitation of the ore and the common dolomite alteration about the ore bodies.

The interested reader is recommended to works by Brown (1970), Snyder (1968) and Roedder (1967b, 1968).

Ores in sediments

INTRODUCTION

The sedimentary realm represents a broad spectrum in which a wide variety of ores have formed and now reside. In spite of the fact that ore-forming processes are more easily observed, studied and duplicated here than in magmatic, hydrothermal or metamorphic realms, intense debate has frequently erupted concerning the origins of many ores in sediments. One problem lies in distinguishing whether the ore mineral formed with the sediment, after deposition of the sediment but before lithification, or long after lithification; obviously there is again the likelihood that processes of ore formation may have extended over long periods of time and spanned the time of lithification. A second problem is the source of the metals and the sulphur which constitute the ore body. A third problem is the mechanism by which sufficient quantities of ore constituents have been transported to their sites of deposition.

The most frequently employed methods of characterizing the nature of the sedimentary realm are those of Eh–pH, log CO_2–pH, or partial pressure diagrams. Among the most commonly cited and useful diagrams in delineating the sedimentary realm is that given in Figure 16. Because of the intimate relationships among Eh, pH, the oxidation state of elements and the solubilities of elements, these diagrams and others like them have proved extremely useful in defining the natural conditions under which many types of ores may form.

For the sake of convenience in considering the diverse nature of sedimentary ores this realm is broken into several related but sometimes overlapping categories:

Weathering—residual deposits.

Erosion: (a) mechanical—placer-deposits; (b) chemical: (sedimentary iron ores, nonmarine manganese ores, manganese nodules, uranium–vanadium ores, oxidation and supergene enrichment).

Chemical precipitates from sources other than weather-

James R. Craig

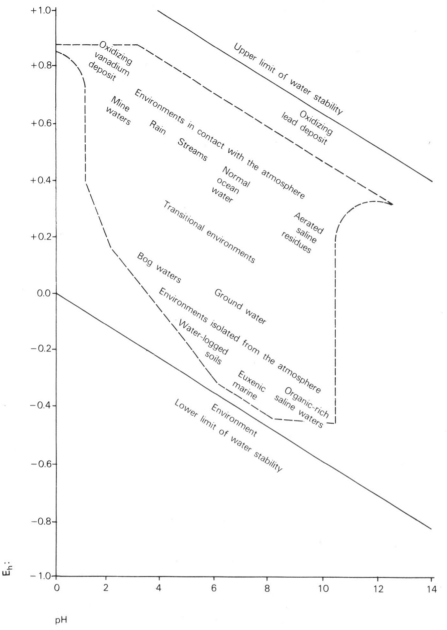

Fig. 16. Approximate positions of some natural environments in terms of Eh–pH as defined by Garrels and Christ (1965), with the limits of natural water analyses (dashed line) as given by Bass-Becking (1960).

ing: (a) some stratiform ores; (b) volcanic exhalative deposits.

WEATHERING—RESIDUAL DEPOSITS

The survival, mobility, and/or enrichment of ores in the weathering realm is determined by the geochemical nature of the metals involved. In this portion we consider only subaereal weathering effects. The formation of bog iron formations and the geochemistry of supergene enrichment, although clearly related to subaereal weathering, are held aside for separate discussions in later sections.

The realm of subaereal weathering varies in terms of pH and Eh (Fig. 15) from rain water to water-logged soils, but generally remains strongly oxidizing because of immediate or near-immediate contact with the atmosphere. As a result the only minerals formed and/or preserved are oxides, carbonates, sulphates and silicates—and more rarely, nitrates and phosphates—and grains sufficiently inert to resist oxidation. Sulphide minerals are readily destroyed by oxidation.

Among the most economically important residual ores are those of aluminium, nickel and manganese. Conversion of the original non-economic rocks (with either low values of metals or metals bound in such manner as to be unavailable) to usable, concentrated ores requires the intense weathering generally active only in a tropical environment.

Although the *in situ* residual weathering process described below is accepted by most workers for most bauxite deposits, Strachen (1958) has suggested chemical precipitation of Al and Fe from normal weathering as the source of some bauxites, and Davidson (1964) has described the debouching and precipitation of large quantities of Fe and Al hydroxides into the sea from volcanic springs in the Kurile Islands. Bauxite and laterite formation have recently been the topic of a paper by Norton (1973) who concluded that the primary controls were (a) climate (temperature above 25 °C such that microflora can destroy humus more rapidly than macroflora generates it, and the amount and temporal distribution of rainfall), (b) topography (high enough to provide a well-drained local); (c) groundwater movement (to provide removal of dissolved materials); (d) low rate of mechanical erosion relative to chemical; (e) vegetation (to generate acid conditions); and (f) bedrock. Under tropical conditions with abundant moisture and low rates of mechanical erosion, the development of bauxites and laterites is primarily a function of the soil pH and Eh because these parameters control the relative solubilities of the Fe, Al and Si oxides. The removal of the more soluble mineral constituents (i.e. Mg^{2+}, Ca^{2+}, Na^+, K^+) is readily accomplished by either congruent solution of the more soluble minerals:

$$Mg_2SiO_4 + 4H^+ \longrightarrow 2Mg^{2+}_{(aq)}H_4SiO_{4aq},$$

or incongruent solution of those less soluble:

$$2KAlSi_3O_8 + 2H^+ + 7H_2O \longrightarrow Al_2Si_2O_5(OH)_4 +$$
orthoclase   kaolinite
$$+ 2K^+_{aq} + 4H_4SiO4_{aq}.$$

Temporary residence of Mg, Ca, Na and K in other soil minerals (i.e. secondary chlorite) occurs but this interim condition is insignificant in long periods of weathering. The problem of bauxite and laterite formation rests then on the selective removal of the 'insoluble' minerals of Al, Si and Fe through reactions of the type:

$$Al_2Si_2O_5(OH)_4 + 5H_2O \longrightarrow 2Al(OH)_3 + 2H_4SiO_{4(aq)}.$$
kaolinite   gibbsite

The solubility of silica (controlled in nature more by the presence of amorphous silica than by quartz alone, Krauskopf, 1957) is greater than that of either aluminium oxides or ferric oxides throughout the pH range of about 4 to 10. Under more acid conditions, the solubilities of both iron and aluminium oxides and hydroxides rise rapidly. The aluminium compounds, being amphoteric, also exhibit a rapid rise in solubility in basic solutions with pH above about 10.3. Armed with

this knowledge and using the computation techniques of Garrels and Christ (1965), Norton (1973) has presented the data (see Fig. 17) which demonstrated the isosolubility lines for gibbsite and the iron oxides. Of particular interest are the dashed line, which indicates the line of equal solubility of the Fe and Al oxides (i.e. along which there would be no concentration of either Fe or Al relative to the other) and the dotted line to the right of which (to pH values of 10.3) there is no significant solubility of either of these metal oxides. Under the extremely acid but oxidizing conditions of point A, the solubilities of silica and alumina are much greater than that of iron and *in situ* soil development tends to be lateritic. On the other hand, under the somewhat less acid and less oxidizing conditions at point B, silica and iron are removed and alumina left behind such that *in situ* bauxite formation occurs. The acid conditions prevailing at point C dissolve both iron and alumina

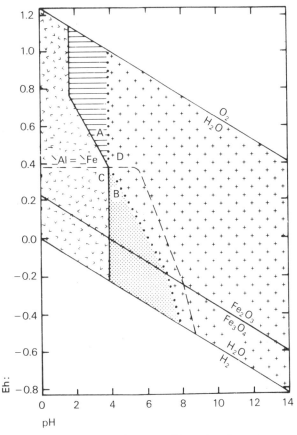

FIG. 17. Schematic Eh–pH diagram illustrating the various regions of Fe_2O_3, Al_2O_3 and SiO_2 enrichment in tropical soils. The dashed line represents equal solubility of Al and Fe; the dotted line represents the highest pH (up to 10.3) at which there is any significant solubility of Al or Fe (10^{-6} m/l). The effects under each set of conditions are: (A) Fe enrichment, Si, Al removed, (B) Al enrichment, Fe, Si removed, (C) Si enrichment, Al, Fe removed, (D) Fe, Al enrichment, Si removed (after Norton, 1973).

leaving silica-enriched soils; and the intermediate pH conditions at D preferentially remove silica but do not effectively change Fe:Al ratios in soils.

Bauxite deposits are world-wide in distribution in tropical areas but are only known as far as 60° N. in the U.S.S.R. and 42° S. in Tasmania. The deposits have been broadly classified as (a) those associated with igneous parent rocks and (b) *terra rossa* bauxites developed on or in close association with carbonate rocks. Patterson (1967) has published an extensive survey of the world's bauxites.

In addition to bauxites, residual weathering has been instrumental in the formation of some extensive deposits rich in nickel and others rich in manganese. The lateritic nickeliferous deposits are of two types: (a) nickeliferous ferruginous laterites containing about 40 per cent Fe and about 1 per cent Ni but with no discernible nickel mineral and, (b) nickel-silicate deposits with relatively low iron content but with the nickel-silicate, garnierite (Ni-antigorite (Mg, Fe, Ni)$_3$Si$_2$O$_5$(OH)$_4$). The first type of deposit is typified by those developed on serpentines in Cuba and the Philippines, whereas the second type are typified by the deposits formed on peridotites in New Caledonia. As shown by Hotz (1964) for the deposits in south-western Oregon, in the United States, the usual geochemical changes attendant during formation of these ores are enrichment in Al$_2$O$_3$, Ni, H$_2$O and total Fe, slight enrichment in Cr and Co, slight loss of Mn, and extreme depletion in MgO, SiO$_2$ and CaO. Similar trends have been observed for other nickeliferous laterites.

The high solubility of nickel under acid to neutral conditions suggests that relatively high pH solutions play a role in the formation of nickeliferous laterites. Such conditions (pH = 9) would result in retention of nickel, aluminium and iron but loss of silica, magnesia and lime as noted above. There remains, however, the unsolved problem of how such alkaline solutions can evolve in moist tropical environments. Krauskopf (1956a) has suggested that the chert and quartz associated with garnierite deposits may have been precipitated colloidally by evaporation. In line with this reasoning Hotz (1964) concluded that variations in the watertable and alternate wetting and drying might have been sufficient to result in the formation of the garnierite deposits. The experimental synthesis of garnierite has been accomplished at temperatures as low as 55 °C in pH ranges of 5.8–6.5 (Henin, 1956; Caillere *et al.*, 1956).

EROSION

Mechanical

Placer deposits

In contrast to the deposits formed by intense *in situ* weathering are the placer deposits which generally represent simple accumulations of relatively unaltered metal-bearing debris. The preservation of native metals to give rise to placer deposits is common only among the noble metals, i.e. gold and the platinum group, and a few metals which develop a sufficiently impervious oxide coating to resist total oxidation (e.g. antimony and tin). In addition, placer deposits are well known of some metal oxides, e.g. the cassiterite (SnO$_2$) placers of South-East Asia, and the rutile- and ilmenite-rich beach sands known in many localities around the world. The development of economic placer deposits of materials such as these depends not so much on geochemical parameters but upon the material having a sufficiently high specific gravity to be concentrated by physical processes, and upon the material being sufficiently 'tough' to prevent its being reduced to such small size as to be readily dispersed in suspension.

Chemical

Sedimentary iron ores

Iron constitutes the principal metal of modern technology because of its high terrestrial abundance, the ease of its extraction, its strength and workability and its low cost. It is nearly ubiquitous in geological occurrence and plays a vital role in both the animal and plant realms. Iron is a typical transition element and exists in the geologic realm in two common valency states, ferrous, Fe^{2+}, and ferric, Fe^{3+}. The abundance of iron (2–3 per cent in sedimentary rocks, 8.5 per cent in basalts, and ~5 per cent of the total crust), and the existence of two valency states lead to a wide variety of occurrences, a proportion of which are economic. The present discussion is confined to iron ores of sedimentary affiliation; those which today provide the bulk of the world's iron.

The geochemical behaviour of iron in the sedimentary realm is summarized fairly well by two rules succinctly put forth by Goldschmidt (1958):

Rule 1. Oxidizing conditions promote the precipitation of iron, reducing conditions promote its solution.

Rule 2. Acid conditions generally promote the solution of iron, alkaline conditions promote its precipitation.

There appear to be two major sources of the iron which may ultimately become incorporated into ore deposits, i.e. weathering of the continents (especially the ferromagnesian minerals), and volcanic activity. Transport of iron from volcanic sources is readily accomplished in the form of ferrous chlorides (Zelenov, 1958, 1959). On the other hand, transport of significant amounts of iron from the continents by means of the normal stream waters presents a problem. Typical surface waters are well oxygenated; as a result, any ferrous iron is rapidly oxidized to the ferric state, which precipitates immediately as the extremely insoluble hydroxide. Only under extremely acid conditions does the ferric ion become soluble. Castano and Garrels (1950) demonstrated, however, that significant quantities of iron may be carried in neutral solutions as organic complexes (especially when tannic-acid levels are high) or as ferric colloids with organic acids. Other organic complexes, including chelates, have been noted (Bjerrium *et al.*, 1957; Hern, 1960) but little of their natural occurrence is known. In addition, iron may be transported as fine hydroxide suspensions and as clays (Carroll, 1958). James (1966) noted in fact that 'bulk of iron produced by present-day erosion is transported to the ocean as colloidal ferric hydroxide, as adhering oxide on clay, or chemically bound ions in clastic, mineral particles'.

The great sedimentary iron ores are of three fairly distinct types: (a) the 'bog' irons; (b) Clinton or minette-type ironstones and (c) the banded Pre-Cambrian ironstones.

The bog irons are typically found in temperate to tundra regions of the northern hemisphere and occur as lenticular to banded masses of goethite, [FeO(OH)], with variable concentrations of clays, sands and organic matter in lakes, along slowly flowing streams, and in marshes. Manganese is generally present (sometimes up to 40 per cent MnO_2), and occasionally also is the blue phosphate vivianite ($Fe_3^{2+}(PO_4)_2 \cdot 8 H_2O$). Some deposits, known as 'black-band ironstones' consist essentially of sideritic accumulations with variable amounts of clastics and organic matter.

The decay of abundant organic matter in saturated soils results in the generation of CO_2-enriched waters, in which iron is readily reduced from insoluble ferric compounds into the soluble ferrous bicarbonate. Subsequent contact and mixing of these groundwaters with oxygenated waters results in re-oxidation of the iron to the insoluble ferric state and reprecipitation of the iron as goethite (see Fig. 18). The formation of the black-band ores is not certain but appears to be a diagenetic effect in brackish waters wherein total oxidation does not occur.

The remaining two types of iron formations, both found over generally broad areas, have been differentiated on the basis of age, texture, mineralogy and, to some degree, chemistry by James (1966) and the reader is referred to his excellent work on iron formation for a more comprehensive analysis.

The minette-, Lorraine- or Clinton-type iron ores consist of extensive, stratigraphically controlled, sedimentary facies, which vary up to 15 m in thickness. They occur in near-shore marine to fluviatile and lacustrine sedimentary facies and often contain abundant fossils and oolites. Their age varies from Middle Pre-Cambrian to Pliocene. The variability of the mineralogy of these deposits—reflected in four distinct ore facies, oxide, silicate, carbonate, sulphide—indicates a wide range of environments of formation. The facies appear to represent the sequential appearance of the stable fields for oxide, carbonate and sulphide with decreasing Eh (Fig. 19). This corresponds to decreasing availability of oxygen in the bottom environments of restricted basin with increasing depth and has been observed in part in the modern Niger delta (Porrenga, 1967).

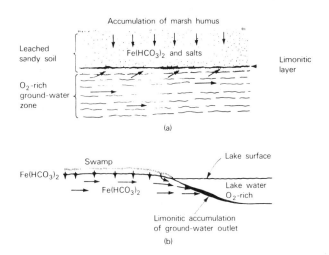

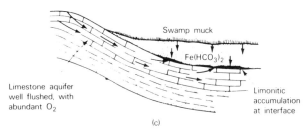

FIG. 18. Development of 'bog iron'-type deposits: limonite deposition: (a) close to water-table in a marsh; (b) in a lake where iron-bearing groundwaters encounter oxygenated lake waters; and (c) at a swamp water, oxygenated aquifer–water interface (from Borchert, 1960, and Stanton, 1972).

James (1966) noted that except for the high iron content, there was little distinctive about the chemistry of the ores. They are massive to poorly banded, rich in normal marine detritus, and occur in basins which are rarely greater than 160 km in length.

The exact mechanisms by which the minette-type ores have formed are not entirely clear, but it appears that iron from multiple sources (weathering, sea-bottom reaction and, at times, from contemporaneous volcanic activity) enters into the sedimentary realm to be precipitated in one of a sequence of facies as a function of Eh, pH and other available constituents (e.g. silica, carbonate, sulphide). The idealized process has been discussed and summarized diagrammatically by Borchert (1960, 1972)—as shown in Figure 20. In the oxygenated regions closest to shore, iron precipitates as oölitic limonite beds. In other neutral to slightly reduced zones, where silica is abundant, chamosite ($(Fe_4Al_2)Si_2Al_2O_{10}(OH)_8$) and/or glauconite ($(K)(Fe, Al)_4Si_7AlO_{20}(OH)_4 \cdot 9 H_2O$) form; in associated areas where sufficient bicarbonate is available, siderite may form. Under stagnant, reducing bottom conditions, H_2S-generated organic decay and bacterial activity precipitates iron in the form of pyrite (and/or other iron sulphide phases).

James R. Craig

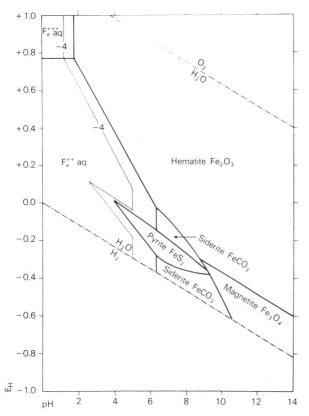

FIG. 19. Eh–pH relations of iron oxides, sulphides, and carbonates with total dissolved sulphur = 10^{-6} M and total dissolved carbonate = 1 M at 25°C. The 10^{-4} M iso-concentration line of Fe^{2+} and Fe^{3+}, at activity of 10^{-1} is also indicated. (From Garrels and Christ, 1965. Reproduced by permission of Harper & Row Inc.)

The third type of sedimentary iron ores, the Pre-Cambrian banded iron formations are by far the most important of all iron deposits. They are restricted to the Pre-Cambrian and Cambrian[1] in age (Goldich, 1973) but occur in several parts of the world under the names of 'taconite', 'itabirite', 'hematite-quartzite', 'banded jaspilite', and 'quartz-banded ore' often in areas measured in tens to hundred of miles. The interested reader is recommended the excellent collection of papers devoted to 'Pre-Cambrian Iron Formations of the World' in *Economic Geology* (Vol. 68, No. 7, 1973), *Survey of the World Iron Ore Resources* (United Nations, 1970) and *Genesis of Precambrian Iron and Manganese Deposits* (Unesco, 1973). These banded iron formations consist of finely laminated beds of iron oxide, carbonate, silicate and sulphide. The most significant features of their geochemistry, which set them apart from the previously discussed iron stones, is their extremely low Al_2O_3 content and their extremely high chert content. In this respect they are very different from normal detrital sed-

1. Boyle and Davies (1973), took issue with this age restriction and suggested that identical ores of Ordovician age occur in the Bathurst—New Castle area of New Brunswick (Canada).

iments. The ore mineralogy consists of magnetite and haematite (geothite is absent), pyrite, greenalite ($Fe_6Si_4O_{10}(OH)_8$), and siderite ($FeCO_3$). The facies concept, as outlined for the minette-type ores, fits the observed field relations of the Pre-Cambrian ores equally well; however, the low Al_2O_3 and high chert contents have posed a problem. The low Al_2O_3 values appear to indicate that the iron was introduced in a relatively pure form—in solution or colloidally—unrelated to clay minerals. The high silica content has prompted several suggestions:

Regular fluctuation in supply due to seasonal climatic changes, i.e., moist seasons with waters more acid and hence iron-containing, and dry seasons with waters more alkaline and silica-containing (Sakamoto, 1950).

Regular fluctuation in biological activity, i.e. seasonal explosive growth of silica-accreting organisms (Krauskopf, 1956a; Huber, 1959; James, 1966; La Berge, 1973).

Regular fluctuations in tectonic activity, i.e. affecting the rates of erosion (Cullen, 1963).

Regular overturn of dimictic lakes, i.e. combination of oxygenation effects with variable stream inflow, biological activity, and convective overturn of lakes (Hough, 1958).

Regular volcanic exhalations of iron solutions, i.e. pulsed exhalative activity with immediate iron precipitation and followed by slower settling of flocculated silica (Goodwin, 1964).

Upwelling of iron-bearing sea waters, i.e. mixing of slightly reducing sea water and slightly oxygenated waters with resultant Fe-oxide and amorphous silica precipitation (Borchert, 1960; Holland, 1972).

Seasonal fluctuations in alkaline lakes, i.e. precipitation of iron hydroxides and carbonates with Fe-silica gels in evaporating saline lakes. Magadiite or Na–Fe silicate gels decomposed under later metamorphic conditions may have provided the source of silica, which later became chert (Eugster and Chou, 1973).

Suffice it to say that the mechanism (or mechanisms) of formation of the banded iron formations is not yet unequivocally established. The apparent absence of any modern analogue of the Pre-Cambrian banded iron ores may well indicate their correlation with the early Pre-Cambrian atmosphere and its low oxygen content.

Non-marine manganese ores

Manganese, an element related to iron, behaves much like it, being soluble in neutral and acid solutions. The problem in understanding sedimentary manganese deposits has been to determine the mechanisms which permit manganese enrichment independent of that of iron. Krauskopf (1957) demonstrated that no significant segregation of Mn from Fe occurred in dissolution at an erosional source; on the other hand, Hewitt (1966) suggested that sub-surface separation must take place since so many hydrothermal waters contain more Mn than

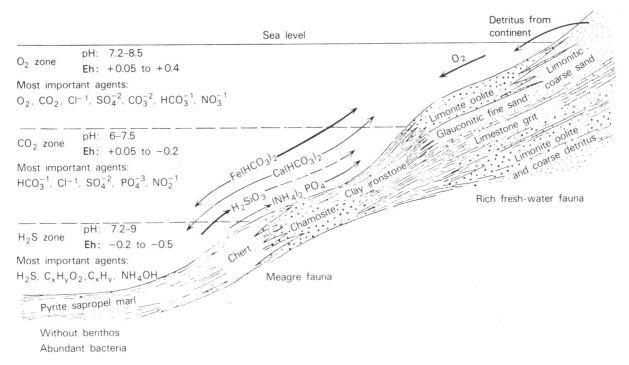

Sea level

Detritus from continent

O_2

O_2 zone

pH: 7.2–8.5
Eh: +0.05 to +0.4

Most important agents:

O_2, CO_2, Cl^{-1}, SO_4^{-2}, CO_3^{-2}, HCO_3^{-1}, NO_3^{-1}

Limonitic coarse sand

Limonite oolite
Glauconitic fine sand
Limestone grit
Limonite oolite and coarse detritus

CO_2 zone

pH: 6–7.5
Eh: +0.05 to −0.2

Most important agents:

HCO_3^{-1}, Cl^{-1}, SO_4^{-2}, PO_4^{-3}, NO_2^{-1}

$Fe(HCO_3)_2$
$Ca(HCO_3)_2$
H_2SiO_3
$(NH_4)_2 PO_4$
Clay ironstone

Rich fresh-water fauna

Chamosite

H_2S zone

pH: 7.2–9
Eh: −0.2 to −0.5

Most important agents:

H_2S, $C_xH_yO_2$, C_xH_y, NH_4OH

Chert

Meagre fauna

Pyrite sapropel marl

Without benthos
Abundant bacteria

Fig. 20. Schematic diagram of the relations between ironstone facies and physico-chemical conditions (from Borchert, 1960).

Fe. Krauskopf (1957) also showed, however, that manganese oxidized less rapidly and less completely than iron and that manganese was slightly more soluble than iron under many natural conditions of E_H and pH. In fact, at any given E_H value, an increase of pH led to preferential precipitation of iron. In contrast to purely inorganic processes, preferential precipitation of manganese by bacteria (Zapffe, 1931) and by mosses (Ljunggren, 1955) has been observed. More recent studies have also demonstrated enzymatic reduction of oxidized manganese (Mn^{3+}, Mn^{4+}) with subsequent solution (Trimble and Erlich, 1968), and bacterial oxidation of reduced manganese (Mn^{2+}) in the oxide form (Silverman and Ehrlich, 1964; Ehrlich, 1968).

Manganese deposits are known to be forming in fresh-water (soil zones and lakes) and salt-water milieux, the latter to be discussed in a later section. Typical of fresh-water deposits is that at Dawson Settlement, New Brunswick (Canada), (Hanson, 1956) in which a 3-m-thick mat of manganese oxides (cryptomelane, psilomelane, pyrolusite) and peat has developed. Formation of the manganese is ascribed to movement as bicarbonates and humates in marshy soils at low pH and Eh, with their subsequent oxidation and precipitation on entering the oxygenated surface waters of the lake. Reducing conditions on the oxygen-depleted lake floor once again permits solution of the manganese and lateral migration until permanently oxygenated regions are again reached, whereupon final precipitation occurs. An excellent review of manganese ore deposits has been presented by Stanton (1972).

Manganese nodules

The occurrence of manganese in marine sediments is sufficiently important to warrant a brief discussion of its own.

Manganese occurs in marine sediments both as dispersed amorphous oxides and hydroxides in concentrations of several hundred p.p.m. and as rich iron–manganese nodules. The total amount of manganese in the ocean sediments requires sources in excess of that which could be ascribed to continental erosion. This excess has generally been attributed to volcanic activity and Stanton (1972) noted a correlation between volcanic activity and manganese content of sediments.

Bostrum (1967) pointed out that Mn had a longer residence time in oceans than iron because iron is more rapidly precipitated through oxidation reactions such as: $4 Fe^{2+} + 6 H_2O + O_2 = 4 FeO(OH) + 8 H^+$. Bonatti and Joensuu (1966) noted that such oxidation reactions should give rise to fractionation of iron and manganese in the vicinity of the emanating source; i.e. higher Fe:Mn ratios close to the source and lower such ratios at a distance—this variation is actually observed in sediments. Furthermore, if both iron and manganese are present in sediments, the Mn is preferentially mobilized under reducing (lower Eh–pH) conditions generated in the anaerobic zones beneath the sediment–water interface.

The most intriguing and important marine occurrence of manganese is that of nodules. These are concretionary or colloform structures averaging 3 cm in diameter and consisting of amorphous hydrated iron–

manganese oxides, goethite α FeO(OH), todorokite [(Mn, Fe, Mg)$_2$Mn$_5$O$_{12}$·3 H$_2$O] and birnessite γ – MnO$_2$. The average contents (Bender, 1972) of nodules include: 16.4 per cent Fe, 14.0 per cent SiO$_2$, 13.8 per cent Mn, 5.4 per cent Al$_2$O$_3$, 4.4 per cent Ti, 4.0 per cent K, 3.3 per cent Ca, 2.3 per cent Na, 0.3 per cent Co and 0.4 per cent Ni. The growth mechanisms of these nodules has been the subject of much conjecture. Cheney and Verdenburgh (1968) suggested that iron-rich protonodules formed under oxidizing conditions at or near the sediment–water interface. Burial permitted encroachment of reducing conditions under which bacteria-produced sulphide reacted with the iron to form pyrite—while manganese, mobilized by the lower Eh–pH, migrated to the sediment-surface oxidizin zone where it precipitated on any available nucleating site (sharks teeth, bones, etc.). Bender et al. (1966) envisaged direct precipitation of the manganese onto deep-sea debris which was kept rolling by 'unknown mechanisms'. On the other hand, some workers (Ehrlich, 1968; Perfil'ev et al., 1965) suggested that micro-organisms were largely responsible for the manganese precipitation in nodule genesis. The textural evidence in addition to isotopic studies clearly indicate that the growth of the nodules has been a very slow accretionary process in the presence of very little clastic deposition.

Whatever the mechanisms of their formation, manganese nodules represent the world's principal reserves of manganese and major reserves of cobalt and nickel (Horn et al., 1973a, b).

Sandstone uranium–vanadium 'roll-type' ores

One of the interesting types of ore deposit which has received considerable attention in the past three decades, as we have entered more deeply into the atomic age, has been the sandstone uranium–vanadium (–copper) ores typified by the 'roll-type' ores of the Colorado Plateau. Similar deposits include those in Wyoming—as described by Fischer (1970)—and some in north and west Australia. The nature and setting of the ores was summarized by McKelvey et al. (1955) as follows:

Most of the deposits are in tabular masses, elongated in the direction of the long axes of the sandstone or conglomerate lenses in which they occur, or in a direction parallel to the orientation of logs or other marks of current lineation. The host rocks are thus interpreted to be fossil stream channel deposits. Most of the deposits occur in or near the thicker parts of the lenses, where mudstone partings or fine debris are present and where logs and other types of carbonaceous matter are abundant. The deposits commonly cut across the bedding, particularly where they form concretion-like structures known as rolls.

The obvious link between these ores and their host rocks is locally modified by faults, anticlines, etc. and is superimposed on a regional trend of higher V:U ratios on the eastern side of the Colorado Plateau to lower V:U ratios on the west, the intermediate zone being appropriately called the Uravan Belt.

The mineralogy of the deposits generally varies with depth, the surficial generally brightly yellow portion containing: carnotite, [K$_2$(UO$_2$)$_2$(VO$_4$)$_2$·3 H$_2$O], tyuyamunite, [Ca(UO$_2$)$_2$(VO$_4$)$_2$·nH$_2$O], autunite, [Ca(UO$_2$)$_2$(PO$_4$)$_2$·nH$_2$O], rutherfordine, (UO$_2$CO$_3$), and the deeper grey-to-black portions containing: uraninite (pitchblende UO$_2$, often closer to U$_3$O$_8$), coffinite, (USiO$_4$·nH$_2$O), montroseite, [VO(OH)], roscoelite, [K(V, Al, Mg)$_3$(AlSi$_3$)O$_{10}$(OH)$_2$].

Typically associated with these later ores are abundant quantities of pyrite and minor amounts of copper sulphides, galena and sphalerite, and the non-metallic gypsum, calcite, dolomite, fluorite, barite and clays.

The mineralogy of these ores, the apparent relationships to past or present water-tables, and the general geological settings have led to extensive examinations of the geochemical behaviour of uranium and vanadium in terms of Eh–pH diagrams (Garrels, 1955; McKelvey et al., 1955; Evans and Garrels, 1958; Hostetler and Garrels, 1962). Such studies have led to an understanding of the geochemical behaviour of uranium and vanadium in aqueous solutions and thrown much light on the origin of the sandstone–uranium–vanadium ores.

Uranium occurs in multiple valency states (+1 to +6) but only the uranous (U^{4+}) and the uranyl (U^{6+}) states are of importance. The U^{1+}, U^{2+} and U^{3+} are such strong reducing agents that they liberate hydrogen from water, and U^{5+} is less stable in the presence of water than U^{4+} and U^{6+}. The transition of U^{4+} to U^{6+} has a standard redox potential within the normal range of geological environments, hence one would expect both ions to occur. The uranous ion, U^{4+}, reacts with water to form an extremely insoluble hydroxide, U(OH)$_4$, which, upon dehydration, forms UO$_2$, pitchblende—a very stable compound under reducing environments (see Fig. 21).

In the presence of carbonate ions, the solubility of the U^{6+} ion is increased considerably from the formation of complexes (Krauskopf, 1967):

$$UO_2(CO_3)_2 = UO_2^{2+} + 2CO_3 =$$
$$= UO_2(CO_3)_3^{4-} = UO_2(CO_3)_2 = +CO_3^{2-}.$$

Similarly, the presence of sulphate ion increases the solubility of U^{6+} by formation of the complex (UO$_2$)SO$_4$.

The work of Hostetler and Garrels (1962) (Fig. 21) revealed that the calculated stabilities of uraninite (under rather reducing conditions) and carnotite (under more oxidizing conditions) were certainly compatible with their observed mode of occurrence. Vanadium, like uranium, occurs in multiple oxidation states, three of which (V^{3+}, V^{4+} and V^{5+}) are stable under normal, near-surface, geological conditions. The dominant aqueous species under conditions of intermediate redox potential is the vanadate ion, VO$_4^{3-}$, which is precipitated

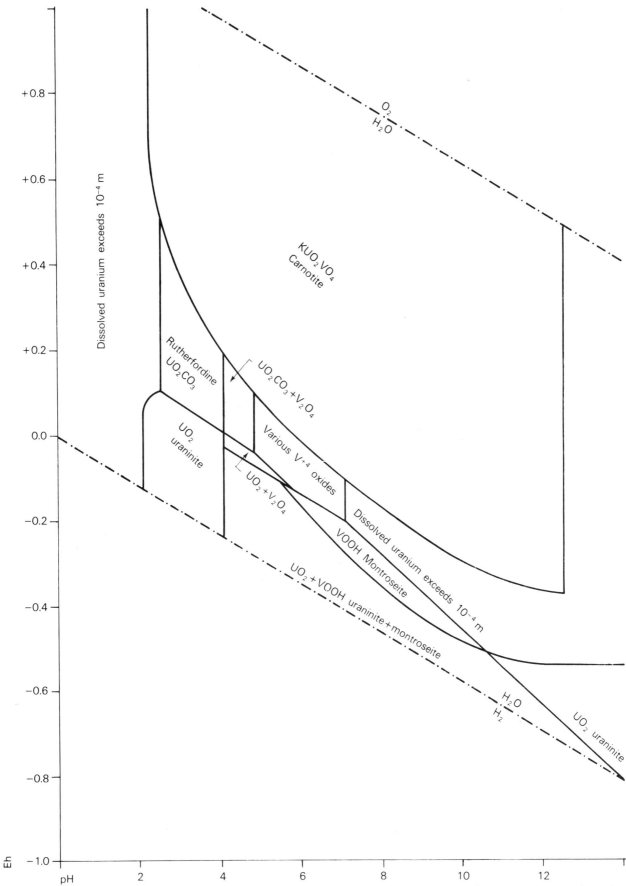

FIG. 21. Eh–pH diagram for uranium and vanadium compounds at 25 °C. Total dissolved V $= 10^{-3}$ M, carbonate $= 10^{-1}$ M (after Garrels and Christ, 1965).

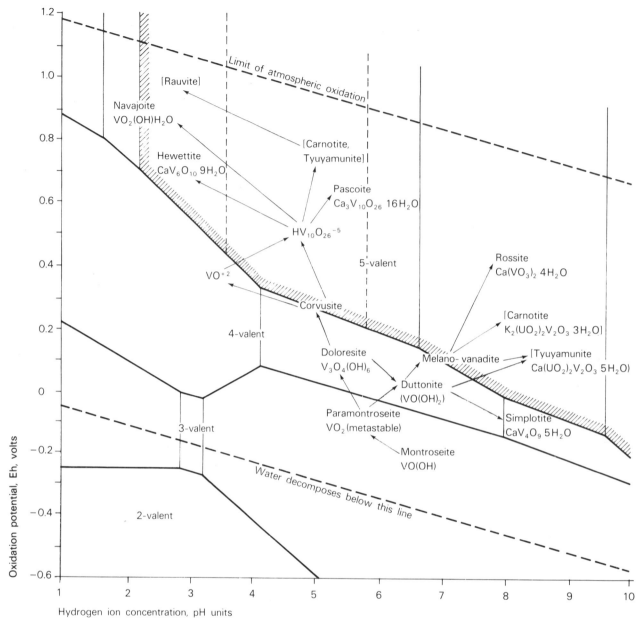

FIG. 22. Alteration sequences of vanadium minerals in terms of Eh and pH. The shaded boundary indicates the stability region of carnotite and tyuyamunite, which supersede all other vanadium species in the presence of the U^{6+} ion (after Evans and Garrels, 1958).

under oxidizing conditions as vanadates [such as carnotite $K_2(UO_2)_2(VO_4)_2 \cdot 3\,H_2O$], and under reducing conditions as montroseite, $VO(OH)$.

The inclusion of this information in Figure 21 shows the complex picture of the formation of sandstone uranium–vanadium ores; under the reducing conditions resulting from the decomposition of organic matter the uranyl ion—possibly mobilized by CO_3^{2-} or SO_4^{2-}—and the vanadate ion are readily precipitated as uraninite and montroseite respectively. Jensen (1958)

has suggested that the reaction of sulphur (in the form of biogenically generated H_2S) with iron oxides may have precipitated the abundant pyrite in these deposits and released the hydrogen to create a reducing environment. The sulphur could also precipitate other metal ions (Pb^{2+}, As^{2+}, Cu^{2+}) as sulphides; these would also remain stable under such reducing conditions. Subsequent influx of more oxygenated waters would result in the formation of the minerals commonly observed near the surface of these deposits, e.g. carnotite, tyuymunite,

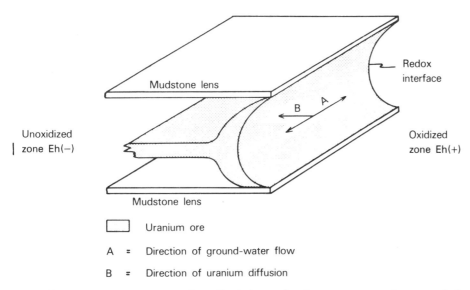

Unoxidized
| zone Eh(−)

Redox
interface

Oxidized
zone Eh(+)

Mudstone lens

Mudstone lens

Uranium ore

A = Direction of ground-water flow

B = Direction of uranium diffusion

FIG. 23. Diagrammatic representation of relations of redox zones to uranium-ore-roll deposits (from Adler, 1964).

etc. The possible alteration processes, as summarized by Evans and Garrels (1958), are shown in Figure 22.

The development of 'roll' structures was apparently the result of what Gruner (1956) termed a 'multiple-migration accretion', i.e. the recycling of uranium by slowly moving groundwater; precipitation occurs on a low Eh (reduced side) or a redox interface—solution on the high Eh (oxidized side). Alder's (1964) concept of this process is shown in Figure 23. Granger and Warren (1969) speculated that the non-biological oxidation-reduction of pyritic sulphur was a major control of the relative Eh–pH conditions developing in the roll front and hence of the site of uranium deposition.

The initial source of the uranium and vanadium is not known with certainty but could easily have been extracted by normal weathering of granitic or volcanic rocks. Sulphur isotopic studies (Jensen, 1958; Warren, 1972) of pyrites in the uranium–vanadium ores indicated that biological activity had been responsible for generation of the pyrite.

Oxidation and supergene enrichment

Subsequent to ore formation, many ores are exposed to surface or near-surface conditions and significant geochemical changes may then occur. The changes may be destructive and dispersive, they may modify ore mineralogy with little effect in ore tenor, or they may actually be concentrative in nature resulting in enrichment of ores. The development of laterites and bauxites as a result of normal weathering has already been discussed.

Although sulphide ore minerals are not stable when exposed to the atmosphere, the rates of oxidation in dry air are extremely slow. In the presence of water, how-

ever, the sulphides begin to alter more rapidly through reactions of the type:

$$H_2O + CO_2 = H_2CO_3 \, ;$$

$$PbS + 2H_2CO_3 = Pb^{2+} + 2HCO_3^- + H_2S;$$

$$H_2S + 2O_2 = SO_4^{2-} + 2H^+, \text{ and}$$

$$Pb^{2+} + SO_4^{2-} = PbSO_4 \, .$$

Sato (1960a, b) noted that Eh values of mine waters were often far below the theoretical maximum for the $O_2 - H_2O$ couple and he suggested that reactions involving peroxide, H_2O_2, might also be important, i.e.

$$2H_2O + O_2 = 2H_2O_2;$$

$$4H_2O_2 + PbS = PbSO_4 + 4H_2O \, .$$

The most potent agent in the alteration of ores is generally held to be sulphuric acid which is generated by the decomposition of pyrite through reactions of the type:

$$FeS_2 + 3.5O_2 + H_2O = Fe^{2+} + 2SO_4^{2-} + 2H^+ \, ;$$

$$2Fe^{2+} + 3SO_4^{2-} + 2H^+ + 0.5O_2 =$$

$$= 2Fe^{3+} + 3SO_4^{2-} + H_2O;$$

$$6Fe^{2+} + 1.5O_2 + 3H_2O = 4Fe^{3+} + 2Fe(OH)_3, \text{ and}$$

$$2Fe^{3+} + 6H_2O = 2Fe(OH)_3 + 6H^+ \, .$$

Such sequences of reactions serve to develop the limonite gossans which cap exposed ore deposits, and to generate ferric-sulphate solutions which serve as sol-

255

James R. Craig

vents in the leaching of other sulphides through reactions like:

$$Fe_2(SO_4)_3 + Cu_2S = CuSO_4 + 2FeSO_4 + CuS;$$

$$Fe_2(SO_4)_3 + CuS + 1.5O_2 + H_2O = CuSO_4 + \\ + 2FeSO_4 + H_2SO_4;$$

$$Fe_2(SO_4)_3 + PbS + 1.5O_2 + H_2O = PbSO_4 + \\ + 2FeSO_4 + H_2SO_4, \text{ and}$$

$$Fe_2(SO_4)_3 + ZnS + 1.5O_2 + H_2O = ZnSO_4 + H_2SO_4.$$

Metals extracted by the generation of acid solutions are transported as long as the acid conditions persist. Re-precipitation may occur as a result of (a) reaction with wall rocks—i.e. neutralization, or (b) by dilution effects on entering the water-table (i.e. rise in pH). Precipitation may also occur as a replacement reaction re-

TABLE 6. Relationships among some common supergene minerals and zones in which enrichment occurs

Zone	Copper minerals	Silver minerals	Lead minerals	Zinc minerals
Gossan or capping	Perhaps traces of some minerals of the oxidized zone	Cerargyrite	Anglesite Cerussite	
Oxidized zone	Malachite Azurite Chrysocolla Ceuprite Tenorite Brochantite[2] Atacamite[2] Andorite[2]	Cerargyrite Native silver	Anglesite Cerussite	Smithsonite[1] Calamine[1]
Supergene	Chalcocite	Argentite		
Sulphide zone	Covellite	Some native silver Pyrargyrite[3] Proustite[3]		

1. Commonly below lead minerals if they are present.
2. Commonly in more arid regions.
3. Generally below argentite.

Source: After Lamey, 1966.

sulting from electrolysis (i.e. $Cu^{2+} + ZnS = CuS + Zn^{2+}$). The general relationships among oxides, sulphides, and dissolved species are best shown in E_h-pH diagrams (e.g. Fig. 16, which outlines the various conditions encountered in the natural realm), detailed discussions of which are given by Garrels (1954), Garrels and Christ (1965), Krauskopf (1967) and Anderson (1955).

The result of the oxidation and supergene reactions is the development of a sequence from the surface downwards: (a) leached Gossan, (b) oxidized zone, (c) supergene zone, (d) primary sulphides. The boundary between the oxidized and supergene zones is commonly attributed to the position of the present or a former water-table. Lamey (1966) summarized the effects on primary copper, silver, lead and zinc sulphides (see Table 6).

Lamey (1966), on the basis of experimental studies and mine-dump leaching, also stressed the possible role of bacteria in promoting oxidation and enrichment of copper deposits; effects on natural ores are unknown.

Although supergene enrichment has traditionally been considered in Cu–Pb–Ag ores, the effects in Ag–Co–Ni ores (Boyle and Dass, 1971b) and even Fe–Ni–Cu ores (Nickel, 1973; Nickel *et al.*, 1974; Watmuff, 1974) have recently been recognized.

CHEMICAL PRECIPITATES FROM SOURCES OTHER THAN WEATHERING

Within the sedimentary realm are many deposits which have formed at the time of sedimentation and/or diagenesis and whose primary features were completed prior to lithification. The ores generally lie conformably within their enclosing sediments and often occur in close proximity to volcanoes or volcanic rocks.

These sedimentary ores appear to constitute a continuum (Fig. 24) between those with obvious and significant volcanic contribution (Volcano Islands, Cyprus pyritic deposits, *kuroko*-type ores of Japan—even pure sulphur flows in Japan), to those or probable volcanic affiliation (i.e. volcanics in close proximity, e.g. Mount Isa), to those with no recognizable volcanic affinities (Kupferschiefer of northern Europe, Zambian Copperbelt). Examples of modern ores of this latter type are the Red Sea deposits which, though forming as a result of hydrothermal emanations, neither contain nor are associated with volcanic rocks. Similar deposits formed in the geologic past would bear no tell-tale volcanic affiliations even if the fluids were purely volcanic in origin. For convenience, these ores are considered in two groups: (a) the stratiform ores not obviously associated with volcanism and (b) the volcanic exhalative deposits directly related to volcanism.

Some stratiform ores

Many strata-bound marine deposits of broad areal extent bear no apparent relationship to metal sources other

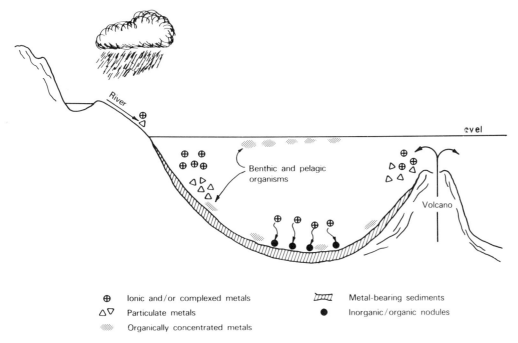

⊕   Ionic and/or complexed metals   ▨   Metal-bearing sediments

△▽   Particulate metals   ●   Inorganic/organic nodules

▨   Organically concentrated metals

FIG. 24. Schematic representation of the sources of incorporated metals in marine sediments. (After Hallberg, 1972. Reproduced by permission of Springer-Verlag New York Inc.)

than sea water or the sediments which enclose them. Two classic examples are the Kupferschiefer of northern Europe (in which are found extensive ores of galena, sphalerite and chalcopyrite with large quantities of pyrite), and the Zambian Copperbelt (in which are found deposits rich in chalcopyrite, bornite, and carrollite and linaeite also with large quantities of pyrite). The understanding of the genesis of these deposits is tied to a determination of the source of the metals and the source of the sulphur.

Several workers (Butlin, 1953; Kimata *et al.*, 1955; Skopintsev, 1961; Berner, 1970) have noted that the primary generators of sulphur in modern sediments (and presumably most ancient ones as well) are the sulphate-reducing bacteria, *Desulphovibrio* and *Desulphotomaculum*, a most abundant group of micro-organisms in many sediments (Simindu and Aiso, 1962). They have also shown that simple organic decomposition releases only negligible quantities of sulphur and that the sulphide content of the sediments is proportional to the concentration of sulphate-reducing bacteria and/or organic carbon. The bacteria facilitate reactions of the type (using the general carbohydrate CH_2O):

$$2CH_2O + SO_4^{-2} \longrightarrow HCO_3^- + HS^- + CO_2 + H_2O .$$

Oxygen is toxic to the sulphate-reducing bacteria, hence sulphide production is limited to regions in the sediments in which excess oxygen has been removed by reaction with metals, sulphide, ammonia, methane, etc. as shown in Figure 25. In these oxygen-depleted regions reaction of available metal and bisulphide can produce

sulphides—as is evidenced by the abundance of disseminated sulphide (nearly always as pyrite since iron is practically ubiquitous and generally in greater abundance than any other metal) in the reduced layers of modern sediments. The problem of the formation of syngenetic ores in most of the ocean today is that of transporting sufficient metal ions other than iron (most iron transport is probably as clastics or as colloidal iron oxyhydroxides) to the sites of HS^- production. Rickard (1973) amongst others has pointed out that (a) metals are in very low concentrations in sea water, and (b) many of the metals would precipitate in other forms (i.e. Cu as malachite) if their concentrations in sea water were to rise. This situation is at least approached on the bed of the Red Sea, where hydrothermal emanations are at present precipitating large quantities of Fe, Mn, Zn, Pb and Cu (Bischoff, 1969; Hackett and Bischoff, 1973)—much of it in the form of oxides because there is insufficient sulphur to react with all of the metal. The sulphide layers that do occur in these sediments may well reflect diagenetic changes, i.e. periods of sulphur introduction from the same sources as the metals, or changes in bottom conditions, as noted below, the reducing, sulphide-rich waters. Near-shore areas also have the additional disadvantage of high rates of clastic sedimentation, which can overwhelm sulphide deposition and drastically dilute metal grades in sediments.

A somewhat unique situation exists in closed euxinic basins such as fjords and the Black and Baltic Seas. Here the oxidized-reduced boundary (Fig. 25) lies well above the sediment–water interface, thereby affording a

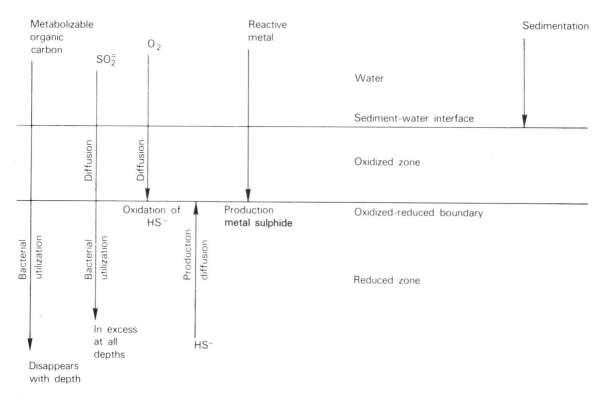

FIG. 25. Schematic representation of a model for sedimentary metal sulphide formation; the oxidized-reduced boundary is within the sediment (after Rickard, 1971).

bottom zone of sulphide-rich waters. In such environments, ore-grade deposits of metal sulphides can be readily formed if sufficient metal is available. Ordinary, organic-rich black shales do show slight metal anomalies (Vine and Tourtelot, 1970) but the values are far below ore grade. Rickard (1973) noted that 'at least 0.1 per cent dry weight organic carbon is required to produce synsedimentary metal sulphide deposits containing more than 1 per cent metal ... [but that] ... an additional metal flux over and above that normally observed in seawater is essential'. He concluded that optimum sites for synsedimentary sulphide ores were organic-rich, fine-grained sediments with contemporary volcanic associations.

The entire topic of sulphide formation by direct or indirect biological activity has generated a voluminous literature; some of the more extensive works to which the interested reader may wish to refer include Bass-Becking and Moore (1969), Love (1962), Temple (1964), Temple and LeRoux (1964) and Nicholas (1967).

In the light of the discussion above, the problem of forming extensive syngenetic ores such as those of the Kupferschiefer (Fe, Cu, Pb, Zn sulphides), or the Zambian Copperbelt (Fe, Cu, Co sulphides), is primarily the difficulty of finding a metal source. Traditionally, two sources have been suggested: (a) terrestrial weathering

with metal transport in rivers and streams, and (b) hydrothermal fluids (springs or volcanic emanations). Streams have been suggested both as the metal source for the Zambian Copperbelt (Garlick, 1961) and for the Kupferschiefer (Gregory, 1930; Deans, 1948). In spite of Gregory's noting that the Kupferschiefer basin would have received drainage from the highly mineralized areas of the Harz, Erzegebirge and Thuringerwald, Krauskopf (1956b) observed that streams, even from the areas of known ores, were notably devoid of significant quantities of dissolved metals. Weiss and Amstutz (1966), Harańczyk (1970) and others have suggested that the metals were not in solution but rather were adsorbed to clay particles and were only released upon contact with sulphide-rich bottom waters of the basin.

Alternative suggestions for the sources of the metal of the Kupferschiefer have been the underlying sandstones (Oberc and Serkies, 1968), the overlying sabkha deposits (Renfro, 1974), or circulating epigenetic fluids (Davidson, 1962). Love (1962) noted that many of the Cu, Pb and Zn sulphides of the Kupferschiefer appeared not to be original precipitates but to be pseudomorphous after pyrite; unfortunately he was unable to pinpoint the time of replacement. Similar observations have been made by Weiss (1973) who postulated diagenetic replacement of pyrite in the Parting shale of the

White Pine deposit by copper sulphides through reactions of the type:

$$8FeS_2 + 7Cu_{aq}^{2+} + 8H_2O = 7CuFeS_2 + Fe^{2+} + 2SO_4^{2-} + 16H^+,$$

and

$$9CuFeS_2 + 11Cu_{aq}^{2+} + 8H_2O = 4CU_5FeS_4 + 5Fe^{2+} + 2SO_4^{2-} + 16H^+.$$

A magmatic source or a volcanic emanation—gaseous or liquid—may leave no record of its presence other than its metals. Certainly this is the case in the Red Sea deposits and the Salton Sea brine, neither of which contains a significant amount of sulphur. Hence, the fluids from which each derived would be difficult to trace to a given source if they were not now in the act of forming.

Volcanic exhalative deposits

In contrast to the aforementioned ores, some large ore deposits bear a clear relationship to volcanic rocks and may thus have adjacent metal sources. Excellent reviews on this subject have been written by Anderson (1969), Hutchinson (1973) and Tatsumi *et al.* (1970). Hutchinson noted that the latest studies of the massive pyritic base metal sulphide deposits in volcanic rocks suggested that they were of volcanic origin, formed in recurrent episodes of sea-floor fumarolic activity during prolonged periods of subaqueous volcanism. The principal metals of the sulphide minerals in these ores are (in order of importance): iron, zinc, lead and copper. Iron, in the form of pyrite, is generally the dominant mineral, but Stanton (1964), Hutchinson (1973) and others have noted that the ores may be subdivided into four classes: iron, iron–copper, iron–copper–zinc, and iron–copper–zinc–lead.

These ores have been generated from the Pre-Cambrian through the Tertiary and are probably still forming, e.g. at Volcano (Honnorez *et al.*, 1973). They are associated with a wide variety of magmas, basaltic to rhyolitic, and include sedimentary rock types ranging from cherts and iron formations to volcano-clastics and greywackes.

Classic examples of this type of ore deposit which have received much attention in recent years are the *kuroko* or black ores of Japan. These are fine-grained, compact masses of pyrite, chalcopyrite, galena and sphalerite, associated with rhyolitic breccias, flows and tuffs (their mode of formation is illustrated in Fig. 26).

The nature of the fluids responsible for formation of the *kuroko* type (and quite likely many other similar volcanic-associated ores) has recently been evaluated by Kajiwara (1970, 1973a, b). He calculated that deposition occurred from near neutral (pH = 5.5) chloride-rich, fluids between 150–200 °C, with the relative activities of CO_2 (10^{-6} atm), S_2 (10^{-10} atm), O_2 (10^{-35} atm).

The lead isotope data for volcanic marine deposits suggests a mantle source for the lead (and perhaps for the copper and zinc also), whereas the sulphur isotopes suggest multiple or mixed sources for this element—at least part of it being derived from sea water (Ohmoto *et al.*, 1970).

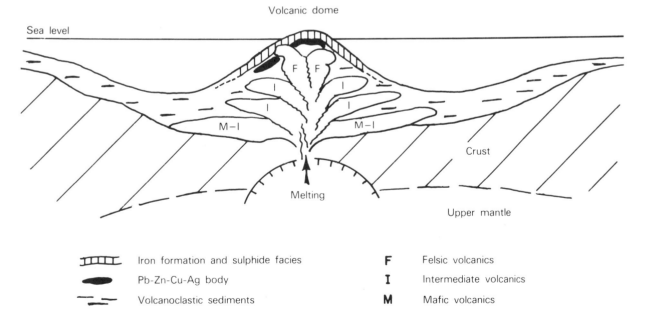

FIG. 26. Diagrammatic illustration of *kuroko*-type volcanogenic massive sulphide deposit formation (after Hutchinson, 1973).

Ores of metamorphic affiliation

Many types of ores occur in terrains that have been subjected to metamorphism. Some ores appear to bear a genetic relationship to the metamorphic episode(s) whereas others appear to have been merely passive, suffering only minor changes in ore textures. The higher the grade of metamorphism, thermal and/or dynamic, the more uncertain is the true origin of the ores due to the metamorphic overprint.

Mookherjee (1970b) pointed out the desirability in differentiating among:
'Metamorphosed' ore deposits—ores which are not genetically related to the metamorphism but which have been subjected to it.
'Metamorphic' ore deposits—ores that 'owe their origin' to metamorphic process(es).
'Mobilized' ores—ores which have been moved—in whole or in part—by metamorphic processes.
Mookherjee suggested that mobilization might be considered in one of three ways: (a) chemical—solution, diffusion, volatile transport; (b) transitional—partial melting and transportation as crystal–liquid aggregate; or (c) physical—plastic flowage, melting.

Generally, metamorphism is discussed in terms of two types (which obviously are not totally exclusive): regional metamorphism, which occurs on very broad sis, often related to orogenesis; and contact metamorphism, which is limited to zones adjacent to igneous rocks. A related activity, which can be of great importance in the development of ore bodies, is 'metasomatism' which Thompson (1959) defined as: '… any process involving a change in the bulk chemical composition of the mineral assemblage'. It may be active under either regional or contact metamorphism.

The manner in which metamorphism results in the ultimate distribution of elements, their dispersal or concentration, is extremely complex and not completely understood. Barth (1962) pointed out that the primary control is variation in chemical potential and that physical variables (e.g. melting-point, sheer strength, etc.) would play only subsidiary roles. He related a fundamental law: '… dispersion of minerals at places where the activity is less and consolidation there, constitute the fundamental processes in metasomatism and metamorphic differentiation'. He noted that activity gradients were created by differences in five main factors: (a) pressure, (b) temperature, (c) chemical compositions of minerals and rocks, (d) the size of minerals, and (e) the surrounding milieu. Detailed discussions of metamorphic processes in general are given by several workers (Fyfe et al., 1958; Thompson, 1959; Vidale, 1969).

Our immediate concern is metamorphism as related to ores. The discussion below is of necessity a brief overview; the interested reader is referred to some excellent works by MacDonald (1967), Vokes (1969), Ramdohr (1969), Stanton (1972) and Kalliokoski (1965).

Furthermore, this discussion will be primarily limited to chemical effects; for details on structural and textural changes in ores the reader is referred to Ewers (1967), Stanton (1972), and Stanton and Gorman (1968).

CONTACT METAPHORMISM

Contact metamorphism of ores, like other rocks, results from baking effects and the introduction of elements from igneous fluids. The extent of the affected aureol depends upon the temperature, size and volatile content of the intrusive, and the nature of the rock into which the intrusive is emplaced. Generally, acidic igneous bodies are emplaced at lower temperatures than are basic bodies; however, the acidic bodies contain much greater amounts of volatile constituents than do basic rocks. The release of these constituents and metal cations in water-rich fluids has resulted in formation of many of the hydrothermal ore deposits. Accordingly, only arbitrary distinctions can be drawn between contact metamorphic and hydrothermal ores.

Although thermal effects around igneous intrusives are common, few ores are directly attributable to them. Barnes (1959), in fact, found that even extreme thermal gradients were insufficient to mobilize Cu, Pb, Zn and Fe in metal-bearing shales. Rather, the thermal effects have generally caused relatively minor oxidation-reduction reactions—especially involving sulphur. Antun (1967) found that emplacement of a granitic body caused loss of sulphur from pyrite disseminated in shales, (i.e. $2 FeS_2 = 2 FeS + S_2$), as much as 3 km from the intrusive contacts. Similarly, intrusion of a basic dike at the Geco Mine, Manitouwadge, Ontario (Canada) has converted adjacent pyrite-bearing ore into pyrrhotite ore (Mookherjee and Dutta, 1970; Mookherjee, 1970a).

On the other hand, the loss of sulphur from pyrite may actually aid in the formation of some ores. Naldrett (1966) described the generation of iron–nickel ores of the Alexo deposit in Ontario as the result emplacement of a peridotite sill adjacent to a pyritic volcanic host rock (Fig. 27). Iron and nickel were extracted from the peridotite by sulphur released through thermal decomposition of the pyrite. Massive sulphide ore was subsequently deposited in a dilatant zone developed along a shear zone. Naldrett (1973) has subsequently concluded that thermally induced sulphidation was not the ore-forming process at Alexo but that this process has been active in other deposits of this type. The sulphurization, or sulphidation, process initially envisaged as active at Alexo has been described in general terms by Kullerud and Yoder (1965) who have pointed out that sulphur, however, supplied or released, is extremely capable of extracting iron or other chalcophile elements from silicates by reactions of the type:

$$4FeMgSiO_4 + S = FeS + Fe_3O_4 + 4MgSiO_3;$$

Fe-Mg olivine   pyrrhotite   magnetite   Mg-pyroxene

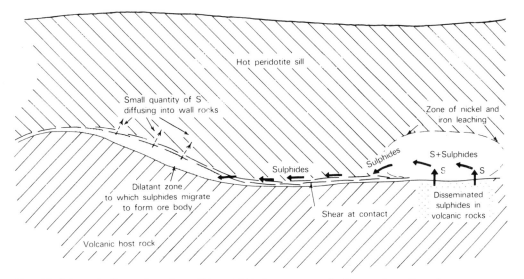

FIG. 27. Schematic illustration of possible sulphidation sequence leading to formation of a iron-nickel sulphide ore (after Naldrett, 1966).

$$4(Mg,Fe)_2Al_2Si_5O_{18} + S = FeS + Fe_3O_4 +$$

Mg, Fe-cordierite   pyrrhotite   magnetite

$$+ 2Mg_2Al_4Si_5O_{18} + 4 Al_2SiO_5 + {_6}SiO_2 .$$

Mg-cordierite   andalusite   quartz

(sillimanite)

Contact metasomatism, with introduction of iron, magnesium and silica, had led to formation of magnetite-rich skarn deposits where carbonate sequences have been intruded by basic rocks. Two rather classic examples are the ore bodies at Cornwall and Morgantown, Pennsylvania in the United States (Sims, 1968). These bodies and others like them (see Park, 1972) have apparently formed through extensive replacement of the carbonate by iron-bearing fluids or solutions. Turner and Verhoogen (1960) suggested that the development of fluoride- and chloride-bearing minerals in the skarns, e.g. fluorite, CaF_2, and scapolite, $[(Na, Ca)_4[(Al, Si)_4O_8]_3(Cl, CO_3)]$, may be indicative of transport of the iron via the volatile species of FeF_3 and/or $FeCl_3$; this contention is supported by the work of Krauskopf (1964). Park (1972) also noted the common association of apatite $[Ca_5(PO_4)_3(Cl, F, OH)]$ with iron ores of both magmatic and metasomatic affiliation and suggested that fluorine and/or phosphorus might play an important role in their formation.

Contact metasomatism around more acidic intrusive has led to development of the rich tungsten ores in skarn (tactite) deposits at Bishop, California (United States) (Gray et al., 1968). The metasomatic activity resulted in replacement of limestones by minerals of silica, alumina, iron, manganese, fluorine and tungsten. Interestingly, Gray et al. (1968) noted that scheelite $[Ca(WO_4)]$ mineralization was almost always associated with quartz, fluorite and apatite.

REGIONAL METAMORPHISM

Regional metamorphism includes the complete gradiation of phenomena from diagenesis to anatexis. Correlation of grade of metamorphism with occurrence of ore bodies on any broad scale has not as yet been successful. Nevertheless, migration of metals and their entrapment in favourable sites and/or the upgrading of deposits by concentration of material has been observed, at least on a small scale (Backlund, 1950). One significant effect of high-grade metamorphism has been the alteration of mineral assemblages. An example appears to be the unique Zr–Mn ores of the Franklin Furance area of New Jersey (United States), which are believed to have been deposited as Red-Sea-type marine deposits (Callahan, 1966) and converted to rare zinc and manganese oxides and silicates (Metsger et al., 1969) during diapiric settling through a plastic marble host.

Hagner and his co-workers (Hagner et al., 1963, 1967) have suggested that high-grade metamorphism was responsible for iron extraction from accessory magnetite and recrystallized pyroxenes and its redeposition in areas of low pressure and high oxygen activity. They envisaged a process of grain boundary migration of a dispersed phase of ions, molecules, or atoms which extracted iron from the areas as much as 200 m from the present positions of the ore bodies.

The remobilization of sulphide ores as melts during metamorphism has been suggested as a possible means of transporting ores to new sites of emplacement. Generation of such a melt has been postulated at Mount Isa and at Broken Hill (Brett and Kullerud, 1966; Lawrence, 1967), though the effects were not large. The generation of melts in the Fe–Pb–S systems as low as 718 °C (Brett and Kullerud, 1966) and in the Cu–Pb–S system as low as 508 °C (Craig and Kullerud, 1969) have supported the suggestion of similar melts in highly metamorphosed ores. At metamorphic temperatures below those of melting it appears that some differential movement of elements may occur by preferential extraction by fluid phases which are generated by metamorphism (Vokes, 1969, 1970).

Walker and Buchanan (1969) found that heating of carbonaceous sediments yielded a host of hydrocarbon, sulphide, chloride and fluoride species which could readily aid the movement of cations in metamorphic mobilization or hydrothermal fluids thus generated.

Just as sulphur is readily mobilized in contact metamorphic zones, it may also apparently be mobilized during regional metamorphism. Thus, Marmo (1960) noted the relationship between Caledonide ores and black sulphide–graphite schists and suggested that mobilization of ore-forming materials from the schists resulted from tectonic deformation and accompanying hydrothermal activity. Williamson and Myers (1969) noted a pyrrhotite and pyrite isograd in Maine, and Carpenter (1974) mapped a similar isograd in the vicinity of the famed Ducktown, Tennessee, ores in the United States, which marked the zone along which pyrite is decomposed to pyrrhotite.

A somewhat similar but more pronounced 'desulphurization' process has been recognized during serpentinization. A good example is that of the Muskox intrusion (Chamberlain et al., 1965), in which sulphides have been reduced to native metals as a result of the hydrogen generated during serpentinization. The reactions by which hydrogen is generated and by which the sulphides are decomposed are of the type:

$$10 Mg_{1.6}Fe_{0.4}SiO_4 + 12 H_2O + 2/3 SiO_2 =$$
olivine   silica

$$= 16/3 Mg_3Si_2O_8(OH)_4 + 4/3 Fe_3O_4 + 8/3 H_2;$$
serpentine   magnetite

$$Fe_3O_4 + 4 H_2 = 3 Fe + 4 H_2O;$$
magnetite   iron

$$FeS + H_2 = Fe + H_2S;$$
pyrrhotite   iron

$$2(FeNi)_9S_8 + 7 H_2 = 9 Ni_2Fe + 9 FeS + 7 H_2S;$$
pentlandite   awaruite   pyrrhotite

$$5 CuFeS_2 + 2 H_2 = Cu_5FeS_4 + 4 FeS + 2 H_2S;$$
chalcopyrite   bornite   pyrrhotite

$$2 Cu_5FeS_4 + H_2 = 5 Cu_2S + 2 FeS + H_2S; \text{ and}$$
bornite   chalcocite   pyrrhotite

$$Cu_2S + H_2 = 2 Cu + H_2S.$$
chalcocite   copper

Metallogenic epochs and provinces

An examination of the nature and distribution of ore deposits cannot help but lead one to the conclusion that the ore-forming processes have not operated uniformly throughout time and space. Such observations have led many workers to the concepts of metallogenic epochs and provinces—concepts which have been applied to all portions of the Earth's crust. Prime examples are: (a) the restriction of banded iron formations to the Pre-Cambrian; (b) the restriction of nearly all porphyry Cu–Mo deposits of the south-western United States to the time span 50–70 my, and (c) the well-defined tin belts of South-East Asia and in Bolivia.

Turneaure (1955) presented an extensive review of metallogenic epochs and provinces on the basis of literature available up to 1955. Since that time numerous new works on these topics have appeared and the concepts of metallogenic epochs and provinces have been

TABLE 7. Articles on geochemical aspects of metallogenic epochs and provinces (since 1955)

Region	References
North America	Burnham (1959); Lang (1969); Tweto and Sims (1963); Schmitt (1966); Callahan (1967); United States Geological Survey (1970); Canadian Institute of Mining and Metallurgy (1966); Naldrett and Gasperini (1971); Sutherland Brown et al. (1971).
Central America	Levy (1970); Ferencic (1971); Gableman and Krusiewski (1968).
South America	Schneider-Scherbino (1962); Stoll (1964); Ahlfeld (1967); Petersen (1970); Goossens (1972); Schuiling (1967); Ahlfeld (1974).
Africa	de Kun (1963); Clifford (1964, 1965, 1966); Fonteilles (1967); Chantraine and Radelli (1970).
Europe	Superecanu (1970); Gableman and Krusiewski (1972); Sierra et al. (1972); Ovtracht and Tamain (1972).
Asia	Magakian (1972); Semenov et al. (1967).
Australia	Solomon et al. (1972); McAndrew (1965).
Japan and Pacific Basin	Tatsumi et al. (1970); Radkevich (1972).
General and/or world-wide	Tatrinov (1957); Goodwin (1961); Brock (1964); Pereira and Dixon (1965); Petrascheck (1965); Stoll (1965); Tugarinov (1966); Bilibin (1968); Rundkvist (1969); Mancher (1972); Guild (1972); Stanton (1972); Gableman and Krusiewski (1969); Naldrett (1973).

extended from obvious, merely close relationships to mobile belts (i.e. McCartney, 1962; McCartney and Potter, 1962) to include global scale tectonics (i.e. Canadian Institute of Mining and Metallurgy, 1966; Kutina, 1969; Guilbert, 1971; Guild, 1972; Sawkins, 1972; Sillitoe, 1973; and Stanton, 1972). Table 7 is an attempt, intentionally incomplete, to provide an updated listing of some of the literature on metallogenic epochs and provinces since Turneaure's work; articles emphasizing structural rather than geochemical aspects are omitted.

The problem of why and where metallogenic belts exist is not known with certainty but appears to be related to the geochemical evolution of the Earth, its continents, its atmosphere and its hydrosphere. Although no single comprehensive statement regarding metallogenics is yet available, it is possible to summarize briefly some aspects of ore deposits (largely from Stanton, 1972).

Banded iron formations. Restricted to shield areas of the Pre-Cambrian; quite possibly indicative of significant differences in the Earth's early atmosphere and oceans.

Nickel–sulphide ores of mafic–ultramafic association. Almost entirely confined to Canadia, Western Australian and Siberian shield areas; most ores are Pre-Cambrian in age and possibly related to evolution of the Earth's mantle (Naldrett, 1973).

Stratiform volcanic ores. Widespread in time and space but most apparent during $3,200-2,500 \times 10^6$, $1,700-1,500 \times 10^6$, and $500-300 \times 10^6$ years (Hutch-

inson, 1973); apparently closely tied to mobile belts.

Anorthositic iron-titanium oxide ores. Major development in North American Pre-Cambrian during $1,400-1,000 \times 10^6$ years.

Porphyry copper–molybdenum ores. Widespread in Circum-Pacific region with major development at about 50–70 my.

Hydrothermal veins. Widespread in time and space but some associations are well defined on regional bases, i.e. Sn provinces of South-East Asia, northern Europe, Bolivia, China.

Stratiform Pb–Zn ores. Post-Pre-Cambrian, northern hemisphere, generally in geosynclinal shelf sediments.

As our knowledge of the evolution of the Earth increases, it is hoped that many of the above-mentioned generalities can be replaced with firm statements. Investigation into the many unsolved problems of ore genesis will no doubt aid in our understanding of the ore-forming process and hopefully in our exploitation of the Earth's storehouse of metals.

Acknowledgements

The writer is much indebted to Paul B. Barton, Jr., and Helmuth Wedow for their critical reviews and comments which have been of great help in the preparation of this manuscript. Completed December 1974.

References

ABRAHAM, K. P.; DAVIES, M. W.; RICHARDSON, F. D. 1960. Sulphide Capacities of Silicate Melts, Pt. I. *J. Iron Steel Inst.*, Vol. 196, p. 309–12.

ADLER, H. H. 1964. The Conceptual Uranium Ore Roll and its Significance in Uranium Exploration. *Econ. Geol.*, Vol. 59, p. 46–53.

AHLFELD, F. 1967. Metallogenetic Epochs and Provinces of Bolivia. *Mineralium Deposita*, Vol. 7, p. 291–311.

——. 1974. Neue Beobachtungen über die Tektonik und die Antimonlagerstätten Bolivens. *Mineralium Deposita* Vol. 8, p. 125–31.

AHRENS, L. H. 1965. *Distribution of the Elements in Our Planet.* New York, Mc-Graw-Hill. 110 p.

ANDERSON, C. A. 1955. Oxidation of Copper Sulfides and Secondary Sulfide Enrichment. *Econ. Geol. 50th Ann. Vol.*, p. 324–40.

——. 1969. Massive Sulfide Deposits and Volcanism. *Econ. Geol.*, Vol. 64, p. 129–46.

ANTUN, P. 1967. Sedimentary Pyrite and its Metamorphism in the Oslo Region. *Norsk Geol. Tidsskr.*, Vol. 47, p. 211–35.

ARNOLD, M; MAUCHER, A; SAUPE, F. 1973. Diagenetic Pyrite and Associated Sulphides at the Almaden Mercury Mine, Spain. In: G. C. Amstutz and A. J. Bernard (eds.), *Ores in Sediments*, p. 7–19. Berlin, Springer-Verlag.

BACKLUND, H. G. 1950. Some Observations on Homogeniza-

tion and on Geochemical Discontinuities in Granite Areas. *18th Int. Geol. Congr. Pt. 3*, p. 31–42.

BARNES, H. L. 1959. The Effect of Metamorphism on Metal Distribution Near Base Metal Deposits. *Econ. Geol.*, Vol. 54, p. 919–43.

——. 1962. Mechanisms of Mineral Zoning. *Econ. Geol.*, Vol. 57, p. 30–7.

BARNES, H. L. (ed.) 1967. *Geochemistry of Hydrothermal Ore Deposits* p. 334–81. New York, Holt, Rinehart & Winston.

BARNES, H. L.; CZAMANSKE, G. K. 1967. Solubilities and Transport of Ore Minerals. In: H. L. Barnes (ed.), *Geochemistry of Hydrothermal Ore Deposits*, p. 334–81. New York, Holt, Rinehart & Winston.

BARNES, H. L.; KULLERUD, G. 1961. Equilibria in Sulfur-containing Aqueous Solutions in the System Fe–S–O, and their Correlation During Deposition. *Econ. Geol.*, Vol. 56, p. 648–88.

BARTH, T. F. W. 1962. *Theoretical Petrology* New York, Wiley, 416 p.

BARTON, P. B. 1959. The Chemical Environment of Ore Deposition and the Problem of Low Ore Transport. In: P. H. Abelson (ed.), *Researches in Geochemistry: 1*, p. 279–300. New York, Wiley.

——. 1967. Possible Role of Organic Matters in the Precipitation of the Mississippi Valley Ores. In: J. S. Brown (ed.),

Genesis of Stratiform Lead–zinc–barite–fluorite Deposits. *Econ. Geol.* p. 371–8. (Mono. 3.)

——. 1970. Sulfide Petrology. *Min. Soc. Am. Spec. Paper 3,* p. 187–98.

BARTON, P. B., SKINNER, B. J. 1967. Sulfide Mineral Stabilities. *In:* H. L. Barnes (ed.), *Geochemistry of Hydrothermal Ore Deposits,* p. 236–333. New York, Holt, Rinehart & Winston.

BASS-BECKING, L. G. M.; KAPLAN, I. R.; MOORE, D. 1960. Limits of the Natural Environment in Terms of pH and Oxidation-reduction Potentials. *J. Geol.,* Vol. 68, p. 243–84.

BASS-BECKING, L. G. M; MOORE, D. 1961. Biogenic Sulfides. *Econ. Geol.,* Vol. 56, p. 259–72.

BASTIN, E. S. 1939. The Nickel–Cobalt–Native Silver Ore Type. *Econ. Geol.,* Vol. 34, p. 1–40.

BATEMAN, A. M. 1951. The Formation of Late Magmatic Oxide Ores. *Econ. Geol.,* Vol. 46, p. 404–26.

BENDER, M. L. 1972. Manganese Nodules. *In:* R. Fairbridge (ed.), *Encycl. Geochem. Environ. Sci.,* p. 673–7.

BENDER, M. L.; KU, T.-L.; BROECKER, W. S. 1966. Manganese Nodules: Their Evolution. *Science,* Vol. 151, p. 325–8.

BERNER, R. A. 1970. Sedimentary Pyrite Formation. *Am. J. Sci.,* Vol. 268, p. 1–23.

BETHKE, P. M.; RYE, R. O.; BARTON, P. B. 1973. Hydrogen, Oxygen and Sulfur Isotopic Compositions of Ore Fluids in the Creede District, Mineral County, Colo. *Geol. Soc. Am. Abstr. With Prog.,* Vol. 5, p. 549. (Abstract.)

BILIBIN, Y. A. 1968. *Metallogenic Provinces and Metallogenic Epochs,* 1–35. Flushing, N.Y., Queens University Press.

BISCHOFF, J. L. 1969. Red Sea Geothermal Brine Deposits. Their Mineralogy Chemistry and Genesis. *In:* E. T. Degens and D. A. Ross (eds.), *Hot Brines and Recent Heavy Metal Deposits in the Red Sea,* p. 368–401. New York, Springer-Verlag.

BJERRIUM, J.; SCHWARZENBACH, G.; SILLEN, L. G. 1957. Organic Ligands. *Stability Constants of Metal Ion Complexes: Lond. Chem. Soc. Spec. Pub.,* 6. Pt. 1, 105 p.

BONATTI, E.; JOENSUU, O. 1966. Deep-sea Iron Deposit from the South Pacific. *Science,* Vol. 154, p. 643–5.

BORCHERT, H. 1960. Genesis of Marine Sedimentary Iron-ores. *Trans. Instn. Min. Metall.* Vol. 69, p. 261–77.

——. H. 1972. Zur Bildung Marin-sedimentäres Eisen- und Manganerze in Verknüpfung mit Splititischen und Keratophyrisch-weilburgitischen Gesteinassoziationen. *Mineralium Deposita,* Vol. 7, p. 18–24.

BOSTRUM, K. 1967. The Problem of Excess Manganese in Pelagic Sediments. In: P. H. Abelson (ed.), *Researches in Geochemistry,* Vol. 2. p. 421–52. New York, Wiley.

BOYLE, R. W. 1968. The Geochemistry of Silver and its Deposits. *Geol. Surv. Can. Bull.,* p. 160.

BOYLE, R. W.; DASS, A. S. 1971a. The Origin of the Native Silver Veins at Cobalt, Ontario. *Can. Min.* Vol. II, p. 414–7.

——. 1971b. The Geochemistry of the Supergene Processes in the Native Silver Veins of the Cobalt–South Lorrain Area, Ontario. *Can. Miner.,* Vol. II, p. 358–90.

BOYLE, R. W.; DAVIES, J. L. 1973. Banded Iron Formations. *Geochim. Cosmochim. Acta,* Vol. 37, p. 1389.

BOYLE, R. W.; LYNCH, J. L. 1968. Speculations on the Source of Zinc, Cadmium, Lead, Copper, and Sulfur in Mississippi Valley and Similar types of Pb–Zn Deposits. *Econ. Geol.,* Vol. 63, p. 421–2.

BRETT, P. R.; KULLERUD, G. 1966. Melting Relationships of Galena–Pyrite–Pyrrhotite Assemblages. A Homogeneous Sulfide Melt at 718 °C. *Econ. Geol.,* Vol. 61, p. 1302. (Abstract.)

BROCK, B. B. 1964. Global Approach to the Problem of Metal-

logenic Provinces. *Chron. Mines Rechn. Min.,* Vol. 334, p. 273.

BROWN, A. C. 1971. Zoning in the White Pine Copper District, Ontonagon County, Michigan. *Econ. Geol.,* Vol. 66, p. 543–73.

BROWN, J. S. 1970. Mississippi Valley Type Lead–Zinc Ores. *Mineralium Deposita,* Vol. 5, p. 103–19.

——. (ed.). 1967. *Genesis of Stratiform Lead–zinc–barite–fluorite Deposits. Econ. Geol.* 443 p. (Mono. 3.)

BUDDINGTON, A. F.; LINDSLEY, D. H. 1964. Iron–Titanium Oxide Minerals and Synthetic Equivalents. *J. Petrol.,* Vol. 5, p. 310–57.

BURNHAM, C. W. 1959. Metallogenic Provinces of the Southwestern United States and Northern Mexico. *New Mexico Bur. Mines and Miner. Resources, Bull.* 65.

BURNIE, S. W.; SCHWARCZ, H. P.; CROCKET, J. H. 1972. A Sulfur Isotopic Study of the White Pine Mine, Michigan. *Econ. Geol.,* Vol. 67, p. 895–914.

BURNS, R. G.; FYFE, W. S. 1966. Distribution of Elements in Geological Processes. *Chem. Geol.,* Vol. 1, p. 49–56.

BUSHINSKY, G. I. 1958. Bauxites, Their Mineralogy and Genesis. In: N. M. Strachan (ed.), *Econ. Geol.,* Vol. 54, no. 5. p. 957–65. Moscow. 488 p.

BUTLIN, K. R. 1953. The Bacterial Sulphur Cycle. *Research* (Lond.), Vol. 6. 184 p.

CABRI, L. J. 1965. Phase Relations in the An–Ag–Te System and their Mineralogical Significance. *Econ. Geol.,* Vol. 60, p. 1569.

CAILLÉRE, S.; HÉNIN, S.; ESQUEVIN, J. 1956. Étude Expérimentale du Mécanisme de la Formation des Antigorites Nickélifères. *Soc. Fr. Mineral. et Crist. Bull.,* Vol. 79, p. 408–20.

CALLAHAN, W. H. 1966. Genesis of the Franklin-Sterling, New Jersey Ore Bodies. *Econ. Geol.,* Vol. 61, p. 1140–41.

——. 1967. Some Spatial and Temporal Aspects of the Localization of Mississippi Valley–Appalachian Type Ore Deposits. In: J. S. Brown (ed.), *Genesis of Stratiform Lead–Zinc–Fluorite Deposits: Econ. Geol.* p. 14–19. (Mono. 3.)

CAMERON, E. N.; DESBOROUGH, G. A. 1968. Occurrence and Characteristics of Chromite Deposits—Eastern Bushveld Complex. In: H. B. D. Wilson (ed.), *Magmatic Ore Deposits: Econ. Geol.* p. 23–40. (Mono. 4.)

CANADIAN INSTITUTE OF MINING AND METALLURGY 1966. *Tectonic History and Mineral Deposits of the Western Cordillera.* (Special Volume No. 8.)

CARPENTER, R. H. 1974. Pyrrhotite Isograd in Southeastern Tennessee and Southwestern North Carolina. *Geol. Soc. Am. Bull.,* Vol. 85, 451–6.

CARROLL, D. 1958. The Role of Clay Minerals in the Transportation of Iron. *Geochim. Cosmochim. Acta,* Vol. 14, p. 1–28.

CASTANO, J. R.; GARRELS, R. M. 1950. Experiments on the Deposition of Iron with Special Reference to the Clinton Iron Ore Deposits. *Econ. Geol.,* Vol. 45, p. 755–70.

CHAMBERLAIN, J. A.; MCLEOD, C. R.; TRAILL, R. J.; LACHANCE, G. R. 1965. Native Metals in the Muskox Intrusion. *Can. J. Earth Sci.,* Vol. 2, p. 188–215.

CHANTRAINE, J.; RADELLI, L. 1970. Tectono-minerogenetic Units of the Basement of Madagascar. *Econ. Geol.,* Vol. 65, p. 690–9.

CHENEY, E. S.; VREDENBURGH, L. D. 1968. The Role of Iron Sulfides in the Diagenetic Formation of Iron-poor Manganese Nodules. *J. Sedim. Petrol.,* Vol. 38, p. 1363–5.

CLIFFORD, T. N. 1964. The Upper Proterozoic–Lower Paleozoic Structural Units and Metallogenic Provinces of Southern

Africa. *Res. Inst. Afr. Geol.* (Leeds). (8th Anniversary Report, 46.)

——. 1965. Structural Units and Tectono-metallogenic Provinces within the Congo and Kalahari Cratons of Southern Africa. *Res. Inst. Afr. Geol.* (Leeds). (9th Annual Report, 30.)

——. 1966. Tectono-metallogenic Units and Metallogenic Provinces of Africa. *Earth and Planet. Sci. Lett. 1.*

COUSINS, C. A. 1968. The Merensky Reef of the Bushveld Igneous Complex. In: H. B. D. Wilson (ed.), *Magmatic Ore Deposits: Econ. Geol.* p. 239–51. (Mono. 4.)

CRAIG, J. R.; KULLERUD, G. 1967. Sulfide Melts in the Cu–Fe–Pb–S System. *Econ. Geol.*, Vol. 62, p. 868–9. (Abstract.)

CULLEN, D. J. 1963. Tectonic Implications of Banded Ironstone Formations. *J. Sedim. Petrol.*, Vol. 33, p. 387–92.

DAVIDSON, C. F. 1962. The Origin of Some Strata-bound Sulfide Ore Deposits. *Econ. Geol.*, Vol. 57, 265–73.

——. 1964. Uniformitarianism and Ore Genesis. *Min. Mag.*, Vol. 110, p. 176–85, 244–53.

DEANS, T. 1948. The Kupferschiefer and the Associated Lead–Zinc Mineralization in the Permian of Silesia, Germany and England. *18th Int. Geol. Cong. Pt. VII*, p. 340–52.

DEGENS, E. T. 1965. *Geochemistry of Sediments.* Englewood Cliffs, N.J., Prentice-Hall. 342 p.

DEGENS, E. T.; ROSS, D. A. (eds.). 1969. *Hot Brines and Recent Heavy Metal Deposits in the Red Sea.* New York, Springer-Verlag. 600 p.

DE KUN, N. 1963. The Mineralogenetic Provinces of Africa. *Econ. Geol.*, Vol. 58, p. 774–90.

DESBOROUGH, G. A.; CZAMANSKE, G. K. 1973. Sulfides in Eclogite Nodules from a Kimberlite Pipe, South Africa, with Comments on Violarite Stoichiometry. *Am. Min.*, Vol. 58, p. 195–202.

DEWING, E. W.; RICHARDSON, F. D. 1960. Thermodynamics of Mixtures of Ferrous Sulphide and Oxide. *J. Iron Steel Inst.*, Vol. 179, p. 147–54.

DICKSON, F. W.; TUNNEL, G. 1958. Equilibria of Red HgS (Cinnabar) and Black HgS (Metacinnabar) and their Saturated Solutions in the Systems HgS, $Na_2S–H_2O$ and $HgS–Na_2O–H_2O$ From 25 °C to 75 °C at 1 Atmosphere Pressure. *Am. J. Sci.*, Vol. 256, p. 654–79.

——. 1968. Mercury and Antimony Deposits Associated with Active Hot Springs in the Western United States. In: J. D. Ridge (ed.), *Ore Deposits in the United States, 1933–1967, 2*, p. 1673–1701. American Institution of Mining and Metallurgy Petroleum Engineers.

DOE, B. R.; DELEVAUX, M. H. 1972. Source of Lead in Southeast Missouri Galena Ores. *Econ. Geol.*, Vol. 67, 409–25.

DOE, B. R.; HEDGE, C. E.; WHITE, D. E. 1966. Preliminary Investigation of the Source of Lead and Strontium in Deep Geothermal Brines Underlying the Salton Sea Geothermal Area. *Econ. Geol.*, Vol. 61, p. 462–83.

ERLICH, H. L. 1968. Bacteriology of Manganese Nodules II. Manganese Oxidation by Cell-free Extract From a Manganese Nodule Bacterium. *Appl. Microbiol.*, Vol. 16, p. 197–202.

EUGSTER, H. P.; CHOU, I.-M. The Depositional Environments of Precambrian Banded Iron Formations. *Econ. Geol.*, Vol. 68, p. 1144–68.

EVANS, H. T.; GARRELS, R. M. 1958. Thermodynamic Equilibria of Vanadium in Aqueous Systems as Applied to the Interpretation of Colorado Plateau Ore Deposits. *Geochim. Cosmochim. Acta*, Vol. 15, 131–49.

EWERS, W. E. 1967. Physico-chemical Aspects of Recrystallization. *Mineralium Deposita*, Vol. 2, p. 221–7.

FAIRBARN, H. W.; AHRENS, L. H.; GARFINKLE, L. G. 1953. Minor Element Content of Ontario Diabase. *Geochim. Cosmochim. Acta*, Vol. 3, p. 34–46.

FERENCIC, A. 1971. Metallogenic Provinces and Epochs in Southern Central America. *Mineralium Deposita*, Vol. 6, p. 77–88.

FINCHAM, C. J. B.; RICHARDSON, F. D. 1954. The Behavior of Sulphur in Silicate and Aluminate Slags. *Proc. R. Soc. Lond.*, Vol. 223, p. 40–62.

FISCHER, R. P. 1970. Similarities, Differences, and Some Genetic Problems of the Wyoming and Colorado Plateau Types of Uranium Deposits in Sandstone. *Econ. Geol.*, Vol. 65, p. 778–84.

FONTEILLES, M. 1967. Appréciation de l'Intérêt Métallogénique du Volcanisme de Madagascar à Partir de ses Caractères Pétrologiques: France. *Bur. Res. Geol. Min. Bull. No. 1*, p. 121–54.

FYFE, W. S., TURNER, F. J.; VERHOOGAN, J. 1958. *Metamorphic Reactions and Metamorphic Facies.* 259 p. (Geological Society of America Memo. 73.)

GABLEMAN, J. W.; KRUSIEWSKI, S. V. 1968. Regional Metallotectonic Zoning in Mexico. *Soc. Min. Engr. Trans.*, Vol. 241, p. 113–28.

——. 1969. Metallotectonic Evidence for Continental Drift. *Union Trans.*, Vol. 50, p. 318. (Abstract.)

——. 1972. The Metallotectonics of Europe. *24th Int. Geol. Congr. Sect. 4*, p. 88–90.

GALLAGHER, D. 1940. Albite and Gold. *Econ. Geol.*, Vol. 35, p. 698–736.

GARLICK, W. G. 1961. The Syngenetic Theory. In: F. Mendelsohn (ed.), London, *Geology of the Northern Rhodesian Copperbelt.* p. 146–65. London, Roan Antelope Copper Mines Ltd and MacDonald and Co. Pubs. Ltd.

GARRELS, R. M. 1954. Mineral Species as Functions of pH and Oxidation-reduction Potentials, with Special References to the Zone of Oxidation and Secondary Enrichment of Sulphide Ore Deposits. *Geochim. Cosmochim Acta*, Vol. 5, p. 153–68.

——. 1955. Some Thermodynamic Relations Among the Uranium Oxides and their Relation to the Oxidation States of the Uranium Ores of the Colorado Plateau. *Am. Miner.*, Vol. 40, p. 1004–21.

GARRELS, R. M.; CHRIST, C. L. 1965. *Solutions, Minerals and Equilibria.* New York, Harper & Row. 450 p.

GARRELS, R. M.; NAESER, C. R. 1958. Equilibrium Distribution of Dissolved Sulphur Species in Water at 25 °C and 1 atm. Total Pressure. *Geochim. Cosmochim. Acta*, Vol. 15, p. 113–30.

GEIJER, P. 1931. The Iron Ores of the Kiruna Type. *Sver. Geol. Unders.*, Ser. C, No. 367.

——. 1967. Internal Features of the Apatite-bearing Magnetite Ores. *Sver. Geol. Unders.*, Ser. C, No. 624, Arsbok 61, No. 9.

GINZBURG, A. I. 1956. Some Peculiarities of the Geochemistry of Tantalum. *Geokhimiya*, p. 74–83.

GOLDICH, S. S. 1973. Ages of Precambrian Banded Iron-formations. *Econ. Geol.*, Vol. 68, p. 1126–34.

GOLDSCHMIDT, V. M. 1937. The Principles of Distribution of Chemical Elements in Minerals and Rocks. *J. Chem. Doc.*, p. 655–72.

——. 1958. *Geochemistry.* London, Oxford University Press.

GOODWIN, A. M. 1961. Some Aspects of Archean Structure and Mineralization. *Econ. Geol.*, Vol. 56, p. 897–915.

——. 1964. Geochemical Studies at the Helen Iron Range (Ontario). *Econ. Geol.*, Vol. 59, p. 684–718.

——. 1966. Archean Proto Continental Growth and Mineralization. *Can. Min. J.*, Vol. 87, p. 57–60a.

——. 1973. Archean Iron-formations and Tectonic Basins of the Canadian Shield. *Econ. Geol.*, Vol. 68, p. 915–33.

GOOSSENS, P. J. 1972, Metallogeny in Ecuadorian Andes. *Econ. Geol.*, Vol. 67, p. 458–68.

GORNITZ, V.; Warde, J. M. 1972. Niobium (Columbium): Element and Geochemistry. In: R. W. Fairbridge (ed.), *Encycl. Geochem. Environ. Sci. IVA,* Van Nostrand. 1321 p.

GOVETT, G. J. S. 1966. Origin of Banded Iron Formatons. *Bull. Geol. Soc. Am.*, Vol. 77, p. 1191–1212.

GRANGER, H. C.; WARREN, C. G. 1969. Unstable Sulfur Compounds and the Origin of Roll-type Uranium Deposits. *Econ. Geol.*, Vol. 64, p. 160–71.

GRAY, R. F.; HOFFMAN, V. J.; BAGAN, R. J.; MCKINLEY, H. L. Bishop Tungsten District, California. In: J. D. Ridge (ed.), *Ore Deposits in the United States, 1933–1967: 2,* p. 1531–54. New York, American Institution of Mining and Metallurgy Petroleum Engineers.

GREGORY, J. W., 1930. The Copper-shale (Kupferschiefer) of Mansfeld. *Trans. Instn. Min. Metall.,* Vol. 40, p. 1–55.

GRUNER, J. W., 1956. Concentration of Uranium in Sediments by Multiple Migration-accretion. *Econ. Geol.*, Vol. 51, p. 495–520.

GUILBERT, J. M 1971. Known Interactions of Tectonics and Ore Deposits in the Context of New Global Tectonics. *Am. Inst. Min. Metall. Pet. Eng., Soc. Min. Eng. Preprint.* 19 p. (71–S–91.)

GUILD, P. W. 1972. Metallogeny and the New Global Tectonics. *24th Int. Geol. Congr. Sect. 4,* p. 25–36.

GUSTAFSON, L. B. 1963. Phase Equilibria in the System Cu–Fe–As–S: *Econ. Geol.*, Vol. 58, p. 667–701.

GUSTAFSON, L. B.; HUNT, J. P. 1971. Evolution of Mineralization at El Salvador, Chile. *Econ. Geol.*, Vol. 66, p. 1266–7.

HACKETT, J. P.; BISCHOFF, J. L. 1973. New Data on the Stratigraphy Extent, and Geologic History of the Red Sea Geothermal Deposits. *Econ. Geol.*, Vol. 68, p. 553–64.

HAGNER, A. F.; COLLINS, L. G. 1967. Magnetite Ore Formed During Regional Metamorphism, Ausable Magnetite District, New York. *Econ. Geol.*, Vol., 62, p. 1034–71.

HAGNER, A. F.; COLLINS, L. G.; CLEMENCY, C. V., 1963. Host Rock as a Source of Magnetite Ore, Scott Mine, Sterling Lake, New York. *Econ. Geol.*, Vol. 58, p. 730–68.

HALL, W. E.; FRIEDMAN, I. 1963. Compositions of Fluid Inclusions, Cave-In-Rock Fluorite District, Illinois, and Upper Mississippi Valley Zinc-Lead District. *Econ. Geol.*, Vol. 58, p. 886–911.

HALL, W. E.; FRIEDMAN, I.; NASH, J. T. 1973. Fluid Inclusion and Light Stable Isotope Study of the Climax Molybdenum Deposits, Colorado. *Geol. Soc. Am. Abst. With Progr.*, Vol. 5, No. 7, p. 649–50. (Abstract.)

HALLBERG, R. O. 1972. Sedimentary Sulfide Mineral Formation—An Energy Circuit System Approach. *Mineralium Deposita,* Vol. 7, p. 189–201.

HANSON, G. 1956. Manganese in Canada. *20th Int. Geol. Congr., Mexico, Manganese Symp.;* Vol. III, p. 9–14.

HARAŃCZYK, C. 1970, Zechstein Lead-bearing Shales in the Fore-Sudetian Monocline in Poland. *Econ. Geol.*, Vol. 65, p. 481–95.

HAUGHTON, D. R., ROEDER, P. L., SKINNER, B. J. 1974. Solubility of Sulfur in Mafic Magmas. *Econ. Geol.*, Vol. 69, p. 451–67.

HEINRICH, E. W. 1966. *The Geology of Carbonatites.* Chicago, Rand-McNally 657 p.

HEINRICH, E. W.; LEVINSON, A. A. 1961. Carbonatic Niobium–Rare Earth Deposits, Ravoli County, Montana. *Am. Miner.*, Vols. 4–6, p. 1424–47.

HELGESON, H. C. 1964. *Complexing and Hydrothermal Ore Deposition.* New York, Pergamon Press. 128 p.

——. 1969. Thermodynamics of Hydrothermal Systems at Elevated Temperatures and Pressures. *Am. J. Sci.,* Vol. 267, p. 729–804.

——. 1970. A Chemical and Thermodynamic Model of Ore Deposition in Hydrothermal Systems. *Min. Soc. Am. Spec. Paper 3,* p. 155–86.

HELGESON, H. C.; GARRELS, R. M. 1968. Hydrothermal Transport and Deposition of Gold. *Econ. Geol.*, Vol. 63, p. 622-35.

HEM, J. D. 1960. Complexes of Ferrous Iron with Tannic Acid. *U.S. Geol. Surv. Wat. Supply Paper 1459–D,* p. 75–94.

HEŃIN, S. 1956. Synthesis of Clay Minerals of Low Temperature. In: A. Swineford (ed.), *Clays and Clay Minerals,* p. 54–60. National Research Council.

HESS, H. H. 1960. *Stillwater Igneous Complex, Montana: A Quantitative Mineralogical Study.* 230 p. (Geological Society of America Memo. 80.)

HEWITT, D. F. 1966. Stratified Deposits of the Oxides and Carbonates of Manganese. *Econ. Geol.* Vol. 61, p. 431–61.

HILTY, D. C.; CRAFTS, W. 1952. Liquidus Surface on the Fe–S–O System. *J. Metals.* New York, Vol. 4, p. 1307–12.

HOLLAND, H. D. 1967. Gangue Minerals in Hydrothermal Deposits. In: H. L. Barnes (ed.), *Geochemistry of Hydrothermal Ore Deposits,* p. 382–436. New York, Holt, Rinehart & Winston.

——. 1972. The Geologic History of Seawater—An Attempt to Solve the Problem. *Geochim. Cosmochim. Acta,* Vol. 36, p. 637–52.

HONNOREZ, J., HONNOREZ-GUERSTEIN, B.; VALETTE, J.; WAUSCHKUHN, A. 1973. Present Day Formation of an Exhalative Sulfide Deposit at Volcano (Tyrrhenian Sea), Part II. Active Crystallization of Fumarolic Sulfides in the Volcanic Sediments of the Baia de Levante. In: G. C. Amstutz and A. J. Bernard (eds.), *Ores in Sediments,* p. 139–66. Berlin, Springer-Verlag.

HORIKOSHI, E. 1969. Volcanic Activity Related to the Formation of the Kuroko-type Deposits in the Kosaka District. *Mineralium Deposita,* Vol. 4, p. 321–45.

HORN, D. R., DELACH, ; M. N.; HORN, B. M. 1973a. *Metal Content of Ferromanganese Deposits of the Oceans.* (Technical Report 3, NSF–GX 33616.) 51 p.

HORN, D. R.; HORN, B. M.; DELACH, M. N. 1973b. *Ocean Manganese Nodules Metal Values and Mining Sites.* 57 p. (Technical Report 4, NSF–GX 33616.)

HOSTETLER, P. B.; GARRELS, R. M. 1962. Transportation and Precipitation of Uranium and Vanadium at Low Temperatures, with Special Reference to Sandstone-type Uranium Deposits. *Econ. Geol.*, Vol. 57, p. 137–67.

HOTZ, P. E. 1964. Nickeliferous Laterites in Southwestern Oregon and Northwestern California. *Econ. Geol.*, Vol. 59, p. 355–96.

HOUGH, J. L. 1958. Fresh-water Environment of Deposition of Precambrian Banded Iron Formations. *J. Sedim Petrol.* Vol. 28, p. 414–30.

HUBER, N. K. 1959. Some Aspects of the Origin of the Ironwood Iron-formation of Michigan and Wisconsin. *Econ. Geol.*, Vol. 54, p. 82–118.

HUFF, L. C. 1952. Abnormal Copper, Lead, and Zinc Content of Soil Near Metalliferous Veins. *Econ. Geol.,* Vol. 47, p. 517–42.

HUTCHINSON, R. W. 1973. Volcanogenic Sulfide Deposits and their Metallogenic Significance. *Econ. Geol.,* Vol. 68, p. 1223–46.

IRVINE, T. N.; SMITH, C. H. Primary Oxide Minerals in the Layered Series of the Muskox Intrusion. In: H. B. D. Wilson (ed.), *Magmatic Ore Deposits: Econ. Geol.* p. 76–94. (Mono. 4.)

JACKSON, E. D. 1968. Chemical Variation in Co-existing Chromite and Olivine in Chromitite Zones of the Stillwater Complex. In: H. B. D. Wilson (ed.), *Econ. Geol.,* p. 41-71. (Mono. 4.)

JACKSON, S. A.; BEALES, F. W. 1967. An Aspect of Sedimentary Basin Evolution: The Concentration of Mississippi Valley-type Ores During Late Stages of Diagenesis. *Bull. Can. Petrol. Geol.,* Vol. 15, p. 383–433.

JAMBOR, J. J. 1971. Origin of the Silver Veins of the Cobalt-Gowganda Region. *Can. Min.,* Vol. 11, p. 402–13.

JAMES, H. L. 1954. Sedimentary Facies of Iron Formation. *Econ. Geol.,* Vol. 49, p. 235–93.

——. 1966. *Chemistry of the Iron-rich Sedimentary Rock.* 61 p. (United States Geological Survey Professional Paper 490 W.)

JEDWAB, J. 1957. Distribution of Certain Trace Elements in Granite and their Usefulness in Geochemical Prospecting. *Freiberger Forschrift.,* C 31, p. 7–10.

JENSEN, M. L. 1958. Sulfur Isotopes and the Origin of Sandstone Type Uranium Deposits. *Econ. Geol.,* Vol. 53, p. 598–616.

——. 1967, Sulfur Isotopes and Mineral Genesis. In: H. L. Barnes (ed.), *Geochemistry of Hydrothermal Ore Deposits,* p. 143–65. New York, Holt, Rinehart & Winston.

——. 1971, Provenance of Cordilleran Intrusives and Associated Metals. *Econ. Geol.,* Vol. 66, p. 34–42.

KAJIWARA, Y. 1970. Some Limitations on the Physico-chemical Environment of Deposition of the Kuroko Ore. In: T. Tatsumi (ed.), *Volcanism and Ore Genesis,* p. 367–80.

——. 1973a. Chemical Composition of Ore-forming Solution for the Kuroko Type Mineralization in Japan. *Geochem. J.,* Vol. 6, p. 141–9.

——. 1973b. A Simulation of the Kuroko Type Mineralization in Japan. *Geochem. J.,* Vol. 6, p. 193–209.

KALLIOKOSKI, J. 1965. Metamorphic Features in North American Massive Sulfide Deposits. *Econ. Geol.,* Vol. 60, p. 485–505.

KELLY, W. C. 1972. Gold: Economic Deposits. In: R. W. Fairbridge (ed.), *Encycl. of Geochem. and Environ. Sci.,* p. 467–70. New York, Van Nostrand & Reinhold.

KELLY, W. C.; GODDARD, E. N. 1969. Telluride Ores of Boulder County, Colorado. *Geol. Soc. Am. Mem.* p. 109, 237 p.

KELLY, W. C.; TURNEAURE, F. S., 1970, Mineralogy, Paragenesis and Geothermometry of the Tin and Tungsten Deposits of the Eastern Andes, Bolivia. *Econ. Geol.,* Vol. 65, p. 609–80.

KIMATA, M.; KADOTA, H.; HATA, Y.; TAJUMA, T. 1955. Studies on Marine Sulphate-reducing Bacteria. I, Distribution of Marine Sulphate Reducing Bacteria in the Coastal Waters Receiving a Considerable Amount of Pulp Mill Drainage. *Japan. Soc. Sci. Fish. Bull.,* Vol. 21, p. 104, 108.

KRAUSKOPF, K. B. 1951. Physical Chemistry of Quicksilver Transportation in Vein Fluids. *Econ. Geol.,* Vol. 46, p. 498–523.

——. 1956a. Dissolution and Precipitation of Silica at Temperatures. *Geochim. Cosmochim. Acta,* Vol. 10, p. 1–26.

——. 1956b. Factors Controlling the Concentration of Thirteen Rare Metals in Sea Water. *Geochim. Cosmochim. Acta,* Vol. 9,

——. 1957. Separation of Manganese from Iron in Sedimentary Processes. *Geochim. Cosmochim. Acta,* Vol. 12, p. 61–84.

——. 1964. The Possible Role of Valatile Metal Compounds in Ore Genesis. *Econ. Geol.,* Vol. 59, p. 22–45.

——. 1967. *Introduction to Geochemistry,* New York, McGraw-Hill. 721 p.

——. 1967. Source Rocks for Metal-bearing Fluids. In: H. L. Barnes (ed.), *Geochemistry of Hydrothermal Ore Deposits.* New York, Holt, Rinehart & Winston, p. 1–33

KRUMBEIN, W. C.; GARRELS, R. M. 1952. Origin and Classification of Chemical Sediments in Terms of pH and Oxidation-reduction Potentials. *J. Geopl.,* Vol. 60, p. 1–33.

KULLERUD, G.; YODER, H. S. 1965. *Sulfide–Silicate Reactions and their Bearing on Ore Formation Under Magmatic, Post Magmatic and Metamorphic Conditions. 22nd Int. Geol. Congr. Pt. 2,* p. 327–31.

KUTINA, J. 1969. Hydrothermal Ore Deposits in the Western United States: A New Concept of Structural Control of Distribution. *Science,* Vol. 165, p. 1113–19.

LA BERGE, G. L. 1973. Possible Biological Origin of Precambrian Iron Formations. *Econ. Geol.,* Vol. 68, p. 1098–1109.

LABORATOIRE de GÉOLOGIE APPLIQUÉE, UNIVERSITÉ DE PARIS, FRANCE, 1973. Some Major Concepts of Metallogeny. *Mineralium Deposita,* Vol. 8, 237–58.

LAMEY, C. A. 1966. *Metallic and Industrial Mineral Deposits.* New York, McGraw-Hill. 567 p.

LANG, A. H. 1961. A Preliminary Study of Canadian Metallogenic Provinces. *Geol. Surv. Can. Paper 60–33.*

LARSEN, L. T. 1973. Textural Study of Polycrystalline Pyrrhotite by Reflectance Measurements and X-ray Pole-figures. *Econ. Geol.,* Vol. 68, p. 671–80.

LAWRENCE, L. J. 1967. Sulphide Neomagmas and Highly Metamorphosed Sulphide Deposits. *Mineralium Deposita,* Vol. 2, p. 5–10.

LEARNED, R. E.; TUNELL, G.; DICKSON, F. W. 1967. The Mutual Solubilities of Cinnabar and Quartz in Na2S Solutions and their Implication for the Genesis of Quicksilver Deposits. *Geol. Soc. Am. Spec. Paper 115,* p. 336. (Abstract.)

LEBEDEV, I. M.; NIKITINA, I. B. 1968. Chemical Properties and Ore Cotent of Hydrothermal Solutions at Chelekem. *Dokl. Akad. Nauk SSR* [Earth Sci. Sect.]. Vol. 183, p. 180–2.

LEVY, E. 1970. La Metalogenesis en América Central in Mapa Metalogenetico de América Central. *Public. Geol. del I CAITI, Guatemala, No. III,* p. 19–57.

LINDGREN, W. 1933. *Mineral Deposits.* 4th ed. New York, McGraw-Hill. 930 p.

LISTER, G. F. 1966. The Composition and Origin of Selected Iron–Titanium Deposits. *Econ. Geol.,* Vol. 61, p. 275–310.

LIVINGSTON, D. E.; MANGER, R. L.; DAMON, P. E. 1968. Geochronology of Emplacement, Enrichment and Preservation of Arizona Porphyry Copper Deposits. *Econ. Geol.,* Vol. 63, p. 30–6.

LJUNGGREN, P. 1955. Differential Thermal Analysis and X-ray Examination of Iron and Manganese Bog Ores. *Geol. Foren. Stockholm Forch.,* Vol. 77, p. 135–47.

LOVE, L. G. 1962. Biogenic Primary Sulfide of the Permian Kupferschiefer and Marl Slate. *Econ. Geol.,* Vol. 57, p. 350–66.

LOWELL, J. D.; GUILBERT, J. M. 1970. Lateral and Vertical Alteration–Mineralization Zoning in Porphyry Ore Deposits. *Econ. Geol.*, Vol. 65, p. 373–408.

MCANDREW, J. (ed.). 1955. Geology of Australian Ore Deposits. *Eighth Commonwealth Mining and Metallurgical Congress, Australia and New Zealand.* Melbourne, Australian Institute of Mining and Metallurgy. 547 p.

MCCARTNEY, W. D. 1962. Mineralization in Mobile Belts. *Econ. Geol.*, Vol. 57, p. 1131–2.

MCCARTNEY, W. D.; POTTER, R. R. 1962. Mineralization as Related to Structural Deformation, Igneous Activity and Sedimentation in Folded Geosynclines. *Can. Min. J.*, April, p. 83–7.

MCDONALD, J. A. 1967. Metamorphism and its Effects on Sulphide Assemblages. *Mineralium Deposita*, Vol. 2, p. 200–20.

MCKELVEY, V. E.; EVERHART, D. L.; GARRELS, R. M. 1955. Origin of Uranium Deposits. *Econ. Geol. 50th Ann. Vol.*, p. 464–533.

MACLEAN, W. H. 1969. Liquidus Phase Relations in the $FeS–FeO–Fe_3O_4–SiO_2$ System and their Application in Geology. *Econ. Geol.*, Vol. 64.

MAGAKIAN, I. G. 1972. The Complexes (Series) of Ore Formations in Different Types of USSR Ore Provinces. *24th Int. Geol. Congr. Sect. 4*, p. 60–4.

MANCHER, A. 1972. Time and Stratabound Ore Deposits and the Evolution of the Earth. *24th Int. Geol. Cong. Sect. 4*, p. 83–7.

MARKHAM, N. L. 1960. Synthetic and Natural Phases in the System Au–Ag–Te, Pt. I. *Econ. Geol.*, Vol. 55, p. 1148–78, 1460–77.

MARMO, V. 1960. On the Possible Genetical Relationship Between Sulphide Schists and Ores. *21st Int. Geol. Congr., Pt. 16*, p. 160–3.

METSGER, R. W.; SKINNER, B. J. BARTON, P. B. 1969. Structural Interpretation of the Sterling Hill Ore Body, Ogdensburg, New Jersey, *Econ. Geol.*, Vol. 64, p. 833. (Abstract.)

MEYER, C.; HEMLEY, J. J. 1967. Wall Rock Alteration. In: H. L. Barnes (ed.), *Geochemistry of Hydrothermal Ore Deposits.* New York, Holt, Rinehart & Winston, p. 166–235.

MEYER, C.; SHEA, E. P.; GODDARD, C. C. 1968. Ore Deposits at Butte, Montana. In: J. D. Ridge (ed.), *Ore Deposits of the United States 1933–1967*, Vol. 2, p. 1374–1416. New York, American Institution of Mining and Metallurgy Petroleum Engineers.

MOOKHERJEE, A. 1970a. Dykes, Sulphide Deposits, and Regional Metamorphism: Criteria for Determining their Time Relationship. *Mineralium Deposita*, Vol. 5, p. 120–44.

——. 1970b. 'Metamorphic' and 'Metamorphosed' Sulfide Deposits. *Econ. Geol.*, Vol. 65, p. 886–93.

MOOKHERJEE, A.; DUTTA, N. K. 1970. Evidence of Incipient Melting of Sulfides Along a Dike Contact, Geco Mine, Manitouwadge, Ontario. *Econ. Geol.*, Vol. 65, p. 706–13.

MUAN, A.; OSBORNE, E. F. 1965. *Phase Equilibria Among Oxides in Steel Making*, Reading, Mass., Addison-Wesley Publishing Co., 236 p.

MUFFLER, L. J. P.; WHITE, D. E. 1969. Active Metamorphism of Upper Cenozoic Sediments in the Salton Sea Geothermal Field and the Salton Trough, Southeastern California. *Geol. Soc. Am. Bull.*, Vol. 80, p. 157–82.

NAGAMORI, M; KAMEDA, M. 1965. Equilibria Between Fe–S–O System Melts and $CO–CO_2–SO_2$ Mixture at 1200 °C. *Trans. Japan Inst. Metals*, Vol. 6, p. 21–30.

NALDRETT, A. J. 1966. The Role of Sulphurization in the Genesis of Iron–Nickel Sulphide Deposits of the Porcupine District, Ontario. *Trans. Can. Inst. Min. Metall.*, Vol. 69, p. 147–55.

——. 1969. A Portion of the System Fe–S–O Between 900 and 1080 °C and its Application to Sulfide Ore Magmas. *J. Petrol.*, Vol. 10, p. 171–201.

——. 1973. Nickel Sulphide Deposits—Their Classification and Genesis, with Special Emphasis on Deposits of Volcanic Association. *Can. Inst. Min. Metall. Trans.*, Vol. 76, p. 183–201.

NALDRETT, A. J.; GASPARRINI, E. L. 1971. Archean Nickel Sulfide Deposits in Canada: Their Classification, Geological Setting and Genesis with some Suggestions as to Exploration. *Geol. Soc. Aust. Spec. Pub. 3.*

NICHOLAS, D. J. D. 1967. Biological Sulphate Reduction. *Mineralium Deposita*, Vol. 2, p. 169–80.

NICKEL, E. H. 1973. Violarite, A Key Mineral in the Supergene Alteration of Nickel Sulphide Ores. Perth Conf. *Aust. Inst. Min. Metall.*, Perth, Australia, May, p. 111–16.

NICKEL, E. H.; ROSS, J. R.; THORNBER, M. R. 1974. The Supergene Alteration of Pyrrhotite–Pentlandite Ore at Kambalda, Western Australia. *Econ. Geol.*, Vol. 69, p. 93–107.

NORTON, S. A. 1973. Laterite and Bauxite Formation. *Econ. Geol.*, Vol. 68, p. 353–61.

OBERC, J.; SERKIES, J. 1968. Evolution of the Fore-Sudetian Copper Deposit. *Econ. Geol.*, Vol. 63, p. 373–9.

OELSNER, O. 1966. *Atlas of the Most Important Ore Mineral Parageneses Under the Microscope.* Oxford, Pergamon Press. 311 p.

OHMOTO, H. 1972. Systematics of Sulfur and Carbon Isotopes in Hydrothermal Ore Deposits. *Econ. Geol.*, Vol. 67, p. 551–78.

OHMOTO, H., KAJIWARA, Y.; DATE, J. 1970. The Kuroko Ores in Japan: Products of Sea Water? *Geol. Soc. Am., Abst. With Prog.*, Vol. 2, Pt. 7, p. 640–1.

OHMOTO, H., RYE, R. O. 1970. The Bluebell Mine, British Columbia. I. Mineralogy, Paragenesis, Fluid Inclusions and Isotopes of Hydrogen, Oxygen and Carbon. *Econ. Geol.*, Vol. 65, p. 417–37.

OL'SHANSKY, Y. I. 1951. The System $Fe–FeS–FeO–SiO_2$. *Izv. Akad. Nauk SSSR Ser. Geol.*, Vol. 6, p. 128–52.

OSBORNE, E. F. 1959. Role of Oxygen Pressure in the Crystallization and Differentiation of Basaltic Magma. *Am. J. Sci.*, Vol. 257, p. 609–47.

OVTRACHT, A.; TAMAIN, G. 1972. La Ceinture Minéralisée Varisque dans le Sud de la Meseta Ibérique. *24th Int. Geol. Congr. Sect. 4*, p. 101–10.

PARK, C. F. 1960. A Magnetite 'Flow' in Northern Chile. *Econ. Geol.*, Vol. 56, p. 431–6.

——. 1972. The Iron Ore Deposits of the Pacific Basin. *Econ. Geol.*, Vol. 67, p. 339–49.

PARK, C. F.; MACDIARMID, R. A. 1970. Ore Deposits. 2nd ed. San Francisco, W. H. Freeman. 522 p.

PATTERSON, S. H. 1967. *Bauxite Reserves and Potential Aluminum Resources of the World.* 176 p. (United States Geological Survey Professional Paper 1228.)

PEREIRA, J.; DIXON, C. J. 1965. Evolutionary Trends in Ore Deposition. *Trans. Inst. Min. Metall.*, Vol. 74, p. 505–27.

PERFIL'EV, B. V.; GABE, D. R.; GAL'PERINA, A. M.; RABINOVICH, V. A.; SAPOTNITSKII, A. A.; SHERMAN, E. E.; TROSHANOV, E. P. 1965. *Applied Capillary Microscopy.* New York, Consultants Bureau. 122 p.

PETERSEN, U. 1970. Metallogenic Provinces in South America. *Geol. Rundschau*, Vol. 59, p. 834–97.

——. 1972. Geochemical and Tectonic Implications of South American Metallogenic Provinces. *Ann N.Y. Acad. Sci.*, Vol. 196, p. 1–38.

PETRASCHECK, W. E. 1965. Typical Features of Metallogenic Provinces. *Econ. Geol.*, Vol. 60, p. 1620–34.

PETRUK, W. 1968. Mineralogy and Origin of the Silver Fields Silver Deposit in the Cobalt Area, Ontario. *Econ. Geol.*, Vol. 63, p. 512–31.

——. 1971a. Mineralogical Characteristics of the Deposits and Textures of the Ore Minerals. *Can. Min.*, Vol. 11, p. 108–39.

——. 1971b. Depositional History of the Ore Minerals. *Can. Min.*, Vol. 11, p. 396–401.

PHILPOTTS, A. R. 1967. Origin of Certain Iron–Titanium Oxide and Opatite Rocks. *Econ. Geol.*, Vol. 62, p. 303–15.

PORRENGA, D. H. 1967. Glauconite and Chamosite as Depth Indicators in the Marine Environment. *Mar. Geol.*, Vol. 5, p. 495–501.

RADKEVICH, E. A. 1972. The Metallogenic Zoning in the Pacific Ore Belt. *24th Int. Geol. Congr., Sect. 4*, p. 52–9.

RADTKE, A. S.; SCHEINER, B. J. 1970. Studies of Hydrothermal Gold Deposition (I) Carlin Gold Deposit, Nevada: The Role of Carbonaceous Materials in Gold Deposition. *Econ. Geol.*, Vol. 65, p. 87–102.

RAMDOHR, P. 1969. *The Ore Minerals and their Intergrowths.* Oxford, Pergamon Press, 1174 p.

RENFRO, A. R. 1974. Genesis of Evaporite-associated Stratiform Metalliferous Deposits—A Sabkha Process. *Econ. Geol.*, Vol. 69, p. 33–45.

RICHARDSON, F. D.; FINCHAM, C. J. B. 1954. Sulphur in Silicate and Aluminate Slags. *J. Iron Steel Inst.*, Vol. 178.

RICKARD, D. T. 1973. Limiting Conditions for Synsedimentary Sulfide Ore Formation. *Econ. Geol.*, Vol. 68, p. 605–17.

RIDGE, J. D. (ed.) 1968. *Ore Deposits of the United States 1933-1967.* Vols. 1 and 2. New York, American Institute of Mining and Metallurgy Petroleum Engineers. 1880 p.

ROEDDER, E. 1965. Evidence from Fluid Inclusions as to the Nature of Ore Forming Fluids. *Symp. Probl. Post Magmatic Ore Deposition, Prague, II*, p. 375–84.

——. 1967a. Fluid Inclusions as Samples of Ore Fluids. In: H. L. Barnes (ed.), *Geochemistry of Hydrothermal Ore Deposits*, p. 515–74.

——. 1967b. Environment of Deposition of Stratiform (Mississippi Valley Type) Ore Deposits: From Studies of Fluid Inclusions. In: J. S. Brown, (ed.), *Genesis of Stratiform Lead–Zinc–Barite–Fluirite Deposits: Econ. Geol.*, p. 349–62. (Mono. 3.)

——. 1968. Temperature, Salinity and Origin of the Ore Forming Fluids at Pine Point, Northwest Territories, Canada: From Fluid Inclusion Studies. *Econ. Geol.*, Vol. 63, p. 439–50.

——. 1971. Fluid Inclusion Studies on the Porphyry-type Ore Deposits at Bingham, Utah, Butte, Montana, and Climax, Colorado. *Econ. Geol.*, Vol. 66, p. 98–120.

——. 1972. Composition of Fluid Inclusions. 164 p. (United States Geological Survey Professional Paper 440JJ.)

ROEDDER, E.; COOMBS, D. S. 1967. Immiscibility in Granitic Melts, Indicates by Fluid Inclusions in Ejected Granitic Blocks from Ascension Island. *J. Petrol.*, Vol. 8, p. 417–51.

ROEDDER, E.; WEIBLEN, P. W. 1970. Lunar Petrology of Silicate Melt. Inclusions, Apollo II Rocks. *Proc. Apollo II Lunar Sci. Conf.* Vol. 1, p. 801–37.

ROSE, A. W. 1970. Zonal Relations of Wallrock Alteration and Sulfide Distribution at Porphyry Copper Deposits. *Econ. Geol.*, Vol. 65, p. 920–36.

RUNDKVIST, D. V. 1969. Accumulation of Metals and the Evolution of the Genetic Types of Deposits in the History of the Earth. *23rd Int. Geol. Congr., Sect. 7*, p. 85–97.

SAKAMOTO, T. 1950. The Origin of the Pre-Cambrian Banded Iron Ores. *Am. J. Sci.*, Vol. 248, p. 449–74.

SALES, R. H.; MEYER, C. 1948. Wall Rock Alteration, Butte, Montana. *AIME Trans.*, Vol. 178, p. 9–35.

SATO, M. 1960a. Oxidation of Sulphide Ore Bodies 1. Geochemical Environments in Terms of Eh and pH. *Econ. Geol.*, Vol. 55, p. 928–61.

——. 1960b. Oxidation of Sulphide Ore Bodies II. Oxidation Mechanisms of Sulfide Minerals at 25 °C. *Econ. Geol.*, Vol. 55, p. 1202–31.

SATO, M.; WRIGHT, T. L. 1966. Oxygen Fugacities Directly Measured in Magmatic Gases. *Science*, Vol. 153, p. 1103–5.

SAWKINS, F. J. 1972. Sulfide Ore Deposits in Relation to Plate Tectonics. *J. Geol.*, Vol. 80, p. 377–97.

SCHILLING, J. G.; WINCHESTER, J. W. 1967. Rare Earth Fraction and Magmatic Processes. In: S. K. Runcorn. *Mantles of the Earth and Terrestrial Planets*, p. 267–83. London, John Wiley.

SCHNEIDER-SCHERBINO, A. 1962. Über Metallogenetische Epochan Boliviens und der Hybriden Charakter der Sog. Zinn-Silber-Formation. *Geol. Jahrbuch* (Hanover), Vol. 81, p. 157–70.

SCHMITT, H. A. 1966. The Porphyry Copper Deposits in their Regional Setting. In: S. R. Titley and C. L. Hicks (eds.). *The Geology of the Porphyry Copper Deposits: Southwestern North America.* Vol. 17, p. 34.

SCHUILING, R. D. 1967. Tin Belts on the Continents Around the Atlantic Ocean. *Econ. Geol.*, Vol. 62, p. 540–50.

SCOTT, S. D.; O'CONNER, T. P. 1971. Fluid Inclusions in Vein Quartz, Silverfields Mine, Cobalt, Ontario. *Ca. Min.*, Vol. 11, p. 263–71.

SEMENOV, A. I.; STARITSKII, Y. G.; SHETALOV, E. T. 1967. Glavnye Tipy Metallogenicheskikh Provintsii i Strukturno-metallogenischeskikh zon na Territorii SSSR. *Zakon Razm Polez Iskop*, Vol. 8, p. 55–78.

SEWARD, T. M. 1973. Thio Complexes of Gold and the Transport of Gold in Hydrothermal Ore Solutions. *Geochim. Cosmochim. Acta*, Vol. 37, p. 379–99.

SHANNON, R. D.; PREWITT, C. T. 1969. Effective Ionic Radii in Oxides and Fluorides. *Acta Crystallogr.*, Vol. 25B, p. 925–45.

SHCHERBINA, V. V.; ZER'YAN, 1964. Paragenesis of Silver and Gold Tellurides as Solid Phases in the System Ag–Au–Te. *Geochem. Int.*, Vol. 4, p. 653–7.

SHIMAZAKI, H.; CLARK, L. A. 1973. Liquidus Relations in the FeS–FeO–SiO$_2$–Na$_2$O System and Geological Implications. *Econ. Geol.*, Vol. 63, p. 79–96.

SHRIVASTAVA, J. N.; PROCTOR, P. D. 1962. Trace Element Distribution in the Searchlight, Nevada, Quartz Monzonite Stock. *Econ. Geol.*, Vol. 57, p. 1062–70.

SIERRA, J.; ORTIZ, A.; BURKHALTER, J. 1972. The Metallogenic Map of Spain 1:200,000. *24th Int. Geol. Congr. Sect. 4*, p. 110–20.

SILLITOE, R.H. 1973. Environments of Formation of Volcanogenic Massive Sulfide Deposits. *Econ. Geol.*, Vol. 66, p. 1321–5.

SILVERMAN, M. P.; EHRLICH, H. L. 1964. Microbial Formation and Degradation of Minerals. *Adv. Appl. Microbiol.*, Vol. 6, p. 153–206.

SIMINDN, V.; AISO, K. 1962. Occurrence and Distribution of Heterotrophic Bacteria in Seawater from Karogara Bay. *Japan. Soc. Sci. Fish. Bull.* Vol. 26, p. 1133–41.

James R. Craig

SIMS, S. J. 1968. The Grace Mine Magnetite Deposit, Berks County Pennsylvania. In: J. D. Ridge (ed.), *Ore Deposits in the United States, 1933–1967.* p. 108–24, New York. American Institute of Mining and Metallurgy Petroleum Engineers.

SKINNER, B. J.; BARTON, P. B. 1973. *Genesis of Mineral Deposits: A. Rev. Earth Planet. Scie.,* Vol. 1, p. 183–211.

SKINNER, B. J.; PECK, D. L. 1968. An Immiscible Sulfide Melt From Hawaii. In: H. B. D. Wilson (ed.), *Magmatic Ore Deposits: Econ. Geol.,* p. 310–22. (Mono 4.)

SKINNER, B. J.; WHITE, D. E.; ROSE, H. J.; MAY, R. E. 1967. Sulfides Associated with the Salton Sea Geothermal Brine. *Econ. Geol.,* Vol. 62, p. 316–30.

SKOPINTSEV, B. A. 1961. Recent Studies of the Hydrochemistry of the Black Sea. *Oceanologiya,* Vol. 1, p. 243–50.

SMITH, C. H. 1972. Chromium: Element and Geochemistry. In: R. W. Fairbridge (ed.), *Encycl. Geochem. Environ. Sci., IV A,* p. 167–70. New York, Van Nostrand, Reinhold.

SNYDER, F. G. 1968. Geology and Mineral Deposits, Mid-continent United States. In: J. D. Ridge (ed.), *Ore Deposits in the United States, 1933-1967,* p. 257–86, New York, American Institute of Mining and Metallurgy Petroleum Engineers.

SOLOMON, M.; GROVES, D. I.; KLOMINSKY, J. 1972. Metallogenic Provinces and Districts in the Tasman Orogenic Zone of Eastern Australia. *Australasian Inst. Min. Metall. Proc.*

STACEY, J. S.; ZARTMAN, R. E.; NKOMO, I. T. 1968. A Lead Isotope Study of Galenas and Selected Feldspars from Mining Districts in Utah. *Econ. Geol.,* Vol. 63, p. 796–814.

STANTON, R. L. 1964. Mineral Interfaces in Stratiform Ores. *Trans. Instn. Min. Metall.,* Vol. 74, p. 45–79.

——. 1972. *Ore Petrology.* New York, McGraw-Hill 713 p.

STANTON, R. L.; GORMAN, H. 1968. A Phenomenological Study of Grain Boundary Migration in Some Common Sulfides. *Econ. Geol.,* Vol. 63, p. 907–23.

STANTON, R. L.; RAFTER, T. A. 1966. The Isotopic Constitution of Sulphur in Some Stratiform Lead–Zinc Sulphide Ores. *Mineralium Deposita,* Vol. 1, p. 16–24.

STOLL, W. C. 1964. Metallogenetic Belts, Centers and Ecocho in Argentina and Chile. *Econ. Geol.,* Vol. 59, p. 126–35.

——. 1965. Metallogenic Provinces of Magmatic Parentage. *Min. Mag.,* Vol. 112, p. 312–23, p. 394–404.

STRACHAN, N. M. (chief ed.); BUSHINSKY, G. I. (ed.). 1958. *Bauxites, Their Mineralogy and Genesis.* Moscow, 488 p. See: *Econ. Geol.,* 1959, Vol. 54, No. 5, p. 957–65.

SUPERECANU, C. I. 1970. The Eastern Mediterranean–Iranian Alpine Copper–Molybdenum Belt. *Int. Min. Assoc.—Int. Assoc. Genesis Ore Deposits: Tokyo–Kyoto, 1970 Proc.,* p. 393–400.

SUTHERLAND BROWN, A.; CATHRO, R. J.; PANTELEYEV, A.; NEY, C. S. 1971. Metallogeny of the Canadian Cordilleras. *Can. Trans. Instn. Min. Metall.,* Vol. 74, p. 121–45.

TATRINOV, P. M. 1957. *General Principles of Regional Metallogenetic Analysis.* Moscow, Gosgeoltekhizdat, 148 p.

TATSUMI, T. (ed.). 1957. *Volcanism and Ore Genesis.* Tokyo, University of Tokyo Press.

TATSUMI, T.; SEKINE, Y.; KANEHIRA, K. 1970. Mineral Deposits of Volcanic Affinity in Japan: Metallogeny. In: T. Tatsumi (ed.), *Volcanism and Ore Genesis.* Tokyo, University of Tokyo Press, p. 3–47.

TAYLOR, H. P. 1971. Oxygen Isotope Evidence for Large-scale Interactions between Meteoric Groundwaters and Tertiary Granodiorite Intrusions, Western Cascade Range, Oregon. *J. Geophys. Res.,* Vol. 76, p. 7855–74.

——. 1973. The Application of Oxygen and Hydrogen Isotope

Studies to Problems of Hydrothermal Alteration and Ore Deposition. *Geol. Soc. Am. Abst. With Prog.,* Vol. 5, 7, p. 834–5. (Abstract.)

TAYLOR, J.; STABO, J. J. 1954. The Sulphur Distribution Reaction between Blast Furnace Slag and Metal. *J. Iron Steel Inst.,* Vol. 178, p. 360–7.

TEMPLE, K. L. 1964. Syngenesis of Sulfide Ores: An Evaluation of Biochemical Aspects. *Econ. Geol.,* Vol. 59, p. 1473–91.

TEMPLE, K. C.; LEROUX, N. W. 1964. Syngenesis of Sulfide Ores. Sulfate Reducing Bacteria and Copper Toxicity. *Econ. Geol.,* Vol. 59, p. 271–8.

THOMPSON, J. B. 1959. Local Equilibrium in Metasomatic Processes. In: P. H. Abelson (ed.), *Researches in Geochemistry,* p. 437–57. New York, Wiley.

TITLEY, S. R.; HICKS, C. L. (eds.). 1966. *Geology of the Porphyry Copper Deposits.* Tucson, Ariz., University of Arizona Press. 287 p.

TRIMBLE, R. B.; EHRLICH, H. L. 1968. Bacteriology of Manganese Nodules III. Reduction of MnO_2 by Two Strains of Nodule Bacteria. *Appl. Microbiol.,* Vol. 16, p. 695–702.

TUGARINOV, A. I. 1966. Causes of the Formation of Metallogenic Provinces. In: A. P. Vinogradov, (ed.), *Chemistry of the Earth's Crust.* Vol. I: Jerusalem, Israel Prog. for Sci. Trans., p. 160–85.

TUNNELL, G. 1964. Chemical Processes in the Formation of Mercury Ores and Ores of Mercury and Antimony. *Geochim. Cosmochim. Acta,* Vol. 28, p. 1019–37.

TUREKIAN, K. K.; WEDEPOHL, K. H. 1961. Distribution of the Elements in Some Major Units of the Earth's Crust. *Geol. Soc. Am. Bull.,* Vol. 72. (D5–A2.)

TURNEAURE, F. S. 1955. Metallogenetic Provices and Epochs. *Econ. Geol. 50th Ann. Vol.,* p. 38–98.

——. 1971. The Bolivian Tin–Silver Province. *Econ. Geol.,* Vol. 66, p. 215–25.

TURNER, F. J.; VERHOOGAN, J. 1960. *Igneous and Metamorphic Petrology.* New York, McGraw-Hill. 694 p.

TUTTLE, O. F.; GITTINS, J. (eds.). 1966. *Carbonatites.* New York, Interscience Publications. 591 p.

TWETO, O.; SIMS, P. K. 1963. Precambrian Ancestry of the Colorado Mineral Belt. *Geol. Soc. Am. Bull.,* Vol. 74, p. 991–1014.

ULMER, G. C. 1968. Experimental Investigations of Chromite Spinels. In: H. B. D. Wilson (ed.), *Magmatic Ore Deposits: Econ. Geol.,* p. 114–31. (Mono 4.)

UNITED NATIONS. 1970. *Survey of the World Iron Ore Resources.* New York, United Nations. 380 p.

UNESCO. 1973. Genesis of Pre-Cambrian Iron and Manganese Deposits. *Proc. Kiev. Symp. 1970.* Paris, Unesco.

UNITED STATES GEOLOGICAL SURVEY. 1970. *Legend for the Metallogenic Map of North America at 1:5,000,000.* Paris, March 1970.

VIDALE, R. 1969. Metasomatism in a Chemical Gradient and the Formation of Calc-silicate Bands. *Am. J. Sci.,* Vol. 267, p. 857–74.

VINE, J. D.; TOURTELOT, E. B. 1970. Geochemistry of Black Shale Deposits—A Summary Report. *Econ. Geol.,* Vol. 65, p. 253–72.

VINOGRADOV, A. P. 1962. Average Contents of Chemical Elements in the Principal Types of Igneous Rocks of the Earth's Crust. *Geokhimiya,* No. 7, 1962, p. 355–571 (In Russian). (Translated in *Geochemistry,* No. 7, 1962, p. 641–4.)

VISSER, D. J. L.; VON GRUENEWALDT, G. (eds.) 1970. *Symp. Bushveldt Igneous Complex and Other Layered Intrusions. Geol. Soc. S. Afr. Spec. Publ. 1.* 763 p.

270

VOGT, J. H. L. 1894. Beitrage zur Genetischen Classification der Durch Magmatische Differentiations Processe und der Durch Previnathloyse Entslandenen Erzvoskommen. *Z. Prakt. Geol.*, Vol. 2, p. 381–99.

VOKES, F. M. 1969. A Review of the Metamorphism of Sulphide Deposits. *Earth Sci. Rev.*, Vol. 5, p. 99–143.

——. 1970. Some Aspects of the Regional Metamorphism Mobilization of Pre-Existing Sulphide Deposits. *Mineralium Deposita*, Vol. 6, p. 122–9.

WAGER, L. R.; BROWN, G. M. 1967. *Layered Igneous Rocks.* San Francisco, W. H. Freeman. 588 p.

WALKER, A. L.; BUCHANAN, A. S. 1969. *Geochemical Processes in Ore for Motion:* I. The Production of Hydrothermal Fluids from Sedimentary Sequences. *Econ. Geol.*, Vol. 64, p. 919–22.

WARREN, C. G. 1972. Sulfur Isotopes as a Clue to the Genetic Geochemistry of a Rolltype Uranium Deposit. *Econ. Geol.*, Vol. 67, p. 759–67.

WATMUFF, I. G. 1974. Supergene Alteration of the Mt. Windarra Nickel Sulphide Ore Deposit, Western Australia. *Mineralium Deposita*, Vol. 9, p 199–221.

WEDEPOHL, K. H. (ed.). 1969. *Handbook of Geochemistry.* Berlin, Springer-Verlag.

WEISS, A.; AMSTUTZ, G. C. 1966. Ion-exchange Reactions of Clay Minerals and Cation-selective Membrane Properties as Possible Mechanisms of Economic Metal Concentration. *Mineralium Deposita*, Vol. 1, p. 60–6.

WEISS, R. G. 1973 Mineralogy and Geochemistry of the Parting Shale, White Pine, Michigan. *Econ. Geol.*, Vol. 68, p. 317–31.

WEISSBERG, B. C. 1970. Solubility of Gold in Hydrothermal Alkaline Sulfide Solutions. *Econ. Geol.*, Vol. 65, p. 551–6.

WHITE, D. E. 1967. Mercury and Base Metal Deposits With Associated Thermal and Mineral Waters. In: H. L. Barnes (ed.), *Geochemistry of Hydrothermal Ore Deposits*, p. 575–631. New York, Holt, Rinehart & Winston.

——. 1968. Environments of Generation of Some Base Metal Ore Deposits. *Econ. Geol.*, Vol. 63, p. 301–35.

——. 1974. Diverse Origins of Hydrothermal Ore Fluids. *Econ. Geol.*, Vol. 69, p. 954–73.

WHITE, D. E.; ANDERSON, E. T.; GRUBBS, D. K. 1963. Geothermal Brine Well: Mile-deep Drill Hole May Tap Ore Bearing Magmatic Water and Rocks Undergoing Metamorphism. *Science*, Vol. 139, p. 919–22.

WHITE, D. E.; MUFFLER, L. J. P , TRUESDELL, A. H. 1970. Vapour-dominated Hydrothermal Systems Compared with Hot Water Systems. *Econ. Geol.*, Vol. 66, p. 75–97.

WILLIAMSON, T. C.; MYER, G. H. 1969. A Pyrrhotite–Pyrite Isograd in the Waterville Area, Maine. *Geol. Soc. Am. Abst. With Prog.*, Pt. 7, p. 237–8. (Abstract.)

WILSON, H. B. D. (ed.). 1968. Magmatic Ore Deposits. *Econ. Geol.* 336 p. (Mono 4.)

WYLLIE, P. J. (ed.). 1967. *Ultramafic and Related Rocks.* New York, Wiley. 464 p.

YAZAWA, A.; KAMEDA, M. 1953. Fundamental Studies on Copper Smelting: I. Partial Liquids Diagram for FeS–FeO–SiO System. *Tech. Rep. Tohoku Univ.*, Vol. 18, p. 40–58.

ZAPFFE, C. 1931. Deposition of Manganese. *Econ. Geol.*, Vol. 26, p. 299–832.

ZELENOV, K. K. 1958. On the Discharge of Iron in Solution into the Okhotsk Sea by Thermal Springs of the Ebeko Volcano (Paramuchir Island). *Dokl. Akad. Nauk SSSR*, Vol. 120, p. 1089–92. (English translation by Consultants Bureau, Inc., New York, p. 497–500, 1959.)

——. 1959. Transportation and Accumulation of Iron and Aluminium in Volcanic Provinces of the Pacific. *Izv. Akad. Nauk, SSSR, Ser. Geol.*, p. 47–59.

271

Geochemical prospecting

Frederic R. Siegel

Department of Geology,
George Washington University, Washington,
D.C. (United States of America)

Introduction

Geochemical prospecting has led to the discovery of important ore finds, some of which are described in the literature (e.g. Hawkes, 1969; Felder, 1974) others being maintained in confidential company files. The technique of geochemical prospecting can also be useful for more complete definition of established deposits, thus indicating directions for their further development.

Classical geochemical exploration techniques for regional and local areas (rock analyses, stream sediment and/or alluvium analyses and soil analyses) have been augmented over the past decade by techniques based on biogeochemical, hydrogeochemical and gasometric parameters; even information from agricultural surveys or epidemiological maps has been used to obtain important data.

The cost efficiency of an exploration geochemistry programme depends on careful and systematic planning from conception of the programme through to fruition. For an exploration geochemistry programme to be successful, an exploitable mineral deposit must be found, or a deposit with a tenor of ore of definite exploration potential should economic conditions change must be targeted. Siegel (1974) laid out an idealized ten-point programme for an exploration geochemistry project, but such a programme must always be planned in terms of a realistic assessment of available resources. This includes an assessment of the facilities available for the programme, budgetary limitations and time requirements.

The omnipresent problem in geochemical prospecting is how to find mineral deposits efficiently and economically. Research has been centred on several different lines.

Evaluating multiple sample types or sample subclasses from single sites in different environments.

Evaluating, modifying and developing analytical techniques.

Identifying pathfinder elements for a given commodity or group of commodities under different geochemical environment conditions.

Understanding how physical, chemical and biological conditions influence both pathfinder and target elements dispersions both vertically and horizontally from an ore source.

Statistically evaluating element concentration data for target selection and indications of regional trends in element distribution.

Prospecting in the marine environment.

Prospecting for petroleum and natural gas.

Using geochemical prospecting data in other fields (and vice versa).

Relating new theories and concepts (e.g. plate tectonics) to the regional localization of potential ore provinces.

There is a great deal of literature on geochemical prospecting (exploration geochemistry, applied geochemistry) and the interested reader is referred to the following texts: Hawkes and Webb (1962), Ginzburg (1960) and another classic early monograph by Hawkes (1957). Other contributions, highlighting later work are Hawkes (1971, 1976), Malyuga (1964), NASA (1968), Brooks (1972), Granier (1973), Levinson (1974), and Siegel (1974); unlike the other texts mentioned, this latter book includes considerations of geochemical prospecting in the marine environment and of geochemistry in health and pollution problems.

Sample evaluation

Geochemical prospecting is based on a thorough consideration of the geological characteristics of an area. Prospecting is then evaluated in terms of a series of chemical element analyses from samples of the area. The geological characteristics are only as good as the prepared geologic/topographic map, ahd the significance of this data will depend on several factors.

Selection of the sample that provides the best contrast with respect to a statistically determined background level.

The representativeness of the sample of the site from which it was taken.

The geochemical laboratory handling, preparation, and analysis of the sample.

Since each type of sample class has its own background level, inherent marked differences in the concentration: background contrast ratio may exist; many recent studies which examined the element concentration differences existing between sample types and/or subclasses for a single site have clearly shown the need for pilot studies before embarking on major programmes.

For example, Wolfe (1974) reported the results of rock, soil and biogeochemical sampling at the Setting Net Lake molybdenum deposit, north-western Ontario (Canada), a single intrusive phase with a mineralized quartz monzonite forming the northern half, and granodiorite forming the barren southern half of the oval 5.7 km^2 area. The small area has two distinct soil environments, which complicated the geochemical sampling. One is of low relief and poorly drained (with water-saturated peat and muskeg) and the other is of upland areas of free drainage (i.e. a more alkaline soil compared with a podzolic soil). Of the elements analysed (Mo, Cu, Hg, S, Pb, Zn, Mn, Ni and Co), Cu and Mo were enriched in the granitic rocks that host the mineralized quartz veins. The Mo of the B zone and the second-year needles of black spruce (by comparison with the bark or second-year twigs) proved to be the most useful techniques in geochemical exploration for low-grade mineralization of the Setting Net Lake type.

The pattern of anomalous Mo in the B soil zone was restricted to bedrock-glacial till fensters in the general clay-muskeg cover for two reasons: (a) it was difficult to take samples below a thick cover and thus evaluations were not fully made; and (b) Mo remains in soils on bedrock-glacial till material whereas it is mobilized in alkaline soils on clay deposits. However, the anomalous Mo concentrations in the second-year black spruce needles extended to both environments and were generally more widespread and well defined than soil anomalies. Because there is a simple relationship between the metal content of soil on bedrock and that in spruce-needle ash, and this relationship is determined mainly by Mo in soil (and only partly by other biochemical requirements), it can be useful in prospecting. Copper is not so good an indicator because Cu levels in the vegetation studied are only secondarily determined by the Cu concentration in the soils, being primarily determined by the specific plant requirements.

Brundin and Nairis (1972), evaluating Swedish methods, made a comparative study of stream sediments, stream waters and organic material collected in a 400 km^2 area in which seven types of mineralization were known (from outcrops, boulders and test drillings) but never exploited. These included:

1. Molybdenite in aplite and as a dissemination in metamorphic, supracrustal rocks.
2. Scheelite in skarn.
3. Chalcopyrite, galena and sphalerite in metamorphic, supracrustal rocks.
4. Galena and chalcopyrite in Cambrian sandstone.
5. Uraninite in basal conglomerates overlying the Pre-Cambrian crystalline basement.
6. Uraninite, disseminated in syenite.
7. Uraninite as one of the minerals filling fissures in metamorphic, supracrustal rocks.

As might be expected from this area situated on the Arctic circle in the western part of northern Sweden, the bedrock is generally covered by glacial till which was transported from the north-west. An evaluation of more than 1,000 samples of each type showed that for regional work, all sample classes were useful for defining patterns of mineralization for Mo and Zn. Organic material and stream waters were best for regional U prospecting. Organic samples were more reliable and sensitive than stream sediments for indications of mineralization containing Cu and Pb. As a result of this research, the Geological Survey of Sweden now uses only organic samples and water samples in large-scale regional prospecting for U sampling close to stream-road intersections (sampling density of 0.05 sample/km^2) and is at present studying the comparative merits of organic material samples versus stream sediment samples for regional Ni prospecting. Many investigations in geochemical prospecting and other fields (e.g. marine geochemistry) have demonstrated the capacity of fine-grained particles: (a) to adsorb and concentrate elements from an aqueous medium (due to the enhanced reactivity of their surface); (b) to be attracted to colloids of opposite charge; and (c) to contain specific minerals of known exchange and adsorption characteristics. In 1972. Perhac and Whelan made a comparison of water, suspended solids and bottom sediment analyses (Cd, Co, Cu, Fe, Mn, Ni, Pb and Zn) for geochemical prospecting in the Zn district of northern Tennessee in the United States. They presented evidence which suggested that the colloidal fraction of the suspended solids (0.01–0.15 μm) gave the best general results in terms of magnitude of anomaly. They suggested that surface adsorption might be significant since their study which investigated three sample sites showed increasing metal content with decreasing particle size and with increasing illite content. However, the colloidal fraction still presents the disadvantages that particulate recovery from water is slow and requires costly equipment.

Such evaluation of the best sample or samples to be used for geochemical prospecting in areas of mono-economic deposits or multieconomic deposits situated where the geology may be relatively simple or rather complex continue to be undertaken, especially in orientation or pilot programmes. The analytical data from these samples are considered in terms of the varying geochemical environments from which the samples derive, through which the samples have passed, and/or in which the samples are taken; this consideration is combined with a judgement of the influence of the character of the different sample types themselves on their element concentrations.

Once the determination of the best sample or samples for a pilot project has been made, a practical assessment of the data, taking into consideration all the other factors, can then determine what parameters should be used in the full geological programme.

Analytical methods

Colorimetry, emission spectrography and atomic absorption spectrometry are the work-horse techniques in geochemical prospecting analysis though many other techniques are used both in the laboratory and in the field.

Improved methods are being constantly reported for specific elements or a wider spectrum of elements. For example, it is well known that mercury is a pathfinder element for economic mineralization of a sulphide character (Jonasson, 1970; McCarthy, 1972, McNerney and Buseck, 1973) and specific detectors have been designed for use in geochemical prospecting (e.g. Vaughn, 1967). Bristow (1972) recently described the quartz crystal microbalance technique for measuring nanogram quantities of adsorbed mercury vapour which is related to the measurement of the frequency shift of a conventional quartz radio crystal as added mass is adsorbed onto its electrode surface. Selective ion electrodes have been developed for several elements, many of which can be especially useful in geochemical prospecting (e.g. Cu^{2+}, Pb^{2+}, S^{2-}, and F^-).

In some cases, the selective ion electrode technique for measuring element concentrations may be used in the field, particularly where water is being sampled, but this technique is usually used in the laboratory only after considerable sample preparation to eliminate interference. In general, the extent of interference depends on four factors: (a) the concentration of interfering ions; (b) the pH of the sample solution; (c) the ionic strength of the solution, and (d) the total concentration of the ion being determined; and these factors will tend to decrease the response of the ion under investigation, i.e. indicate a lesser concentration of the element than is actually present. Friedrich et al. (1973) illustrated this clearly in their investigation on the cupric-sensitive electrode. They reported that certain ions such as hydroxide and sulphide which precipitate Cu^{2+}, and acetate and ammonia, which complex Cu^{2+}, decreased the amount of free cupric ions in a solution. Also, some cations such as Hg^{2+} and Ag^+ may poison the sensing element and, if present, must be eliminated from the sample solution. The results of this study indicate that

cupric-sensitive electrodes could be recommended for stream-water surveys if rough estimates of the copper content were all that were necessary. They emphasized that only low levels of interfering ions should be present and found that for field measurement, a 1:100 Tri-trisol buffer could be used to avoid the effect of copper precipitation by phosphate ions. The best results were when the cupric ion concentration was between 0.6 and 60 p.p.b. at water pH around neutral.

Fluoride and chloride studies with the selective ion electrodes are especially favoured since they can save a great deal of time. Kesler et al. (1973), analysing fluoride in rocks, noted that the interference caused by hydroxyl ions and fluoride complexing with Si^{4+}, Al^{3+} and Fe^{3+} (which tended to remove free fluoride from interaction with the electrode, thus yielding lower fluoride concentration values) could be overcome by buffering the sample solutions to a pH of between 5 and 7. Because tin is often associated with fluorine, a pilot study was made in a zone of tin mineralization in the Maya Mountains of Belize, using unweathered host granite as the sampling medium. Three out of four anomalous values for fluoride were found in the zone of tin mineralization. Also, areas of granite associated with known alluvial cassiterite yielded samples with high fluoride values. Kesler et al. (1973) found that whole rock fluoride values (especially when fluoride values were high) varied directly with water-leachable fluoride values so that the less expensive, more rapid water-leach method they described (after Van Loon et al., 1973) may have field applications.

In addition to rock, stream sediment and stream-water samples, soils have been studied using the fluoride-selective electrode. Farrell (1974) analysed fluoride in soils as a direct indicator of fluorite mineralization in Derbyshire (United Kingdom). He used both a partial (cold extractable) extraction and a total fluoride method and found that both sets of values could indicate areas of vein fluorite mineralization concealed below residual soil and soil developed on transported overburden.

Mahaffey (1974) has reported on a spectrophotometric method that concentrates and isolates rhenium from samples (of rock, sediment, water, brine and mineral concentrates) by using carbon adsorption and solvent extraction. The technique agrees well with neutron activation analysis. Hoffman and Waskett-Myers (1974) noted that molybdenum determination with dithiol in soils and sediments was difficult when an emulsified organic layer formed and interfered with colorimetric analysis. They thus modified the dithiol method by adding small quantities of acetone for clarification of the organic phase and found no loss of molybdenum from standards; some samples in fact indicated that acetone did not appreciably deplete the molybdenum complex from the organic phase. Thus acetone addition is recommended for samples which do not yield an adequately separated organic phase or where test-tube walls are rendered opaque by coatings. Similarly, Kothny (1974) described a direct, sensitive and simple spectrophotometric method for determination of sub-microgram amounts of platinum in acid-leachates from geological materials.

The leach that is used in geochemical prospecting is most important and must be carefully studied. Olade and Fletcher (1974), for example, showed that a $KClO_3 - HCl$ leach was sulphide-selective and had advantages over other procedures (e.g. ascorbic acid-hydrogen peroxide, aqua regia or nitric-perchloric acids) in estimating the copper sulphide content of granodiorites. Bradshaw *et al.* (1974) investigated three different chemical extractions on sediments and soils for metals: (a) 'total metal'—perchloric acid extraction ($HClO_4$); (b) 'weak acid'—hydrochloric acid extraction (HCl), and (c) 'cold-extractable' extraction (EDTA).

They demonstrated that the stronger chemical attack accentuated the effects of mineralization, i.e. when variations in metal content of different rock types were large with respect to the anomalies that might result from mineralization. Also, if the metal was tightly bound, the EDTA-extractable metal would not give good results. Pilot studies must thus be carried out in which extraction procedures can be evaluated. One of the better examples of the effects of different sample preparation and extraction methods on analytical results was given by Bondar (1969) (see Table 1).

Instrument techniques have been adapted or are now being studied, for the purposes of *in situ* marine geochemical prospecting. Friedrich *et al.* (1974) described a radioisotope energy-dispersive X-ray fluorescence technique (EDX) designed for use on land which could be adapted to be used aboard ship and could be used for the analysis of Mn, Fe, Co, Ni, Cu and Zn in manganese nodules.

The technique is practical, analytically excellent, and can be carried out without weighing or special preparation of a sample apart from drying. However, although the method provides data rapidly so that evaluations can be made on board to guide further exploration, it does not provide immediate *in situ* analysis.

A system for *in situ* analysis which is at present undergoing testing is based on the thermal neutron activation of material with ^{252}Cf and interpretation of the resulting capture gamma spectrum. This analysis has been successful in tests with the transition elements, chlorine, gold, manganese and mercury; Senftle (personal communication) further noted that nickel was readily detectable under about 140 cm of overburden. He proposed the use of an *in situ* marine system for activating a large volume–area sample (as against a single nodule, for example), reading the capture gamma emissions with a Ge–Li detector, using a computer for rapid data reduction (five-minute lag) and subsequent mapping; this would facilitate immediate programme operation planning and changes. In 1973, Battele–North-west Laboratory reported successful testing of the first self-contained unit for *in situ* analysis of up to 30 elements in sea-bed minerals by such a probing technique which is lowered from a surface ship for operation. Senftle (1970) suggested that the system be included on an ocean bottom submersible unit that could be directed by onboard scientists.

Dogs, too, may be conditioned to detect as little as 10^{-18} g of gases such as SO_2 and H_2S. Where the dog indicates the presence of these gases, there may be sulphide-bearing minerals at or below the Earth's surface. The technique has been successfully used in Scandinavia and research is going on to determine its usefulness in areas of western Canada (Brock, 1972). For a discussion of the detection limits in instrumental analyses the reader is referred to the article by Karasek and Laub (1974).

Pathfinder elements

Research continues into the identification of guides to or indicators of mineralization. Since these guides or indicators may each respond differently to varying physical, chemical and biological conditions, an understanding of their responses to the different environments in which they are being studied is essential. When discussing elements in this context, either as free ions, molecules, compounds or complex ions, the designation 'pathfinder' is often used.

A pathfinder element, rather than being the element actually sought, is really an element whose geochemical characteristics (and hence affinities) result in it being intimately associated with the principal object of any search. However, the only proof of an accumulation of an element is the element itself. The geochemical

TABLE 1. Comparison of six analyses of the same stream sediment for nickel using the atomic absorption technique on aliquots prepared or extracted in different ways[1]

	Preparation		Chemical attack	Content of nickel (p.p.m.)
	Ground	Fraction		
Laboratory A	No	−80 mesh	70 per cent $HClO_4$	20
Laboratory B	No	−80 mesh	1:3 HNO_3	60
Laboratory C	No	−80 mesh	$HNO_3 - HCl$ mixed acid	150
Laboratory D	No	−80 mesh	1:1 HCl	320
Laboratory E	Yes	−100 mesh	1:1 HCl	14
Laboratory F	No	−100 mesh	1:1 HCl	1,120

1. *Sample description:* stream sediment with 0.5 per cent magnetite; magnetite contains an average of 0.28 per cent Ni; 80 per cent of the magnetite is in the −100 mesh size fraction; 99 per cent of the sample falls in the +100 mesh size fraction; and 96 per cent is in the +80 mesh size fraction.
Source: Bondar, 1969.

TABLE 2. Affinities of the chemical elements for the principal phases that compris the Earth

Siderophile phase	Chalcophile phase	Lithophile phase	Atmophile phase	Biophile phase
Fe, Ni, Co, Ru, Rh, Pd, Re, Os, Ir, Pt, Au, Ge, Sn, Sb, (Pb), C, (As), P, Mo, W, (Nb), Ta, Se, Te, Cu, Ga	((O)), S, Se, Te, Fe, Cr, (Ni), (Co), Cu, Zn, Cd, Pb, Sn, Ge, Mo, (Os), As, Sb, Bi, Ag, (Au), He, Ru, (Pt), (Rh), Ga, In, Tl, (Pd)	O, (S), (P), (H), (C), Si, Ti, Zr, Hf, Th, Li, Na, K, Rb, Cs, F, Cl, Br, I, B, Al, (Ga), Sc, Y, RE (La–Lu), Be, Mg, Ca, Sr, Ba, (Fe), V, Cr, Mn, Nb, Ta, W, U, (Tl), (Ge), (Zn), (N)	(H), C, N, O, F, Cl, Br, I, Ar, He, Ne, Kr, Xe	C, H, O, N, P, S, Cl, I, (Ca), (Mg), (K), (Na), (V), (Mn), (Fe), (Cu)

Source: Compiled from several sources but based on the original presentation by Goldschmidt, from Siegel, 1974.

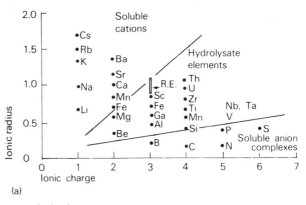

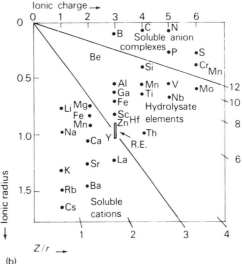

FIG. 1. Diagrammatic representations that show the Z/r relation by plots of ionic charge against ionic radius, and the responses of elements in hydrogeological sedimentary environments (from Siegel, 1974).

characteristics that define most element associations are dominated by ion·size and charge (related directly to the number and arrangement of electrons about an atomic nucleus). These characteristics determine to a great degree the many 'rules' and concepts that have been evolved to explain element substitution (e.g. Goldschmidt's rules, electronegativity, bonding energy and crystal field stabilization energy concepts). The close relationship between ion size, charge and element affinity is clearly indicated in Table 2, in which we see element affinity by phase (siderophile, chacophile, lithophile, biophile and atmophile); Figure 1 shows the response of some elements in a weathering environment by relating their ionic potential (Z/r) to the forms in which they may predominate, i.e. as free ions, as elements bound to hydrolysates or as complex anions. Both of these relationships are used by the exploration geochemist to assist in prospecting programmes. This also implies another useful property of a pathfinder, i.e. association with a wide range of elements so that its utility in exploration geochemistry is not restricted to only one type of economic deposit.

The usefulness of a pathfinder is influenced not only by these associations, but also by dispersive haloes either in primary deposits or in secondary ones—and by any subsequent activity of a physical, chemical and biological nature. This is, of course, a function of the element's response to the conditions under which the pathfinder was deposited and dispersed. An element

may be essentially immobile under one set of pH conditions but significantly mobilized under another; e.g. because it is often associated with porphyry Cu deposits, Mo can be used as a pathfinder to such deposits and provide a wider target to prospecting than would the Cu itself. Thus in order to use pathfinders effectively the exploration geochemist must be aware of how the different elements and their associated forms react under given environmental conditions. Practical analytical methods must also be available for the use of any element as a pathfinder. Finally the analytical technique must be economically viable for the exploration programme.

One of the better studied pathfinder elements is arsenic and its geochemistry and use as an indicator element in geochemical prospecting has been reviewed recently by Boyle and Jonasson (1973). They have emphasized that arsenic was a good indicator for several elements of economic value because it is found combined with many elements (Cu, Ag, Au, Zn, Cd, Hg, U, Sn, Pb, P, Sb, Bi, S, Se, Te, Mo, W, Fe, Ni, Co and

TABLE 3. Principal type deposits in which As is a constituent

Deposit type	Example	Mineralogy	As content
Copper shale, argillite, and schist	Kupferschiefer-White Pine-Zambia type	As in pyrite, copper sulphides, and various other sulphides,	t–5,000 p.p.m.
Copper (U, Vd, Ag) sandstone deposits	'Red Bed' type	In pyrite and various other sulphides	5–5,000 p.p.m.
Lead-zinc deposits in carbonate rocks	Mississippi Valley-Pine Point type	Mainly in pyrite	< 5 p.p.m.
Pyritiferous quartz-pebble conglomerate and quartzites enriched in one or more of Au, Ag, U, Th, and R.E.	Witwatersrand–Blind River type	In pyrite and arsenopyrite with small amounts in minerals such as pentlandite, pyrrhotite, and cobaltite	50–8,600 p.p.m.
Native copper deposits in or associated with amygdaloidal basalts	Keweenaw, Michigan type	In copper arsenides, in a few Ni-Co arsenides, in the native Cu	5 p.p.m.
Skarn	Nickel Plate Mine, British Columbia	In arsenopyrite	High, may reach about 20 per cent in some ore shoots
	Bau District, Sarawak	In arsenopyrite, native As; generally with Cu, Pb, Zn, Cd, Au, Ag, Mo, and Fe and less commonly with W, Sn, Bi, Sb, Te, Ni and Co	
Pegmatites and pegmatite-like bodies enriched in one or more of Sn, W, Nb, Ta, Bi, Mo, and U		In arsenopyrite and pyrite; rarely in native As and loellingite	< 500 p.p.m.
Massive, bodies, mantos, lodes, veins, and stockworks	Massive Ni–Cu sulphide bodies commonly with basic rocks (Sudbury) associated with Fe, Ni, Co, Cu, Ag, Au, Pt metals, Te, Se, Bi, S	In niccolite, maucherite, gersdorffite, arsenopyrite, sperrylite	t to low per cent
	Massive Cu–Zn–Pb sulphide bodies in volcanic and sedimentary areas (Noranda–Flin Flon–Bathhurst type) associated with Cu, Pb, Zn, Cd, Ag, Fe, and Sb; sometimes with Sn, Bi, Sc and Te	In arsenopyrite, pyrite, tennantite, enargite, and various other arsenic-bearing sulpho-salts	~800 p.p.m. (with wide range)
	Veins, lodes, and stockworks mainly in granitic rocks and associated sediments; very enriched in one or more of Sn, W, Bi and Mo	Sometimes with cassiterite and wolframite with Cu, Fe, Pb and Zn sulphides; sometimes W–Mo veins, scheelite–quartz–Au veins, W–Ag–Au veins, W–Cu veins, W–Ag–Cu–Au–Zn veins; elements concentrated are many and varied: Li, Rb, Cs, Be, B, F, U, Sc, R.E., Mo, Re, W, Mn, Fe, Cu, Ag, Au, Zn, Cd, Ga, In, Tl, Ge, Sn, Pb, As, Bi and S. Many with wallrock alteration zones with extensive tourmalinization and greisenization. Arsenopyrite and pyrite	t to several per cent s
	Various polymetallic veins, stockworks, mantos and bodies with Cu, Ag, Pb, Zn and Fe sulphides and sulpho-salts	In arsenopyrite, pyrite and variety of sulpho-salts. Enargite may be important	t to percent s
	Veins greatly enriched in Ni Co, As, Sb, Bi, and U (Cobalt–Great Bear Lake–Jackymov type)	Ni, Co, Fe arsenides and native Ag, sometimes native As, arsenopyrite, and various As-bearing sulphosalts. Pitchblende may be characteristic	2– 50 per cent

TABLE 3 *(continued)*

Deposit type	Example	Mineralogy	As content
	Gold–quartz veins, stockworks, and silicified zones	Arsenopyrite, As-bearing pyrite and variety of sulpho-salts	Few p.p.m. to per cent s (in ore shoots)
	Realgar-orpiment	Rare; associated with S, Sb, Ag, Pb, Cu, Fe, Au and rarely Hg	
	Mercury deposits—fissure veins, stock-works, disseminations, impregnations, and replacement bodies along faults or brecciated zones; cinnabar	With realgar, orpiment, arsenopyrite, and arseniferous pyrite and/or marcasite	Low to high As contents
	Disseminated bodies in various rocks: Au, mainly in sedimentary rocks	Arsenopyrite, arseniferous pyrite	480 p.p.m. to > 1 per cent
	Porphyry Cu and Mo deposits	Pyrite, enargite	Low (especially Mo type) ~100 p.p.m.

Source: Compiled from Boyle and Jonasson, 1973.

Pt metals) either as a primary component of a mineral with other elements, or as a trace element substituting for one in the aforementioned list (e.g. for S in pyrite). In addition to its occurrence in many classes of mineral deposits (Table 3), arsenic may be mobilized into secondary environments (e.g. as the arsenate ion) and can provide an extensive dispersion halo away from the primary deposit, thus presenting the exploration geochemist with a larger target area than the deposit itself does. Let us examine the mobilization of arsenic as reviewed by Boyle and Jonasson (1973).

The primary mobilization of As and its compounds, mainly as mobile anionic complexes and polymers with O and S, takes place in the gaseous and solution states as well as by diffusive processes. Mobilization is through fractures, fissures and faults (which may result in an offset anomaly) until deposition takes place because of changing physical and chemical conditions in the transporting medium and/or by contact with reactive host rocks. Arsenic has two oxidation states, $+3$ and $+5$; in primary environments with low oxidation potentials the $+3$ (arsenite) state would be dominant, but with high oxidation potentials in the environment, the $+5$ state (arsenate) would be dominant. In the primary environments, As is mobile in both acidic and alkaline conditions, but neutral or alkaline hot spring waters do tend to carry and precipitate rather large amounts of the element, suggesting that As-containing deposits may have been deposited from neutral and alkaline media.

In the weathering environment, rocks and deposits containing As yield soluble arsenates ($+5$ state). In a strongly acidic environment in which supergene solutions come into contact with minimum carbonate material, the As is mobile and may migrate considerable

distances from its source, thus amplifying the exploration geochemistry target. However, in weak acidic, neutral or basic solutions, soluble As is rather immobile because of hydrolysis and reactivity with various natural compounds and the target size may be limited. In soil-forming processes, then, As may be mobilized as the arsenate ion and be thus adsorbed or absorbed by hydrous iron oxides in a secondary environment or may be subject to chelation and/or metal-organic binding depending on the organic component of the secondary environment. Obviously, whether arsenic is present in a secondary environment depends on its availability in waters supplied to a basin of deposition or in waters passing through a soil zone. In soils, As is generally found enriched in the B zone because of adsorption and absorption of the arsenate ion by the hydrous oxides associated with the zone. However, the As may also be found enriched in the A zone and this is often attributed to chelation or organic binding in the zone. In basin situations, arsenic minerals may occur in the heavy mineral concentrates of both eluvial and alluvial phases especially near arseniferous deposits. Another immobilizer of As is reaction of other elements in solution (for example, Fe, Co, Ni, Pb and Zn) with As as soluble arsenates to form secondary minerals. Where the oxidation potential may be low in this environment, the As in a $+3$ state is relatively mobile in both acid and alkaline solutions. Thus, with deep and prolonged oxidation, much As may be removed from the oxidized parts of ore bodies and rocks and dispersed in ground and surface waters slowly. Fortunately, there is no toxic accumulation of As in these waters because of reaction and sequestering of the element by hydrous iron oxides, humus, and natural colloidal materials.

As was pointed out by Boyle and Jonasson (1973),

the solution chemistry and migration behaviour of As is complex. The element may migrate as uncharacterized complexes, as a variety of chelated or other compounds (originating from plant or animal matter decay), and in colloidal state or adsorbed onto hydrous iron oxides, silicates, and humus colloids.

Also there is the expected precipitation of small amounts of As from surface waters contacting gossans, soils, stream sediments, lake sediments, bogs, and other environments mainly by adsorption onto mineral particles, clays, and gels, and by adsorption and/or chelation or other forms of organic binding by humic constituents. The scavengers for As are hydrous iron oxides and natural precipitates containing hydrous Fe oxides, Mn oxides, and humic constituents. Geochemically, some of the associations and non-associations may be explained by the characteristics of ions or ion complexes; the negatively charged arsenate ions are attracted by positive charged colloids (or gels) such as might be the case for iron and hence the arsenic–iron positive correlation association. On the other hand, Mn oxides are negatively charged and would repel the arsenate negative complex ion so that arsenic accumulation would not be associated with Mn oxides presence unless the manganese oxides were ferriferous. Another concentration factor may be associated with ion replacement; for example, As is often associated with phosphatic materials and this may be explained by the isomorphic replacement of $(PO_4)^{3-}$ by $(AsO_4)^{3-}$ in the environment being studied. Concentration may also be the result of H_2S generation by bacterial action in the decay of plant and animal matter; this would result in the precipitation of As as sulphides and as arsenopyrite and/or arseniferous pyrite or marcasite in a reducing environment and given pH conditions.

Once precipitated in soils, sediments and bogs, for example, As may be remobilized into the environment by micro-organisms in a variety of forms, either inorganic or organic, dependent on any change in the geochemical environment which would allow for the remobilization.

In colder climates where the environment may be dominated by muskeg, peat, and peaty mucks, As may be concentrated by precipitation by H_2S and other sulphide species, and as As in phosphates.

Because of its geochemical characteristics which are similar to those elements for which it is a pathfinder element, or because of its geochemical responses to a given geochemical environment, As has been found to be a particularly good pathfinder for Au but less useful as a pathfinder for Ag. Arsenic is intimately associated with Sb and Bi deposits and hence is a good indicator element for those deposits. Arsenic is also used as a reliable pathfinder to mineral deposits containing varying associations of Cu, Zn, Hg, Sn, Pb, Mo, W, Fe, Co, Ni, and the Pt metals, as well as for Se and S. In some cases, it has been found that As is present as an enrichment in U deposits, especially those containing Ni, and

Co arsenides, and may be a useful pathfinder element to such deposits. In these latter cases, where the association is not always clear cut, a pilot or orientation study would be in order to define the utility of As in the particular situation.

Arsenic then is a good indicator of mineral deposits for a wide range of elements and deposit types. The selection of the sample to be used in a given exploration zone must be determined after a pilot study because selection will vary with the mobility or fixation of As in the host rocks of a deposit, in the overlying soils or glacial drift, in the ground, spring and stream waters and their precipitates, in stream and lake sediments, and in vegetation.

Mercury is another excellent pathfinder to an important suite of elements in mineral deposits of varying types. Like arsenic, mercury is a chalcophilic element but with some atmophilic characteristics. It has been used as a pathfinder for deposits containing Au, Ag, Cu, Ni and Zn, in addition to itself as native mercury or cinnabar. Although mercury concentrations in rocks and soils have been used in the search for ores, the present emphasis is on using mercury vapour as a pathfinder since mercury has a high vapour pressure and may be emitted from its own minerals or those with which it is associated (McCarthy, 1972; McNerney and Buseck, 1973). Thus, with the development of instruments capable of detecting and quantitatively determining mercury-vapour concentrations in (a) soil gases, (b) the atmosphere immediately above the soil, or even (c) far above the soil (e.g. 90 metres), many areas of mercury-containing mineralization in the subsurface were established. This technique presumes that the mercury emanating from concealed mineralization is able to rise vertically through an overburden. If the mercury has been laterally displaced from the mineral deposit—which may well be the case—this may be reflected as a halo-type dispersion pattern in the soil gases or atmospheric gases analysed. Such displacement of an anomalous mercury-vapour accumulation can be due to various considerations, including displacement by heating of a dome-like accumulation, erosion of the upper part of a deposit, and concealment by overburden deposition, or the presence of an impermeable overburden that forces the displacement of emanating mercury vapour towards ascending pathways (channels) such as fissures, fractures, joints and faults. The degree of this last type of displacement obviously depends on the distance of the pathways for mercury vapour from the deposit, the attitude of these channelways, and the depth of the deposit with respect to the samples taken. This displacement is an oft-cited feature when a gaseous pathfinder is employed for mineral location.

In the same way as arsenic and mercury have been used as pathfinders to sulphide mineralization, the sulphur-containing gases have the same function, especially SO_2, which is produced by the oxidation of mineral sulphides. Although techniques exist for rapid, real-time analysis of SO_2 in air or soil gases (adapted from pollution monitoring and control instrumentation) an exciting field method for detecting 'abnormal' SO_2 *in situ* has been successful: 'canino-detectors'. Dogs can detect sulphureous gases at quantities as low as 10^{-18} g; this is far better than any existing equipment. They may be thus conditioned to detect and point out emanating SO_2 and have been directly responsible for the discovery of base-metal deposits in Finland (Ekdhal, 1976). Research in western North America into the usefulness of dogs as geochemical prospecting assistants also yielded preliminary results (Brock, 1972) which were most encouraging.

Many other volatiles (and/or their dissolved species) are being used or studied as potential pathfinders. Fluorine, for example, has been used as a direct indicator of fluorite mineralization in Australian geochemical soil surveys (Farrell, 1974). Although the fluorine content of the samples was determined by the selective ion electrode after extractions for rapid, partial and total fluoride were made, there is no reason to believe that fluorine in a volatile phase would not have given the same results. An examination of the associations of fluorite or other fluorine-bearing compounds in ore deposits of other minerals shows that fluorine could serve as a pathfinder to deposits of tin, tungsten or molybdenum. Thus, Graham *et al.* (1975) used fluorine in groundwater (from wells) as a guide to Pb–Zn–Ba–F mineralization in flat-lying carbonate rocks of middle North America. This they described as a good technique for blind deposits beneath hundreds of feet of overburden.

McCarthy (1972) has cited in his review on gaseous pathfinders, other relationships that have been found between gases and mineral deposits: (a) anomalous concentrations of fluorine and chlorine in air over gold veins; (b) bromine and iodine in air over a porphyry copper deposit; (c) carbon dioxide in soil-gas measurements to map buried mineralization; (d) SO_2 in air over silver and gold veins (containing pyrite), in soil gas or air over disseminated sulphide deposits, and in soil gas or air over roll-type uranium deposits (from associated pyrite). He also emphasized the many factors that can influence the use of volatiles as pathfinders in geochemical prospecting; although it is tempting to consider the use of helicopters or other aircraft carrying instruments for immediate analysis, this type of sampling has met with mixed success owing to limitations imposed by dilution, season, climate and other meteorological factors.

Dispersion

The distribution of elements in ore deposits is primarily dictated by the genesis of the mineralization: of igneous, sedimentary, or metamorphic origin, or combinations. Such an accumulation may suffer subsequent alteration and dispersion of components by physical, chemical and/or biological processes.

Research into physical (or mechanical) dispersion, although rather simplistic in nature, has shown that the mechanisms of physical dispersion can combine together to cause either greater or lesser overall effect on mineral dispersion. For example, Bugrov (1974) found that in the Eastern Desert of Egypt, where there is much wind transport of small particles of detrital matter, geochemical mineral anomalies were lessened, i.e. their presence was 'diluted' by the addition of this fine fraction. However, by sieving out this fine fraction the geochemical anomaly could be effectively regained. Conversely, Bugrov found that there could be an accumulation of heavy or oxidation-resistant minerals in holes of rock debris and within small depressions on the surface of outcropping bedrock. Such a false geochemical anomaly can occur in zones of little or no mineralization.

Perhaps the simplest type of physical dispersion of geological material (with its geochemical indicator) is eluvial in nature and will be primarily affected by the topographically plane surface and gravity. As material is removed or undergoes disintegration and soil formation is initiated, an eluvium would normally remain as float on the lowered weathered soil surface. If, however, the weathering and soil-forming processes take place on an incline, dispersion will be amplified in a down-slope direction and the resulting deposit is termed 'colluvium'. When water which, as rivers and streams, is an important dispersal agent, aids such dispersion, detrital material is also transported and deposited; the deposition may be temporary—as, for example, on a flood plain—and is called 'alluvium'. Where there is a significant lessening of gradient at the base of a topographically high system, any anomaly in alluvium may spread out and assume a dispersion pattern similar to the alluvial fan.

Physical dispersion of geological material also results from glacial activity. Glacial dispersal is characterized by geochemical data plots in which the mineral distributions may be linear or broad tear-shaped in form and like striations and/or 'chatter marks' give a clear indication as to the direction from which the glacier moved; this provides an excellent clue to further exploration. The glacial dispersion trains may show discontinuous development if obstructions caused a halt in the path of ice movement. The superimposition of glacial activity on other effects will cause pre-existing anomalies to be 'smeared out' in the direction of ice flow.

Wind is another agent of physical dispersion. A wind of 36 km/h can move particles up to 1 mm in diameter for short distances; at a wind velocity of 7–11 km/h, only very fine detrital material (less than 0.25 mm in diameter) is moved significant distances. If the wind blows across a mineral deposit, any anomaly factors will tend to be found downwind and concentrated either at physical barriers to continued movement (scrubs, cacti, etc.) or at points where the kinetic energy of the wind is unable to sustain further movement of material. Obviously, this latter influence will vary regionally and seasonally but an essentially unidirectional wind will cause a linear dispersion; with lateral vectors of wind flow added to the system, there will be corresponding widening or spreading of wind-transported material.

Although data on physical dispersion of geochemical elements continues to be collected, most contemporary research into element dispersion is focused on that caused by chemical and biological factors. Such investigations are complex and must often be combined to allow for the best possible interpretation; for example, in one soil an ionic complex may be mobile and thus available for plant uptake, whereas in another environment the same complex, in the absence of a mobilizing medium, may not.

We know much about the response of minerals and their component elements to solution processes under varying experimental conditions, especially pH and Eh. However, in vitro data may not apply in environments where competing reactions are not simply chemical in character, but biogeochemical as well.

The utility of recognizing the role of pH–Eh relations in mineral prospecting has been treated in relatively simple but excellent form by Hansuld (1967) and much recent research has been directed towards quantifying pH–Eh effects in many different environments. Studies on Cu and Mo have been especially prominent. For example, Hoffman and Fletcher (1972) described the distribution of Cu in the semi-arid region of south-central British Columbia, and showed that in the brunisol profiles samples, the Cu content increased with depth. In such environments, Cu generally increases with pH which rises from 5.5 at the surface of such profiles to 8.1 at the calcareous horizon of the profiles. The Cu is leached from the surface and precipitated and accumulated under progressively stronger alkaline conditions. Thus, where alkaline horizons are present in soil profiles, Cu anomalies in the related hydromorphic regime should not be expected since the upward and lateral dispersion of dissolved Cu would be small.

However, the low concentrations of dissolved Cu (as Cu^{2+}, $CuCO_3$, and $Cu(CO_3)_2^{2-}$) in river water can be explained in other ways. For example, by dilution of anomalous waters with barren water, or by equilibrium reactions. Hoffman and Fletcher (1972) calculated that for the Rayfield River (pH = 7.5–7.8, SO_4^{2-} and CO_3^{2-} activities $\sim 10^{-4}$), the total dissolved Cu should be less than 10 p.p.b. Thus, if greater-than-equilibrium concentrations entered the river, Cu would be precipitated or adsorbed by the sediments. At pH = 7.8, the Rayfield River waters had a Cu concentration of about 2 p.p.b.

Wolfe (1974) employed the pH–Eh concept of Hansuld (1967) to explain Cu and Mo dispersion in his research on porphyry-type Mo–Cu deposits. Under oxidized, acidic conditions, Cu compounds are fairly soluble and Cu is mainly present as the simple bivalent cation in aqueous solution; under similar conditions Mo is immobile and may be present as ferrimolybdate. In oxidizing, alkaline, aqueous systems, however, Mo forms a soluble, mobile molybdate complex anion $(MoO_4)^{2-}$. Such a solubility–pH relationship was evident from the unusually high Mo concentrations in trees growing in alkaline soils—where the mobility of the molybdate anion complex is high—and anomalous Cu levels in high pH soils, which apparently reflected the precipitation of Cu^{2+} in the alkaline environment.

When living material is being used as evidence for the existence of mineral deposits, its ability to create a micro-environment—at the tips of a root system, say—must always be considered, as well as the varying requirements of different plants and plant tissues to concentrate specific elements. Most recently, for example, Quin et al. (1974) showed during a biogeochemical exploration programme for tungsten in New Zealand that trunks contained more tungsten than did leaves, and leaves contained more tungsten than did twigs. They also found that for predicting the concentration of tungsten in soils they were studying, ferns were more effective than shrubs or large trees; this was thought to be due to their different root systems.

Coope (1973) reviewed geochemical prospecting for porphyry-copper-type mineralization and emphasized the pH–Eh relationship between Cu and Mo. However, he noted that the dispersive behaviour of Cu and Mo in aqueous systems might not be exclusively related to redox or the pH conditions. The Cu can be more extensively dispersed due to the complexing of Mo by Fe—as the ferrimolybdate. Also, bacteria can mobilize Cu under certain pH–temperature conditions and abet its dispersion aureole. Local conditions of climate, topography, mineralogy, geology and vegetation will all influence the magnitude and extent of any dispersion; hence only after the correlation of data from areas where such local conditions vary will one be able to establish the degree of influence of the interrelated factors.

Research is also progressing into the role of humus matter in soils and organic matter in sediments as agents that can immobilize and concentrate ions with which they come in contact. In a study of the formation of metal dispersions in organic lake sediments, Timperly and Allan (1974) found that the accumulation of metals in reducing sediments was dominated either by organic complexing or by metal sulphide precipitation. They found that a guide to the order of stability of organo-metallic complexes was:

$$Cu^{2+} > Ni^{2+} > Pb^{2+} > Co^{2+} > Fe^{2+} > Zn^{2+} > Mn^{2+}.$$

Chowdhury and Bose (1971) had earlier investigated the influence of humus matter on the formation of geochemical anomalies in a series of laboratory experiments. The stability constants for the chelates of metal humates they worked with were in the order: $Pb>Cu>Zn>Ni>Co$. They also found that the equilibrium concentration of the metals in solution required for maximum retention by humus matter was in the reverse order; i.e. $Co>Ni>Zn>Cu>Pb$. The stability of the organo-metallic complexes and the chelates of the corresponding metal humates are quite different. This highlights the need for further research on well-characterized organics and their individual relationships to selected ore metals. They also suggested that metal absorption (w/w) by the humic fraction of soil was governed by the atomic weight and valency of the elements, and hence that the relative adsorption of two elements of equal valency was proportional to their atomic weights.

Timperly and Allan (1974) have pointed out that although we may obtain evidence of the role of organic complexing in the laboratory and even in real environments, competing reactions are always important; for example, in gyttja, which often contains free S^{2-} ion, the S^{2-} concentration will control whether the accumulation of some cations (for example, Cu^+, Cu^{2+}, Zn^{2+}, Fe^{3+}) is by sulphide precipitation, organic complexing, or both. The order of metal sulphide precipitation in their study was: $Cu^{2+}>Pb^{2+}>Co^{2+}=Ni^{2+}>Zn^{2+}>Fe^{2+}>Mn^{2+}$. The S^{2-} ion may be generated by bacteria during decay of organic matter and during anaerobic activity on sulphate under reducing conditions.

The bond strength of a complex can be studied by examining the effects of different solvents possessing varying capabilities of separating selected ions from their bonded sites either in or on mineral lattices. Thus, we can categorize element response to extraction: first, those elements that are 'cold-extractable', i.e. loosely attached elements adsorbed only on clay-sized particles or other mineral units; second, 'weak acid extractable', i.e. loosely bonded and absorbed metals, precipitated salts, and possibly some of the less resistant silicates, such as layered silicates, that respond to this attack; third, 'strong acid extractable' or 'total metal extractable', i.e. all loosely bound or absorbed metal and metals from alkaline silicates (feldspar, layered silicates, such as mica, and sulphides); and fourth, total metals present, i.e. the total sample is solubilized. The extractions are, in essence, chelation, weak acid leaching, strong acid leaching and wet ashing, respectively, of the samples and, in any of the extractive categories except the last, there are degrees of relative effectiveness.

Although research continues in this area of geochemical prospecting, it may not be of critical importance to the execution of effective prospecting as long as consistent techniques are used throughout a programme. For further reading, and for a detailed study of conceptual models of this branch of exploration geochemistry, the reader is referred to the volumes edited by Bradshaw (1975) and Kauranne (1976).

As a final aspect of element dispersion from its point of origin, the offset or displaced anomaly, which

is often a combination of physical and chemical dispersion mechanisms, must be considered. Such displacement may result, for example, from the attitude of a deposit and the degree of weathering of soil; or a combination of this with a compaction or slide activity of the material bearing a detected anomaly. Displacement by gravity is simple to visualize—and in the field may sometimes be discerned by an experienced observer—but if the attitude of a blind deposit is not known, especially with respect to its depth below the surface, the vertical drilling from an anomaly will prove negative. Such displacement can also take place via dipping channelways from the original major deposit. Faults, fractures and joint systems all provide pathways for element mobilization and, depending on their attitudes with respect to a blind deposit and the depth of a deposit with respect to the surface sample taken for prospecting, can result result in an offset or displaced anomaly.

Statistical evaluation of target

Statistical evaluation of element concentration data for geochemical target selection and indications of regional trends in element distribution continues to be refined and developed. Siegel (1974) reviewed some fundamental concepts that are necessary to geochemical prospecting and gave a general critique of several of the techniques used by geochemists from the late 1960s onwards for the evaluation and selection of apparent anomalies. In an earlier article, Nichol et al. (1969) presented an excellent study of the roles of trend surface and factor–vector analyses in the interpretation of regional geochemical exploration data.

Basically, all of the statistical methods study single or multiple populations of data and establish arbitrary limits for regional fluctuation values of element concentrations (often designated 'background'), local fluctuation values (often designated 'threshold'), and values greater than the threshold upper limit (often designated 'anomalies'). In many evaluations of geochemical data, these values are calculated from the mean and the standard deviation. However, the standard deviation, σ is classically determined for a population in which measured values have a normal distribution, i.e. they are symmetrically distributed about an arithmetic mean, $\bar{X}$, when plotted as a frequency distribution curve. In such a system, the mean $\pm$ one standard deviation, i.e. $\bar{X}\pm1\sigma$ will include 68.26 per cent of the values in the population and $(\bar{X}+1\sigma)$ may be considered as the upper limit of the background value; $\bar{X}\pm2\sigma$ will include 95.44 per cent of the values and $(\bar{X}+2\sigma)$ may be considered as the upper limit of the threshold value; beyond this, a measurement may be considered anomalous. Thus in a population of 1,000 values, 26 measurements may be statistically expected to exceed the $(\bar{X}+2\sigma)$ limit. [It has been suggested that $(\bar{X}+3\sigma)$ should be the limit beyond

Frederic R. Siegel

which we have anomalies. However, this seems rather restrictive for geochemical prospecting use since only one or two values in population of 1,000 might be expected statistically to exceed $(\bar{X}+3\,\sigma)$.]

The previous figures refer to normal distribution in a population. In geochemistry, it seems that element concentrations in most populations approximate to log-normal distributions, i.e. a plot of frequency against class value on an arithmetic scale would show an asymmetrical curve skewed either to one side or the other—with the positive skew being the most common. A plot of frequency against class value on a logarithmic scale (log transform), will yield a symmetrical curve about a *geometric* mean. In geochemical prospecting, the $\bar{X}$ and σ values are usually derived from a log-normal plot in the same way as they would from a normal plot, and can be used directly for the establishing of background, threshold and anomaly limits. Statistically, this may not be entirely justifiable, but empirically, the technique has proved most useful.

The characterization of the log-normality of a population may also be established by plotting cumulative frequency against class values in logarithmic form on probability paper on which, with a log-normal distribution, the plot will approximate to a straight line. The values for $\bar{X}$, $(\bar{X}+\sigma)$ and $(\bar{X}+2\,\sigma)$ can be easily derived from such a plot by relating them to the cumulative 50, 84 and 97.5 percentiles, respectively. This graphical method has been treated in fine detail by Lepeltier (1969).

Bolviken (1971) considered the distinction between background, threshold and anomaly and suggested that because the concepts of background and anomaly values could each be treated as individual statistical distributions, each should give a straight line when plotted on probability paper. In essence, he was saying that because each category forms a statistical distribution itself, a single value for mean and standard deviation for the combined distributions was of limited value and that a mean and standard deviation should be determined for each distribution. He also felt that the concept of threshold was incorrect and that only two concentration groups should be used: (a) background—the distribution to which the majority of observations in a set of random samples from a zone belong; and (b) anomaly—any distribution differing from the background.

In the Geological Survey of Norway, an atlas of nomograms is used to determine the best combination of distributions that can be fitted to the plotted empirical distribution and the background and anomaly are estimated as the two single distributions of the nomogram (from which there may be an indication of the relative proportions of background and anomaly in the combined distribution).

Threshold does have significance, however, to most exploration geochemists, who consider it as a concentration band that separates a lower concentration data set (the background) from a higher concentration data set

(the anomalous). From an economic aspect, these concentration sets represent either a non-mineralized or a mineralized geological environment. Whether or not the mineralized environment is economically exploitable can be determined only after test drilling to investigate the tenor of the ore and the tonnage reserves.

Recently, Sinclair (1974) described the use of probability graphs for the selection of threshold values between background and anomalous values in geochemical data sets. His method was based on the partitioning of a cumulative probability plot of the data, resulting in fundamental grouping of data values. That probability plots can be used is based on the premise that element analyses approximate to log-normal density distributions. Very often a sigmoidal curve results from such a plot due to the overlap of background and anomalous populations. Sinclair defined the threshold as the point of inflexion (or maximum curvature) on the cumulative probability plot and, from the position of the inflexion point on the cumulative percentile, was able to determine the contribution of each population to the curve. Once this proportion has been established, the individual contributing populations can be extracted and presented as straight-line plots representing the idealized populations. The geometric mean may be read off (50 percentile) as can one standard deviation (84 and 16 cumulative percentiles), two standard deviations (97.5 and 2.5 cumulative percentiles) or indeed any limits that the researcher chooses to set as thresholds for a study.

Shortly thereafter, Parslow (1974) reviewed the graphical methods of analysing bi-modal frequency distributions for background and threshold values and proposed a method for extracting the populations contributing to a natural geochemical data set, i.e. background and 'anomalous' distributions. As did Sinclair, Parslow assumed normality or log-normality for the distributions and used the probability scales graph in his approach. However, he emphasized that the majority of trace element data sets from exploration geochemistry projects were sigmoidal and analysed the sigmoidal curve by trying to determine the contributions of the two overlapping populations. In fact, the graphical methods proposed by Sinclair and Parslow are very similar in nature and differ from each other only in the manner in which linear distributions for background and anomalous distributions are fixed.

In geochemical prospecting, statistical analysis has an important role in establishing target areas for detailed follow-up and perhaps subsequent drilling. However, what Hawkes and Webb wrote in 1962 is still true today:

It should be emphasized that although statistics may help in presenting and analyzing geochemical data, it can not provide the interpretation. A reliable interpretation of anomalies in terms of ore requires a combination of complex human experience and a capacity to recognized significant geometrical correlations. Pure mathematical analysis, therefore, is not likely to replace the subjective interpretative talents of the exploration geologist for some time to come.

Marine geochemical prospecting

In marine geochemical prospecting, much of the more recent work has been directed towards the development of *in situ* analytical equipment (Siegel, 1974) or simple, rapid shipboard analysis; these advances were mentioned in a previous section. Although most interest in marine prospecting (apart from hydrocarbons and minerals that can be dredged from near-shore waters) is centred around metal recovery from manganese nodule accumulations on the deep-sea floor (Horn, 1972) or from hot brines or sediments (with metal concentrations) in selected areas of the deep sea (Tooms, 1972; Holmes and Tooms, 1973), there is evidence of renewed activity in the near-shore and/or continental shelf environment.

As geoscientists attain more knowledge of Quaternary geomorphology, geologic agents and processes, and their extensions into and regressions from what we now identify as the continental shelf, economic geologists and geochemists are investigating palaeoenvironments for river channelways, deltas and channel gradient changes, and the analyses of samples therefrom as guides to old placer accumulations in the continental shelf environment. Existing placer accumulations have been mapped and presented by McKelvey and Wang (1970) as *World Sub-sea Mineral Resources*; none the less, the recognition of older and palaeosystems may prove beneficial to onshore programmes as well. For example, Fairburn (1973) completed a programme of regional stream sediment sampling in the south-west of Western Australia and noted that in one area, greater than background Pb and Sn contents in the -80 mesh fractions of samples were related to unusually high Nb contents in the samples. Subsequent research showed that the anomalous samples were from streams draining eroded sand dunes related to a former shore-line. These dunes contained heavy mineral sands, differentiates of which gave values up to 0.12 per cent Nb in samples of altered ilmenite. It is essential to note that this particular programme was set up partly because of the geological environment of the region and partly upon the implications of global tectonics, a topic to be discussed later.

The use of pathfinders or guides to economic deposits of selected metals or non-metals, rather than using a major component of the commodity itself, is limited in the marine environment—we should emphasize here the great costs that are involved in taking samples from the marine environment, especially in the bathyl and abyssal depths—however, this problem does not always exist to the same degree. Summerhayes *et al.* (1970) basing their argument on the fact that uranium is often present in phosphate deposits in trace quantities, proposed that the radioactivity of samples could be used as guides to submarine phosphate deposits. Preliminary tests were carried out with a submersible scintillometer in water up to 1,000 m in depth. The data obtained were promising enough to warrant the development of equipment for continuous recording which could be towed by a boat. The excessively high price of petroleum products that are required for fertilizer production has made this discovery of potentially enormous importance to developing countries that do not possess oil deposits.

Recently, Siegel *et al.* (1976, 1977) proposed that trace metal analysis of marine mineral suspensates from inshore environments could be useful as a new geochemical prospecting technique for relatively near-shore land regions. This would be especially effective for areas with difficult land access because of physical and logistic problems and also because of the costs involved in supporting a total land programme in such areas (e.g., the Chilean archipelago, the Antarctic peninsula, Greenland fjords, and others). The ability to target a reasonably small area with maximum mineral potential within a large region is cost effective in regional geochemical exploration and thus desirable. The premises for using the mineral suspensate as a sample type for geochemical evaluation of mineral resources potential are: first, that the suspended matter sample is derived from the region sampled and that it is representative of the suspensate from that region; second, that if it is derived from a mineralized zone in that region it will carry a chemical signal that will indicate its anomalous nature with respect to suspended matter that has come from non-mineralized zones; and third, that the geochemists have an indication of direction and velocity of current movement in the off-shore sampling area so as to be able to trace suspensates back to their probable zone of injection into the marine system, giving proper consideration to other factors that could influence suspensate transport, mineralogy, and chemistry (oceanographic and tidal conditions, climate, biology, atmospheric input and others).

Finally, it should be mentioned that although, at present, marine exploration—academic or applied—is not generally restricted by national, artificial geographical limits of control, exploitation of natural resources (inorganic and organic) from the marine environment certainly is—according to the claimed sovereignty limits of different countries—limits which range from about 5 to 300 km, or to waters at the edge of the continental shelf with a depth of approximately 200 m. One would hope that this freedom of research will be maintained in all national waters, for to restrict such research would be stifle man's quest for the support which may help his continued existence on our planet.

However, what authority is to control mineral harvesting beyond the limits of national jurisdictions? Who may exploit deposits such as manganese nodules? Technologically, systems are already available for harvesting these nodules from the deep ocean and economic metallurgical techniques are said to exist for the processing of their contained metals. At present, this technology is only available to a few nations. This is equally true for the exploitation of potential petroleum and/or natural-gas accumulations in the open ocean. However, the marine resources in the open sea should not belong to any one nation or group of nations whose industry and technology permit it; rather, the marine resources belong to *all* the nations of the world—this wealth is man's wealth. It is evident that a world organization or agency *must* be responsible for apolitical control and leasing of claims to companies that can be allowed to make a fair and acceptable profit on investment according to an agreed-upon formula. But much of the profit must be injected back into man; to provide, for example, funds for natural or man-made disaster victims, funds for paying increased commodity prices, funds for social services, and funds for continual environmental research in and around the ocean areas so that an unforeseen physical/chemical/biological alteration or mutation will not overtake and unbalance the ecological niche of localized or distant oceanic zones.

This is, of course, work for the Conferences on the Law of the Sea that have been taking place regionally and internationally over the past several years. Hopefully, these international conferences will terminate the class of thinking in terms of national or regional interests and face up to the reality that decisions have to be made, not delayed.

Geochemical prospecting for hydrocarbons

Prospecting for hydrocarbons still tends to be directed away from geochemistry towards geology and geophysics. This is because the target being sought is the container rather than what it may contain. Geophysical techniques are principally used to find appropriate traps for hydrocarbon accumulation that can be defined and located, e.g. sedimentary domal features, anticlines, faults, salt domes, etc. Geology is used to try to establish likely geological sites for stratigraphic-type traps. Methods and models have become more refined and sophisticated, especially when used in conjunction with computer functions. Obvious manifestations of the hydrocarbons may be found as seepages of crude oil and/or natural gas and, in some cases, they may be evident as a type of 'eternal' flame. Geochemistry is concerned with the location of the less obvious results of hydrocarbon accumulation in the subsurface.

Geochemical prospecting for petroleum and/or natural gas is not new, being used as far back as 1933, when Laubmeyer measured the concentrations of methane in soil atmospheres and attempted to relate

'anomalous' concentrations of the methane to hydrocarbons in the subsurface. Subsequently, Soviet and American scientists continued the development of 'gasometric' techniques and improved upon them considerably both in precision and in the spectrum of gases identified and analysed (Sokolov et al., 1971; McCrossan and Ball, 1971). The basis of the gasometric technique is that the light gases associated with petroleum or natural gas deposits rise and provide a soil gas field greater than that which might be considered the normal background for an area. In some cases, a particular gas (e.g. propane) may not even have a natural soil gas field and hence must represent gas generated from a hydrocarbon deposit.

Recent advances in gasometric prospecting have been made by Soviet scientists in their deep-well, gas-logging projects (Sokolov, 1971; Sokolov et al., 1971; Sokolov and Dadashev, 1972; Geodekian and Stroganov, 1973). Instead of degassing the gases from soils at or near the surface of the Earth, they drill test wells 200–600 m deep and study gases from drilling chips representing given depths, drilling muds that are allowed to equilibrate with formational gases at known stratigraphic intervals, and core sequences. This technique is supposed to give better definition of gaseous anomalies related to deep hydrocarbon accumulations. Their most recent technical advance permits them to seal cores hermetically *in situ* and bring them to the surface for analyses; this enhances anomalies and hence allows even subtle anomalies to become recognizable. However, the costs involved in drilling wells for the purpose of obtaining deep subsurface gas samples are high, calling for considerable government subsidy. In addition, for private companies, the financial risks involved may not justify this method over present, conventional means of locating petroleum and natural gas.

In the marine environment, however, private companies are using gasometric techniques for geochemical prospecting. Specially designed shipboard systems are being used either for continuous gas chromatographic analysis of water brought up from close contact with the sea floor, or for analysis of sediments after they are obtained from the bottom. The use of encapsulated gas chromatograph units, towed by a ship, and used to make continuous gas analysis profiles of bottom waters *in situ,* has also been reported. Significant research is also being carried out on gases, dissolved in subsurface water, which are associated with formations known to contain hydrocarbon accumulations (Zorkin, 1969; Zarella, 1969; Zarella et al., 1967).

Although gasometrically determined anomalies may be able to predict the presence of hydrocarbons in the subsurface, they cannot show whether the hydrocarbons are collected in an economically exploitable quantity. This can be demonstrated only by drilling test wells and determining the flow characteristics. Siegel (1974) provided a summary of geochemical prospecting for hydrocarbons and a recent volume of the *Journal of Geochem-*

ical Exploration (Hitchon, 1977) has been devoted to the application of geochemistry to the search for hydrocarbons.

Use of geochemical prospecting data in other fields

Geochemical concepts, interpretative capabilities and prospecting data have all been useful in other disciplines, notably in environmental fields dealing with health and pollution, The use and relationships of geochemistry with these fields have been treated in earlier articles of this volume and the interested reader is referred to that and texts by Warren (1973), Cannon (1974), Cannon and Hopps (1971, 1972), Hemphill (1973) and Hopps and Cannon (1972).

Plate tectonics in geochemical prospecting

Finally, we shall consider how recent concepts of global tectonics are being used to establish the relationship between plate tectonics and ore emplacement under given geological conditions so as to form metallogenic provinces; if such situations can be reconstructed from studies of the geological column, it is possible that areas whose mineral potential had not hitherto been considered would be brought to light. In general papers, Sillitoe (1972a) presented a plate tectonic model for the origin of porphyry copper deposits, and Sawkins (1972) much the same for general sulphide deposits; Guild (1974a, b) has described the distribution of metallogenic provinces in relation to major Earth features and the application of global tectonic theory to metallogenic studies. In papers on specific regions, Sillitoe (1972b) showed the relationship of metal provinces in western North America with subduction of oceanic lithosphere and, in 1974, discussed the implications for magmatism and metallogeny of the tectonic segmentation of the Andes; also, Mitchell and Garson (1972) described the relationship of porphyry copper and circum-Pacific tin deposits with Palaeo-Benioff zones. The evolution of the island arc and its relation to mineral deposits has been treated by Mitchell and Bell (1973). Both these and other papers have alerted geochemists and economic geologists alike to the importance of relating the plate tectonics knowledge with the plate systems that must have existed in the past. If, as they seem, plate systems are related to metallogenesis, the recognition of positions of palaeoplate systems should be an outstanding guide to exploration zones for geochemical prospecting teams.

In an excellent paper, Rona (1974) summarized the relationships between plate tectonics and mineral depos-

its. He proposed that a major source of metals (sulphide ores) could be generated either by convergent or divergent plate systems. In the convergent system, two adjacent plates come together and either collide, or one plate subducts downwards under the other and is absorbed into the earth. Sawkins (1972) and Sillitoe (1972a) have shown that most of the major sulphide deposits in the world are located along existing or older convergent plate boundaries where an oceanic plate plunged beneath a continental margin (including the continental platform) or beneath an island arc chain. It is possible that the sulphide deposits are present at convergent plate boundaries because mineralizing solutions were generated from the plunging plate mass which underwent fusion or anatexis (melting) as it was absorbed into the earth.

In the case of the divergent system, two adjacent plates move apart and new lithosphere is added to each plate during sea-floor spreading. Sulphide minerals in existing divergent plate systems (e.g. the Red Sea) may be disseminated in sediments in an expanding basin developing from the spreading process. Such deposits may be saturated with and overlain by brines carrying the same metals in solution as those present in the sulphide deposits. The origin of the metals in the brines is not known but it is possible that the brines are being charged with metals from undersea volcanic emanations or from the halmyrolisis of submarine extrusions. Thus, if present-day knowledge can direct us to ancient analogues of the existing plate systems, focuses for mineral exploration can possibly be established.

Rona (1974) also attempted to relate conditions favourable for the accumulation of petroleum and natural gas to convergent and divergent plate boundaries. These conditions include the preservation of organic matter (favoured by a toxic environment in which organic matter is not utilized by living forms), and lack of oxygen (so that organic matter cannot be decomposed to H_2O and CO_2). In a convergent system, the subduction zones are characterized by the existence of deep-sea trenches running parallel along the length of the continental margin being subducted. Also, in addition to the trench systems, volcanic island arc chains may divide an ocean basin into smaller units partially enclosed between the islands and the adjacent continent. Both the trenches and the island arc systems can create basins favourable to the accumulation and preservation of petroleum progenitors and petroleum itself. These basins entrap sediment and organic matter from the continent and oceanic areas; their shapes are such as both to restrict circulation of the ocean and, hence, oxygen replenishment; thus organic matter will be preserved. Finally, the accumulations and the geological structures that develop as a result of the deformation of sediments and sedimentary rocks by tectonic forces provide reservoirs and traps for the accumulation of petroleum. Some marginal, semi-enclosed basins may therefore be promising areas meriting future exploration.

In a divergent plate system, we are concerned with the deep-sea environment and hydrocarbon accumulations in that environment. The rift into two of the continent allows for the formation of a sea between the spreading sections with circulation being impeded by the continents themselves. The result is preservation of organic matter. In a climatic–meteorological condition where evaporation exceeds precipitation or recharge of basin sea waters, the precipitation of halite (salt) strata may take place along with the deposition of organic matter. With continued spreading and possible concurrent subsidence, the sea evolves from a restricted to an open circulation system. If the strata of organic matter and salt are buried under sediments, two processes can follow: first, the formation of petroleum from organic matter; and second, the salt developing into domes or diapir masses that can provide traps for hydrocarbons at their contacts with the intruded sediments.

Therefore, petroleum and natural gas may be found in deep ocean basins that have grown through a stage of a restricted sea by sea-floor spreading.

Sillitoe (1972a) used the concept of lithosphere plate tectonics to develop a speculative model for the origin and space–time distribution of porphyry copper and porphyry molybdenum deposits. From already published chemical and isotope data, he first showed that partial melting of oceanic crustal rocks on underlying subduction zones at the elongate compressive boundaries between lithospheric plates could generate the components of porphyry ores and associated calc-alkaline igneous rocks. His model proposes that metals in the porphyry ore deposits were mantle-derived at divergent plate boundaries and were transported laterally to subduction zones as components of basaltic-gabbroic oceanic crust and small amounts of overlying pelagic sediments. He suggested that the time-space distribution of porphyry ore deposits depended on three factors: (a) the

erosion level of an intrusive volcanic chain; (b) the time and location of magma generation, and (c) the availability of metals, on an underlying subduction zone.

His theory is that porphyry ore deposits formed in a series of relatively short, discrete pulses, perhaps correlated with changes in the relative rates and directions of motion of lithospheric plates. This type of motion is supported by the work of Anderson (1975).

From a distribution of Mesozoic–Cenozoic subduction zones and compilations of ages of magmatism in island arcs, Sillitoe (1972a) also suggested as potential porphyry provinces, Japan and New Zealand; the Aleutians, Izu-Bonin, Sumatra–Java, Banda, north and south Celebes, and Afghanistan; and Morocco and Algeria.

Geochemical prospecting is an excellent adjunct to geology and, if the data of geochemical prospecting are used correctly, they can be of fundamental importance in resolving problems of basic research or of applied natural resource evaluations. Recognition is the key to planning and carrying out a good geochemical exploration project; recognition of the distribution of geological units and their relationship with recent conceptual advances, recognition of element associations with rock types, recognition of ore-forming processes and physical, chemical and biological processes that may subsequently affect the ores, recognition of anomalies, and recognition of ore minerals. It has been written that: 'Geochemical reconnaissance is attractive from the point of view of rapidly and cheaply evaluating the economic potential of an area and of outlining new and extending old boundaries of metallogenic district.' In reality, geochemical prospecting is a highly specialized field in which complex processes and many interrelated factors may influence results: recognition of these and the ability to use them well may make the difference between a successful operation and an ordinary one.

References

ANDERSON, D. L. 1975. Accelerated Plate Tectonics. *Science,* Vol. 187, p. 1077–9.

BOLVIKEN, B. 1971. A Statistical Approach to the Problem of Interpretation in Geochemical Prospecting. In: R. W. Boyle (ed.), *Geochemical Exploration,* p. 564–7. Canadian Institute of Mining and Metallurgy. (Special Volume No. 11.)

BONDAR, W. F. 1969. Some Principles of Geochemical Analysis. In: Canney (ed.), International Geochemical Exploration Symposium. *Colo. Sch. Mines Q.,* Vol. 64, p. 19–22.

BOYLE, R. W.; JONASSON, I. R. 1973. The Geochemistry of Arsenic and its Use as an Indicator Element in Geochemical Prospecting. *J. Geochem. Explor.,* Vol. 2, p. 251–96.

BRADSHAW, P. M. D. (ed.). 1975. Conceptual Models in Exploration Geochemistry—The Canadian Cordillera and Canadian Shield. *J. Geochem. Explor.,* Vol. 4. 213 p.

BRADSHAW, P. M. D.; THOMSON, I.; SMEE, B. W.; LARSSON, J. O. 1974. The Application of Different Analytical Extrac-

tions and Soil Profile Sampling in Exploration Geochemistry. *J. Geochem. Explor.,* Vol. 3, p. 209–26.

BRISTOW, Q. 1972. An Evaluation of the Quartz Crystal Microbalance as a Mercury Vapour Sensor for Soil Gases. *J. Geochem. Explor.,* Vol. 1, p. 55–76.

BROCK, J. S. 1972. The Use of Dogs as an Aid to Exploration for Sulphides. *West. Miner.,* December, p. 28–32.

BROOKS, R. R. 1972. *Geobotany and Biogeochemistry in Mineral Exploration.* New York, Harper & Row. 209 p.

BRUNDIN, N. H.; NAIRIS, B. 1972. Alternative Sample Types in Regional Geochemical Prospecting. *J. Geochem. Explor.,* Vol. 1, p. 7–46.

BUGROV, V. 1974. Geochemical Sampling Techniques in the Eastern Desert of Egypt. *J. Geochem. Explor.,* Vol. 3, p. 67–76.

CANNON, H. L. 1974. *Geochemistry and the Environment.* Vol. 1: *The Relation of Selected Trace Elements to Health and Disease.* Washington, D.C., National Academy of Sciences. 113 p.

CANNON, H. L.; HOPPS, H. C. 1971. *Environmental Geochemistry in Health and Disease.* 230 p. (Geological Society of America Memo. 123.)

—— (eds.). 1972. *Geochemical Environment in Relation to Health and Diesease.* 77 p. (Geological Society of America Special Paper 140.)

CHOWDHURY, A. N.; BOSE, B. B. 1971. Role of 'Humus Matter' in the Formation of Geochemical Anomalies. In: R. W. Boyle (ed.), *Geochemical Exploration,* p. 401–9. Canadian Institute of Mining and Metallurgy. (Special Volume No. 11.)

COOPE, J. A. 1973. Geochemical Prospecting for Porphyry Copper-type Mineralization—A Review. *J. Geochem. Explor.,* Vol. 2, p. 81–102.

EKDAL, E. 1976. Pielavesi: The Use of Dogs in Prospecting. In: L. K. Kauranne, (ed.), *Conceptual Models in Exploration Geochemistry. Geochemical Exploration,* Vol. 5, p. 296–8.

FAIRBURN, W. A. 1973. Distribution of Niobium in Stream Sediments Related to Heavy-mineral Sand Dunes, near Pemberton, Western Australia. *J. Geochem. Explor.,* Vol. 2, p. 403–10.

FARRELL, B. L. 1974. Fluorine, A Direct Indicator of Fluorite Mineralization in Local and Regional Soil Geochemical Surveys. *J. Geochem. Explor.,* Vol. 3, p. 227–44.

FELDER, F. 1974. Shawinigan Nickel–Copper Property—A case History of a Reconnaissance Geochemical Discovery in the Grenville Province of Quebec, Canada. *J. Geochem. Explor.,* Vol. 3, p. 1–24.

FRIEDRICH, G. H. W.; KUNZENDORF, H.; PLUGER, W. L. 1974. Ship-borne Geochemical Investigations of Deep-sea Manganese Nodule Deposits in the Pacific Using a Radioisotope Energy Dispersive X-ray System. *J. Geochem. Explor.,* Vol. 3, p. 303–18.

FRIEDRICH, G. H.; PLUGER, W. L.; HILMER, E. F.; ABU-ABED, I. 1973. Flameless Atomic Absorption and Ion Sensitive Electrodes as Analytical Tools in Copper Exploration. In: Jones (ed.), *Geochemical Exploration 1972,* p. 435–43. London, Institution of Mining and Metallurgy.

GEODEKIAN, A. A.; STROGANOV, V. A. 1973. Geochemical Prospecting for Oil and Gas Using Reference Horizons. *J. Geochem. Explor.,* Vol. 2, p. 1–10.

GINZBURG, I. I. 1960. *Principles of Geochemical Prospecting.* New York, Pergamon. 311 p. (Translated from the original Russian, 1957, edition.)

GRAHAM, G. S.; KESLER, S. E.; VAN LOON, J. C. 1975. Fluorine in Ground Water as a Guide to Pb–Zn–Ba–F Mineralization. *Econ. Geol.,* Vol. 70, p. 396–8.

GRANIER, C. 1973. *Introduction à la Prospection Géochimique des Gîtes Métallifères.* Paris, Masson. 143 p.

GUILD, P. W. 1974a. Distribution of Metallogenic Provinces in Relation to Major Earth Features. *Schr Reihe. Erdwiss. Kamm. Osterr. Akad., Wiss.* (Vienna), Vol. 1.

—— 1974b. *Application of Global Tectonic Theory to Metallogenic Studies. Presented at Symp. Ore Dep. of Tethys Region by Commission on the Tectonics of Ore Deposits, IAGOD and CGMW at Varna, Bulgaria.*

HANSULD, J. A. 1967. Eh and pH in Geochemical Prospecting. *Proceedings, Symp. Geochem. Prospecting, Geolog. Surv. of Canada, Dept. of Energy, Mines and Resources,* p. 172–87. (Paper 66–54.)

HAWKES, H. G. 1957. Principles of Geochemical Prospecting. *U.S. Geol. Surv. Bull. 1000-F,* p. 225–355.

—— 1969. A Geochemical Case History from New Zealand.

In: Canney (ed.), *Int. Geochem. Explor. Symp. Colo. Sch. Mines Q.,* Vol. 64, p. 509. (Abstract.)

——. 1971. *Exploration Geochemistry Bibliography. Period January, 1965 to December, 1971.* Toronto, Association of Exploration Geochemists. 118 p. (Special Volume No. 1.) (With Supplement 1972 in *J. Geochem. Explor.,* Vol. 2, 1973, p. 41–76, and Supplement 1973 in *J. Geochem. Explor.,* Vol. 3, 1974, p. 89–128.)

HAWKES, H. E., WEBB, J. S. 1962. *Geochemistry in Mineral Exploration.* New York, Harper & Row. 415 p.

——. (comp.). 1976. *Exploration Geochemistry Bibliography, Period January 1972 to December 1975,* Rexdale, Ontario, Association of Exploration Geochemists. 195 p. (Special Volume No. 5.)

HEMPHILL, D. D. (ed.). 1973. *Trace Substances in Environmental Health.* Columbia, Mo., University of Missouri. 399 p. (1972, 559 p.; 1971, 456 p.; 1970, 391 p.; 1969, 318 p.; 1968, 236 p.)

HITCHON, B (comp. and ed.). 1977. Application of Geochemistry to the Search of Crude Oil and Natural Gas. *Geochem. Explor.,* Vol. 7, 293 p.

HOFFMAN, S. J.; FLETCHER, K. 1972. Distribution of Copper at the Dansey-Rayfield River Property, South-central British Columbia, Canada. *J. Geochem. Explor.,* Vol. 1, p. 163–80.

HOFFMAN, S. J.; WASKETT-MYERS, M. J. 1974. Determination of Molybdenum in Soils and Sediments with a Modified Zinc Dithiol Procedure. *J. Geochem. Explor.,* Vol. 3, p. 61–6.

HOLMES, R; TOOMS, J. S. 1973. Dispersion from a Submarine Exhaltive Orebody. In: Jones (ed.), *Geochemical Exploration 1972,* p. 193–202. London, Institution of Mining and Metallurgy.

HOPPS, H. C.; CANNON, H. L. (eds.). 1972. Geochemical Environment in Relation to Health and Disease. *Ann. N.Y. Acad. Sci.,* Vol. 199. 352 p.

JONASSON, I. R. 1970. *Mercury in the Natural Environment: A Review of Recent Work.* 39 p. (Geological Survey of Canada, Paper 70–57.)

KARASEK, F. W.; LAUB, R. J. 1974. Detection Limits in Instrumental Analysis. *Research/Development,* Vol. 25, p. 36–8.

KAURANNE, L. K. (ed.). 1976. Conceptual Models in Exploration Geochemistry. *Geochem. Explor.,* Vol. 5. 420 p.

KESLER, S. E.; VAN LOON, J. C.; BATESON, J. H. 1973. Analysis of Fluoride in Rocks and an Application to Exploration. *J. Geochem. Explor.,* Vol. 2, p. 11–18.

KOTHNY, E. L. 1974. Simple Trace Determination of Platinum in Geological Materials. *J. Geochem. Explor.,* Vol. 3, p. 291–300.

LAUBMEYER, G. 1933. A New Geophysical Prospecting Method, Especially for Deposits of Petroleum. *Petroleum,* Vol. 29, p. 1–4.

LEPELTIER, G. 1969. A Simplified Statistical Treatment of Geochemical Data by Graphical Representation. *Econ. Geol.,* Vol. 64, p. 538–50.

LEVINSON, A. A. 1974. *Introduction to Exploration Geochemistry,* Alberta, Applied Publishing Ltd. 612 p.

MCCARTHY, J. H., Jr. 1972. Mercury Vapor and Other Volatile Components in the Air as Guides to Ore Deposits. *J. Geochem. Explor.,* Vol. 1, p. 143–62.

MCCROSSAN, R. G.; BALL, N. L. 1971. An Evaluation of Surface Geochemical Prospecting for Petroleum. In: *Geochemical Exploration,* p. 529–36. Canadian Institute of Mining and Metallurgy. (Special Volume No. 11.)

MCKELVEY, V. E.; WANG, F. F. H. 1970. *World Sub-sea Mineral Resources (Preliminary Maps).* United States Geological

Survey Miscellaneous Geologic Investigations, Map I-632 (second printing plus discussion to accompany Map I-632.

MCNERNEY, J. J.; BUSECK, P. R. 1973. Geochemical Exploration Using Mercury Vapor. *Econ. Geol.*, Vol. 68, p. 1313–20.

MAHAFFEY, E. J. 1974. A Spectrophotometric Method for the Determination of Rhenium in Geologic Materials. *J. Geochem. Explor.*, Vol. 3, p. 53–60.

MALYUGA, D. P. 1964. *Biogeochemical Methods of Prospecting*. New York, Consultants Bureau. 205 p.

MITCHELL, A. H.; BELL, J. D. 1973. Island Arc Evolution and Related Mineral Deposits. *J. Geog.*, Vol. 81, p. 381–405.

MITCHELL, A. H.; GARSON, M. S. 1972. Relationship of Porphyry Copper and Circum-Pacific Tin Deposits to Palaeo-Benioff Zones. *Trans. Inst. Min. Metall.*, Sec. B, Vol. 81, p. 10–25.

NASA. 1968. *Application of Biogeochemistry to Mineral Prospecting*. National Aeronautics and Space Administration. 134 p. (Special Publication 5056.)

NICHOL, I.; GARRETT, R. G.; WEBB, J. S. 1969. The Role of Some Statistical Mathematical Methods in the Interpretation of Regional Geochemical Data. *Econ. Geol.*, Vol. 64, p. 204–20.

OLADE, M.; FLETCHER, K. 1974. Potassium Chlorate–hydrochloric Acid: A Sulphide Selective Leach for Bedrock Geochemistry. *J. Geochem. Explor.*, Vol. 3, p. 337–44.

PARSLOW, G. R. 1974. Determination of Background and Threshold in Exploration Geochemistry. *J. Geochem. Explor.*, Vol. 3, p. 319–36.

PERHAC, R. M.; WHELAN, C. J. 1972. A Comparison of Water, Suspended Solid and Bottom Sediment Analyses for Geochemical Prospecting in a Northeast Tennessee Zinc District. *J. Geochem. Explor.*, Vol. 1, p. 47–54.

QUIN, B. F.; BROOKS, R. R.; BOSWELL, C. R.; PAINTER, J. A. C. 1974. Biogeochemical Exploration for Tungsten at Barrytown, New Zealand. *J. Geochem. Explor.*, Vol. 3, p. 43–52.

RONA, P. A. 1974. Plate Tectonics and Mineral Resources. *Scient. Am.*, Vol. 231, p. 86–95.

SAWKINS, F. J. 1972. Sulphide Ore Deposits in Relation to Plate Tectonics. *J. Geol.*, Vol. 80, p. 377–97.

SENFTLE, F. E. 1970. Mineral Exploration by Nuclear Techniques. *Min. Congr. J.*, Vol. 56, p. 21–8.

SIEGEL, F. R. 1974. *Applied Geochemistry*. New York, Wiley. 353 p.

——. Marine Exploration Geochemistry. In: Fairbridge and Finkl (eds.), *Encyclopedia of Earth Sciences*, Volume on: *Applied Geology and Soil Science*. New York, Van Nostrand.

SIEGEL, F. R.; PIERCE, J. W.; BLOCH, S.; HEARN, P. P. 1976. Mineral Suspensate Geochemistry, Argentine Continental Shelf (48°–53°30′ s), R/V Hero Cruise 75–3. *Antarctic J. of the U.S.*, Vol. XI, p. 230–1.

SIEGEL, F. R.; PIERCE, J. W.; SAYALA, D. 1977. Reconnaissance Geochemical Prospecting With Marine Suspensates. *VIII Caribbean Geol. Conf.*, Curaçao, *Abstract Book*. p. 188–9. (Abstract.)

SILLITOE, R. H. 1972a. A Plate Tectonic Model for the Origin of Porphyry Cu Deposits. *Econ. Geol.*, Vol. 67, p. 184–97.

——. 1972b. Relation of Metal Provinces in Western America to Subduction of Oceanic Lithosphere. *Bull. Geol. Soc. Am.* Vol. 83, p. 813–18.

SINCLAIR, A. J. 1974. Selection of Threshold Values in Geochemical Data Using Probability Graphs. *J. Geochem. Explor.*, Vol. 3, p. 129–50.

SOKOLOV, V. A. 1971. The Theoretical Foundations of Geochemical Prospecting for Petroleum and Natural Gas and the Tendencies of its Development. In: R. W. Boyle (ed.). *Geochemical Exploration*, p. 544–9. Canadian Institute of Mining and Metallurgy. (Special Volume No. 11.)

SOKOLOV, V. A.; DADASHEV, F. G. 1972. Deep Gas Survey of the South Caspian Basin. *Abstracts, 4th Int. Geochem. Explor. Symp.*, p. 60. London, Institution of Mining and Metallurgy.

SOKOLOV, V. A.; GEODEKYAN, A. A.; GRIGORYEV, G. G.; KREMS, A. Y.; STROGANOV, V. A.; ZORKIN, L. M.; ZEIDELSON, M. I.; VAINBAUM, S. J. 1971. The New Methods of Gas Surveys, Gas Investigations of Wells and Some Practical Results. In: R. W. Boyle (ed.), *Geochemical Exploration*, p. 538–43. Canadian Institute of Mining and Metallurgy. (Special Volume No. 11.)

SUMMERHAYES, C. P.; HAZELHOFF-ROELFZEMA, B. H.; TOOMS, J. S.; SMITH, D. B. 1970. Phosphirite Prospecting Using a Submersible Scintillation Counter. *Econ. Geol.*, Vol. 65, p. 718–23.

TIMPERLY, M. H.; ALLAN, R. J. 1974. The Formation and Detection of Metal Dispersion Halos in Organic Lake Sediments. *J. Geochem. Explor.*, Vol. 3, p. 167–90.

TOOMS, J. S. 1972. Potentially Exploitable Marine Minerals. *Endeavor*, Vol. 31, p. 113–17.

VAN LOON, J. C.; KESLER, S. E.; MOORE, C. M. 1973. Analysis of Water-extractable Chloride in Rocks by Use of a Selective Ion Electrode. In: Jones (ed.), *Geochemical Exploration 1972*, p. 429–34. London, Institution of Mining and Metallurgy.

VAUGHN, W. W. 1967. A Simple Mercury Vapor Detector for Geochemical Prospecting. *U.S. Geol. Surv. Circ. 540.* 8 p.

WARREN, H. V. 1973. Some Trace Element Concentrations in Various Environments. In: How and Loraine (eds.), *Environmental Medicine*, p. 9–24. London, Heinemann Medical Books.

WOLFE, W. J. 1974. Geochemical and Biochemical Exploration Research Near Early Precambrian Porphyry-type Molybdenum-copper Mineralization, Northwestern Ontario, Canada. *J. Geochem. Explor.*, Vol. 3, p. 25–42.

ZARELLA, W. M. 1969. Applications of Geochemistry of Petroleum Exploration. In: Heroy (ed.), *Unconventional Methods in Exploration for Petroleum and Natural Gas*, p. 29–41. Dallas, Tex., Southern Methodist University Press.

ZARELLA, W. M.; MOUSSEAU, R. J.; COGGESHALL, N. D.; NORRIS, M. S.; SCHRAYER, G. J. 1967. Analysis and Significance of Hydrocarbons in Subsurface Brines. *Geochim. Cosmochim. Acta*, Vol. 31, p. 1155-66.

ZORKIN, L. M. 1969. Regional Regularities of Underground Water Gas Contents in Petroleum-gas Basins. *Geol. USSR*, No. 2.